COMPUTATIONAL
FRAMEWORK FOR
KNOWLEDGE

COMPUTATIONAL FRAMEWORK FOR KNOWLEDGE

Integrated Behavior of Machines

Syed V. Ahamed

WILEY

A JOHN WILEY & SONS, INC., PUBLICATION

Copyright © 2009 by John Wiley & Sons, Inc. All rights reserved.

Published by John Wiley & Sons, Inc., Hoboken, New Jersey.
Published simultaneously in Canada.

No part of this publication may be reproduced, stored in a retrieval system, or transmitted in any form or by any means, electronic, mechanical, photocopying, recording, scanning, or otherwise, except as permitted under Section 107 or 108 of the 1976 United States Copyright Act, without either the prior written permission of the Publisher, or authorization through payment of the appropriate per-copy fee to the Copyright Clearance Center, Inc., 222 Rosewood Drive, Danvers, MA 01923, (978) 750-8400, fax (978) 750-4470, or on the web at www.copyright.com. Requests to the Publisher for permission should be addressed to the Permissions Department, John Wiley & Sons, Inc., 111 River Street, Hoboken, NJ 07030, (201) 748-6011, fax (201) 748-6008, or online at www.wiley.com/go/permission.

Limit of Liability/Disclaimer of Warranty: While the publisher and author have used their best efforts in preparing this book, they make no representations or warranties with respect to the accuracy or completeness of the contents of this book and specifically disclaim any implied warranties of merchantability or fitness for a particular purpose. No warranty may be created or extended by sales representatives or written sales materials. The advice and strategies contained herein may not be suitable for your situation. You should consult with a professional where appropriate. Neither the publisher nor author shall be liable for any loss of profit or any other commercial damages, including but not limited to special, incidental, consequential, or other damages.

For general information on our other products and services or for technical support, please contact our Customer Care Department within the United States at (800) 762-2974, outside the United States at (317) 572-3993 or fax (317) 572-4002.

Wiley also publishes its books in a variety of electronic formats. Some content that appears in print may not be available in electronic formats. For more information about Wiley products, visit our web site at www.wiley.com.

Library of Congress Cataloging-in-Publication Data:

Ahamed, Syed V., 1938-
 Computational framework for knowledge : integrated behavior of machines / Syed V. Ahamed.
   p. cm.
 Includes bibliographical references and index.
 ISBN 978-0-470-44686-7 (cloth)
1. Data mining. 2. Web databases. 3. Knowledge acquisition (Expert systems) I. Title.

QA76.9.D343A338 2009
006.3'12–dc22

2008052043

Printed in the United States of America

10 9 8 7 6 5 4 3 2 1

CONTENTS

Foreword   xvii

Preface   xxi

Introduction   xxv

1 New Knowledge Environments   1

   Chapter Summary / 1
- 1.1 The Need to Know / 2
  - 1.1.1 Global Power of Knowledge / 6
  - 1.1.2 Scientific Aspects / 9
  - 1.1.3 Wealth Aspects / 10
- 1.2 Role of Technology / 12
  - 1.2.1 Three Major Contributions / 12
  - 1.2.2 A String of Secondary Contributions / 15
  - 1.2.3 Peripheral Contributions / 17
- 1.3 Knowledge and Wealth / 17
- 1.4 Evolving Knowledge Environments / 18
  - 1.4.1 Components of Knowledge / 20
  - 1.4.2 The Processing of Knowledge / 22
- 1.5 Structure and Communication of Knowledge / 22
  - 1.5.1 Velocity of Flow of Knowledge / 23
  - 1.5.2 Truisms in the Knowledge Domain / 29
  - 1.5.3 Philosophic Validation of Knowledge / 29

　　　　1.5.4　Scientific Principles in the Knowledge Domain / 31
　　　　1.5.5　Aspects of Knowledge / 33
　　1.6　Intelligent Internet and Knowledge Society / 34
　　　　1.6.1　Four Precursors of Modern Wisdom / 34
　　　　1.6.2　Knowledge Bases to Derive Wisdom / 35
　　　　1.6.3　Role of National Governments / 36
　　　　1.6.4　Universal Knowledge-Processing Systems / 41
　　　　1.6.5　Educational Networks / 50
　　　　1.6.6　Medical Networks / 52
　　　　1.6.7　Antiterrorism Networks / 55
　　1.7　Knowledge Networks / 58
　　　　1.7.1　Evolution of Knowledge Networks / 58
　　　　1.7.2　Knowledge Network Configuration / 59
　　1.8　Conclusions / 59
　　References / 61

2　Wisdom Machines　　　　　　　　　　　　　　　　　　　　　　64

　　Chapter Summary / 64
　　2.1　Many "Flavors" of Wisdom / 65
　　2.2　Three Orientations of Wisdom / 67
　　　　2.2.1　Absolute Wisdom / 68
　　　　2.2.2　Materialistic Wisdom / 68
　　　　2.2.3　Opportunistic Wisdom / 69
　　　　2.2.4　Needs and Wisdom / 70
　　　　2.2.5　What Are Wisdom Machines? / 71
　　2.3　Optimization of Wise Choices / 73
　　　　2.3.1　Derived Axioms / 73
　　　　2.3.2　Priming of Machine Wisdom for Directional Axioms / 74
　　2.4　Three-Level Functions / 75
　　　　2.4.1　Level I: Access and Administrative Functions / 76
　　　　2.4.2　Level II: Linkage, Scientific, and Statistical Functions / 76
　　　　2.4.3　Level III: Human Authentication / 77
　　2.5　Knowledge Machine Building Blocks / 77
　　　　2.5.1　What Are Knowledge Machines? / 78
　　　　2.5.2　Knowledge-Machine-Based Wisdom Machines / 79
　　　　2.5.3　Sensor-Scanner-Based Wisdom Machines / 81

            2.5.4   Bus Configurations and Switch Locations / 84
    2.6   Machine Clusters / 86
            2.6.1   Single-Wisdom Single-Machine Systems / 86
            2.6.2   Single-Wisdom Multiple-Machine Systems / 89
            2.6.3   Multiple-Wisdom Single-Machine Systems / 91
            2.6.4   Multiple-Wisdom Multiple-Machine Systems / 93
    2.7   From Wisdom to Behavior / 95
    2.8   Order, Awareness, and Search / 97
    2.9   Conclusions / 98
    References / 99

3   **General Theory of Knowledge**   **101**

    Chapter Summary / 101
    3.1   A Basis for the Theory of Knowledge / 102
    3.2   Comprehension, Nature, and Knowledge / 102
            3.2.1   A Functional Approach / 103
            3.2.2   Incremental Changes / 104
            3.2.3   Elemental Convolution and Knowledge Operations / 108
    3.3   Central Processing and Knowledge Processing / 108
    3.4   Accumulation of Information, Knowledge, and Wisdom / 111
    3.5   The Enhanced Knowledge Trail / 112
    3.6   Sequencing of Events at Nodes / 114
    3.7   Transitions at I, K, and C Nodes / 116
    3.8   Transition Management at Nodes / 117
    3.9   An Inverse Universe / 120
    3.10 Origin and Destination / 122
            3.10.1   Nature, Origin of Knowledge Trail / 122
            3.10.2   Two Destinations of Knowledge Trail / 124
            3.10.3   Multiple Feedbacks along the Knowledge Trail / 126
            3.10.4   Dynamics of Knowledge in Societies / 130
            3.10.5   I, K, C, W, and E Bases to Replace Nodes / 132
            3.10.6   Nodal Forces in Societies / 137
            3.10.7   Summary of Sociological Forces on Nodes / 139
            3.10.8   Multiplier Effects at Intermediate Nodes / 140
            3.10.9   Implications of Knowledge Equations / 141
    3.11 The Entropy of Knowledge / 141
            3.11.1   The Change of Entropy / 142

    3.11.2 The Duration for Change of Entropy / 142
    3.11.3 No Prior Knowledge: No Change in Entropy / 143
    3.11.4 No Knowledge Process: No Change in Entropy / 143
    3.11.5 Examples of Prevalent Knowledge Processes / 143
    3.11.6 Generalization of Knowledge Processes / 144
  3.12 Conclusions / 145
  References / 146

4 Verb Functions and Noun Objects   148

  Chapter Summary / 148
  4.1 Positive and Negative Social Forces / 149
    4.1.1 Forces in Society and Operations on Objects / 150
    4.1.2 Assistive, Neutral, and Resistive Objects / 152
  4.2 Framework of Knowledge / 153
    4.2.1 Hyper-Dimensionality of Knowledge / 153
    4.2.2 Intertwined Spaces of Knowledge / 154
    4.2.3 Contours of Knowledge / 155
    4.2.4 Funicular Polygon of Concepts / 155
  4.3 Compilation of Knowledge / 156
    4.3.1 Lexical Analysis for Knowledge / 157
    4.3.2 Parsing Bodies of Knowledge / 157
    4.3.3 Syntactic Analysis of Knowledge / 161
    4.3.4 Semantic Analysis for Knowledge / 161
    4.3.5 Knowledge Machine Code / 163
  4.4 Derivation of Knowledge / 163
    4.4.1 Microscopic or Basic Assembly-Level Knowledge Instructions / 164
    4.4.2 Application-Level Knowledge Programs / 165
    4.4.3 Universal-Level Knowledge Programs / 168
  4.5 Knowledge Machine Software Hierarchy / 171
  4.6 Knowledge Hardware and Software Systems / 172
  4.7 Classical Migration Path of Knowledge / 175
    4.7.1 Knowledge Traps / 176
    4.7.2 Transition at Generic Nodes / 176
  4.8 Conclusions / 179
  References / 180

5 Humanistic and Semi-Human Systems — 182

Chapter Summary / 182
5.1 Humanistic Chip Sets / 183
5.2 Essence of Human Activity / 185
    5.2.1 Varieties of Noun Objects / 186
    5.2.2 Types of Verb Function Commands / 188
5.3 Smart Verbs and Intelligent Nouns / 189
    5.3.1 A Qualitative Synopsis of the Response / 190
    5.3.2 Nonlinear Response / 191
    5.3.3 An Intelligent Response / 192
5.4 Algorithmic Representation / 193
    5.4.1 Step 1: Compute Change Index x / 193
    5.4.2 Step 2: Compute Change Δ after $kopc$ / 195
5.5 Effects of Intelligent Response / 197
5.6 Gradation of Knowledge Processor Unit Responses / 197
    5.6.1 Traditional Computer Response / 198
    5.6.2 Life Form Response / 198
    5.6.3 Rational Response / 198
    5.6.4 Artificial Intelligence Response / 199
    5.6.5 Humanistic Response / 199
    5.6.6 Human Response / 200
    5.6.7 Other Types of Responses / 201
5.7 Fragments of Overall Human Activity / 202
    5.7.1 Minor Tasks and Microcodes / 203
    5.7.2 Major Ordeals and Complex Programs / 203
    5.7.3 The Microcode of the Humanistic Chip Set / 203
5.8 Dual Knowledge Processor Unit Human Interaction Model / 204
    5.8.1 Bilateral Human Relations / 205
    5.8.2 Models of Human Interaction / 206

References / 208

A.1 Nonlinear Responses / 212
    A.1.1 Exponential Representation / 212
    A.1.2 Memory Effects / 212
    A.1.3 Saturation / 212
    A.1.4 Peaky Response / 213
    A.1.5 Hysteresis Effects / 213
A.2 Oscillatory Response of Noun Objects / 215

B.1 Knowledge Machine for Human Interactive Processes / 219
References / 221
B.2 Knowledge Machine for Labor–Management Negotiation / 222
    B.2.1 Background of Labor–Management Negotiations / 222
    B.2.2 Constituents of Model for Labor–Management Negotiations / 223
    B.2.3 Demand Intensities of Labor and Management / 224
    B.2.4 Structure of the Model for Lasor–Management Negotiations / 225
    B.2.5 Negotiation Process and Validation / 225
    B.2.6 Successful Negotiation / 227
    B.2.7 Impasse in Labor–Management Negotiations / 227
    B.2.8 Conclusions for Labor–Management Negotiation / 228
References / 229
B.3 Knowledge Machine for Corporate Interactions / 229
    B.3.1 Background of Economics-Based Interactions / 230
    B.3.2 Expert System Concepts / 231
    B.3.3 Computational Aspects in Negotiations / 233
    B.3.4 Mathematical Representations for the Cost of Agreement, Cost of Disagreement, and Demand Intensity / 235
    B.3.5 Systems Model of Interaction / 237
    B.3.6 Pruning and Predicting the Outcome / 241
    B.3.7 Conclusions for Corporate Negotiations / 243
References / 244

6 Information and Knowledge Filters   246

Chapter Summary / 246
6.1 Junk Information and Hype Knowledge / 247
6.2 Design of Conventional Signal Filters / 249
    6.2.1 Types of Traditional Signal Filters / 251
    6.2.2 Active Filters / 251
    6.2.3 Digital Filters / 251
6.3 Signal Waves and Knowledge Flow / 252
6.4 The $(I \ll \gg K)$ Filters / 253
    6.4.1 Symbols as Objects and Functions / 254
    6.4.2 Role of Humans / 256
    6.4.3 Precise Solutions and Confident Results / 257
6.5 The Design of $(I \ll \gg K)$ Filters / 258

	6.5.1	Knowledge Filter Concepts / 259

- 6.5.1 Knowledge Filter Concepts / 259
- 6.5.2 Noun Objects and Verb Functions / 265
- 6.5.3 Scanning for Operative Nouns and Verbs / 266

6.6 Configuration of $(I \ll \gg K)$ Systems / 266
- 6.6.1 Application Constraints / 266
- 6.6.2 Design Concepts for $(I \ll \gg K)$ Systems / 266
- 6.6.3 Application for Filtering and Comparing / 270
- 6.6.4 Low-Pass and High-Pass Information Filters / 271
- 6.6.5 Design Steps for $(I \ll \gg K)$ Filters / 272
- 6.6.6 Design Methodology / 272

6.7 Selection of Noun Objects and Verb Functions in Samples / 273

6.8 Systems for $(I \ll \gg K)$ Filters / 275

6.9 Two-Port Knowledge Network / 277
- 6.9.1 Two-Port Electrical Network Theory / 278
- 6.9.2 Configuration of a Two-Port Knowledge Network / 279
- 6.9.3 Mathematical Implications of the Two Port Network / 283
- 6.9.4 Change of Operators $(+, \times, /, \text{ and } =)$ / 283
- 6.9.5 Intelligent Two-Port Knowledge Networks / 286
- 6.9.6 Needs, Knowledge, and Innovations / 286

6.10 Knowledge Gates / 286
- 6.10.1 Knowledge-Based AND Gate / 289
- 6.10.2 Knowledge-Based OR Gate / 290
- 6.10.3 Knowledge-Based NOT Gate / 291
- 6.10.4 Practical Use of Knowledge Gates / 292

6.11 Contamination of $(I \ll \gg K)$ / 295

6.12 Decontamination of Knowledge / 297

6.13 Conclusions / 300

References / 302

- 6.A.1 T-Zero Filtering and Output $BOK_1\text{-}T$ / 303
- 6.A.2 R-Zero Filtering and Output $BOK_1\text{-}R$ / 304
- 6.A.3 TR-Zero Filtering and Output $BOK_2\text{-}TR$ / 305
- 6.B.1 S-Zero Filtering and Output $BOK_1\text{-}S$ / 310
- 6.B.2 Generality of Information Filtering / 310
- 6.C.1 Application to Three Positive Noun Objects / 312
- 6.C.2 Filter Template Based on TVB / 314

xii CONTENTS

        6.C.3   New Filter Template Based on TVB and Soul / 316

7 Process and Change of Entropy   **320**

Chapter Summary / 320
- 7.1 Actions and Entropy / 321
- 7.2 Knowledge-Centric Objects / 321
  - 7.2.1 Passive Knowledge-Centric Objects / 322
  - 7.2.2 Reactive or Intelligent Knowledge-Centric Objects / 322
- 7.3 Classification of Knowledge-Centric Objects / 326
  - 7.3.1 Single Simple Knowledge Object (SSKO) / 327
  - 7.3.2 Single Simple Knowledge Object, Single Attribute (SSKO-SA) / 328
  - 7.3.3 Single Simple Knowledge Object, Multiple Attributes $(1 - n)$ / 328
  - 7.3.4 Single Complex Knowledge Object, Single Dependent Object / 329
  - 7.3.5 Single Complex Knowledge Object, Single Dependent Object, Single Attribute / 330
  - 7.3.6 Single Complex Knowledge Object, Single Dependent Object, Multiple Attributes n / 330
  - 7.3.7 Single Complex Knowledge Object, Multiple Dependent Objects ℓ / 331
  - 7.3.8 Single Complex Knowledge Object, Multiple Dependent Objects ℓ, One Attribute / 332
  - 7.3.9 Single Complex Knowledge Object, Multiple Dependent Objects ℓ, Multiple Attributes n / 333
- 7.4 Clusters of Complex Knowledge-Centric Objects / 333
  - 7.4.1 Multiple Complex Knowledge Objects m, Multiple Dependent Objects ℓ, Single Attribute / 334
  - 7.4.2 Multiple Complex Knowledge Objects m, Multiple Dependent Objects ℓ, Multiple Attributes n / 335
- 7.5 Single-Process *Kopcodes* for Generic Knowledge-Centric Objects / 336
  - 7.5.1 Single-Process *kopcodes* for Simple Objects / 336
  - 7.5.2 Single-Process *kopc* for Single Complex Object / 337
  - 7.5.3 Single-Process *kopc* for Multiple Complex Objects / 339
- 7.6 Multiple-Process Instructions / 340
  - 7.6.1 MP-MCKO-MDO-NA, Class-16 *kopc*16 Instruction / 340

- 7.6.2 MP-MCKO-MDO-SA, Class-17 *kopc*17 Instruction / 340
- 7.6.3 MP-MCKO-MDO-MA, Class-18 *kopc*18 Instruction / 340
- 7.7 Passive *Kopcs* / 341
  - 7.7.1 Report Generation *kopcs* on Passive Knowledge-Centric Objects / 342
  - 7.7.2 Report Generation *kopcs* on Dynamic Knowledge-Centric Objects / 343
- 7.8 Active *Kopcs* / 343
  - 7.8.1 Active *kopcs* for Passive Knowledge-Centric Objects / 344
  - 7.8.2 Active *kopcs* for Reactive Knowledge-Centric Objects / 345
  - 7.8.3 Active *kopcs* for Intelligent Knowledge-Centric Objects / 345
- 7.9 Execution of Multiple-Process Instructions / 346
- 7.10 Iterative and Reflexive Processing / 347
- 7.11 Macroinstructions for Knowledge Processor Units / 347
- 7.12 Knowledge Processor Unit Architectures for *kopc*01 − 12 Instructions / 348
  - 7.12.1 Single-Process Single-Object Processors and Machines / 348
  - 7.12.2 Multiple-Process Multiple-Object Processors and Machines / 350
- 7.13 Conclusions / 353
- References / 354

8 Knowledge System Architectures   355

Chapter Summary / 355
- 8.1 From the Central Processor Unit to Knowledge Processor Unit / 356
- 8.2 The Philosophic Dimension / 357
- 8.3 Iterative Cyclic Social Processes / 359
  - 8.3.1 Awareness and Knowledge / 359
  - 8.3.2 Communications and Interactions / 362
  - 8.3.3 Bargaining and Negotiation / 363
  - 8.3.4 Compromise During Transactions / 363
  - 8.3.5 Adjustment and Relationships / 363
- 8.4 The Scientific Dimension / 364

8.4.1 The Search for Precision and Infallibility / 364
8.4.2 Machine Accuracy and Human Adaptability / 365
8.5 Varieties of Processors and Machines / 365
8.5.1 Central Processor Units and Generic Computers / 366
8.5.2 Communications Processor and Electronic Switching Systems / 373
8.5.3 Input/Output Processors / 379
8.5.4 Display/Graphics Processor Units / 380
8.5.5 Numeric Processor Units and Array Processors / 382
8.5.6 Digital Signal Processors / 382
8.5.7 Digital Object-Processing Environments / 383
8.5.8 Digital Knowledge Environments / 386
8.5.9 Digital Medical-Processing Environments / 394
8.6 The Trend of Concepts / 401
8.6.1 Bridge between Science and Wisdom / 401
8.6.2 Concept Processing / 406
8.6.3 Human Concept Machines / 406
8.7 Recent Trends Toward Wisdom / 406
8.7.1 Artificial Wisdom / 407
8.7.2 Old and New Wisdom / 409
8.7.3 Wisdom in Machines and Instinct in Humans / 411
8.8 The Trend of Values and Ethics / 412
8.8.1 Ethical Issues and Role of Machines / 412
8.8.2 Conflictive Roles of Humans and Machines / 413
8.9 Conclusions / 413
References / 414

9 Humans, Machines, and Networks — 416

Chapter Summary / 416
9.1 Ethics and Values in High-Tech Society / 417
9.2 Needs that Drive / 419
9.3 Networks that Transport / 420
9.4 Overlap of Needs and Networks / 423
9.4.1 The Human and Social Aspects / 423
9.4.2 Individual Needs / 424
9.4.3 Individuals, Machines, and Networks / 424
9.4.4 Corporate Needs / 427
9.4.5 Corporations, Computers, and Networks / 428

    9.4.6   Societal Needs / 429
    9.4.7   Societies, Machines, and Networks / 430
9.5 Rationality of Humanistic Machines / 431
    9.5.1   The Human Element / 435
    9.5.2   The Open Human Inter-Connect Model / 438
    9.5.3   Role of Open Human Inter-Connect in National/Social Settings / 440
9.6 Wisdom Domain and Knowledge Rush / 440
    9.6.1   Pitfalls in a Knowledge Society / 442
    9.6.2   Humanistic Systems that Refuse / 443
9.7 Needs Pyramid of a Society / 443
    9.7.1   Individuals as Reputatuble Writers / 445
    9.7.2   Machines as Processors of Information / 445
9.8 Self-Perpetuating Power Loops / 447
9.9 Convergence of Knowledge and Motivation Hierarchies / 450
    9.9.1   The Hierarchy of Knowledge / 451
    9.9.2   The Hierarchy of Motivation / 453
9.10 Knowledge Machines on Knowledge / 455
9.11 Conclusions / 456
References / 458

10 Architecture of Knowledge   460

Chapter Summary / 460
10.1 A New Breed of Knowledge / 461
10.2 The Knowledge Loop and Its Stability / 462
10.3 Continuity of Knowledge / 465
10.4 Three Orientations of Wisdom / 466
10.5 Long-Term Movements within the K-Loop / 468
10.6 Conflictive Roles of TVB<>DAH Orientations / 469
10.7 Tracking of The K-Loops / 469
    10.7.1   Expansionary Upward Movement (TVB) / 470
    10.7.2   Recessive Downward Movement (DAH) / 471
10.8 Details of Upward Migration / 472
10.9 Details of Downward Migration / 473
10.10 Knowledge Loops in the *X*-And *Y*-Dimensions / 473
10.11 Four Ports of Super-Node K / 476
10.12 Construction of A Knowledge Plane / 477

10.12.1 Embedded Noun Objects and Verb Functions / 477
10.12.2 Primary K-Plane in the XY-Dimensions / 478
10.13 Knowledge Bases in the Primary K-Plane / 479
10.14 Successive K-Loops in the Time Dimension / 479
10.14.1 Stacking of K-Loops / 479
10.14.2 Lessons from Knowledge Lost / 483
10.14.3 Cosmic Space, Knowledge Space, and Intellectual Space / 484
10.15 The Nebula of Knowledge: Knowbula / 486
10.16 Hindsight and Foresight / 489
10.16.1 Coarse Survey of Human Struggle / 489
10.16.2 Scan of Modern Methodologies / 490
10.16.3 Structured and Predictive Knowledge / 491
10.17 Knowledge, Freedom, and Creativity / 493
10.17.1 Knowbula Almost Fills the Cube of Knowledge / 493
10.17.2 Knowbula Loosely Fits the Cube of Knowledge / 495
10.17.3 Knowbula Becomes Insignificant in the Cube of Knowledge / 496
10.18 Knowledge and Money / 496
10.19 Pinnacle of Mind and Infinity of Thought / 498
10.20 Social Migration and Political Agenda / 500
10.21 Conclusions / 502
Reference / 503

Acronyms   **505**

Index   **519**

Author Biography   **537**

FOREWORD

This is a very unusual book dealing with a very unusual subject.

We are all familiar with the filtering and processing of *signals* according to certain given criteria, such as selecting a frequency band out of a communications signal, reducing unwanted noise or other artifacts from a given electrical signal, or filtering out certain colors from a light signal. Information theory, invented by Claude Shannon some 60 years ago, went a step further. Shannon defined mathematically the *information* contained in communications signals by formulating a mathematical measure for information that he called entropy (reduction of uncertainty in a given information message). With this mathematical formulation, he was able to process and quantify *information* and to formulate the maximum quantity of information that a given channel can transmit. It is important to note here that Shannon's concept of information is entirely independent of the "knowledge" that is contained in his information signal.

Professor Syed Ahamed has now gone a decisive step further, and suggests something that has not been suggested in so much detail ever before: the "filtering" of *knowledge*, according to criteria postulated by the receiver of a stream of information conveying this knowledge. Since the "quality" or "value" of knowledge is by its very nature subjective, Professor Ahamed introduces methods of formulating and quantifying the parameters necessary to filter knowledge *according to the value system of the person at the receiving end of that knowledge*. In the hands of a lesser expert this endeavor may well seem utopian. However, Professor Ahamed's credentials are most impressive, and it is no exaggeration to say that if anyone can tackle this project with some hope of success, *which may essentially mean starting an entirely new discipline*, then Professor Ahamed is that person.

Syed Ahamed is a well-recognized communications and systems expert with many years of experience at Bell Labs; he is also a university professor in the same field. He is well versed in computer science and mathematical methods that he has used in numerous purely technical areas, and documented in several textbooks. What makes him unusual, and indeed exceptional, is that he is also a quasi-philosopher and humanist, who has devoted much of his life to contemplating problems of society and how to alleviate them. His intentions in this respect are very pure and idealistic; however, they are tempered by the practical and realistic view of an engineer and scientist.

This latter point, namely, the genuine and benevolent intentions of Syed Ahamed, is very important, because the "filtering of knowledge"—if it can be done successfully—may, of course, be misused by every propagandist, tyrant, and dictator anywhere on this planet. This becomes clear when we realize that the object of this book is to go far beyond what we know today as simple "SPAM filtering" and the like. However, I feel that if the filtering and manipulation of knowledge *can* be done—and Professor Ahamed's book will convince us that there is a high likelihood that it can—then it *will* be done by someone, and possibly by someone much less qualified to do so than he. Thus, the apprehensions that may be expressed by a critic of the subject of this book, namely, the *manipulation of knowledge*, are similar to the apprehensions that might have been expressed at the time of the invention of matchsticks, razor blades, or any other items that can be used for the benefit or the detriment of mankind—depending on who lays their hands on them

Among the highlights of this book, which define the novelty of its contents, are the following: (a) It combines philosophical, societal, and artificial intelligence concepts with those of computer science and information theory, and (b) knowledge is treated as a mathematical object-based entity. Furthermore, the book tries to cope with the "knowledge explosion" (as exemplified, e.g., by the Internet, Google, and Wikipedia) by letting machines (i.e., computers) do the processing, classifying, and prioritizing of available information. Because of the immense amount of available information, these tasks are considered by the author as too formidable and insurmountable by humans without the help of machines.

This realization is, in fact, one of the core motivations of the author for writing the book, namely, in his words, to "disentangle the gigantic knot of information and knowledge available on a global scale via the Internet and computer communications." Being an engineer and scientist, and more specifically an expert in computer science and electrical communications systems, Professor Ahamed is also a humanist, philosopher, and poet. These are unique qualifications that entitle him to tackle the formidable task at hand. To do so, he introduces new concepts such as that of "wisdom machines" that interact with humans. He claims that "information is in the domain of machines, knowledge in the domain of humans." The "forces of society" are broken down into a "triangle of knowledge" consisting of the three corners of *science, economics*, and *technology*, much as the "color triangle" used in the theory of color (as utilized, e.g., in a color television) has the corners *red, green*, and *blue*. He introduces these forces of society in order to quantify, classify, and combine the characteristics of knowledge in terms of verb functions and noun objects. With newly defined terms such as these, he feels confident that "purifying the contents of embedded knowledge suddenly appears as a distinct scientific feasibility".

This book has the potential of becoming a must read for a vast readership, including all scientists and engineers in the information technology field, but also psychologists, sociologists, people studying public communications, journalism, advertising, and more. The subject is both very original and highly ambitious.

Eventually, with the ongoing information revolution as typified by the Internet age, someone was bound to tackle the problem of "knowledge filtering", and there could not be a better choice than Professor Ahamed to start the process going.

PROFESSOR GEORGE S. MOSCHYTZ
Professor Emeritus
Swiss Federal Institute of Technology (ETH)
Zurich, Switzerland

Currently, Head, School of Engineering,
Bar-Ilan University, Israel.

PREFACE

In the knowledge society, computational processes dominate the intellectual space. Computers supplement human thoughts as much as automobiles facilitate transportation, or as much as fine arts stimulate the imagination. The role of computers and networks in scientific space is too ingrained to be retrieved. From an intellectual and philosophic perspective, knowledge is the vehicle of social progress. Knowledge-based computer systems serve the dual role of modifying the fuel for progress and the mode of transportation. From internal combustion to jet propulsion, from automobiles to spacecraft, new knowledge has taken human beings from being earthlings to astronauts.

This book deals with knowledge and society. In a confined sense, knowledge is considered as a mathematical entity that can be scientifically processed by minds and machines within a predefined boundary of knowledge. On a social and global basis, knowledge is considered as an Internet-based yet finite and dynamic entity that is processed by super-computers and intelligent networks to serve the society.

Environmental conditions constantly reshape the resources for human needs. At least three pyramids of human needs become apparent: personal, social, and global. Human thinking by itself cannot resolve the complexity of the inter-twined processes in the resolution of such needs within the three pyramids. Machines become helpless without the intuition of humans and their collective wisdom. Process-based wisdom becomes as crucial as an integrated wisdom-based process. In an attempt to cut through this gigantic puzzle, a computational rather than purely mathematical approach is proposed in this book. *Four* dominant trends from the past few decades now fully established in society are abundant and fully harnessed to develop the framework of knowledge.

First, the recent strides in information, knowledge, and computer technologies essential to nudge society forward are evident in almost all network-based information and knowledge systems. Electronic governments, medical systems, educational institutions, research centers, etc., deploy some level of information technology, knowledge processing, computational facilities, and network communication. The technology within these applications provides the first cornerstone for the computational framework for knowledge.

Second, the integration of the global IP-based addressing capability of the Internet and application-oriented processing of data that is the forte of computer systems is already available from worldwide knowledge bases. TCP/IP has provided a harmonious blend of OSI functionality in existing computer and communications facilities. Distance learning, remote hospitals, robotic arms, and sensing systems provide realistic applications for the seamless integration of computers and communication. The networks behind these applications provide the second cornerstone for the computational framework for knowledge.

Third, knowledge and its processing facilitate mathematically accurate, socially benevolent, and economically viable optimal decisions for localized and global environments. The algorithms, specialized hardware, and user-friendly interface embedded in corporate decision support systems, remote medical services, safety nets, and emergency response systems provide the third cornerstone for computational framework for knowledge.

Fourth, combined switching/routing technologies and data communication are features of most traditional digital networks, and the precise computing of massive mainframe computers is already being deployed in global facilities. Space centers, tsunami warning centers, newsrooms, financial markets, etc., provide the fourth cornerstone for the computational framework for knowledge.

Recent groups of clustered innovations have made society dynamic, intensive, and powerful. However, throughout the four systems that serve the three human needs, (personal, social, and global), the lack of a unified theory of knowledge is evident. At least two fundamental issues remain unaddressed in order (1) to make knowledge processing as precise and powerful as numerical and logical processing, and (2) to build a scientific structure for knowledge ware from knowledge operation codes to knowledge application programs.

If we look back at the evolution of societies, the correlation between knowledge within any society and its achievement becomes evident. Human creativity and social environment enhance each other as they contribute to knowledge. Knowledge is chosen as the focus of this book to offer *two* windows: *first*, to look for creativity inside the microorganism of an individual human agent of change; and *second*, to look outside into the macro organism of society, culture, or the surrounding social entity. The derived knowledge offers scientific "connectivity" with data and information, and an "extrapolation" into the goals and ambitions of future generations. Humanistic systems (Chapter 5) function as computer systems driven by human goals and ambitions. These computer systems are primed with knowledge bases to achieve individual, social, and national goals and secured with tamper-proof layers of code to protect the hardware, middleware, software, and firmware. Together, knowledge-based humanistic systems will tackle the problems that face human beings for many generations.

There are ten chapters in this book and the first two offer the platform of technology platform for implementing general-purpose knowledge-processing environments. The book develops a mathematical (Chapters 3 and 4) and a computational framework (Chapters 6–10) for dealing with knowledge. The book explores and deploys knowledge to extend it based on the universality and ruggedness it has developed in the past. Traditional filter theory is ushered into the knowledge domain to purify and decontaminate knowledge. Knowledge filters (Chapter 6) systematically verify the retained authenticity of knowledge in any "body of knowledge" and evaluate the extent of contamination as information passes through any social or broadcast media. The content-to-contamination ratio (in dB) in knowledge is evaluated as a numerical measure of its authenticity. Conversely, such filters can also block serious errors and misconceptions in the representation of knowledge. Knowledge machines that deploy such filters have the numerical estimation of the robustness of a graph of knowledge embedded in the body of knowledge. Broken graphs can thus be mended by these new machines. This book presents a unique theory of knowledge that permits machines and humans to scientifically decompose complex structures of knowledge (in any specified direction classified by the Dewey Decimal System or the Library of Congress classification) and to rearrange knowledge-centric objects in a fashion that suits the specific goal(s) of society, the culture, or the surrounding social entity.

In a sense, the genesis for the perpetuation of knowledge is proposed based on native instructions (Chapters 7) for knowledge-processing machines. Raw knowledge is converted into processed knowledge in order to comply with the specific needs of society at any instant of time (Chapter 8–10). Knowledge never dies; it is manifest in another vein, in another shade, and perhaps in a different mathematical framework. The development of quasi-mathematical formulations is presented by a general theory of knowledge followed by knowledge operation codes for machines that cause incremental changes in the entropy of existing knowledge. The dynamic nature of knowledge is implied as restless minds think and tireless machines process. Only timeless knowledge lingers on as oracles of wisdom, benevolence to society, shrines of beauty.

SYED V. AHAMED
Holmdel, New Jersey

INTRODUCTION

Knowledge is the precursor of human activity and a prime mover of social progress. In its pristine form, knowledge defies the boundaries of physical sciences and leaps over the constraints of space and time. It can extend into universe or break into eternity. All the same, knowledge can also enter the neurons in the brain and the neural nets of reasoning. The constrained representation of knowledge leaves the mind wondering about the origin and use of knowledge. However, when appropriately segmented and linked, knowledge discloses its beginnings but not its end; it is dynamic and continuously evolving. It is alive and being modified by the very thoughts that makes each of us alive and vibrant. In order to be practical and realistic, knowledge still needs to be channeled into gates, processed by circuits, and carried by communications media. Computers and networks now start to appear as knowledge systems rather than semiconductor chips and silica glass. In this book, we propose that knowledge be treated as a structured, abstract entity with a set of well-articulated rules for processing, authenticating, purifying, and predicting its course and contents, eventually to identify the means of generating cohesive and self-sustaining entities of pristine knowledge and axioms of profound wisdom.

Computers and networks serve individuals and societies. User preferences influence the invention and design of systems. At the top of the intellectual chain resides human creativity as an undaunted frontier for all activities. Well-balanced novelty in computer- and network-based systems brings economic rewards and becomes the key to corporate survival. The older inventions are displaced by devices that are more powerful and gadgets that integrate newer services in fewer and more user-friendly devices, leaving more time for creativity and the search for newer gadgets. This cyclic momentum has been building for the last few decades. Society now moves at a faster pace, and the creativity has increased.

Productivity is higher but at the expense of greater human involvement. The newer machines proposed in this book take over some of the human functions. The typical computer systems have not kept pace with society demanding more and more from human beings. In Chapters 3 and 4, we propose a new breed of knowledge-, concept-, and wisdom-based humanistic machines to relieve the human mind from performing the processing of knowledge, isolation of concepts, and derivation of wisdom. These new machines perform well at an intermediate level and imitate the human intellect as humanistic machines

Worldwide acceptance of the OSI model, fiber optics, cellular technologies, broadband networks, TCP/IP protocol, and the Internet have contributed to the demise of the older service networks. The traditional ISDN of the 1980s and 1990s is now replaced by the newer and more powerful frame relay networks that dominate the digital highways. From global networks to household-sensing networks, fiber optics for the transportation of high-speed digital data and asynchronous mode (ATM) for communication have become the present and the future of most networks, just as silicon had become the basis for delivering VLSI technology. More recently, plastic fiber optic technology delivers data within small networks at a fraction of the cost of other technologies, further enhancing the role of frame relay networks.

In Western societies, Marshall's laws govern microeconomic behavior in dealing with human and corporate needs and the Keynesian perspective governs macroeconomic behavior in dealing with domestic and international needs. The social needs and legal framework are addressed by politicians and ministers who constitute the driving force in implementing a secure environment. The wisdom and ethical aspects are left for philosophers, priests, and pastors with very little power to sense, maintain, and monitor these aspects. In a sense, a vacuum at the nucleus of human personality where wisdom and ethics reside starts to appear.

This book also introduces the concepts of white, black, and green knowledge based on monetary wealth having three classifications as white, black, and green monies. In the knowledge domain, information and knowledge derived and compiled at authenticated and classical institutions and propagated for socially beneficial reasons may be considered white knowledge. The main emphasis is that of validity and universality conveyed by that body or occult of knowledge. Scientific, ethical, and generally benevolent use of such knowledge is implicit. On the other hand, information and knowledge from questionable institutions, war camps, military colleges, and torture training institutions carry the label of black knowledge. Like black art, black knowledge has an aura of questionable ethics and abuse of knowledge. The existence of gray or murky knowledge becomes real in dealing with white knowledge and black knowledge. Selfishness and arrogance of users lurk behind individuals and nations that deploy such black knowledge.[1]

---

[1] In most situations, the existence of black knowledge is generally overlooked. In practice, the abuse of such knowledge is much too abundant in politics, corporations, and even at the interpersonal level.

The most frequent form of abuse of knowledge occurs during coercion and cruelty of individuals and nations.

It is clear that when a machine pursues the ethical foundations and/or implications of such gray or murky knowledge, the motives of the abusers are quickly exposed. In a sense, the science and methodology in pursuing knowledge are directed toward the foundations and implication of its use. The machine thus acts as a filter to block the spread of virus-ridden knowledge much as the anti virus programs block the spread of virus-ridden information.

In many cases, the extent of murky and irrelevant information overwhelms an unprepared mind. The immediate response of the Internet challenges the freedom of human thought to be free and creative and thus soar above ignorance. The human mind that is destined to reign free and leap over apathy is often trapped in trivia. It becomes a paradox that the information society which offers freedom also drowns the senses. It almost blocks the mind from being socially responsible and astute. Global harmony and peace are sold for an Internet song of gray knowledge. The long-term effects are the pursuit of trivia and a bias in the communications media.

Modern machines can process knowledge as easily as computers can process numbers. Fiber optic networks offer (almost) unlimited bandwidth. Context-related information processing is as easy as a UNIX-based search. It becomes a paradox that the information society which offers freedom also corrupts the senses to be socially responsible, astute, and mindful of global harmony and peace. The trait and bias in the communications media that disseminate the murky information become insidious within it.

Industrial societies drive the modern world. Here, the profit motive is dominant. Many times, that motive is camouflaged and almost invisible to unsuspecting individuals and social entities. Social viruses such as deception in advertising, the misrepresentation of facts, and the willful distortion of truth are taken casually in most societies. The open practice of swaying public opinion (especially during elections and public speeches) challenges the foundations of honesty and integrity in the audience. In an urgent sense, the tolerance to such practice is a willingness to slide down into greater deception. The defense against such a downward spiral lies in the education and reinforcement of ethical conduct in business, education, government, and community.

On the one hand, the truth, virtue and beauty (TVB) embedded in information and knowledge facilitate the mind to spiral toward peace, social justice, and wisdom. On the other, the deception, arrogance, and hate (DAH) that lie hidden in information and media clips push the mind deep into the plots of self-preservation, selfishness, and arrogance. Both attitudes and accompanying behavior are common in individuals, corporations, cultures, and nations. The reactions to perceived threats become so real that nations are ready to initiate wars at the drop of a hat.

CHAPTER 1

NEW KNOWLEDGE ENVIRONMENTS

All men by nature desire knowledge.
— **Aristotle** (384–322 B.C.), *Metaphysics*

CHAPTER SUMMARY

This Chapter deals with the synergy between computer and communications systems that makes the new knowledge environments urgent and feasible. Redundant information appears to overflow beyond all constraints and boundaries. A basic methodology to contain, organize, and deploy the overabundance of information in academia, business, medicine, electronic media, governments, and financial institutions is proposed for quick and effective separation and the subsequent deployment of sensitive information, valuable knowledge, key concepts, and lasting wisdom.

Key concepts necessary for building specialized computers and networks are introduced. Architectures of subsystems and standardized platforms for software are summarized from computer sciences. Specialized switching platforms and organizations are borrowed from communications systems engineering. The last three decades of fiber-optics development for the transmission of binary data and advent of packet switching technology for routing information are examined to incorporate intelligence in the networks. The novel strides in both computer and communication disciplines are interwoven to develop and design intelligent medical, educational, security, tsunami warning, and government systems.

Computational Framework for Knowledge. By Syed V. Ahamed
Copyright © 2009 John Wiley & Sons, Inc.

The historic landmarks in hardware, software, and firmware are now firmly embedded in these machines. Putting things in perspective, one realizes that knowledge processing is as feasible now as data processing that appeared in the late 1940s, or as feasible as very large-scale integration (VLSI) that appeared in the early 1950s. The ultimate success of many enduring methodologies depends on the collective will of the industrial community. Major trends are initiated by potential financial gains, provided the technology exists. For example, global information networks prevail because of fiber-optics technology, and modern diagnostic cancer and surgical procedures exist because of magnetic resonance imaging (MRI) techniques. Social demand plays a role in increasing potential rewards. In the same vein, if knowledge processing can solve some of the greater and global intellectual needs of human beings, then it is conceivable that machines will be built simply to address our curiosity at first and then to establish the foundations of wisdom for later generations.

1.1 THE NEED TO KNOW

In the modern world, precise and incisive knowledge is essential for accurate decision making. Maps of knowledge networks in the Internet domain are not unlike road maps in commercial centers or neural maps in the human mind. Knowledge environments influence intellectual settings in academia, business, and the financial markets. In the knowledge domain the confrontation between the knowledgeable and the ignorant leads to the survival of the deftest. Skill in the Internet society is based on possessing a mental map of knowledge bases around the globe and their access methodology. These concepts are well exploited by communications engineers. Telephone numbers, web addresses, and even satellites addresses are deployed effectively to reach out to anyone in any location. Implementation in the field of communications occurs via these numerical addresses and is based on the routing algorithms programmed. Such programs are decoded and executed in the VLSI circuits of large electronic switching systems [1], routers [2] and wireless stations [3].

The synergy between the intelligent Internet, computer, and network technologies has accelerated the growth of the information and knowledge society. This unprecedented growth calls for new knowledge based technologies for the powerful elite in society, thus enhancing their power.[1] The full synergy of knowledge and power acting in concert is being retraced now as it was in the days of Francis Bacon. *Human intelligence*, now supplemented by machine intelligence, *society*, now supplemented by global transportation, and *networks*, now supplemented by trans-oceanic fiber-optic information highways, form a tightly confined triad of forces that keeps the global economy moving in a long and sustained period of

---

[1]Knowledge and human power are synonymous, since the ignorance of the cause frustrates the effect. Francis Bacon (1561–1626), *Aphorism III*.

expansion. These three prime movers are discussed next in the context of the current social setting.

First, consider human intelligence. A gift of heredity in humans, intelligence flourishes in the lap of society. Sustained intelligent decisions propel positive social movements. This symbiotic relationship develops early and lasts for long periods of spiraling growth for peoples and nations alike. Nature and humans coexist in harmony as a gift of intelligence in human beings and intelligent societies composed of human beings. Such societies cultivate the healthy natural processes resulting in the social adaptations of humans. In a majority of cases, the cycle is broken by natural and self-imposed disasters. The symbiotic relationship between intelligence and society needs unbiased mentoring and maintenance for a positive cycle of growth.

In the information age, information and knowledge-based machines can maintain the stability of a triangular balance between intelligence, society, and nature. It appears impossible to maintain stability within the vertical axis of this triangular balance without premeditated and complementary changes to all the three mitigating factors and the newer society driven machines and networks can find the optimal blend for an intelligent society.

Far from being stationary, this compounded pendulum of intelligence, society, and nature swings continuously. Swings of the socioeconomic balance can be gentle and easy, or violent and brutal. On the one hand, appropriate human attitude nurtures societies, fosters intelligence, and preserves nature. On the other, improper attitude kills societies, forbids intelligence, and destroys nature. Enhancement of positive intelligence is social responsibility. The continued revival of negative intelligence is a human process fueled by greed, selfishness, and hate in those who suppress society.

In the present context, the mentoring and monitoring of the entire scenario dealing with the deployment of intelligence, movements in society and preservation of values are best accomplished by the precision of machines and by interconnectivity of networks. Numerous parallelisms exist. The Federal Reserve system and economic institutions mentor and monitor[2] the flow of monies, production of durable goods, and employment in nations. In monitoring patients in hospitals, machines perform well under human supervision. When the processes become sophisticated and complex, collaboration and synergy between machines and humans become desirable. Robotic production lines, corporate management, and petrochemical industries deploy such a strategy of collaborative and synergistic effort.

Second, consider society. Composed of individuals and leaders, society becomes a mirror. The collective self of a group and prolonged, well-accepted

---

[2]The econometric models of developed countries involve many hundreds of parameters and a set of interdependent equations that govern the economic activity. The equations embedded in the models are generally solved by matrix inversion methods and time series analysis by econometricians to make projections for the future. Such methodologies to govern and stabilize the social values and ethics of a nation can be developed and solved by the sociologists of the future.

attitudes of its constituents are reflected. When the mirror is concave and focused, it can integrate human activity toward a path of progress for the entire society. If the mirror becomes warped, the human activity clutters local hot spots, causing social unrest and violence in the community. When the mirror becomes rough and convex, individual activity is splintered and defused without bringing the due social rewards to the community. In reality, neglected societies experience every type of ill. Such sick societies are restored by the polished and geometric profiles by disciplined and charismatic leaders. Social progress follows. The intellect within society and the capability of leaders become the precursors to success.

Human beings playing a collaborative role do provide an environment for social progress, but machines can design and construct the tools for social change. Knowledge machines provide a basis for systematically cumulating the conditions and circumstances for such social change. In the medical field, MRI systems provide the demarcation between healthy and cancerous tissues in the human body for surgeons to perform delicate surgery. In the VLSI environment, simulation programs offer the circumstances when pulse shapes deteriorate beyond repair.

The monitoring of all the components for desirable social change is a full-time and relentless task. If it is human to err, then the errors are compounded much too fast to sustain steady social growth. If it is mechanistic for computers to be accurate, then accuracy is safeguarded, at least for a while. If it is human to be creative, then creativity can lead to a sustained and wise social goal. In this vein, it is our contention that accurate machines and creative humans be intertwined such that the combined system is better than machines or humans. It is possible to combine the positive qualities of machines and humans toward a steady social pattern for growth over relatively long periods. On the other hand, it is equally possible to couple killer machines with brutal humans.

Under ideal conditions, a balanced hybrid system of humans (creative, selfless, and well-intentioned) and machines (information, knowledge, and well-primed wisdom) will outperform the current mostly human and little machine system. It is apparent that the little machines presently in use are the plain computer systems of the past. It is also our contention that the humanistic machines proposed in this book are wisdom machines [4] when primed with human values and ethics.

Third, consider networks. Telecommunications is the backbone of commerce. Financial transactions drive humans and nations to search for the means to satisfy human and social needs. Society churns and life goes on. Telephone networks sufficed during the early days. The digital era made began with circuit-switched digital capability (CSDC) within the telephone networks during the late 1970s, during the 1980s, digital services were implanted in simple telephone networks such as the plain old telephone service or public service telephone network (POTS or PSTN) by deploying the integrated services digital network (ISDN). But a impact occurred dramatic and sustained since the inception of fiber-optic technology for the local and global networks. More recently plastic fiber optics have held the promise to making fiber networks robust, cheap, and universal for carrying digital data on local networks.

Apart from the strides made in for physical media, the switching technology has evolved dramatically. From human operators to the programmable electronic switching systems (ESSs) of the mid-1960s, progress has been slow. The switching speeds for modern communication systems needed the innovations of the transistor and its integration to be imported as fully evolved computer systems. Numerous versions of these ESSs served as the major building blocks in the networks of 1970s to 1990s.

Packet switching rather than circuit switching made an impact during the 1950s and 1960s. Having been developed for computer networks rather than telecommunications, packet switching offered a new approach to the design and deployment of switching protocol. The universal form of this transfer control protocol is the TCP/IP, used for Internet switching. Most modern networks deploy high-speed switches and routers.

Optical switching devices brought a dramatic impact in networking technology. Interconnection between globally dispersed locations occur via a network command and fibers transmit information at mega-rates—too much for the human senses to absorb and assimilate. The glut in the optical capacity of networks does not trigger faster human senses. Evolutionary by nature, optic and auditory nerves retain their capacities. Additionally, the processing of information into retained knowledge in the mind is much slower.

Minds retain wisdom and networks carry bits. Perceptually, the information divide between bits and wisdom is deep and wide. The mind is lost in this divide that it cannot take a quantum leap into the wisdom domain from the photonic data in optical fibers. This book presents a methodology to systematically traverse the divide along the knowledge highway that spans the two distant nodes. This book also extends this well-trodden span into the social norms and ethics that are the pinnacle of wisdom. A beautiful human face starts to appear in the cyber space of sharp binary pulses of data.

The simplest form of a graph is feasible with seven nodes (bits, data, information, knowledge, concept, wisdom, and ethics) and six links in between. A progression of order and reason is suggested by installing the six links between the seven nodes, with data and ethics at the extremities. Accordingly, these six links are:

1. Low-level logic circuits to respond to the binary bits
2. Computers to process the structured binary bits as data
3. Information technology platforms to instill order in the data
4. Knowledge machines to generalize and categorize information
5. Concept machines to construct a framework of rationality
6. Wisdom machines to derive universal axioms of wisdom for human beings to "plug and play" in other social settings

In reality, all the links between all the nodes are not ready, even though the links (1) through (3) and partly into (4) are well entrenched in technology. In the modern era, networks have evolved significantly to keep pace with the growing collections of human needs. Since the age of fiber optics, networks have grown at a pace faster than the financial activities to sustain society. Additional bandwidth has brought about the gadgets and devices that suit most whims and fancies of society rather than its needs. One of those needs is to continue to search, even if the search is for the unknown or the search itself is an unknown. The urge to search is a search for the unknown. Before the machine becomes lost in a wonderland (labyrinth) of fantasies, its human counterpart can stop the futile search for humanistic machines and realign to solve realistic social problems.

1.1.1 Global Power of Knowledge

On the positive side, information-rich societies, like networked nations, benefit from quick and global access to numerous knowledge bases around the world. A similar phenomenon occurred at the beginning of the Industrial Revolution, when access to technology provided a basis for the wealth of nations. On the negative side, lagging societies are likely to suffer the ill-effects of the growing knowledge divide between the two sectors within the same society. From an economic perspective, the wealth of information for networked nations can build a crag between knowledge-rich and knowledge-poor nations. However, the laws of economics governing the distribution and flow of wealth of knowledge are distinctly different from the laws of economics that govern the distribution and flow of money. The impact of the knowledge revolution is very likely to be different from that of the Industrial Revolution.

The struggle of nations is ridden with wars and deprivation. Yet, most nations have weathered the wave of the Industrial Revolution around the globe. Such a revolution is likely to reappear again in the future but in the knowledge domain. In the earlier era, the lower or more basic needs of societies were met by the products of the Industrial Revolution of the sixteenth century. In the current scenario, the higher-level needs of societies are likely to be chess pieces in the game of knowledge revolution now unfolding. The writings of Karl Marx who predicted the exploitation of labor (*Economic and Political Manuscripts* in 1844, *Theory of Surplus Value* in 1862, and *Das Capital* in 1867) are most likely to become relevant again during the exploitation of the ignorant, the illiterate, and undernetworked nations in the next two decades. Exploitation appears to be the only commonality between the two revolutions.

Wealth and power are built on a foundation of knowledge. Devices and machines have slowly taken over the functions of humans. A snapshot of the current and futuristic technologies to process knowledge and to extract concepts and wisdom from it is depicted in Figure 1.1. The forward pace is accelerating, and machines have already displaced the simpler functions of human thinking. More complex thinking falls in the domain of artificial intelligence (AI) embedded in machines. The roles of machines as the means for wealth by industrialization and power by military confrontation has become evident.

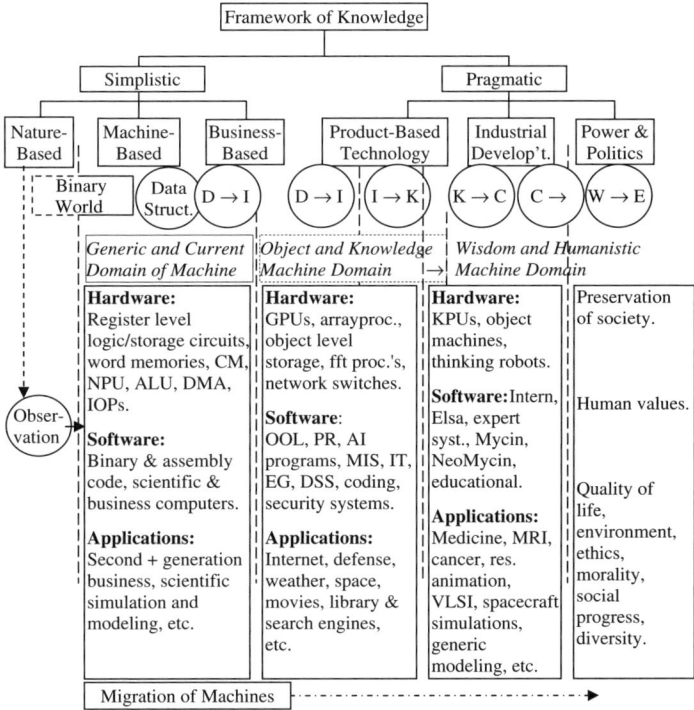

Figure 1.1 Overview of current and futuristic technologies to process knowledge and to extract concepts and wisdom from it. The letters B, D, I, K, C, W, and E stand for binary, data (and data structures), information, knowledge, concepts, wisdom, and ethics.

For underprivileged nations and societies, astute and accurate self-analysis in the early stage of knowledge revolution will prevent a life of intellectual slavery in the decades that follow. After all, the slave trade flourished for a century. Guns and ships allowed British and Portuguese sailors to exploit the social conditions in Africa. In the knowledge society, the strategic algorithms in AI and well-guarded knowledge banks are likely to become the new weapons and technology. The unfolding knowledge games that people and nations play are the early warnings of the knowledge wars to follow. It is proposed here that machines sense, maintain, and monitor an ethical framework and balance for the distribution of the wealth of knowledge.

Societies like humans have needs. In the Utopian knowledge society, monetary wealth becomes subservient to the wealth of knowledge. Knowledge (and wisdom) about monetary wealth will lead to its optimum utilization, and the marginal utility theory applies for the distribution of wealth. The classic writings in Sanskrit emphasize that wisdom without greed is the gateway to *unselfish* wealth. In the modern context, wealth is a transition but wisdom is a state. Knowledge, wisdom, and wealth make up a triangle of hope and fear for the struggles of

8 NEW KNOWLEDGE ENVIRONMENTS

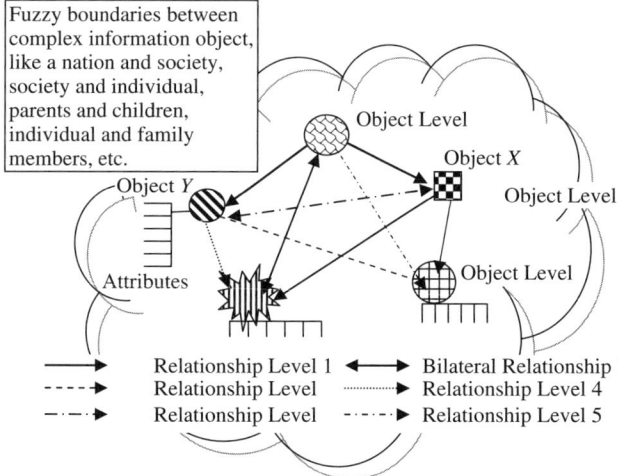

Figure 1.2 Representation of a complex information/knowledge object with five lower-level information objects. Strong, weak, casual, unilateral, and bilateral relationships are shown.

humans and nations alike. To capture the essence of this, we present the roles of humans and machines into two Figures 1.1 and 1.2.

Internet is the prime mover of information for the resolution of needs. In most instances, relevant information causes the mind to blossom but a flood of information drowns creativity. A rapid and unrestrained glut of information floods the senses and drenches thought. In fact, all the irrelevant information obtained via high-speed networks burdens the mind. In the post-information revolution, the mind has become specifically sensitive to the validity and pertinence of information. A tunable information filter serves the need to separate out junk from germane information.

In most cases, human beings are ill equipped for these tasks and the result is distraction and lack of focus to solve the immediate problem. The purpose of this book is to introduce methods of implanting some of the low-level human functions, such as selective filtering, pattern recognition, enhancing, and relevancy checking, in machines and knowledge-processing systems. Humans function at a higher level in terms of innovation, invention, creation, transformation, transmutation, etc. The interaction between a human being and a machine will be that of rider and an intelligent horse in the connected terrain of knowledge, concepts, and wisdom.

This capacity to carry through human concepts of communication in hardware is a skill of human beings. Human beings program, that is, fragment any complex function repeatedly until it becomes a hardware-realizable VLSI or photonic process, as a very systematic and scientific step-by-step task. Programming occurs at numerous layers, ranging from invoking a sequence of powerful global Internet commands to the most rudimentary binary-level instructions to invoke micro-functions in logic circuits. In many instances, it becomes impossible to "see through" the thought process of a human being represented as a sequence of logical processes in circuits. The fragmentation of tasks, and the layering and organization of software functions (encoded as routines) offer the ultimate methodology to build complex intelligent machines and networks.

Since the inception of the central processing unit (CPU) in early computers of the 1940s and 1950s, machine instructions have remained in extensive use. However, in order to make the computer more popular, languages for numerous applications were introduced. As the applications have become more universal, the languages and their compilers have also become sophisticated. In a sense, the evolution of higher-level languages (HLLs) has made the use of computers easy for the user. The introduction of application programs with a graphical user interface (GUI) has made any end-users capable of executing intricate software programs on blocks of input user data. The final manifestation is the ease and comfort with which Internet users may access and deploy the information and knowledge stored at any World Wide Web (WWW) address.

Logic circuits and their operations have not changed drastically for some time, but the organization and layering of hardware have evolved dramatically. The skill to compose large complex functions involving the fewest high-level end-user commands (including the clicking of a mouse) has now shifted to the designers of Internet and applications-level software, compilers, and assembly-level programs. The software to execute the user functions in the VLSI circuits embedded in sophisticated hardware is as much a part of the knowledge environment as automobiles are a part of human society.

1.1.2 Scientific Aspects

From the days of the Industrial Revolution, significant changes in society came about from some form of a scientific breakthrough. The foundation of social and materialistic progress generally rests on the well-founded law(s) of physics. Such scientific achievements may be singular or cascading. Once the origin of new scientific thought is defined, progress is generally fueled by society to deploy the invention and exploit its novelty. Current opportunities brought to bear by information technology only enhance the momentum of change.

Over earlier generations, science and wealth did not bear a direct correlation. However, the exploitation of guns and ammunitions to generate vast amounts of wealth by plundering natives was the proven strategy of thugs and pirates. British slave traders and Spanish conquistadors all left deep scars on the face of human civilization. Still practiced in the knowledge and information domain, this artifact of violating the privacy of humans and nations is not a scientific achievement. To some extent, it is indicative of the thievery and piracy of information. However distant or late secure network technology may appear, it offers a glimmer of hope against the abuse of information.

The pinnacle of information and knowledge machine performance is the ability to be ethical in transacting business and executing programs that are not detrimental to society. Initially, implanting this type of intelligence in machines appeared formidable, but the operating systems of present machines do not permit the execution of malicious code. Machines do not transmit virus-ridden information over the Internet. Current medical information code prevents the prescription of dangerous combinations of interacting drugs.

A different scenario emerges when scientific pursuits and human greed are blended in one human mind. Over the last few decades, the greed for wealth emanating from commercial enterprises has not diminished. In a sense, scientists who have acquired a taste for wealth have become greedier. The alarming result in the knowledge domain is the reincarnation of the exploitation (in Marxist terms) of the ignorant. Ill-informed humans stand to lose a lot in a deluge of print and fine print. A new breed of machines that retain a global perspective on the problem and the nature of the proposed solution will provide better solutions and warn users about all the positive and negative aspects from an unbiased perspective.

1.1.3 Wealth Aspects

The deployment of any scientific innovation for the pursuit of wealth is tolerated and sometimes desired by nations. In most cases, the conduct of humans and nations in the pursuit is generally less than ethical. The stringent enforcement of laws is expensive and time-consuming. When the choice lies between being wealthy or ethical, morality is generally shortchanged at an individual and national level.

1.1.3.1 Industrial Development Industrial development is an offshoot of technology and technology is based on innovative ways to satisfy human needs. National development occurs as a result of a chain of cause–effect relationships that lead to an enormous accumulation of wealth. By the same token, human opportunism is also a chain of economic events that govern the price of essential goods. As long as essential needs exist, the opportunity for the accumulation of wealth also exists. In a sense, price as it is determined by supply and demand leads to the accumulation of wealth and that is, in most cases, a precursor to the greed of individuals and nations.

When driven by an obsession for wealth, the manipulation of conditions for maximizing it becomes a preferred strategy for industrialists and industrialized nations. Adam Smith was the first to write about the wealth of nations and Alfred Marshall was the first to enunciate the laws of supply and demand. The superposition of Smith's notion about the wealth of nations on Marshall's law of supply and demand leads to the combined strategy for maximizing the wealth of nations that can curtail supply to maximize price. This drama has been played out by the OPEC countries (or by DeBeers Corporation) over the last few decades in growing wealthier. Such an obsession with profit or wealth by energy companies becomes evident every winter as the need for home heating reaches a peak.

In the knowledge domain, the derivation of a viable strategy for the accumulation of lucrative knowledge bases (KBs) is more devious. However, knowledge as a resource to generate technology based on inventions still prevails in its own niche. The pursuit of knowledge rarely becomes an obsession or greed for knowledge, except for a few scientists and philosophers. Knowledge wars are as unlikely as saintly killings, even though Bell and Edison, Tesla and Marconi fought many scientific duels in their claim to fame. Glory and fame are the objects of greed that conveniently spill over into greed for wealth.

1.1.3.2 Materialistic Achievement Scientific and materialistic achievements have a very fragile boundary except for the most ethical scientists. In the hierarchy of achievements, materialistic wealth provides the foundation for scientific progress. The wealth necessary to satisfy lower-level needs is the necessary and sufficient precursor for individuals and nations to develop the scientific methodology to satisfy higher-level needs and thus cross the barrier of knowledge. It becomes a personal challenge for an intellectual to resist the pull that drags one into the quicksand of materialistic greed.

1.1.3.3 Global Power of Nations Power is distinct from knowledge or wealth, even though wealth by itself may offer some power. However, knowledge is power, and the knowledge of deadly weapons is a basis for deadly power. For nations, an arsenal of knowledge on deadly weapons represents deadly power. In a somewhat surreptitious way, nations fund nuclear technology to build an arsenal of bombs, fund missile technology to instigate star wars, fund communications satellite technology to make spy satellites, etc. In most cases, the knowledge gained from funded research becomes a by-product that may have other marginal value.

In a deceptive game with intellectuals, nations siphon and store deadly knowledge for the supremacy of global power. Along the knowledge trail that is meant to benefit society and humankind, offshoots toward the national greed for power start to appear. The global superpowers have practiced this strategy with no uncertainty many times over. The greed of nations needs an open articulation of the wealth of nations as expressed by Adam Smith. Karl Marx had openly rallied against the greed of capitalistic society and the exploitation of labor. It remains to be determined if the historic voices of Smith and Marx, will rise again for the greed of nations for global supremacy and the exploitation of the human knowledge and intellect.

Knowledge and intellect are more illusive resources than money and economic goods. The laws of mirco- and macroeconomics governing knowledge and intellect have yet to be formulated. Virtual and distant as knowledge and intellect appear, more abstract and remote are the new breeds of economics for knowledge and intellect. In a sense, a viable theory of knowledge (Chapter 3) is only an entry into the domain of knowledge.

1.2 ROLE OF TECHNOLOGY

Technology advances human ambition and eases the burden of the mundane. From primitive tools to the spacecraft, inventions achieve immediate and concrete goals. Technology becomes self-perpetuating because human goals keep increasing and grow dynamic. The hierarchy of technology almost reflects the pyramid of human needs. Human beings have needs and inventions resolve such needs; the technologies to provide optimal solutions then to follow. The laws of economics emerge and a systematic collection of knowledge is accumulated. The expansion of needs-based knowledge is inevitable. In the current knowledge society, the expansion of knowledge accelerates further because of the expansion of needs, of innovations, and of the numerous technologies to resolve the needs. Human efforts to maintain a continuum of knowledge, is made feasible by a new breed of knowledge machines rather than conventional computer systems.

1.2.1 Three Major Contributions

1.2.1.1 Stored Program Control and Conventional Machines Eckert and Mauchly proposed the Electronic Numerical Integrator and Computer (ENIAC) in 1943, and von Neumann proposed the Electronic Discrete Variable Automatic Computer (EDVAC) in early 1945. The Institute of Advanced Study (IAS) unveiled its machine in late 1945 at Princeton. The IAS machine evolved amid a flurry of innovations in vacuum tube engineering and the inception of semiconductors. A rigorous systems approach to the overall design of the IAS computer system was launched rather than assembling different gadgets. Some of the basic gating functions of logic had been already implemented.

The quantum jump that von Neumann proposed was the design of the entire system. The IAS machine architecture established the concept of stored program control that permits the main programs as well as the data to be stored in the main memory as binary bits. The format and representation of these two entities are different. The program and data could thus occupy segments of main memory, and the execution of machine instructions could take place systematically according to binary/assembly-level programs. The program counter contains the address of the next executable instruction.

The design of the CPU was the basis of the breakthrough in the machine. Many modifications have followed since the 1950s, but the five-step sequence (fetch instruction, decode instruction, operand fetch, execute operation code or *opc*, and store interim results) in the execution of each binary instruction has not changed drastically. The main theme of the von Neumann machine is presented since some architectures of the knowledge-processing unit are derived from von Neumann's original proposals and the subsequent enhancements. such as the single instruction multiple data, (SIMD), multiple instruction multiple data (MIMD) and multiple instruction single data (MISD), pipeline, array processing, etc.

The concepts in von Neumann's earlier Electronic Discrete Variable Computer (EDVAC) also deployed the stored program control (SPC) concept for controlling the functions within the machine and used the binary arithmetic logic circuits within the CPU. It also used a two-level memory system for the computer: (1) a 1,024-word faster memory block, and (2) a second block about 20 times this capacity (but slower) as the secondary memory. Some of these concepts are still in use today.

The most distinguishing and innovative features of the IAS machine are its memory size, word length, operation code (OPC) length, and addressable memory space in binary bits. The machine has 4,096 addressable locations in its memory, and the word length at each location is 40 bits, holding two instructions each with one 8-bit byte for the OPC and a 12-bit address field. The machine also uses a fixed bit binary format (with 40-bit representation), with the first bit denoting the sign of the number in the next 39 bits. Thus, two instructions or one data word would occupy one memory location and the computer could execute one instruction after another and keep computing like any other modern machine.

1.2.1.2 Unix and C Environments The UNIX operating system and C language from Bell Laboratories brought the operating systems concept to a new level of freedom for the user. In 1965, when electronic switching systems were introduced [1], the need to control the ESS due to the growth and long-distance nature of telephone calls became apparent and a whole new language was essential to meet the needs of the communications industry. Essentially, a new operating systems (OS) environment (UNIX) and the C language [5] made computers more powerful to serve sophisticated and demanding users. The concepts introduced in the UNIX OS and their deployments in the C language have been enhanced further in the C++ language for midlevel users. For more sophisticated users, LISP [6], PROLOG [7], and more recently introduced Java [8] languages offer greater flexibility in programming complex tasks on machines.

In essence, the new operating systems and languages provide users with the flexibility to interact with the CPU functions (arithmetic, logical, branching, and switching), which was denied by HLL software designers. The UNIX OS simplified the numerous layers of older operating systems and the C languages made HLL software more universal and powerful. The ease of programming in HLLs was slightly compromised to achieve more powerful and optimal instruction sets to solve a more complex and generic breed of problems.

1.2.1.3 The Internet The Internet has brought about a profound revolution in modern society. General-purpose computing facilitated the popularity of computer systems in the 1950s and 1960s. The introduction of inexpensive personal computers in the 1980s and 1990s brought the revolution home. Since the early 1990s, Internet has provided access to information highways within the minds of the younger generation. In modern society, the Internet is the venue to satisfying the most natural instinct of curiosity by surfing and searching cyberspace. It has opened the conceptual dimension and shown that, like the human mind, knowledge derived from the intelligent Internet is infinite—well at least semi-infinite.

There is no generic patent for the Internet. It is free from any gains to be claimed by any commercial entity. Internet ownership does not exist even though current Internet service providers (ISPs) charge a fee for the access and services they provide. In a sense, the Internet concept freely floats from any country to another, any socio-intellectual circle to the next. When the backbone telecommunications network of any country can host the rather rudimentary requirements of packet switching, Internet access is readily provided, offering every user in one country an international gateway to other users and knowledge bases.

The Internet is a willful and determined consolidation of the inventions from many contributors over many decades. During the 1960s and 1970s, the federal government sponsored loosely related research topics as independent projects based on the explosive growth of computational systems. These projects generated islands of knowledge that were founded on sound scientific basis developed at universities and corporations. Later, these islands were bridged and made into a larger body of knowledge by additional investigations lasting for two or three decades, from the 1970s through 1990s. Information highways became a manifest reality.

The Internet is a perfect example of the evolution of two intertwined scientific and business strategies. Most of the significant contributions in the telecommunications arena from the 1960s through the 1990s were systematically accumulated on the Internet platform. One the one hand, the Internet shines brightly on a scientific and mathematical basis because of major inventions in communications. On the other, the Internet is fueled by the economic motives of its participants. Synergy still prevails. Inventions came about, and the Internet sustained its own growth in a socially beneficial and an economically viable fashion. The rewards of the intelligent Internet are already altering the direction of human thought, toward the possibility of processing knowledge. The processing of numbers, symbols, and objects is a well-established technology. Over the last decade, the pace has accelerated and global learning in the knowledge society further adds to Internet technology as it drives society toward greater intelligence, deeper wisdom, and beautiful ethics.

Two federal sponsors [the Office of Aerospace Research (SRMA) and Advanced Research Projects Agency (ARPA)] facilitated the fast track of the Internet. The Aloha system, initiated in the late 1960s, resulted in a novel way to share most transmission media. Although it was developed for radio networks linking numerous campuses of the University of Hawaii and other research facilities on the Hawaiian Islands, it paved the way for random access techniques based on CSMA/CD. Next, under the sponsorship of ARPA, the ARPANET evolved to effectively use computer resources. During the mid-1970s, ARPANET had gained full operational status serving the needs of the Defense Communications Agency. Later, it was used by the Department of Defense (DOD) to transfer both classified and unclassified information and to meet most of the DOD data communication needs. The seminal blending of information processing and mass communication started in the 1970s, and it was funded until the concept gained maturity in the 1980s. During the 1990s, most generic concepts and algorithms were firmly established. The packet network technology is now a cheap reality and has gained immense popularity.

1.2.2 A String of Secondary Contributions

The triadic effects among fiber optics, semiconductors, and the quantum technologies of the 1990s are also an example of both the symbiosis and synergy that have facilitated global gigabit nets. The entire evolution of technology and its implementation are quite recent. A platform of earlier inventions was essential for the string of recent innovations. The role of accelerated inventions and innovations is obvious through the 1980s (for VLSI technology) and 1990s (for the network and fiber-optics technology), and in the present decade at a more moderate rate. Perhaps the saturation that curtails the duration of the creative phase in any discipline is a social phenomenon.

1.2.2.1 VLSI and Ensuing Computing Systems In the early days of computers, gating was accomplished by vacuum tubes. The odds against building large computing systems were immense. The chain of inventions (transistor, planar transistor, integration, SSI, MSI, LSI, and finally VLSI) that followed quite quickly from 1948 to 1965 was indeed the pathway to sophisticated computing systems. The series of computational architectures (SPC, SIMD, MIMD, multiprocessing, parallel, array, distributed, and networking) have further accelerated the progress toward more efficient, widely distributed computer facilities. The series of programming, algorithmic, and software strategies have converged, giving computer technology an edge over any other technology. The closest competition comes from VLSI technology that is a direct beneficiary of computer technology because VLSI is totally dependent on computers (computer-assisted design or CAD programs) that are driven by the VLSI industry of the prior generation. The symbiotic cycle has continued for several generations and is likely to endure further. Quantum computing may well be the next frontier.

Computer devices function with their gates performing logic (and, or, ex-or, etc.) and arithmetic (add, subtract, multiply, divide, trigonometric, etc.) functions in the CPU environment. In an input/output (I/O) environment, the communicating devices are assigned and activated. In a storage environment, the bus access and store/retrieve functions are performed by a variety of digital devices. These devices may also provide signal processing functions by digital techniques, thus leading to digital signal processors. Microprocessors emerge when these processors are further integrated to perform generalized and programmable logic, plus control and arithmetic functions. And, with additional bus access and random access memories, or read-only memories, microcomputers emerge. Size, capacity, and facilities lead to larger and larger systems.

1.2.2.2 Switching and Communication Systems Electronic switching systems (ESSs) are computers in their own right. Massive parallelism (to handle many millions of call simultaneously) dominated the architecture. Dependent on

VLSI technology, ESSs have been the beneficiaries of a string of inventions in the semiconductor industry and the computational power of modern computing systems. The growth of massive switching systems was stymied by the viability of fiber-optics technology, by the global acceptability of the SONET frame relay systems, and then by the asynchronous transfer mode (ATM) cell relay technology. Much less expensive routers, bridges, and gateways had gained prominence for communicating packets through the national backbone networks.

The switching function in a computer system, although essential, is not dominant. The input/output channel, direct memory access, and cache memory channels exist and are switched under the operating system and users' commands. Akin to the virtual channels in communications systems, they are set up on demand and disbanded after the exchange of information occurs. In network environments, the switching function is essential and dominant. Real-time, concurrent control of many thousands of communications channels is essential at most of the nodes within the network. For this reason, the design and architecture of switching systems are a major area of specialty and can become as intricate as the design and architecture of any major mainframe computer system.

Modern digital network devices also function with gated logic. However, the channel management aspects of the network dominate, rather than digital and binary arithmetic. Switching and allocation of channels (switched or packet) and their monitoring become a major network function; the network devices, though digital, are designed for channel control. Control of the network to perform under dynamic conditions needs more adaptation than the control exerted by computer operating systems. Switching of communications channels needs specialized hardware (electronic switching systems), which closely parallels mainframe computers. Both are driven by stored programs, and both have similar architectures and track individual tasks. The communications tasks are shorter but much more numerous: One mid-sized ESS office may service up to 100,000 lines with as many as 10,000 independent channels being active at any one time. Specialized switching also occurs at channel banks that support these ESSs.

The more recent ESS environments handle up to 200,000 lines for every possible combination of services, such as local/toll operator services, wireless, intelligent networks, integrated services digital network (ISDN), and packet and all switching with ATM front-end processors. With the advent of ISDN, the functional demands on these switching systems are expected to increase dramatically, and the complexity of the tasks is also expected to increase. Furthermore, any elaborate national communications network may have hundreds of these switching centers or network nodes. Hence, the network hardware deviates from the conventional computer hardware and follows an expansive architecture to suit the increasingly demanding and intelligent network functions. Fortunately, the hierarchical steps followed by computer scientists in complex computer designs can be, and have been, adapted to network systems. The transmutation of the logical steps traced in both the hardware and software applies to the network architecture and its driving software. Complex computers become basic network building blocks, interspersed with transmission and access facilities, in the same way that general-purpose processor chips have become the basic building blocks of most complex computers. The software driving these computers differs, just as the microcode that differs between processor chips.

1.2.3 Peripheral Contributions

Business and financial environments also have specific needs. Over a period of time, numerous software configurations have come and gone. One of the languages for business applications developed during the 1950s and 1960s was ALGOrithmic Language (ALGOL). The extended use of ALGOL for rapidly increasing business needs and programming occurred in the 1960s and 1970s. In a broad sense, business users have not been as demanding and sophisticated as scientific and telecommunications users. Consequently, the effects of their use have been noticeably different. In most cases, the basic hardware for processing of data for business applications has not changed drastically except for the size and interconnectivity of peripheral devices.

When the demands on the applications processes have become intense [such as in magnetic resonance imaging (MRI), signal processing, radar tracking, missile technology, etc.], the hardware has undergone significant changes. New inventions and innovations have kept pace with the sophisticated demands. Battlefield robots were as much a fiction in the early 1990s as electronic switching systems (ESSs) during the early twentieth century. Yet, both are realities in the early twenty-first century.

Unimaginable progress has been made. Machines routinely assemble MRI scans [9] and generate three-dimensional images of cancerous malignancies in human beings, autopilots [10] navigate airplanes, robotic controls [11] make landings on planets, and drone planes [12] chart enemy territory. It may appear inconceivable that a machine will be able to process the knowledge embedded in information, much less be able to extract wisdom from information. It appears there is no problem unless it is defined. The definition of a problem is half its solution. The conceptual barriers are meant to be broken down, the paper tigers are meant to be burnt down, dark fibers are meant to be deployed, and every Everest is there to be climbed.

The origin or source of incisive knowledge appears immaterial, provided it is accurate, scientific, and deducible. The processing of knowledge can be machine-based, human, or a blend of the two. In the final analysis, the distillation of wisdom is a series of cyclic and coherent processes between the logical deductions of machines and the inherent creativity of humans.

1.3 KNOWLEDGE AND WEALTH

The domain of knowledge is more encompassing than that of wealth and materials. For dealing with the utility of knowledge, all factors (its scarcity, its total utility, its marginal utility, specifically its diminishing marginal utility, its utilitarian value, its exchange value, etc.) that influence the evaluation need to be considered. From a communications perspective, knowledge can be traced backward and extrapolated forward, much like scientific parameter(s). From a structural perspective, we propose that the processing of knowledge be based on the most basic and fewest truisms. These truisms are, in turn, based on reality and they permit the characterization of information and knowledge. To this extent, computational processing does not depend on the philosophic writings of earlier economists. However, the truisms are validated by a longer-term philosophic interpretation of how these truisms have survived so that they can be expanded and reused in scientific and computational environments. This approach permits machines to process knowledge based on the content of a particular piece of information and to enhance the content, presentation, and wealth of knowledge that the information communicates.

To face the reality of the current information revolution, the monumental advances in computer, communications, and information technologies need firm functional integration. Modern machines can process knowledge as easily as computers can process numbers. Fiber-optic networks offer (almost) unlimited bandwidth. Context-related information processing is as easy as a Unix-based search. With such strides in technology, an information glut almost drowns an ill-prepared mind. The Internet challenges the freedom of the human soul to remain free and to soar across oceans of ignorance. The human mind, destined to reign free and leap over the mountains of apathy, becomes trapped in the weeds of trivia. It becomes a paradox that the information society which offers freedom also corrupts the senses, preventing humankind from being socially responsible and astute and safeguarding global harmony and peace. The biases within the communications media that disseminate the information become insidious within it.

On the one hand, the truth, virtue, and beauty embedded in information and wisdom enable the mind to grow and migrate toward peace, social justice, and wisdom. Yet, on the other, the deception, arrogance, and hate that lie hidden in information and media clips cause the mind to spiral deep within the prehistoric caves of self-preservation, even in the absence of any threat. Fear of fear has an exponential growth. Such attitudes and accompanying behavior are common in individuals, corporations, cultures, and nations. The reactions to perceived threats become real in well-armed nations initiate wars for no reason expect to deplete the arsenal.

The role of the Internet in bringing information and knowledge to every member of an educated society has increased rapidly over the last decade. Over the last few years, the four directions of human achievement have been intelligence (both human and artificial), the Internet (both conventional and semantic), wisdom (both traditional and web-based), and finally, networks (both modern and broadband). These four directions appear unrelated at first but soon become interwoven: They blend into each other in a synergistic and harmonious way without bounds. The four-dimensional space and our way of thinking are unified freely, encompassing intelligence, the Internet, knowledge, and networks. These concepts are explored further in Section 1.5.

1.4 EVOLVING KNOWLEDGE ENVIRONMENTS

The evolution of society is based on the systematic collection, validation, and deployment of gainful knowledge. Knowledge can range from gossip to well-guarded national secrets. Gossip and rumor that have little value are filtered out of the computational processes. On the other hand, knowledge that is rare or unique enters the computational domain to be examined, refined, and enhanced. Knowledge is collected systematically (from Internet traffic), validated extensively (from Web knowledge banks), and deployed widely (from the dictionary of axioms available from Web wisdom bases). The true wealth of knowledge (if there is any) is thus evaluated rather than the raw format in which it was presented. Knowledge processing becomes a precursor to the enrichment of knowledge or the distilling of wisdom.

In a symbolic sense, knowledge has two major components: (1) the embedded "objects" around which knowledge is gathered and (2) the process that such objects have undergone to modify the knowledge surrounding them. If there are attributes assigned to such objects, then such attributes need to be measured, documented, and/assigned. Hence, any object forms a nucleus of some knowledge around it. Nothing without any history has no knowledge associated with it, and the nothingness of nothing is no attribute. However, in a more positive sense, every object has some history and some attribute(s).

Knowledge is inclusive of objects around which knowledge is gathered. If objects are classified as nouns, then such objects undergo change through the verbs operating on and around them. Invariably, a verb causes a change in the status of a noun object and corresponds to a *kopc* (verb) executed on a *kopr* (noun). In this content, *kopc* stands for knowledge operation code and *kopr* stands for knowledge operand.

A sentence that is a part of any textual message in any language has at least one verb and one noun. Traditionally, most of the messages processed by the human mind enhance knowledge or understanding in the context (vectorial direction) in which the message is perceived.

In the same vein, every executable statement of a knowledge program needs at least one *kopc* and one *kopr*. The format of multiple *kopc*s and *kopr*s for the command structures of knowledge processors is discussed in [13]. Invariably, a certain duration of time is essential for the verb function (force) to complete any change in the status of the object. If the extent of change thus accomplished is ΔS in Δt seconds, then the symbolic representation of the incremental change in status is as follows:

$$\Delta(\text{Status of a noun object}) = (\text{Intensity of the verb function}) \cdot \Delta t$$
$$\Delta S = \text{Force} \cdot \Delta t$$

An object undergoes a change in its status ΔS depending on the intensity of the verb function exerting a force over an interval Δt of time.[3]

If status (energy) is considered entropy in a particular discipline, then the change of energy is the scalar product of force and displacement. The change of status ΔS of an object (i.e., the change of entropy for that object) is a vector in any given discipline and equals the intensity of verb function times the extent of change over an interval of time to achieve the desired change. For example, if status is wealth, then the change of wealth equals the earning rate (i.e., /hr) times wealth saved over the duration of the time worked.

In reality, such an interpretation becomes more complex in the knowledge domain. If status is the knowledge of a learner in a particular discipline, then the change of knowledge is the scalar product of intensity (e.g., the concentration of the learner dedicated to that subject equivalent to "force") and the duration is time of study.

In such cases, concentration toward study should be considered a vector in the direction of the discipline, and the knowledge gained is the scalar product of concentration toward study in that subject matter and the duration of time studied. The "object" is the learner; the "verb function" is the scalar product of concentration of the learner for that subject and the time spent. Once again, if the concentration varies with time, then the scalar product becomes a time span integral of scalar products at any instant t times dt.

The status (or entropy) of any one object or a group of objects becomes akin to the knowledge retained by a human being or group. In a sense, when an object group is formed, then the status of the combined knowledge undergoes change. Such changes can be slow and insidious or sudden and explosive depending on the intensity of the verb functions and nature of the noun objects.

In knowledge machines, the object environment is tuned and optimized much like the adjustments of social groups in society. The processing of objects within the knowledge machines thus leads to the optimality of performance and achievements of object groups. To this extent, the knowledge machine is capable of performing some preliminary managerial functions based on objects and their attributes.

---

[3]This equation is *similar* to the equation (work = force · displacement) in particle dynamics. If force and displacement are considered as vectors, then a scalar product is implied. In the knowledge domain, time is of greater essence. A measurable change ΔS occurs in the status of an object (or a collection of objects) by exerting one (or a series of) verb function(s) for a duration Δt. Stated alternatively, the equation states that the rate of change of status of noun object is the intensity of the verb function. The equations germane to particle dynamics quickly lose their analogy in the knowledge domain. The equations for rigid dynamics hold some similarity, but even these equations lose their true analogy. It remains to be seen if the entire set of rigid dynamics equations (that include translation and rotation) may have a better correspondence with entropy and its incremental change in the knowledge domain over a given period of time.

1.4.1 Components of Knowledge

There are at least three main components of knowledge: (a) the information regarding a group of n objects around which that knowledge is clustered, (b) the interrelation ships within the group, and (c) the status of each of the n objects. Other secondary knowledge classified below as (d) pertaining to the relationships between the attributes of objects, and attributes of the attributes also exist. It is unclear which of the components may significantly change the character of the clustered group. However, a knowledge machine is well equipped to determine (by scanning the existing knowledge banks) which objects, attributes, and attributes of attributes influence the traits of the group and their relationships. Such estimations may become necessary as the knowledge machine composes and recomposes new objects as it processes knowledge. Thus, knowledge about anything can be written down as:

(a) The n objects in the group that constitutes "anything"
(b) The web of implicit or explicit verbs (bondages) that retain the status of each of the n objects, and their structural relationships with each other to be constitutes as "anything"
(c) The status of each of the attributes, their relationships with objects, and their own interrelationships
(d) The secondary relationships

For developing a methodology for the machine representation of knowledge, if the list of the components is terminated at (b), then knowledge can be represented as an array of n objects and a matrix of the status of each object. The matrix can become cumbersome because each object may have numerous attributes, and each object and/or each attribute (at any given instant of time) would have undergone a series of verb functions to change its status.

To be totally accurate, the cumulative effect of each verb function from $-\infty$ to now ($t = t_1$) at this instant is necessary. However, the simplification and standardization of objects can be deployed to write down the (b) components of the n objects. For example, an "object" car is a vehicle with all the standard secondary objects/attributes (wheels, engine, steering wheel, etc.). If the application of this object is transportation, the machine does not need to have any more information. But, if the need arises, the machine can look up the database for the particular make of the car and find its secondary attributes and the attributes of attributes.

A symbolic representation of knowledge can thus be written with two terms representing the (a) and (b) constituents of knowledge:

$$(\text{Knowledge})_{t_1} \equiv \sum_{i=1}^{i=n}(\text{Group of NO}_i) \cdot \text{plus} \cdot \int_{t=0}^{t=t_1} \sum_{j=1}^{j=v_1}(\text{Series of VF}_j) \cdot dt$$

This equation should be considered representational rather than mathematical. The symbol ·plus· indicates that different kinds of elements are involved. The first term denotes noun objects NO_i and the second term indicates the integrated effects of a series of j verb functions VF_j for each ith of the n noun objects. This relationship is explored and enhanced further in Section 3.2.3.

In the context of allocating memory to a body of knowledge, two linked arrays are necessary for (a) and (b). The matrix for (b) may be numeric (integer, fixed point or floating point data), alphanumeric (to determine object identification), logical (linkage, types) relational (with degree of influence), and/or qualitative (e.g., excellent, very good, good, acceptable, unacceptable, etc.). Storage allocation becomes an integral part of programming knowledge-based problems.

1.4.2 The Processing of Knowledge

The processing of knowledge assumes two components: (1) the filtering of raw information to create "knowledge-rich" information by selecting appropriate "noun objects" with most dominant structural relationships with other noun objects and (2) the processing of "noun objects" with machine-executable "verb functions" or knowledge operation codes on such operand noun objects. The execution of a series of verb functions on a collection of interrelated noun objects thus modifies the structural relationship between the noun objects and creates new noun objects. The combined effect of a series of verb functions v operating on a group of n noun objects from time $t = 0$ to t_1 yields the net entropy of that body of knowledge (in the "noun object" group) and is recorded in Section 1.5.1. The implication is that:

1. Knowledge is dynamic.
2. v events (verb functions or VFs) operating on the n objects will alter the entropy of the body of knowledge encompassing the n objects.
3. The knowledge at any instant of time t_1 is a result of all the knowledge functions that the n objects have experienced over the *entire* history of that knowledge.

The well-designed knowledge program will enhance the entropy of information to suit a social, individual, or scientific need. It becomes desirable to make the new noun objects and their structural relationships practical, viable, and realistic. The new knowledge derived from the old knowledge and new "objects" (situations, scenarios, institutions, societies, etc.) is thus "computed" (and conceived) as the knowledge processing takes place. Knowledge processing becomes a process of creating new objects and enhancing their structural relationships. The distillation of wisdom becomes a series of cyclic and coherent processes between the logical deductions of machines and inherent creativity of humans. These concepts are explored further in Chapters 3–5.

1.5 STRUCTURE AND COMMUNICATION OF KNOWLEDGE

Material and monetary wealth was discussed Adam Smith and has evolved as the basis for national and international trade and commerce. John Maynard Keynes (1883–1946) and his fiscal policy issues are still held in esteem in monitoring the growth of nations [14]. Unlike monetary wealth, combined information and knowledge $(I \ll \gg K)$ have many facets and implications. Whereas the measurement of wealth is scalar and has a numeric measure as the currency value, the wealth of knowledge has more numerous measures. After all, the evolution of society is based on the systematic collection, validation, and deployment of gainful knowledge. Knowledge can range from hearsay to well-guarded national secrets. Unfounded information and gossip have only marginal value and such information has no significance. On the other hand, if knowledge discloses a rare discovery, an invention, or a trade secret, then its value is at a premium. If the information has social significance, is rare, and is still not disclosed, then the value of that information is high. However, information kept in total secrecy has no value unless it is derogatory or damaging. Even long and extended periods of torture are justified for prisoners of war who supposedly have "information" about the enemy! Rare and damaging information has only blackmail value. For these reasons, the economics and strategy for dealing with knowledge and information require different considerations from those established in typical economics or game theory [15].

A certain commonality exists in the economics of knowledge and traditional macroeconomics. Money that becomes stagnant and does not get invested leads to the liquidity trap [14]. The business community is unable to invest and grow because economic opportunities are too few even though interest rates may be low. Valuable technological information that does not find its way into production lines remains dormant as paper in patent offices. In a sense, the possibility of an information-rich but stagnant society starts to become real, somewhat like Japanese society in the 1980s. Valuable knowledge and information (like money) need deployment. Like savings that are invested (savings = investment in classic macroeconomic theory), knowledge (knowledge = production in a knowledge economy) distilled from information needs to be channeled into corporations. Channeling such knowledge into institutions of learning creates a multiplier effect (like that in the national economy) in the $(I \ll \gg K)$ domain.

1.5.1 Velocity of Flow of Knowledge

A certain *velocity in the flow* of *information and* knowledge $(I \ll \gg K)$ is necessary for either information or knowledge to be productive. Information that grows too stagnant (like money during liquidity trap conditions) or too fluid (like money during rampant inflationary conditions) loses its potential to be socially valuable. A certain *viscosity in the flow* of $(I \ll \gg K)$, like money flow, makes the activity rewarding and economically justified. Information that finds no channel(s) for communication has exhausted its life cycle.

A limited commonality also exists in the economics of information and traditional microeconomics. The value of information and knowledge $(I\ll\gg K)$ that is transacted is initially comparable to the value of goods or assets that are transacted. However, $(I\ll\gg K)$ does not get depleted like goods or assets that are physically exchanged. The depletion of the value of $(I\ll\gg K)$ involves exponential decay rather than a sudden change. The rate of decay of the worth of information can be quite sudden (high exponent) for some types of $(I\ll\gg K)$ (e.g., weapons and warfare technologies) compared to others (e.g., educational or medical technologies). The sharing of $(I\ll\gg K)$ may bring down the value as exponential decay, but it still retains some utility for both parties. Both parties benefit from the economic rewards yet retain the wealth of information. Monetary and material wealth that is shared loses value and utility simultaneously. The value of $(I\ll\gg K)$ varies with server-client relationships. Conflictive and cooperative roles are both feasible, thus altering the laws of economics of knowledge and information.

Mainly, $(I\ll\gg K)$ that has social, financial, ethical, or moral implications is a resource that is not as immediately exhaustible as monetary or materialistic wealth. Like any other resource, $(I\ll\gg K)$ can be accumulated, enhanced, stored, or even squandered; however, this resource has special properties. The enhancement of $(I\ll\gg K)$ is a mental/machine activity differing from the enhancement of material wealth, which is a production/robotic activity. For the differences cited above, $(I\ll\gg K)$, "objects" are treated as hyper-dimensional objects that follow the laws of processing but are not quite aligned with the processing of numbers, scalars (such as currency values), or text. Modern computers are capable of processing vectors and graphical objects. Current software packages that handle complex number $(x + iy)$, two-dimensional space for electrical engineers and mathematicians perform as smoothly as those that handle three-dimensional (X, Y, Z) space for graphic designers and movie makers. In dealing with $(I\ll\gg K)$ "objects," special compilers are necessary. Such compilers should perform lexical, syntactic, and semantic analyses of information objects that can identify other information objects and relate them to the newly found objects by variable and adaptive role-based linkages. A recursive compiler can handle such a scenario.

The processing of graphic entities [16] starts to assume the initial flavor of the processing of information objects. Some of the steps suggested in this chapter are initial and rudimentary, but they can be modified and enhanced[4] to suit different types of information object(s) and their interactions. Processing of information objects depends on the application. On the one hand, mechanical and routine transactions of information objects are akin to data processing in banking and commerce. On the other, when information has human and social implications, then a new software layer that emulates human processes (such as love, hate, needs, feelings, education, counseling) becomes necessary. Generally, human interactions follow the underlying economic framework of the exchange

---

[4]Algebraic (multiply, divide, matrix, etc.) operations for complex numbers, the development of software routines followed much later, after the development of assembly-level programs for processing real numbers.

of resources. On a very short-term basis, the marginal utility theory [17] starts to unfold in most transactions. Perceived fairness and valuation are of essence in most cases. In dealing with information, most humans follow a fairness and value judgment analysis unless it is willfully transgressed.

The rational component of human processes follows simple programming approaches. The emotional component is tackled by suggesting (and adapting) a series of statistical paths ranging from common to rare reactions. Such reactions are documented in knowledge bases around the world, and steps are adapted in neural networks. In such instances, the machine-generated resolution of information can be superior to an all-human solution because machines can evaluate every type of emotional response in every culture and can suggest a customized response closer to the tastes of the humans involved.

While machines are communicating or exchanging information, they strictly abide by the I/O commands of humans or of the core operating system. While human beings process information, the value and worth of the information are initially assessed and modified by learning, clarification, and negotiation. While machines are processing information, the information-processing units[5] (IPUs) alter the structural relationships between objects and other objects, objects and their attributes, and relationships between object X attributes with object Y attributes. This scenario is depicted in Figure 1.2. The alteration and redistribution of relationships are not altogether random (unless they are the last resort).

Instead, they are based on laws of probabilities as to which of the relationships are most common and likely to form secure bonds (e.g., information about gasoline and octane values, or information about hang gliders and the wing span of birds). In the process, the machines also investigate unusual and uncommon relationships (e.g., information about the design of hang gliders for Australian coasts and those for Scandinavian coasts), giving rise to novel and unique information, knowledge, or scientific principles (if any).

Machines have an advantage in processing vast amounts of information quickly and accurately. The incremental changes in information are tallied to the incremental changes in external conditions to optimize and predict the information for a given set of new conditions. Incremental changes over any of the parameters (such as time, attributes, or environmental conditions) are accurately tracked and labeled. Processing the key information object(s) that forms the nucleus (nuclei) of the raw information and then reconstituting the information object(s) identify opportunities for possibly new and valuable information, knowledge, or scientific principles. This is the fundamental clue to crossing from the information mode to the knowledge mode.

---

[5]Information processing and knowledge processing are used interchangeably in this paper since the forward processing (distilling) of information leads to knowledge and the backward processing (parsing) of knowledge leads to information. The same machine may be able to process in either direction, perhaps by changing its control memory chip sets. At this stage, it is premature to speak specifically of the many possibilities that still lie ahead.

Unlike monetary wealth and the wealth of nations [18] that are depleted, the wealth of information is shared. Unlike monetary wealth, $(I \ll \gg K)$ has significantly different attributes. Whereas universal and numerical values can be assigned to monetary wealth, information has overlapping qualities and fuzzy parameters to transact information.

Complexity theory [19] starts to resemble knowledge theory because of the highly variable nature of $(I \ll \gg K)$ "objects" and their interrelationships. Most of the precepts of complexity theory become applicable when dealing with information and knowledge. However, in dealing with $(I \ll \gg K)$, we limit the processing to a confined number of objects that will not make the information processing chaotic. The self-contained structure is statistically prioritized with weighted relationships between the objects that are considered valid for the processing of $(I \ll \gg K)$ "objects." In addition, the limitations of the computer system (accuracy, memory size, speed, and possible switching capability) define the size and "body of knowledge" (or "complex initial object") that the machines will handle.

The knowledge-processing system filters out any "objects" that are likely to cause chaotic and unstable oscillations in the processing. It refuses to process inconsistent information, much like computers that refuse garbled data. During the execution phase, irrational requests to process information are terminated and the error condition introduced, just like computers that refuse to execute impossible numeric operations. Unlike complexity theory, knowledge theory will perform legitimate functions on objects for which some earlier statistical information is available in world-wide knowledge banks. If the extent of information is too restrictive, the learn mode [20] is invoked to build a knowledge base for the unknown object. The machine guards itself from being drawn into an execution mode that will end in catastrophe by establishing noncircular forward and backward pointers. Even though recursion is permitted, the depth of recursion is made consistent with machine capacity. Rationality is given higher priority than the task of executing a knowledge program. These bounds of rationality contain the fuzzy bounds of knowledge that is under process. To this extent, the machine regains its own stable operating condition, just as a human being would attempt to do. Thus, overall knowledge-processing systems have a fair chance of solving complex knowledge problems that human beings by themselves cannot attempt.

The knowledge-processing system limits the size of the body of knowledge processed by a quantitative measure of the capacity of the machine in relation to the requirement of a "complex initial object." No such limitation is imposed in complexity theory. For this reason, knowledge theory is based on the computer systems that will attempt to solve a knowledge problem. Knowledge theory is a valid tool in initially formulating a problem and adopting a strategic solution to it. The system resources expended to change the status of information and knowledge (see T2 in 1.5.2 also see P3 in 1.5.3) during the course of the problem's solution will be (in most instances) the bottleneck. In essence, complexity theory is an open-ended theory, but knowledge theory works in the context of machines having discrete (binary or hyperspace) representations, limited in their memory, I/O, switching capacities, and speed of operation.

To this extent, knowledge theory is like information theory that works in most nonchaotic but extremely noisy environments. Knowledge theory does not violate any of the principles (such as auto-organization, edge of chaos, power of connections, circular causality, try and learn, and olograммatic principle) set forth by complexity theory. To some extent, auto-organization and try and learn are based on a survey of the world-wide knowledge bases on the Internet to find out how other complex knowledge objects have accomplished auto-organization and adaptation. To this extent, quantification within knowledge theory (like that within information theory) becomes feasible.

Shared information loses value at a relatively low rate. Whereas there is a suggestion of the strict zero-sum game [21] in transacting the wealth of nations and individuals, there is an impression of the *elastic zero-sum game* as two parties share knowledge and information. Wealth (i.e., all the derived utilities combined together) and value rather than just the price of information are perceived at the time of sharing information. The sale price of a commodity or an asset can only rise in a free-market environment. The price for sharing information is arrived at by both buyer and seller and not determined by market forces. Sometimes, the value of information in a document or book far exceeds the price of the written work, sometimes, the converse can be the case.

In the knowledge domain, an approximation of the scarcity, value, and life of the information is feasible. In terms of scarcity, value, and life (a three-dimensional curve), five coordinate points may be readily identified: (1) totally unshared and secret information that has no value and an indeterminate life, (2) guarded information that has high value and a relatively long life, (3) information shared with a select clientele that has the highest value until it starts to leak and slowly erodes in value, (4) media information that has a media-established price and a short life, and finally, (5) gossip and trivia that have junk value and dissipate without a trace. The value of information in a socioeconomic setting has at least three additional dimensions: the truth contained, the elegance or appeal conveyed, and the social benefit that can be derived from the information.

To deal with the complex nature of information from a computational and processing perspective, we propose four dimensions or senses (truism, philosophic, scientific, and economic), as shown in Figure 1.3, in which information can be characterized. In dealing with information as an object, the truism of all information objects (not their content) states the truth (as well as it is known) about the entire object class. Similarly, the philosophic characterization of all information objects (not their content) states the philosophic nature (as well as it is known) of the entire object class, etc.

Processing an information object can alter its four characteristics (T, P, S, and E). In fact, constructive processing will make marginal information objects into significant information objects, if any significance exists. Worthless information

28 NEW KNOWLEDGE ENVIRONMENTS

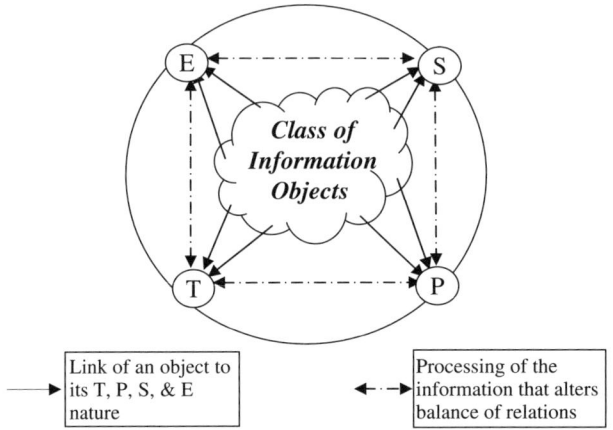

Figure 1.3 Representation of the class of knowledge and information objects with four characteristics: truism, philosophic, scientific, and economic nature of knowledge.

is filtered out from any scientific knowledge processing.[6] The process may be to deduce, interpret, derive, systematize, analogize, categorize, conceptualize, rationalize, and generalize any other process that has a scientific basis.

In order to initiate the information processing that will search out new information or new knowledge from vast amounts of information, three steps are proposed: observation of reality, philosophic validation, and scientific principles that can be generalized and deployed elsewhere. The observation of reality is fundamental to all sciences. Because information has illusive boundaries and flexible formats, the concept resides in the content of information and goes deeper than a statement or representation of information. In a sense, information is the water that can be poured into any vessel. The water represents a real information object with its own properties and the vessel the secondary information object. Together, they form a (partially) stable object group.

Philosophic validation is necessary to provide the long-term continuity and stability of information, such that any inference/scientific conclusion can be drawn. To continue with the earlier example, if the water is poured into a vessel carved out of ice, neither water nor ice will form a stable[7] object group (i.e., water in a vessel). It then becomes necessary to probe the wealth of knowledge and information (water) in society (vessel) to validate their reality as a stable object group.

---

[6] Most compilers block programs (program objects) from proceeding to the execution phase unless they are free of all syntactic, semantic, and linkage errors. In a similar vein, information that is inherently false or malicious, or ridden with pornography, will not gain access to information-object-processing systems.

[7] Unless the situation is adiabatic at 32°F, which becomes too specific to draw any general conclusions.

The derivation of scientific principle(s) becomes an act of courage: standing firm in the observation of reality and a philosophic validation to make a universal statement or hypothesize about the object–object relationship (H and OH in H_2O, or electrons beams and electromagnetic fields in cathode ray tubes) as a scientific truth. The bold steps of scientists (e.g., Hamming, Shannon, Bose, Hocquenghem, Ampere, Gauss, Maxwell) as they formulated their significant conclusions, based on their own human information-processing capabilities, are still valid. To continue with the earlier example, the scientific principle that water, which can be poured, has low viscosity is universal and it can be used in other instances and situations. In the information domain, very fluid and fast-flowing gossip would not have value, because it is not stable enough to become useful as a scientific principle.

We attempt to follow the three-step procedure (reality, philosophy, and science) to get a computational handle on processing information. It becomes possible to extend the three-step procedure into a fourth step and derive an economic basis for dealing with knowledge and information. Both knowledge and information are known to have value. The rules for dealing with them as object-based abstract commodities start to become significantly different from the economics of materialistic wealth.

1.5.2 Truisms in the Knowledge Domain

The observation of reality over long periods leads to generality or truism. In dealing with knowledge, three important concepts are suggested:

T1: *Knowledge has a life cycle.*
T2: *Knowledge can be altered, but any alteration of it requires an expenditure of energy.*
T3: *Knowledge has impact.*

This list is short and other dependent truisms can be derived from the three listed above. The truism layer is shown at the top of Figure 1.4. The T1 to T3 list is kept deliberately short in the hope that a listing of derived scientific principles will also be elementary and short. This would reduce the basic operations that a computer system will have to perform while processing knowledge.

1.5.3 Philosophic Validation of Knowledge

Only four philosophic validations ($P_1 - P_4$) are suggested and the list is deliberately kept short to reduce the instruction set for the machine to process knowledge. It is depicted as the middle layer of Figure 10.3.

Based on T1, the justified philosophic validation (at this time) is as follows:

P1: Knowledge is timely or obsolete, and it can change its characteristics over time. Typically, human or machine processing changes derived knowledge.

30 NEW KNOWLEDGE ENVIRONMENTS

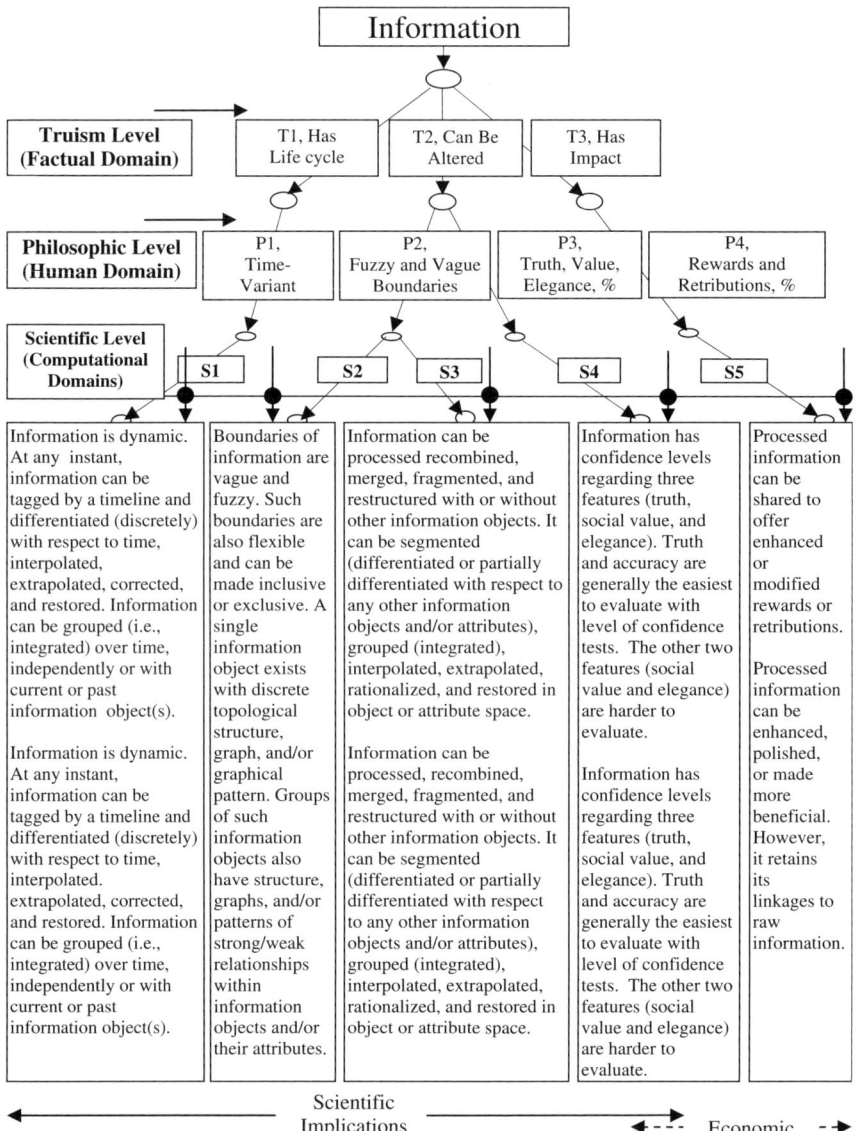

Figure 1.4 Separation of the factual, human, and computational realms of information to derive knowledge, wisdom, and concepts from it.

It ranges from mere gossip to a scientific principle or an equation in physics. When the linkages to raw knowledge are not retained, the processed knowledge may assume the identity of a new knowledge object. Hence, Aristotle's concept of beginning, middle, and end becomes subjective in the knowledge domain. We refer to this particular validation as P1.

Based on T2, two philosophic validations (P2 and P3) are feasible and presented separately because their implications are different:

P2: *The boundaries of knowledge and information are vague and fuzzy.* Returning to the example of water in a vessel, information is blended in human perception. Much like the features of beauty that lie in the eye of the beholder, the boundaries of knowledge lie in the mind of the receptor. Human perception becomes a fading memory (or a leaky bucket) to hold knowledge (water). When a machine receives knowledge, the knowledge objects, their structure, and their relationships are analyzed and stored with the timeline for that specific "body" of knowledge. In a sense, the "knowledge compiler" performs a lexical, syntactic, semantic, and timeline analysis on "knowledge inputs" and identifies the knowledge objects, their structures, and their relationships.

Implication P3 based on T2 is:

P3: *Knowledge has three qualitative features;* truth contained, social value conveyed, and the inherent elegance of content in variable proportions. Knowledge can also have the opposite features (falsehood, social malice, and ugliness) in variable proportions. An equally important principle is that the change of status of knowledge implies an effort (equivalent to force) to bring about the change sustained over the displacement of the status, thus invoking a concept of psychological or social energy or the deployment of resources.

To fall back on the example of water in a vessel, if the water carries three partially dissolved solutes (sugar, sweetener, and honey), then the viscosity changes, thus altering the fluid mechanics and concentration levels in different sections of the vessel. Furthermore, any alteration of the concentration level, after an equilibrium condition is reached, needs energy for the change (such as stirring, shaking, vibrating, or adding more water). The scientific basis for predicting the concentration contours becomes quite complex and even unpredictable (like the weather). However, when a machine has a basis for estimating the truth (sugar), social value (sweetener), and elegance (honey) independently (based on a statistical sampling of other knowledge objects and their relationships), then the raw/processed knowledge can be scientifically evaluated with appropriate confidence levels.

Based on T3, the validation for P4 is stated as follows:

P4: *The sharing of knowledge can bring rewards or retributions in any variable proportion.* This particular implication carries little impact in the scientific domain but becomes significant in the social and economic domains. In the socioeconomic realm, it is a generally accepted practice to exchange items of similar value (including knowledge, patents, techniques, and ideas).

It is also frequent to find the extent of damage inflicted as retribution. In the knowledge domain, litigation and penalties are imposed when negative knowledge and bad publicity are purposely circulated.

1.5.4 Scientific Principles in the Knowledge Domain

Five scientific principles (S1–S5) are derived from the four philosophic validations (P1–P4). The first principle, S1, results from P1 and is stated as follows:

> *S1: It is implied that knowledge is dynamic.* At any instant, knowledge can be segmented (differentiated with respect to time, culture, nations, etc.), encoded, communicated, corrected, interpolated, extrapolated, restored, and even reconstituted. Knowledge can be grouped (i.e., integrated over time, etc.), independently or with current or past knowledge object(s). If knowledge objects are treated as dynamic and continuous in the time domain, then differentiation and integration become possible. The analog and closed-form operations are irrelevant, but finite and event-driven changes are sensed from information and knowledge bases. For instance, every scientific meeting or conference adds or subtracts from the collective knowledge base of a community. Human beings and/or machines can process new knowledge objects continuously. When finite changes are necessary, then the commitment of resources becomes essential. Hence, the concept of (expected) incremental or marginal costs is evaluated and equated to the (expected) incremental or marginal benefit that is gained.

The second principle, S2, is derived as an extension of P2 and is stated as follows:

> *S2: The boundaries of knowledge and information are vague and fuzzy.* Such boundaries are also flexible and can be made inclusive or exclusive of other knowledge objects. Single knowledge objects exist as topological structures, "graphs," and/or graphical patterns. Groups of information objects also have structure, graphs, and/or patterns of relationships within the information objects and/or their attributes. Structures, graphs, and patterns (see Figure 1.4) can have scientific implications for stability. In the domain of knowledge, the knowledge objects need reasonable bonds to remain in existentence for any length of time. Insecure bonds between objects only result in short-lived rumors and gossip.

The third principle, S3, also results as an extension of P2 and is stated as follows:

> *S3: A knowledge object can be processed, corrected, recombined, merged, fragmented, and restructured by itself or in conjunction with other knowledge objects.* It (they) can also be segmented (differentiated or partially differentiated with respect to other knowledge objects or attributes), grouped

(integrated), interpolated, extrapolated, rationalized, and restored in object or attribute space. The basic tools of discrete mathematics become applicable in dealing with the continuity of knowledge over time and the continuity of structural relationships or discrete contours with respect to other objects or their attributes.

The fourth principle, S4, results from P3 and is stated as follows:

S4: Knowledge has confidence levels regarding three features (truth, social value, and elegance). Truth and accuracy are generally the easiest to evaluate in the context of other similar single or multiple knowledge objects with level of confidence tests. Generally, (local and global) knowledge bases that contain knowledge about similar objects can provide a basis for confidence tests. The other two features (social value and elegance) become harder to evaluate.

The fifth principle, S5, results from P4 and is stated as follows:

S5: Processed knowledge can be shared to offer enhanced or modified rewards or retributions. Processed knowledge retains its linkages to raw knowledge. The human processing of knowledge has taken a firm hold in society. Transitory knowledge processed by the human mind is dispersed as conversation. Knowledge that is more important is documented and retained for further reference. In the realm of processing by intelligent machines or systems, knowledge can provide more value (truth, social significance, or elegance) in the processed mode, especially if the processing is done on a scientific basis by following principles S1–S4. For example, segmentation and recombination offer a slightly different form of truth (that is equally valid) than the original truth. Similarly, the mere rearranging of the words can sometimes make a hidden context or idea become more apparent, etc. From a computational perspective, simple differentiation tests (i.e., event analysis and correlation studies) can reveal the more sensitive knowledge objects with a complex knowledge structure.

1.5.5 Aspects of Knowledge

In the comprehension of knowledge, two aspects need to be addressed. *First*, in dealing with knowledge, the truism tempered by long-term philosophic validation leads to scientific principles. These principles are formulated as qualitative and statistical relationships to start a basis of knowledge theory by which the differentiation, integration, and sensitivity of knowledge can be estimated. Primary and secondary knowledge objects are introduced to offer knowledge structure and dependence. The quantitative basis and content of a body of knowledge are established by the number of secondary objects, their structural relationships, the number of attributes of each secondary object, and their own relationship matrices.

34 NEW KNOWLEDGE ENVIRONMENTS

Granularity of the knowledge space is defined as the smallest prism formed by the numerical precision of the computer systems, the lowest Hamming distance between the code words that the networks can carry at their maximum speed, and the perception of human beings who will sense the microprism of knowledge. At least one dimension of this prism is personality-dependent, even though the numerical precision of the computers and lowest Hamming distance through the network can be accurately quantified for that particular human–machine system.

Second, in dealing with the theory of knowledge, the comparison with complexity theory shows that knowledge theory is closely intertwined with the quantity of knowledge (see the paragraph above) in any primary knowledge object. However, knowledge theory is always retractable and (almost) never becomes chaotic for three reasons:

1. The linkage (forward and backward pointers, depth of recursion, size of memory) built in the operating systems of computers will prevent tail-chasing loops through the many knowledge objects.
2. The seven OSI layers will automatically prevent networks from getting trapped in endless send–resend cycles of packets, sessions, blocks, knowledge objects, etc.
3. The human beings who monitor the machines are capable of preventing them from senseless and silly pursuits in the knowledge domain.

Knowledge-processing systems are based firmly on the triad of machines, networks, and humans working in conjunction and cooperation. Three mechanisms—the machines (their architectures and operating systems), the networks (their layering and protocol), and the human beings (their natural intelligentsia)—work synergistically in making a complex but manageable knowledge environment.

1.6 INTELLIGENT INTERNET AND KNOWLEDGE SOCIETY

In some industrial societies that drive their local environments, the profit motive is dominant. Sometimes, the motive is well camouflaged and almost invisible to unsuspecting individuals and social entities. Social viruses such as deception in advertising, misrepresentation of the facts, and willful distortion of the truth are taken casually in most societies. The open practice of swaying public opinion (especially during elections and in public speeches) challenges the foundations of honesty and integrity in the audience. In an urgent sense, the tolerance of such a practice represents a willingness to slide down into greater deception. The defense against such a downward spiral lies in education and the reinforcement of ethical conduct in business, education, government, and community integrated as the modern society.

1.6.1 Four Precursors of Modern Wisdom

Modern wisdom is an important by-product of computer-driven communications systems and human beings who constantly strive for perfection in (almost) everything they do. In the domain of overlap lie the axioms of modern wisdom. Although not eternal and infinitely encompassing, these axioms stretch beyond common knowledge and daily events. We approach this domain of overlap from four directions to identify and use these four precursors of modern wisdom.

First, consider human and artificial intelligence. At the pinnacle resides natural intelligence ready to tackle any problem of which the mind can conceive. At the intermediate level, the role of artificial intelligence (AI) becomes dominant in carrying out slightly higher-level tasks, blending machine intelligence with previously programmed human intelligence. At the lowest rung resides encoded machine intelligence that is cast in silicon and Pentium and that carries out mundane tasks synchronized to the cycles of a cesium clock.

Second, consider the conventional and semantic Internet. At the (current) pinnacle resides the Internet, ready to interface with any Web site in its validated TCP/IP format. Intelligence and the Internet will be fused in the next-generation internets lurking beyond the end of the decade. At the intermediate level, all digital networks are well synchronized and properly interfaced to accept information in an orderly and hierarchical format and to take on the optical paths or ether space. At the lowest rung of networks reside analog networks, cast in copper, space, and any conceivable media in an unsynchronized and imprecise fashion just to get information across.

Third, consider traditional and conventional wisdom. At the pinnacle reside values, ethics, and social benevolence. Here, the process of unification starts to have significance. Intelligence, the Internet, knowledge, and networking commingle to produce a composite fabric of human bondage and long-lasting human values. At the lowest rung of knowledge are binary bits, somewhat like the raw cells in a body. Bits soon blend to become bytes and data blocks, data become information, information turns into knowledge, knowledge gives rise to concepts, concepts lead to wisdom, and wisdom offers values and ethics.

Finally, consider modern and broadband networks. Computer applications exist to serve human beings, but machines do have their limitations. If the traditional seven-layer OSI model is downsized to three layers and then topped with four additional layers (global concern, human ethics, the national environment, and an economic/financial basis), then the new seven-layer network model would encompass the networks as they exist now and include human values in its makeup. The convergence of the four dimensions (intelligence, the Internet, knowledge, and networks) starts to become evident at the top. Currently, the application layer (AL) of the OSI model falls short of satisfying any human needs. At the lowest rung is the physical layer (PL) with six higher layers to make the new OSI model and discussed further in Chapter 10.

1.6.2 Knowledge Bases to Derive Wisdom

Knowledge bases (KBs) are relatively new but becoming [22, 23] increasingly popular. As they are relatively few, still evolving, and proprietary, the algorithms and techniques for knowledge base management systems (KBMSs) have not become as popular as DBMSs. KBs provide logical and textual linkages between the objects stored in databases or object bases, in addition to storing basic information about the database or object entities. Objects and their attributes are structured in fixed or flexible formats. Both information and the structural relationships between the pieces of information or the attributes of objects are retained for consistency and inter-operability. Tight and compact structure results in efficient KB functions such as searches, comparisons, and evaluation of objects/attributes stored. If the run length of entities such as objects, attributes, or linkages becomes fixed, then KB functions become highly programmable.

In many instances, any subject matter is assigned a unique identifier number. This methodology is common in the Library of Congress (LoC) classification and the Dewey Decimal System (DDS) used in most standard libraries. The electronic libraries deploy the KB methodology with an almost unlimited depth of search algorithms. Close identifier numbers indicate subject/logical linkage and can imply the physical proximity of storage space in massive storage facilities. Most Internet KBs are highly organized for quick and efficient searches. For instance, IP network addressing is a hierarchy of the physical location of servers and the computer nodes accessible on the Internet. This standard makes Web addressing, switching, and access very straightforward and quick.

Peripherals, networks, databases (specifically those used in recent executive information systems such as SAP environments [24]), and knowledge bases (specifically those used in the PeopleSoft environments [25]) are expanding at an alarming rate in the current information-based society. The dramatic fall in the price of IC chips and customized chip sets coupled with the cost of communicating very high-speed data locally and globally has permitted society to enjoy an enhancement of in quality of life and literacy. However, the truest contribution of the knowledge worker appears to be solving the issues of wisdom and values at a global level rather than rehashing the invention issues at a local level.

1.6.3 Role of National Governments

One of the roles of a national government is to provide authentic and significant information to its citizens. After the impact of the digital revolution and then the emergence of the knowledge society, that information has been most effectively provided and disseminated by information technology (IT) platforms. Numerous advantages follow in order to manage almost all aspects of the government and its functionality.

One such methodology for the implementation of electronic governments and a secure national network to communicate and monitor the flow of information

INTELLIGENT INTERNET AND KNOWLEDGE SOCIETY **37**

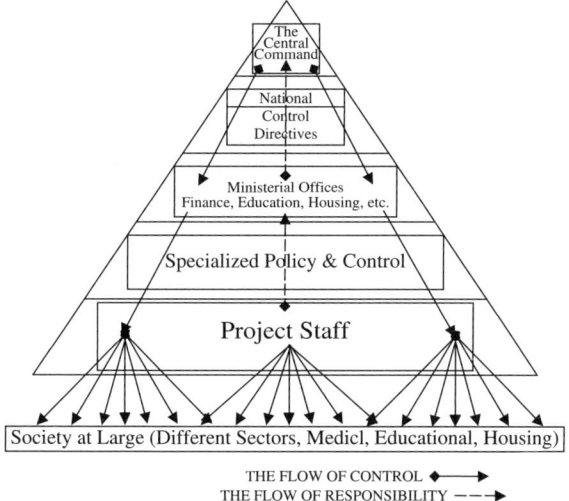

Figure 1.5 Hierarchy of a typical government structure before the introduction of an electronic government platform.

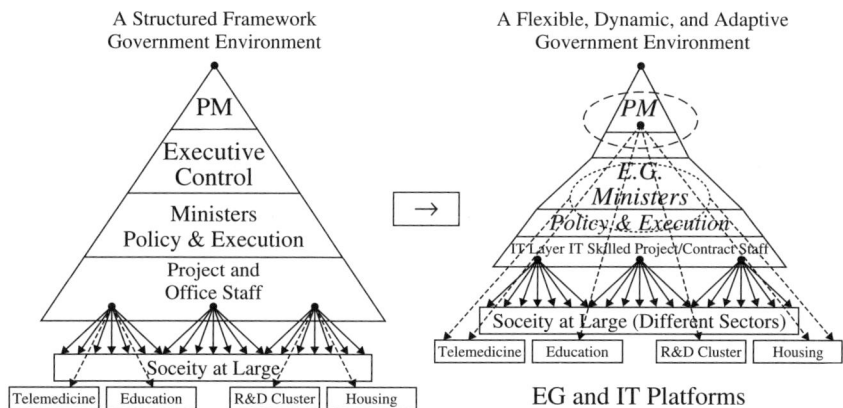

Figure 1.6 Use of an electronic government structure and an information technology platform to facilitate quick social and government changes in a nation.

is shown in Figures 1.5 through 1.7. The conventional pyramid of human power by authority and control is depicted in Figure 1.5.

Generally, a cabinet or the democratic process formulates national policy and provides the general direction for a nation. The executive office (head of state) executes the policy as well as assuming full-time managerial and executive responsibilities. For this reason, the electronic government-IT (EG-IT) architecture for any developing nation should have the capacity to retain extensive information on all managerial functions, such as planning, organizing, staffing,

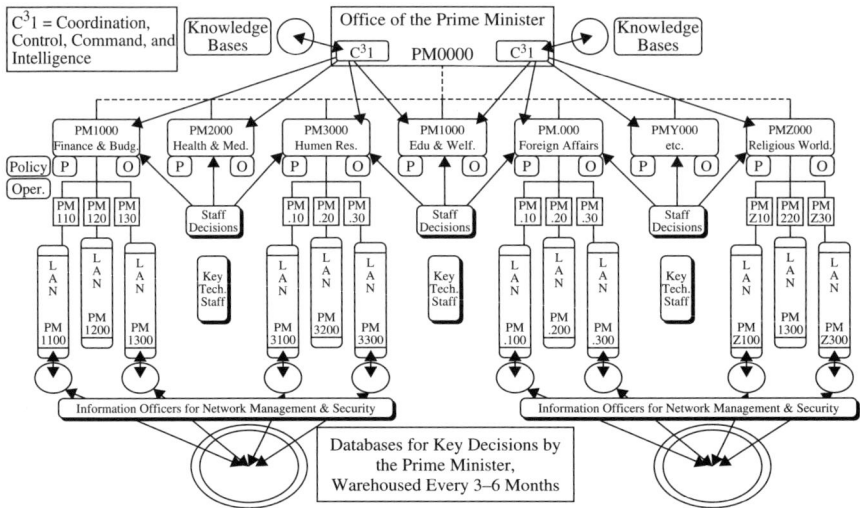

Figure 1.7 An information technology platform at a prime minister's office to support C^3I (coordination, control, and communication with inherent intelligence) to facilitate a quick and efficient response from the government sector. This enables a country to move quickly, dynamically, and decisively.

implementing, and controlling most critical and significant projects, as well as commanding, controlling, and coordinating other ongoing projects of national importance.

The introduction of the most rudimentary IT platforms in the government makes the completion of functions within the government more efficient and offers greater control over the utilization of governmental staff and resources. The attendant organizational change is illustrated in Figure 1.7.

Typically, an EG needs to retain the most recent information for all sectors of the government and the history of that information. The funding for and coordination of the other divisions of the government (such as defense, education, commerce, communication, or social welfare) occur at one physical location, and a dedicated, secure, and centralized data warehouse is essential to track, update, justify, and explain important changes in the government on a regular basis. These same requirements also exist in most large corporations and multinational organizations. The difference is the size, security, and importance of programs that national governments have to address, guarantee, and deliver. Some of these considerations are discussed in [26]

1.6.3.1 Control, Coordination, and Flow of Information A centralized EG is unique in terms of funding and the coordination of various divisions of government (such as defense, education, commerce, communication, and social welfare) and projects arising within the government (such as telemedicine, distance learning, and electronic commerce). The organization and structure of

IT within the EG must reflect this rather unique focus and concentration on management and administration (of itself) as well as the other branches of government.

To accommodate the flow of information based on the nature of functions specific to the executive office and other ministries, a suitable EG architecture is proposed in Figure 1.7. Whereas the government's departments, ministries, and project office may have a unified blueprint for local ITs, the IT of the centralized executive office (head of state) needs special consideration. It is strongly suggested that the network and IT engineers designing the EG of any country incorporate the different functions of the executive office in the initial design of the IT system and the networks that support these functions (coordination, control, and command with embedded intelligence referred to as C^3I and discussed further below). The network is equipped to interface well with the other departments or ministries. If the trend is to use generic programmable network components, then the network and netware that control the operations of these components (to facilitate the EG functions for the executive office) should be endowed with enough intelligence to perform the IT-EG functions in conjunction with those of the other ministries.

To some extent, this matter is not trivial, unless the network designers have facilitated these functions as being programmable or already have assigned paths and addresses unique to the offices and databases within the ministries.

This architecture is neither optimized nor customized to any particular country. A possible blueprint of one of many configurations is documented here. Several rounds of optimization of architecture, components, links, and server (databases) characteristics are essential before embarking on the direct application of this architecture. The customization of the network to a particular nation is as essential as the funding of the project. All the necessary aspects of the detailed planning must be undertaken in the most conscientious way.

Divisions, agencies, or departments of the government handle projects arising within the government (such as telemedicine, distance learning, or electronic commerce). The organization and structure of IT for the various departments start to differ from those of the centralized government of the executive office (president, prime minister, etc.). The architecture and configuration for the EG for the executive office reflect this rather unique focus and concentration on management and administration.

Figure 1.5 depicts the hierarchy of a typical government structure before the introduction of an EG platform. The EG platform shown in Figure 1.7 enhances the connectivity between the central command (the Prime Minister, in this case) and the various ministries and reduces the workforce over time. The transition to an EG platform is time-consuming and the government remains functional during it. Entirely self-sufficient electronic government systems have not been successfully implemented in any country. Sometimes, in organizing EG platforms, it becomes desirable to allocate a numbering system to the many branches of the government. Numbering allocation facilitates the allocation of storage, networks, and processes to the documents, interactions, and activities in the computer systems that house the IT for the EG.

When the document tracking of references to prior or similar tasks becomes necessary, machines can perform most of the routine or AI-based intelligent functions. The design of an electronic government information system (EGIS) can be made akin to a typical large-scale management information system (MIS) or executive information system (EIS) such that a scaled version of current software systems may be used as the modular building blocks of the EGIS.

Examples of such numbering systems are the Dewey Decimal System and Library of Congress classification system. These systems draw a boundary (though vague) between the branches of knowledge. A similar numbering system that broadly identifies the particular branch of government would facilitate classification of all activities within it. Funding, activities, and projects could be tracked and controlled much like the tracking systems in typical human resource applications such as PeopleSoft or SAP.

In the section, a simplified number tracking system is suggested. The alpha numbers are two alphabet letter, four-digit entities, where the first two letters denote the particular office, such as PM for the prime minister's office, EW for the education and welfare office, HR for the human resources office, etc. The next four digits signify the organizational segment, such as 0000 for the Office of the Prime Minister, 1000 for the Office of Finance and Budget, or 2000 for the Office of Health and Welfare, etc. The tree structure continues. The numbering system is quite flexible and can be adjusted as the structure of the government changes.

In most cases, the role of EG platforms is essentially that of a communications provider and provider of routine transactions processing, where there are very few policy decisions. In the support facilities for the government, decision support systems help human beings to acquire better insight into intelligent and informed decision making. When the role of decision making becomes sensitive and crucial, wisdom machines [4] become more desirable, with EG platforms providing the intelligent control coordination and communication within the government. Figures 1.6 and 1.7 depict two stages in the transition of a hierarchy of any typical EG to that of a typical service-based EG.

In the design of EGs, the location of the center for coordination, control, and command needs precise identification. In the particular example presented in this book, the logical concentration of power lies in the Office of the Prime Minister. The physical location can be moved at a later date over wide-area broadband networks. If the databases and servers are not physically moved, then the access time for critical information is enhanced by the transit time between the old and new locations. The logical identification of the center of power serves to strategically place the routers and high-capacity links. Architecture for an IT platform for the prime minister's office supporting the C^3I (coordination, control, command, and intelligence) approach is shown in Figure 1.6. The communications aspect is built in to the hierarchy of the prime minister's own office network in conjunction with the horizontal layout of the local-area networks for the various ministries.

Data warehousing becomes necessary to maintain continuity in the operation of the government over an extended period. The long-term economic and growth trends of a small nation can be planned, organized, staffed, implemented, and also monitored and controlled in the EG framework for the country. The three facets of intelligent coordination, control, and command start with the Office of the Prime Minister, feedback derived from the offices of the ministries and the 13 key indicators measuring economic activity within the country.

1.6.4 Universal Knowledge-Processing Systems

The architectural concepts of a complex knowledge-based system dubbed the universal knowledge-processing system[8] (UKPS) are viewed as wisdom machines (WM) capable of solving both simple and complex problems. Knowledge bases accessible at any Web location offer broad and generic solutions. The UKPs is, in turn, formed of several knowledge-processing systems (KPSs) connected through any backbone or global network. In this section, we present the key components that make up the architecture of this sophisticated system. Some of the concepts presented here are abstract and require further investigation to complete the entire cycle of evolution of machines and development of systems.

1.6.4.1 Knowledge-Processing and Wisdom Machines The architecture of knowledge-processing systems over time has changed just as the philosophy of computing within the organization has: from mainframe-dominated, centralized computing systems to network-based distributed computing systems. Fundamental features that distinguish a distributed system from a previous system are the distribution of resources and functions among several computers at workstations that are networked together using present-day communications protocols such as TCP/IP and RISC processors. The operating systems could be UNIX, Windows NT, or any other adaptations of UNIX. Internet/Intranet technology has undergone lot of changes and, as a result, we see it being applied to various fields. These technologies have reached far into our office and home environment.

The important issue to be resolved for Intranet-based KPSs is real-time performance and the reliability of decision-based control. A brief introduction to the client–server architecture is given, after which the model of the UKPs or wisdom machine is proposed. The features of this model that distinguish it from distributed systems are the use of a browser, a wide-area network (WAN), a wide-area cluster of servers on the WAN, and multicast communication based on IP. The Intranet-based UKPS provides the basic underpinning for improved flexibility and expandability. Several of the KPSs could be integrated to form the UKPS or wisdom machine. The architecture also enables the knowledge-processing centers that contain these KPSs to be rearranged as necessary and when required.

---

[8]The concepts presented in this section were presented by Ajit Reddy and Syed Ahamed at the 2006 International Conference on Information & Knowledge Engineering, Las Vegas, NV, June 26–29, 2006. See *Proceedings of IKE 2006*, Worldcomp'06 Publications, 2006 pp. 65–70.

1.6.4.2 Client–Server Architectures The client–server software architecture is a versatile, message-based, and modular infrastructure that is intended to improve usability, flexibility, interoperability, and scalability as compared to centralized, mainframe, time-sharing computing. A client is the requester of services and a server is the provider of services. A single machine can be both a client and a server depending on the software configuration. Commonly used client–server architectures are either two-tier or three-tier.

Two-Tier Architecture In the two-tier architecture, the client machine has a user interface and the server machine is a powerful machine serving a number of clients at a time. The server could be offering database management services, in which case the server provides stored procedures and triggers. In this case, process management is split between the user system interface environment and the database management server environment. The server maintains a connection with its client through keep-alive messages, even at time when no useful work is done. The limitation on its flexibility to move program functionality from one server to another is seen as a liability of the two-tier architecture system.

Three-Tier Architecture In this architecture, the limitations observed in two-tier architectures are overcome, and a middle tier is introduced between the user system interface environment and the database management server environment. The middle tier can perform queuing, application execution, scheduling, prioritization, and database staging. The middle tier can be implemented in the form of transaction-processing monitors, message servers, or application servers.

In addition to computers and their operating systems, programming languages and their compilers play an important role in writing code in a simple and efficient way such that even complex programs could be developed in a reasonable amount of time with the required manpower needed for development. With this grows the complexity of the code and software that is developed, and hence, different methodologies must be adopted to design, validate, and maintain the software.

1.6.4.3 Knowledge-Processing System A simplistic view of a knowledge-processing system is that it is a system that can make or provide a decision based on certain input information or query together with the knowledge base. In order to process information and arrive at a decision, these systems should be equipped with the necessary knowledge stored within it.

From an architectural perspective, the basic components of such a system are as follows:

- Hardware
- Operating system
- Storage
- Network
- Knowledge processing

INTELLIGENT INTERNET AND KNOWLEDGE SOCIETY   43

```
┌─────┬─────────────────────┬─────┐
│     │     KNOWLEDGE       │  N  │
│     │     PROCESSING      │  E  │
│  S  ├─────────────────────┤  T  │
│  T  │                     │  W  │
│  O  │     OPERATING       │  O  │
│  R  │      SYSTEM         │  R  │
│  A  ├─────────────────────┤  K  │
│  G  │                     │     │
│  E  │     HARDWARE        │     │
└─────┴─────────────────────┴─────┘
```

Figure 1.8 Basic components of a knowledge-processing system.

Figure 1.8 provides a block diagram view of how these components are assembled together to form the knowledge-processing system. The hardware component of the KPS could be a chemical, optical, or silicon-based platform. As the technology evolves, platforms could be based on some other technology as well. Operating systems like UNIX and, Windows XP in the evolution path should be able to work independently of the platform technologies, storage and communication networks. In the evolution path, the operating system, hardware, storage, and network could change, but the knowledge-processing functions and data developed and collected over a period of time should be able to function irrespective of the changes in the basic components of the KPS.

If we apply the three-tier model to this architecture as shown in Figure 1.9, applications such as the monitoring and decision control functions are installed in the second layer. Data in the form of real-time input are stored in the third layer. The application and database layers are installed in the server. The graphical user interface (GUI), command line interface (CLI), or Web browser installed in the first layer enables the operator to monitor and display pages and to control the KPS if required. To control the KPS, an operator connects the GUI or browser to the corresponding system, then requests authority to control the decision process

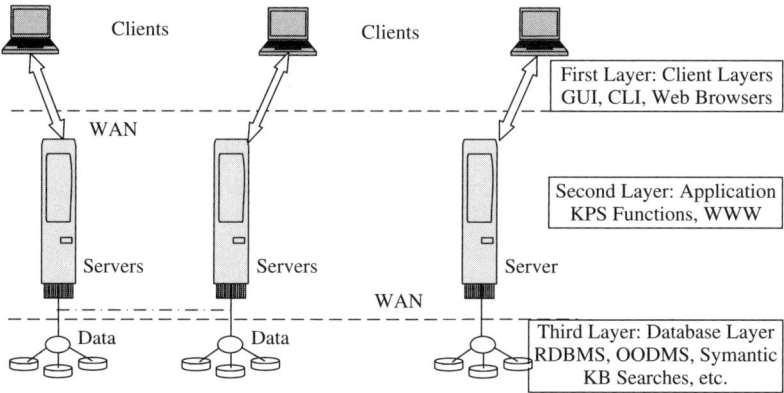

Figure 1.9 A three-tier model of a knowledge-processing system.

through the pages displayed on the browser. Since the servers of the KPS could be located anywhere, the operator needs to know their location. In this model, an authorized operator could use the application in any KPS connected to the WAN and suitably control the decision process if necessary in a positive manner.

Using the wide-area cluster architecture, the servers of the KPS are separated geographically and connected through a WAN. If the server, or servers, in the cluster fails, then the other healthy servers back up the functions automatically. Once the failed server is brought up again, the server is put back in the cluster as a healthy server, thereby providing fault tolerance. Each of the KPSs could have a single server or multiple ones representing the second and third layers.

The architecture considered here exists for high performance, provideing the high-speed response of the human–machine interface. Whenever an application such as the monitoring function or decision control function in the second layer updates the database layer, the high-performance architecture distributes the updated data to the relevant KPS. Multicast communication based on IP is applied to this replication process.

1.6.4.4 Universal Knowledge- or Wisdom-Processing System

The wisdom-processing systems architecture depicted in Figure 1.10 shows the different levels of knowledge-processing systems that are required to arrive at a decision or solution based on wisdom. The systems at all these levels have the same set of basic components as illustrated in Figure 1.8. However, the knowledge contained at each of these levels together with the processing power of these systems could vary, the systems at the top being the most powerful in terms of computing power, storage, network bandwidth, operating systems, and other capabilities and those at the bottom level being the least powerful. The systems at all these different levels are networked together either by a dedicated network or through a shared network.

Level 1 The systems at this level are knowledge-primitive machines with the ability to store user-entered data or data received from external sources, process data, analyze/model data, suggest or make a decision for less complex problems, thus serving as decision support systems. They should also be able to compile data for higher-level systems. The systems at this level have the necessary intelligence

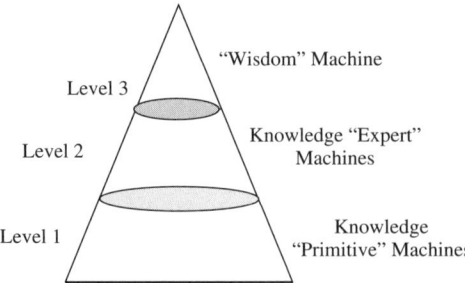

Figure 1.10 Conceptual representation of a wisdom-processing system.

and knowledge to process the preliminary data for any given task by extracting the necessary information and sending the compiled results to the next-level machines for a more thorough processing for the given task or problem.

Level 2 At this level, the systems are knowledge-expert machines that will act as expert systems with access to a knowledge base consisting of the views of experts, sets of rules, and sets of relationships. They contain a database rich with information related to a specific task. They simulate an expert thought process using knowledge-processing library functions and methods for logic and reasoning for the tasks to be analyzed and for the decisions to be provided based on the query or input information. On the development front, they have the means of programming that will enhance the knowledge base and inference process. With this capability, the level 2 machines have the ability to provide recommendations or decisions, solve complex problems, adapt to problem solutions, learn and continually refine the methods in the inference engine. Since these machines reside at a level between that of the level 1 machines and the level 3 machines, they have the necessary knowledge and intelligence on how to interface with these levels. These systems also have the ability to compile data and transmit it to the higher levels.

Level 3 This level consists of a group of very powerful systems connected together using a dedicated network. From the basic knowledge-processing component perspective, they are the same except that they are much more powerful in terms of processing power, storage capabilities, and network capabilities. These systems can independently solve a complex problem or get the preprocessed information compiled together from the levels below, extract the information, and then decide. The compiled results obtained from the different expert or experts for the same problem are collected and further analyzed by giving a certain weight to each of them. It could be that given the nature of the complex problem, there may not be any experts so these systems are in a position to reach a partial decision by using the available expertise and making a judgment. During this process, the level 3 systems could use some other expert system or systems, or possibly the same expert systems as before. By exchanging information back and forth between this network of systems at all levels, it is quite possible to arrive at a conclusive result for a problem of such complexity. If a knowledge base does not exist for the given problem, then it could be obtained by unleashing several queries to the experts that could be systems or humans, in which case the unavailable knowledge base is compiled in real time.

Solving problems in this manner could introduce delays. However for time-critical tasks, decisions must be arrived at without any delays, in which case the system will continue with the decision that would be based on the decisions of the expert systems with the current knowledge base with a certain confidence level. Human interaction is necessary at this critical juncture, either to go ahead with the decision made or to do something different. These systems will also oversee the decision process of the systems at the levels below if the systems at levels 1 and 2 are granted access.

Although in this present architecture only three different levels are shown, in reality each of these levels could be further subdivided in sublevels. The systems at level 2, which are experts, could be classified into different sublevels based on their degree of expertise and systems at level 1 could be divided into sublevels based on their storage, processing, modeling, and analysis capabilities. At level 3, we assume that all the systems are of the same degree in terms of wisdom: They may not be experts in the related area, but given the available information and problem at hand, they have the necessary wisdom to reach a suitable conclusion. Because at this level there is a network of such systems, the final decision is based on the majority of decisions of these systems, so a check on arriving at a wrong decision is maintained. The postprocessing of extracted wisdom is discussed in [4].

Since the Intranet-based UKPS could be accessed from many terminals, a secure socket layer (SSL) should be applied to any communication between the GUI, CLI, or browser and the WWW server, and each of the operators should be allowed to access only predefined contents.

1.6.4.5 Methods of Reasoning Decisions at all levels are arrived at by using different methods of reasoning, such as reasoning by induction, reasoning by both deduction and prediction, and logical reasoning. Inductive reasoning is usually based on observation whereby the premises of inductive arguments are bits of evidence gathered either directly or indirectly, and its conclusions are tentative generalizations about groups, relationships, or predictions. So with inductive reasoning, conclusions based on premises however accurate they may seem could possibly involve other conclusions as well, which is to say that the conclusion drawn is a possible but not necessarily reasonable one for the problem at hand. Thus, inductive conclusions do not follow of necessity from the premises. Nevertheless, inductive reasoning helps us to start building a knowledge base with the general statements and principles we have learned thus far.

In deductive reasoning, necessary conclusions are generated based on the premises of the arguments. If the argument is a syllogism, then it draws conclusion about the member of a group from generalizations about the group and the relationship between the member and the group. If the argument is a hypothetical chain, then it draws conclusions from a general prediction and the given situation. Although the deductive method puts the generalizations of the inductive method to the test, it has to be adaptive so that the generalizations can be finetuned for a specific task or problem.

Critical thought process requires both inductive and deductive reasoning, whereby through inductive methods we get to identify and formulate certain generalizations that need to be put to test by deductive methods to see if they hold.

1.6.4.6 Applications Two applications for knowledge- and wisdom-processing systems are (1) power outage detection and prevention and (2) a tsunami early warning system. The details of both the electrical power outage example and its consequences, and the tsunami warning system fall in the public Internet domain.

Power Outage Detection and Prevention The power outages that have occurred in several major cities in recent times across the globe have affected trains, elevators, and the normal flow of traffic and life. In some cities, water supplies were affected because water is distributed through electric pumps. Airports across the affected region experienced delays and some shut down temporarily. The outage slowed the Internet due to rerouting of its traffic, and Web sites powered by servers in affected cities were unable to respond to requests from their clients. In order to detect and prevent a disaster such as major power outage, KPS could make a determination in advance by isolating the defective power center (s) that is likely to trigger the tripping of all the power grids, thus minimizing the damage caused by overload.

The proposed KPS-based model and architecture are shown in Figure 1.11. The model consists of data acquisition and control units (DACU) connected to the KPS, which sends the acquired data of the power systems to the servers of the KPS, in which the collected data are processed using the knowledge-processing functions. The data from each power center are collected by means of a client used by the operator either manually or automatically from the DACU and sent to the KPS via the WAN. Multicast communication is used between the DACU and KPS servers, and all servers can receive data from each and every DACU.

Each server and each client are connected directly to the WAN. Browsers installed in each client enable the operator to monitor and control the process manually, and allow access to any server that is transparent to the client. Critical functions installed in each of the KPS servers along with the incoming data are constantly backed up. The UKPS, which is an Intranet, connects to the Internet through a firewall. A satellite link is provided so that critical data are sent to another geographic location where a similar KPS setup exists and also to transmit collected data and receive control signals from KPS in another geographic location in the event of a natural calamity that could paralyze the local KPS setup.

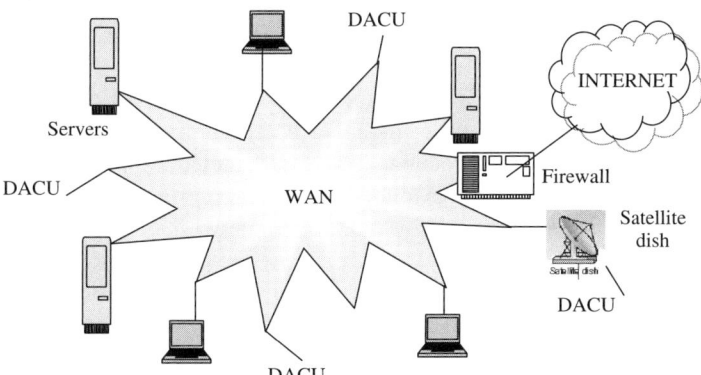

Figure 1.11 Knowledge-processing system model for power outage detection and control. DACU = data acquisition and control unit.

The decision-making process occurs at the UKPS, which connects several KPS networks. In spite of all this, one notes that there is some delay in the DACU collecting the data and sending them to the KPS servers, where the data are processed and forwarded to the UKPS for a final decision after it has analyzed the processed data from the other KPS. The processed data could be a partial decision made by the local KPS, or it might be that no decision is taken by the local KPS or that it suggests a list of remedies to the problem, one over which the UKPS would be the final decision maker. The decision made by the UKPS could be over-ruled in the manual mode by the operator or operators monitoring the power centers. In the auto mode, the decision would be made by the UKPS and the control signals sent to the DACU for the appropriate action.

Tsunami Early Warning System There are so many variables and uncertainty to consider in the detection of a natural disaster like a tsunami. In some cases, we might have little or no information about the situation; in others, the available information is vague or incomplete. In almost all these cases, the system has to operate under severe time constraints, with little or no time to make a critical decision and sound an alarm about the catastrophe that is likely to occur. The variability and inevitability that result from complexity may be traced to the interaction of numerous independent actors. Functioning in an environment of uncertainty, complexity, and variability requires sound judgment, creativity, and initiative—all of which the UKPS should be able to handle.

The tsunami early warning system consists of two stages: the detection stage and the dissemination stage. As part of the detection process, the objective of the KPS is to detect, locate, and determine the magnitude of potentially tsunamigenic earthquakes occurring in any part of the sea floor throughout the world. Earthquake information is provided by seismic stations operating in various nations. If the location and magnitude of an earthquake meet the known criteria for the generation of a tsunami, a warning is issued to advise of an imminent hazard. The warning includes predicted tsunami arrival times at selected coastal communities within the geographic area defined by the maximum distance the tsunami could travel in a few hours.

A tsunami watch with additional predicted tsunami arrival times is issued for a geographic area defined by the distance the tsunami could travel in a subsequent time period. If a significant tsunami is detected by sea-level monitoring instrumentation, the warning is extended to other countries and regions. Sea-level (or tidal) information is provided by monitoring networks and other participating nations of the tsunami warning system (TWS). This effort encourages the most effective data collection, data analysis, tsunami impact assessment, and warning dissemination to all TWS participants. The dissemination process consists of sending bulletins and warnings to concerned government and emergency officials and the general public of participating countries and even those nations that do not officially participate but are likely to be hit by the tsunami. A variety of communications methods are used in disseminating the tsunami warning.

INTELLIGENT INTERNET AND KNOWLEDGE SOCIETY **49**

The system consists of several tsunameter mooring systems (TMS) positioned along the coastal line of various countries. The TMS consist of transducers, RF modems, and a bottom-pressure recording device that detect the changes in pressure caused by the tsunami and transit them to the surface buoy via an acoustic link that the transducers pick up and send to the satellite by means of the RF modem, as shown in Figure 1.12.

The satellite that collects the data from these TMS sends them to processing centers where the collected data are processed and the likelihood of a tsunami occurring will be determined. In addition, the path and magnitude of the tsunami are determined, as well as which countries are likely targets of this disaster. The necessary action is then taken, with tsunami alerts disseminated to affected countries or regions. In the two applications discussed above, a clear picture of a chaotic environment filled with variables and unknowns emerges.

Here, possible configurations of universal knowledge-processing systems are presented and applications arising from the implementation of such a system discussed. The architectures and applications are extensive. Other complex problems can also be solved. Numerous complex problems arise during routine situations;

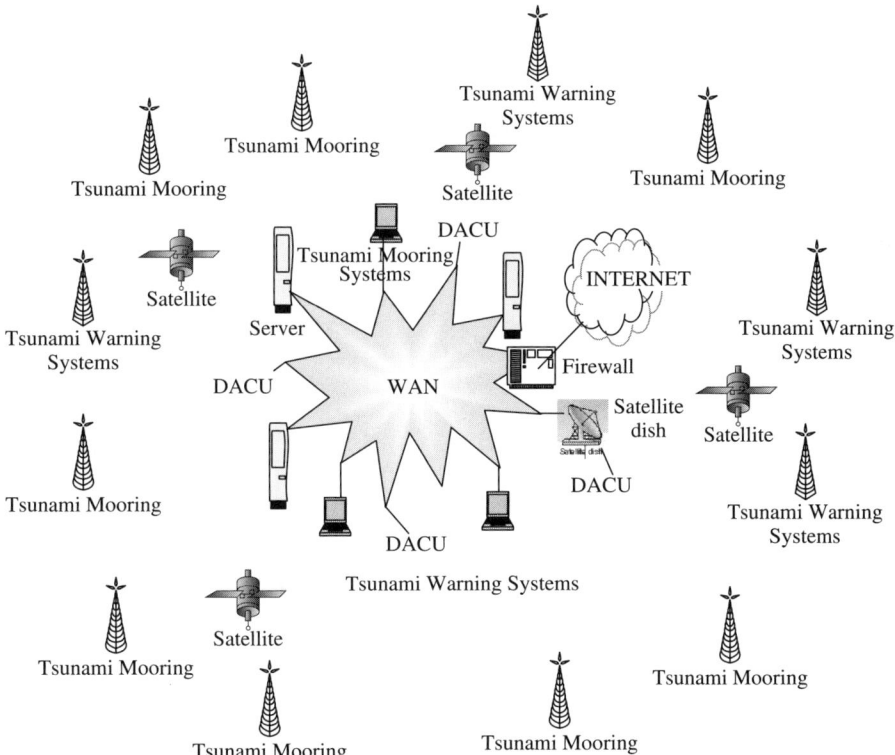

Figure 1.12 Universal Knowledge-Processing System–based tsunami early warning system. DACU = data acquisition and control unit.

50 NEW KNOWLEDGE ENVIRONMENTS

they are frequently issues that cannot ordinarily be solved by the existing computing infrastructure. These problems require the knowledge-processing systems that lead to wisdom machines, those presented in this section, to make decisions that are time-critical in nature.

1.6.5 Educational Networks

Educational networks are becoming universal. From an economic perspective, educational networks are more desirable for the routine and mundane tasks of teaching students the basic (and low-level) skills in any particular subject. The system is able to interpret independently and in conceptual rhythm with the learner, and yet provide comprehensive, dialogue type answers to queries on almost any subject. Web-based learning offers unique advantages for students, and faculty and economic rewards to administrators. This capability is achieved by incorporating the dynamic growth of high-speed networks already integrated in most existing telecommunications networks. Language to interpret a query, artificial intelligence to comprehend the global context of that query and student capabilities, and knowledge processing to uniquely construct a reply make educational networks [13] as desirable as MIS facilities [27] in monitoring most corporate functions.

The basic configuration of a single- or multicampus educational network is shown in Figure 1.13. Campus networks (CNs) may be classified by a numbering system akin to the Dewey Decimal System (DDS) or based on the Library of Congress (LoC) classification. Multinumeral campus network addresses with combined DDS and IP addressing will substantially facilitate the access of any knowledge base internally or on the Web.

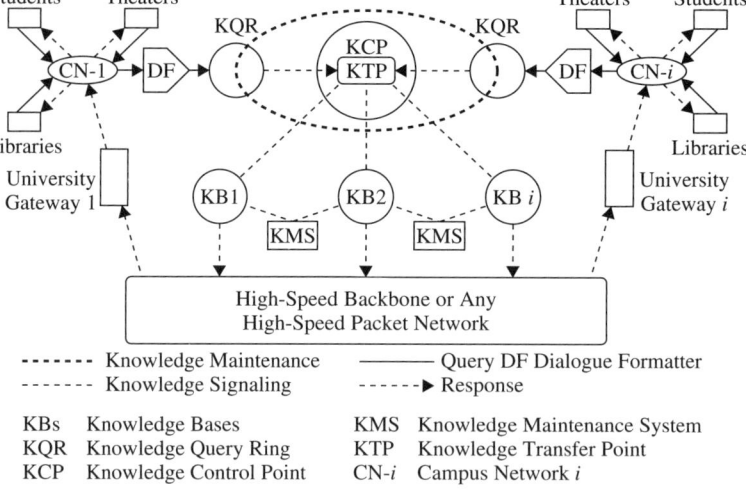

Figure 1.13 Configuration of a basic multicampus educational network.

INTELLIGENT INTERNET AND KNOWLEDGE SOCIETY   51

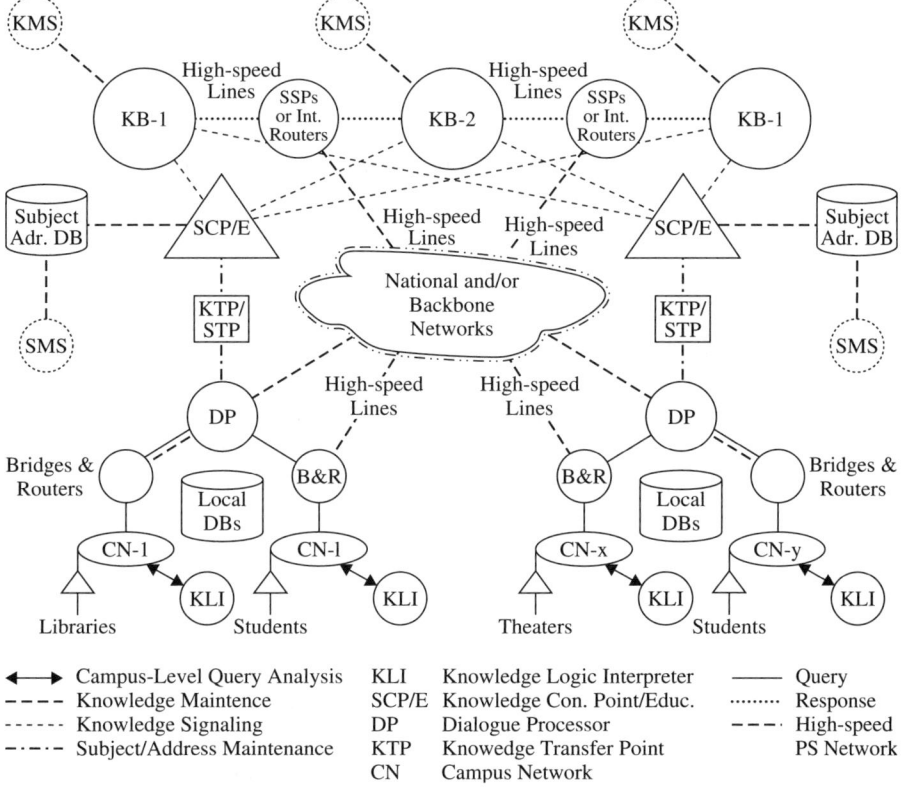

Figure 1.14 Configuration of an educational network for a developing nation.

A slightly modified educational network for a national university is shown in Figure 1.14. This particular network deploys the national, WAN, or MAN backbone network for access to and from distant cities or campuses. The typical service switching points (SSPs) of an intelligent network architecture may also be replaced by intelligent routers that can switch the ATM cells of a high-speed packet-switched network. Specific intelligent routers based on DDS classification have not been developed for educational networks.

Query structure and analysis are performed by dialogue formatters (DF) that interface the campus queries and also signal the KTP/STPs for services from the knowledge control points to obtain the appropriate response from the knowledge bases (KBs). The response time of such a network should not exceed the typical response times of intelligent networks. Simulation and design studies of such well-designed, intelligent network-based, educational network responses have been reported in considerable detail [28]. Extensive simulations indicate that such massive educational systems can be implemented with typical high-speed campus and backbone networks in most developed nations.

A well-designed infrastructure for packet-switched and ATM networks for developing nations will pave the way toward an informed and educated public. World-wide trans-oceanic fiber-optic-based information highways [13] are already providing access to global knowledge bases. The enhancement of any national education infrastructure for global universities requires the careful design of switching centers, link capacities, and the distribution of knowledge centers around the nation. Quality of service (QoS) considerations become crucial in the early design of networks.

1.6.6 Medical Networks

Intelligent medical network (IMN) configurations and designs were initiated at the City University of New York as far back as 1990 by Krol [29]. Numerous other researchers such as Mollah [30], Lueng [31], and Waraporn [32] have extended this methodology over the last decade. In an attempt to blend wisdom machines architecture into IMNs, we present two configurations that utilize intelligent Internet architectures and national medical expert teams that constantly scan the activities in hospitals and medical centers in order to filter out routine activity and identify unusual circumstances to add to medical knowledge bases around the country or the world.

Medical networks can range from small hospital networks to multinational global specialty-based networks for special condtions and rare diseases. In the current environment, search engines provide access to information but lack the human consultative platform for interactive sessions between experts. The financial and economic aspects of transactions could not be handled by former search engines. The development of medical service providers (see Figure 1.15) that do maintain extensive information electronically, ranging from patient ailments to medical insurance coverage and prescribed medications, will result in the online existence of the treatment program for each of their customers, in much the some way that credit card companies can track the financial activities of their clients.

General patient databases, physician access point units, patient access point units, and service facilities are connected to medical data banks and processors via several buses. In an alternative integrated medical computer system, numerous processors are included, each with their own memories and modules, and linked together to establish a processor net unit. This system can be used in a campus environment, where several buildings make up the hospital or several hospitals are interlinked over local-area networks. One such configuration is shown in Figure 1.16. The nature of components and their capacities are matched to the variety of services provided, and the sizes of links and routers are matched to the expected client base served by the medical facilities.

Large groups of independent hospital networks may be integrated (and programmed) to coexist within one larger medical network as independent intelligent networks sharing a network infrastructure. Functionality may also mean performing as a group of localized medical networks [33] for a particular hospital rather than as one massive medical network for a region or country. The role of

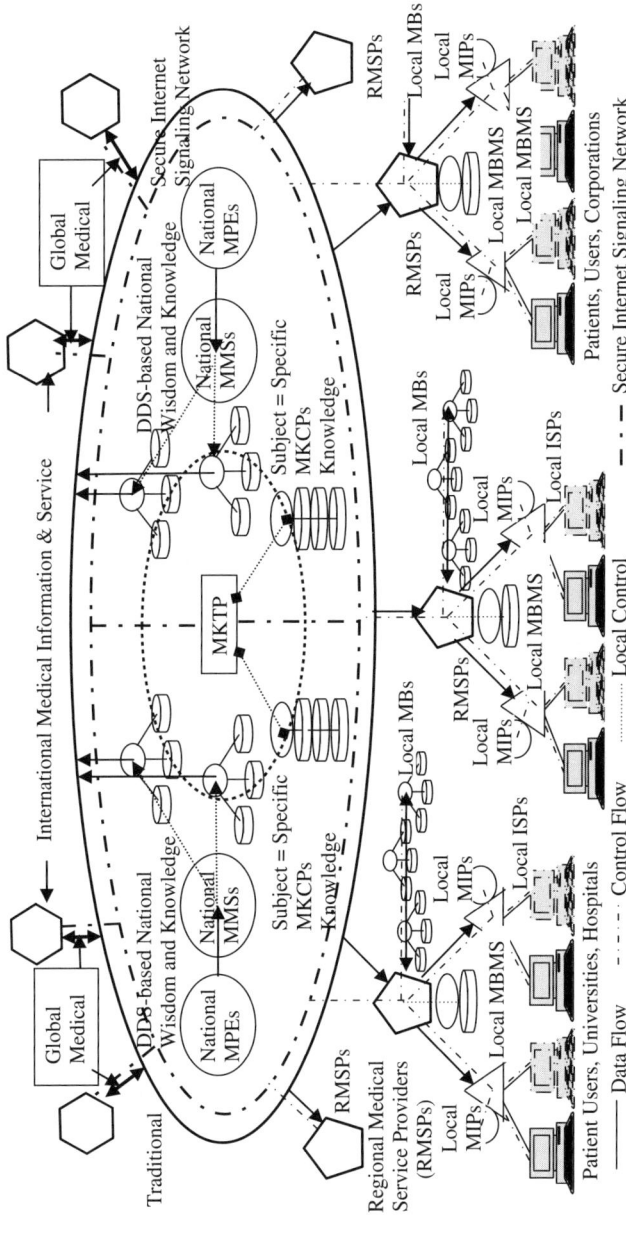

Figure 1.15 Framework of an Internet-based intelligent medical network with traditional intelligent network components. MSPs = medical service providers; MIP = medical intelligent peripheral attached to ISP local network; RMSPs = regional medical service providers; MBs = medical knowledge bases; MBMSs = MB management systems; MPEs = medical services provisioning environments; MKCP = medical knowledge control point; MKTP = medical knowledge transfer point; MMS = medical information management system; MPE = medical services provisioning environment; DDS = Dewey Decimal System/Library of Congress classification system bases.

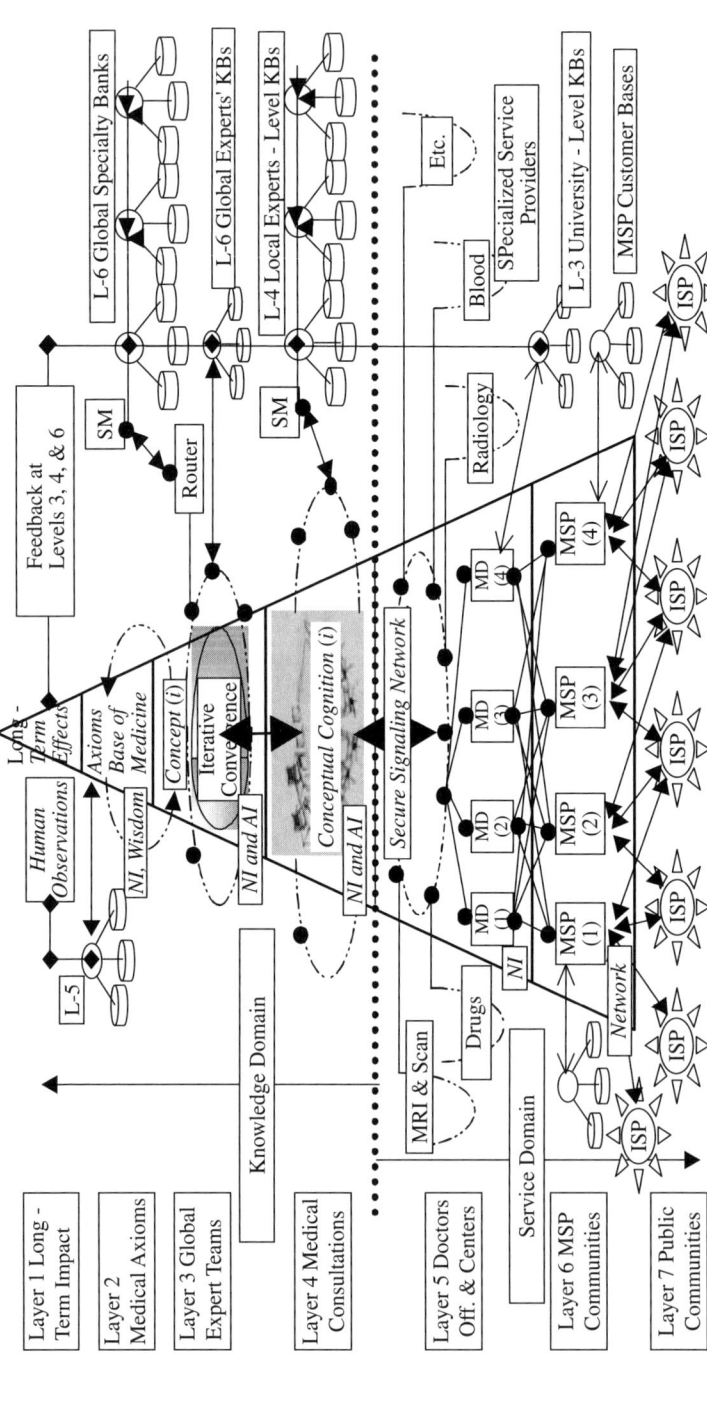

Figure 1.16 A seven-level intelligent hospital environment. This arrangement serves public medical needs via Web access (level 7) to the hospital. Routine physician–patient transactions are handled at the lower levels (1–4). Facilities at the higher levels (5–7) provide venues for physicians and hospital staff to consult with local medical communities and interact with subject matter specialists around the globe. Network security is provided at all seven levels. SM = switching module; MSP = medical service providers; MD = doctors.

numerous medical service providers (MSPs) will thus be confined to local-area services and information access. The main advantage of having a large number of smaller and localized MSPs is that the cost and complexity are much lower than those incurred by a large MSP. The new competition in terms of localized MSP services will reduce medical expenses and overall customer costs.

More local MSPs and businesses can offer medical information and services such as discounted prescription drugs, hospital and nursing services, or physical therapy. Such an increase on the supply side of medical information will help the medical field become more competitive and thus contain medical costs in the long run.

1.6.7 Antiterrorism Networks

An IT platform for an antiterrorism (AT) office within a country is shown in Figure 1.17. The blueprint can be customized to suit any office reporting to a prime minister or executive body. The numbering system provides for organization, document tracking, and continuity among the functions in the different branches of government.

The special PM-AT network AT1000 facilitates linkage with a centralized command facility. This segment can be designed to be physically or logically independent of other networks. Security and network connectivity under fault conditions become important considerations designing this network. The EG computer and network configuration are thus fundamental in monitoring the policy, goals, and project completion in any branch of the government.

The architectures presented in this section offer added benefits to the knowledge community by interleaving computer-based antiterrorist methodologies with the quick response of high-speed networks that serve the Internet and Intranet industries. When artificial intelligence (AI) techniques are used to detect and possibly predict emergency conditions in a country, a network-based sensing and monitoring environment can be integrated with its high-speed backbone networks.

The execution of the policies set forth by the leader(s), and the roles of the minister(s) and the staff, are scanned and documented accurately by the filing system built into the EG infrastructure. Typically, the traditional concepts of planning, organization, staffing, implementation, and control can be achieved efficiently and optimally in any country. The network for the entire EG is depicted in Figure 1.18. The configuration can be customized, and its numbering system makes the possible organization, document tracking, and continuity among the functions in the different branches of government.

Configuration of the routers and links is also shown in Figure 1.8. When the numerous levels of ministerial offices are dispersed throughout the country, the dependability of the backbone network becomes an issue. Secure and trustworthy connections via independent or leased public domain channels sometimes provide intra-office capability.

An intelligent decision support system can be ported in the network environment by appropriately arranging the functions of the three switching systems

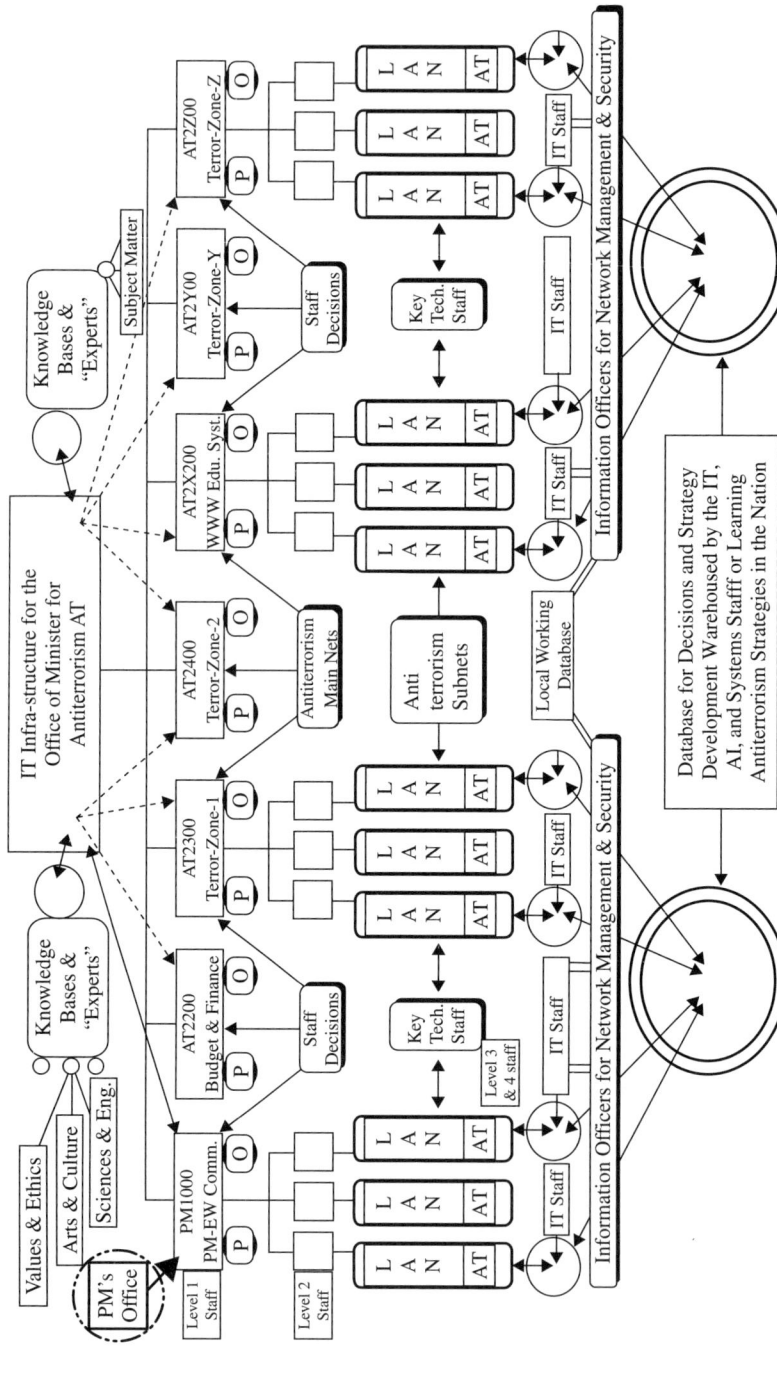

Figure 1.17 Integrated antiterrorist platform for terrorism-related information processing and management. The platform allows the police and antiterrorist special units to learn and teach specific strategies.

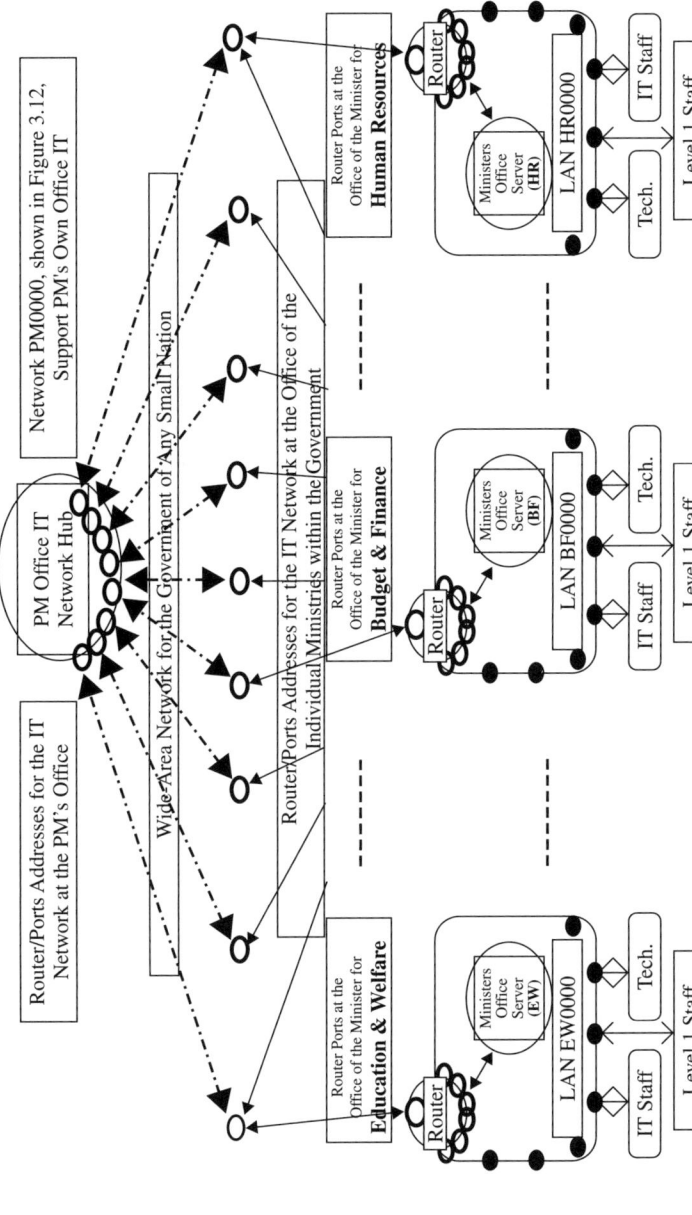

Figure 1.18 A layout of routers and the network incorporating the electronic government backbone network and the functioning of the police, defense, and antiterrorist staff members to develop a unified and coordinated strategy for responding to emergency situations.

or switching the modules of any of those systems. One can simulate all the functions of a national economy, a corporation, a society, or any midsized social entity by distributing "sensors" in the environment and then regulating the "climate" of the entity with intelligence modules distributed throughout the network. One can program the response to each sensed scenario as a quantitative measure of the attributes of a set of programmable objects. One can separate the three basic components of any intelligent MIS (the dialogue processor, analytical/simulation module, and data manager) into three interdependent machine clusters.

1.7 KNOWLEDGE NETWORKS

Knowledge bases and networks are becoming increasingly common in the Internet environment. Knowledge processing to monitor and maintain the specific information that resides in such knowledge bases becomes necessary to be able to provide knowledge, news, and information-specific services to satisfy the public needs.

In this section, we present an architecture in which knowledge-based storage functions are compatible with processing, management, and knowledge services. Intelligent network (IN) environments where customized service provision is common in both second-generation IN (IN/2) and advanced intelligent network (AIN) environments provide a basis for the design of knowledge services provision. Whereas service logic programs offer specific service functions that can be executed in traditional computer environments, knowledge logic programs need very specific architectures to perform knowledge functions. The overall architecture of generic knowledge machines is presented in this section. Such machines perform programmable functions and execute application programs that entail knowledge-based micro- and macrofunctions.

1.7.1 Evolution of Knowledge Networks

In an extremely rudimentary sense, computer networks can minimally serve as knowledge networks. These networks perform knowledge functions such as processing, management, storage, and retrieval of critical objects that constitute the key elements within any body of knowledge, such as an article, a paper, a book, etc. The parsing of knowledge to uncover the crucial objects that play a decisive role in shaping any body of knowledge is akin to parsing for symbols within the compliers of higher-level languages (HLLs). In essence, knowledge implies objects, around which webs of relationships accumulate. The systematic study, analysis, and implication of these relationships give rise to incremental or derived knowledge, which in turn becomes a part of mainstream knowledge. Even though this definition appears to be self-perpetuating, it contributes to the explosion of knowledge on an ever-expanding scale. It also leads to the discovery and invention of new objects and concepts that are added to the dictionary in an unending process.

1.8 CONCLUSIONS

The nature of objects and the relationships between them (such as the keywords of a paper, or chapter, or an index in a book) constitute a segment or body of knowledge. These objects and their interrelationships become disoriented during knowledge processing. Only when there is consistency in the processing of all the objects that are implied in a body of knowledge is the overall processing of knowledge logical, coherent, consistent, and legitimate. Knowledge environments vary in their adaptation and intelligence. Service intelligence in communications systems has been successfully deployed since the 1980s [34, 35]. In the knowledge domain, the three basic categories of intelligence are feasible. They facilitate (1) connectivity to high-speed backbone networks and the Internet, (2) address and service update capability via the service control point database (SCP) of the intelligent networks (IN), and (3) knowledge-based intelligence (KI) look-up capability via the Dewey Decimal System (DDS) or Library of Congress (LoC) classification system [36, 37].

1.7.2 Knowledge Network Configuration

In Figure 1.19, when the adjunct knowledge service processor (AKSP) detects a request from the ISP user, a trigger condition invokes the participation of the appropriate knowledge service provider (KSP-i), The knowledge control point (KCP) unit of the KSP will start the attribute search for KBs around the world and does an initial match. Second-level matching takes place based on the client's attributes. After the raw information for the digital libraries and KBs is processed, the knowledge is again processed to obtain the exact information required by the Internet service provider (ISP) client, and it is dispatched from the KSP (for additional processing knowledge) over the global, national, or local backbone networks.

The limited access, high level of security, and logical separation from other Internet traffic will ensure the high quality of the information, knowledge, concept, or wisdom that the return KSPs to the ISP client. If we look back at the early telecommunications industry, in-band signaling systems were a cause of serious concern in the mid-1980s and early 1990s. The common channel inter-office signaling system (CCISS) and ITU-based standards for international signaling networks (or the CCISS network for local signaling) have evolved with a distinct and standard protocol to alleviate the inband signaling issues.

As the Internet starts to permeate every branch (medical, educational, commercial, business, financial, etc.) and phase of activities, the need for a distinct knowledge-signaling network and the participation of knowledge-processing systems and providers is foreseen. We give an example of such next-generation systems, one that appears to be financially viable and technologically feasible, in Figure 1.19.

1.8 CONCLUSIONS

This chapter introduces the current socioeconomic setting for the accelerating development of computer and communications systems that provide users with

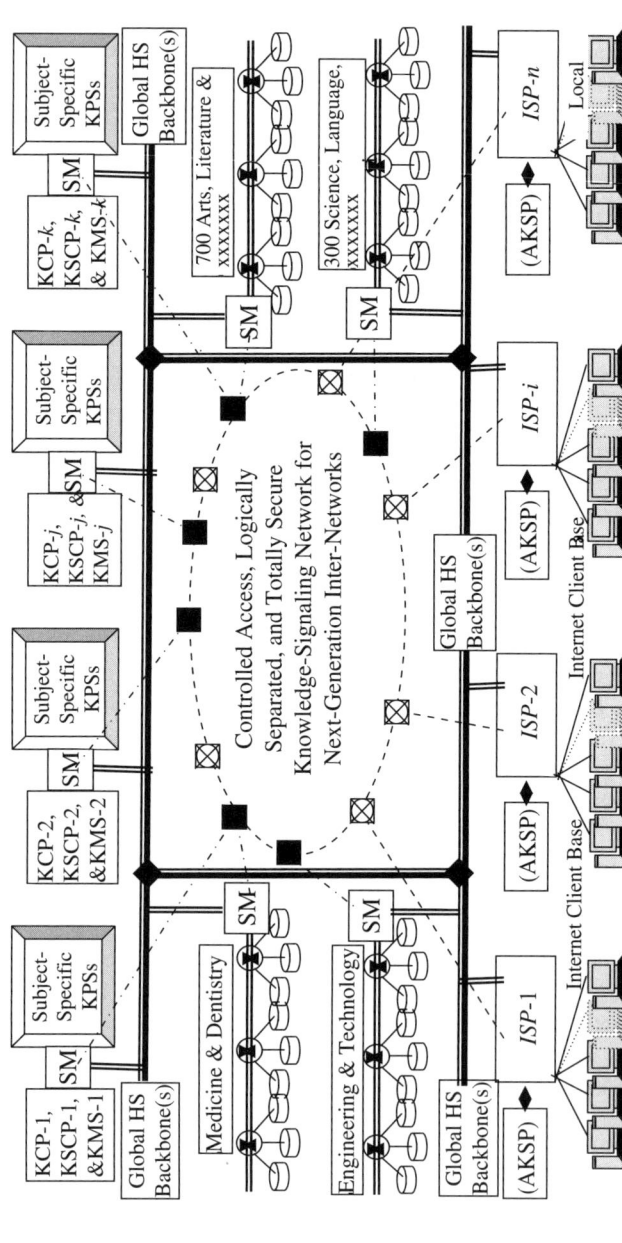

Figure 1.19 Network layouts for two-layer services of the next-generation Internet systems. Intelligent Internet and intelligent networks for building a peaceful and secure future. ISP = Internet service provider; AKSP = adjunct knowledge services processor; SM = switching module; KPS = knowledge service provider; KCP = knowledge control point; KSCP = knowledge services control point; KMS = knowledge management services.

precise and pertinent information. The machines preprocess the information furnished to an extent that the user can make sound and accurate decisions in solving a local problem. Such preprocessing is evident in the current Internet search programs when delivering search results. These current processes are rooted in basic logic rather than intelligence. The configurations presented are derived from the processing of objects contained in the information processed rather than the words used to generate that information. The processing capabilities and algorithms of the networks are readjusted to suit needs the user.

The architectures presented in this chapter cover a wide variety of applications, ranging from intelligent hospital environments to electronic governments for nations. The commonality in all the networks presented is their access to Web knowledge bases that validate and furnish global objects for comparison and inference. The intelligence embedded in such objects also interpreted by machines and processors makes the results returned to the user consistent with the explicit or implicit needs of the user. In a sense, the networks interpret user needs and the methodology to find solutions to such needs from the global and universal objects stored in the knowledge bases.

The machine configuration for most environments offers added benefits to the professional community. In most networks, such as medical or educational ones, the computer-based methodologies for the best solutions to specific situations in localized problems are customized by network intelligence in routing and by artificial intelligence in processing. In specialized networks such as tsunami warning, antiterrorist, or emergency response systems (ERS) networks, the methodologies with the quickest response are coupled with artificial intelligence (AI) techniques to detect and predict possible emergency conditions in the country and are deployed. These techniques can be deployed by any socioeconomic organization (such as a corporation or nation) to predict and monitor stability and continued survival. Machine response moderated by human intuition is generally quick, and the corrections are optimized to prevent oscillations and repeated corrections. A smooth and damped transition occurs from one stable operating condition to the next.

A network-based sensing and monitoring environment can be integrated with the high-speed backbone networks of the country. An intelligent decision support system can be ported in the network environment by appropriately arranging the functions of the three switching systems or switching the modules of any of those systems.

REFERENCES

1. Bell Telephone Laboratories, *Engineering and Operations in the Bell System*. Western Electric Co., Winston-Salem, NC, 1982.
2. S. V. Ahamed and V. B. Lawrence, *Design and Engineering of Intelligent Communication Systems*, Kluwer Academic Publishers, Boston, 1997.

3. Rogier Noldus, CAMEL: *Intelligent Networks for the GSM, GRPS and UMTS Network*, John Wiley & Sons, Hoboken, NJ, 2006.
4. S. V. Ahamed, "The Architecture of a Wisdom Machine," *International Journal of Smart Engineering Systems Design*, London, 5(4), 2003: 537–549.
5. B. Stroustrup, *The C++ Programming Language*, 3rd ed., Addison Wesley, Boston, 2000.
6. P. Seibel, *Practical Common Lisp*, Apress, Berkeley, CA, 2005.
7. S. J. Russell and P. Norvig, *Artificial Intelligence*, Prentice-Hall, Upper Saddle River, NJ, 1995.
8. E. Castro, *HTML for the World Wide Web with XHTML and CSS: Visual QuickStart Guide*, Peachpit Press, Berkeley, CA, 2002.
9. C. P. Poole and H. A. Farach (Eds.), *Handbook of Electron Spin Resonance: Data Sources, Computer Technology, Relaxation, and Endor*, Springer-Verlag, New York, 1999.
10. L. Buckwalter, *Avionics Training: Systems, Installation, and Troubleshooting*, Avionics Communications, Inc., Leesburg, VA, 2005. (auto pilots for planes)
11. K. Williams, *Build Your Own Humanoid Robots: 6 Amazing and Affordable Projects*, McGraw-Hill/TAB Electronics, New York, 2004.
12. B. Yenne, *Attack of the Drones*, Zenith Press, 2004. Also see G. West, "Drones Ahead: Pilotless Planes Reduce Danger and Increase Efficiency," *Alaska Business Monthly* (HTML), April 1, 2007 (digital). Farmington Hills, MT.
13. S. V. Ahamed, *Intelligent Internet Knowledge Networks*, Wiley-Interscience, Hoboken, NJ, 2006.
14. J. M. Keynes, *The General Theory of Employment, Interest, and Money*, Prometheus Books, Buffalo, NY, 1997. Also see R. M. O'Donnell, *Philosophy, Economics, and Politics: The Philosophical Foundation of Keynes's Thought and Their Influence on His Economics*, Palgrave Macmillan, New York, 1989.
15. J. Neumann and O. Morgenstern, *Theory of Games and Economic Behavior*, Princeton University Press, Princeton, NJ, 2004.
16. J. R. Parker, *Algorithms for Image Processing and Computer Vision*, John Wiley & Sons, Hoboken, NJ, 1996.
17. R. Arena and M. Quere (Eds.), *The Economics of Alfred Marshall: Revisiting Marshall's Legacy*, Palgrave Macmillan, New York, 2003.
18. A. Smith, *The Wealth of Nations*, Prometheus Books, Buffalo, NY, 1991.
19. D. S. Byrne, *Complexity Theory and Social Sciences*, Routledge, New York, 1998.
20. E. Alpaydin, *Introduction to Machine Learning (Adaptive Computation and Machine Learning)*, MIT Press, Cambridge, MA, 2004.
21. J. Neumann and O. Morgenstern, *Theory of Games and Economic Behavior*, Princeton University Press, Princeton, NJ, 2004; see "Zero Sum and Non-Zero Sum Games." Also see A. R. Washburn, "Two-Person Zero Sum Games (Topics in Operations Research Series)," Military Applications Social Research, Elsevier Science & Technology, "Handbook in OR" Burlington MA., 2003.
22. E. G. Sable and H. W. Kettler, "Intelligent Network Directions," *AT&T Technical Journal*, Summer 1991: 2–9. Also see W. D. Ambrose, A. Maher, and B. Sasscer, *The Intelligent Network*, Springer-Verlag, New York, 1989.

23. J. O. Boese and R. B. Robrock, "Service Control Point: The Brains Behind the Intelligent Networks," *Bellcore Exchange*, Navesink, NJ, December 1987.
24. *SAP Solutions and eLearning*, SAP (UK) Limited, Feltham, Middlesex, England, 2002.
25. R. Rowley, *PeopleSoft v.8: A Look under the Hood of a Major Software Upgrade*, White Paper, PeopleSoft Knowledge Base, 2005. Also see R. Gillespie and J. Gillespie, *PeopleSoft Developer's Handbook*, McGraw Hill, New York, 1999.
26. S. V. Ahamed and V. B. Lawrence, "Evolving Network Architectures for Medicine, Education and Government Usage," Pacific Telecommunications Council, Hilton Hawaiian Village, January 14–18, 2002.
27. K. C. Laudon, *Management Information Systems*, Prentice-Hall, Englewood Cliffs, NJ, 2003.
28. M. S. Pinto, "Performance Modeling and Knowledge Processing in High-Speed Interconnected Intelligent Educational Networks," Ph.D. dissertation, May 2007.
29. M. Krol, "Intelligent Medical Network," Ph.D. dissertation, City University of New York, 1996. Also see M. Krol, and D. L. Reich, "Development of a Decision Support System to Assist Anesthesiologists in the Operating Room," *Journal of Medical Systems*, Springer Verlag, New York, NY, 24, 2000: 141–146.
30. N. Mollah, "Design and Simulation of International Intelligent Medical Networks," Ph.D. dissertation, City University of New York, 2005.
31. L. Leung, "Design and Simulation of International Intelligent Medical Networks," Ph.D. dissertation, City University of New York, 2005.
32. N. Waraporn and S. V. Ahamed, "Intelligent Medical Search Engine by Knowledge Machine," in *Proceedings of 3rd International Conference on Information Technology: New Generations*, IEEE Computer Society, Los Alamitos, CA, 2006. See also N. Waraporon, "Intelligent Medical Databases for Global Access," Ph.D. dissertation, City University of New York, 2006.
33. S. V. Ahamed et al., "Hospital-Based Integrated Medical Computer Systems for Processing Medical and Patient Information Using Specialized Functional Modules," U.S. Patent 6,272,481 B1, issued August 7, 2001.
34. W. D. Ambrose, A. Maher, and B. Sasscer, *The Intelligent Network*, Springer-Verlag, New York, 1989. Also see *IEEE Communications Magazine*, Special Edition on Intelligent Networks, February 1992. See also J. Meurling, "An Intelligent Network Perspective," *Bellcore Symposium on Intelligent Networks*, Brussels, Belgium, 1988.
35. S. V. Ahamed and V. B. Lawrence, *Intelligent Broadband Multimedia Networks*, Kluwer Academic Publishers, Boston, 1997.
36. M. Mortimer, *Learn Dewey Decimal Classification*, 21st ed., Scarecrow Press, Lanham, MD, 1999. Also see *Dewey Decimal Classification and Relative Index*, 22nd ed., OCLC, Dublin, OH, 2003; J. P. Comaroni, *Dewey Decimal Classification*, 18th ed., Forest Press, Albany, NY, 1976.
37. J. Ganendran, *Learn Library of Congress Subject Access*, Scarecrow Press, Lanham, MD, 2000. Also see U. S. Government, *Library of Congress Classification*, http://catalog.loc.gov.

CHAPTER 2

WISDOM MACHINES

CHAPTER SUMMARY

The information age and knowledge society have evolved in synergy and synchrony. The information revolution has closely trailed the expansion of broadband networks. Broadband networks are the result of the judicious and wise choices of network developers around the world. Looking back over the last three decades, we see that the origin of the information age stems from the introduction of fiber optics for the transmission of binary data and advent of packet-switching technology for routing information.

Wisdom has played a profoundly ingrained role in synergizing the numerous events that preceded the knowledge society. Modern-day wisdom machines are a by-product of the knowledge revolution. The wisdom behind the synergy serves society much like computers have served the business community. The architectures of wisdom machines suited to the current era come from the evolutionary trail of computing systems and the ever-expanding capacity of VLSI chips. The inclusion of many social demands and technological capabilities to build viable national backbone networks is by itself an act of wisdom well suited to the information age.

During the earlier Aristotelian era, logical reasoning rather the scientific methodology dominated the basis of wisdom. From a philosophic point of view, it is debatable if the writings of Aristotle cannot be stretched to suit the

---

Computational Framework for Knowledge. By Syed V. Ahamed
Copyright © 2009 John Wiley & Sons, Inc.

information age. Such an inference can only be force-fit with perfect hindsight. To separate the scientific and philosophic perspectives, it appears more logical to draw an imaginary line of demarcation between the two facets of wisdom—namely, ancient classical wisdom and modern neoclassical wisdom.

Whereas Aristotelian wisdom finds its origins in truth, virtue, and beauty (TVB), modern wisdom is founded on science, economics, and technology (SET).

The two are not mutually exclusive but their relative emphases can be different. Ancient wisdom is based on an illusive triangle of TVB. Ancient wisdom is long ranged and stable. The goals remain abstract. Short-term fluctuations in society and norms do not substantially dictate the rules of ancient wisdom. The focus of TVB seems to lie along the path toward immortality, utopia, and sainthood tempered by the reality of existence. Human perceptions and judgment are loosely (and qualitatively) tied to social norms and degree of freedom within those norms.

On the other hand, modern wisdom is based on SET concepts. It has a quantitative basis. The focus of modern wisdom, with its origins in SET, is optimality, efficiency, and goal orientation. Modern wisdom becomes dynamic. Society and situation can temper the rules of modern wisdom. Timing and social setting are of the essence. This short-term wisdom can lead to the relentless pursuit of immediate goals, and even at the expense of long-term ethics and values.

Initially, the two types of wisdoms appear to occupy two opposite perspectives. Soon enough, they can be combined into an infinite spectrum of wisdom spanning the six dimension in the TVB and SET sets. Typically, read-only memory (ROM) chip sets that drive knowledge processor units are primed with either[1] the three dimensions of TVB or SET. Much as human thought can be influenced by preconditioning, wisdom machines can be embedded with memory chips primed with the TVB or SET combination. The relative emphasis on the two has altered wisdom to suit the many civilizations that have come and gone between the Romans and modernists. The abuse of such machines to manipulate the output poses an imminent danger.

2.1 MANY "FLAVORS" OF WISDOM

Aristotle's wisdom focused on truth, virtue, and beauty (TVB) and modern wisdom is built on science, economics, and technology (SET). At least two personifications for wisdom are evident: a saint and an executive. Each has a code of ethics to follow. The two codes retain some overlap, even though they may initially appear distant. They are both well contained in two semi-parallel universes with well-defined domains of overlap and intersection. For example, if science is based on the laws of universality, then truth is a manifestation of

---

[1] In practice, it is appropriate to have two chip sets and let the user decide which flavor of wisdom (ancient or modernistic) the solution should favor. Any admixture of the flavors is also feasible by forcing the wisdom machine to attach prechosen weights to either flavor or the three components of each.

universality; if mathematics is based on a universally accepted set of numeric and algebraic rules then truth is the result of a set of elegantly formulated ethical rules, etc. Hence, a framework of underlying and intertwined rationality exists between these two distant "flavors" of wisdom.

In the computational domain, machines have greater freedom. Consider the vector T in the TVB flavor of wisdom. If only two extremes are permitted as T and t, then the truth dimension has two steps. Similarly, if the virtue and beauty dimensions also have V and v steps and B and b steps, this leads to 8 (2^3) possibilities. Pushing the discrete steps in the SET dimension leads to another eight possibilities. The flavors of wisdom at the two extremes become an 8×8 matrix.

The user of the wisdom machine has the option of forcing any one of the four steps (T, t, s, or S) for the T/S dimensions. Similarly, four steps (V, v, e, or E) for the V/E dimension and four more steps[2] (B or b, t or T) for the B/T dimension become evident. The machine has 64 [i.e., $(2 \times 2)^3$] flavors of wisdom it can impart to the solution of any specific problem. The machine can also generate 64 (slightly overlapping) axioms[3] or solutions to any problem from the 64 perceptions of wisdom.

When the steps are made more discrete (finely spaced), then the flavors of wisdom become exponentially larger. For instance, three steps ($T^+, T\ T^-; V^+, V, V^-; B^+, B, B^-$) in each dimension yield $(3 \times 2)^3$ or 216 flavors, and four steps in each dimension lead to $(4 \times 2)^3$ or 512 flavors, etc.)

If the basic colors [red, green, or blue (RGB)] represent the three dimension TVB for classical wisdom or SET for modern wisdom, and the number of gradations in each color represents the discrete steps in each dimension for the perception of wisdom, then the flavors of wisdom become consistent with the number of hues that the human eye can perceive. This number is quite large and individualistic and in essence, the number of flavors of wisdom also tends to become (infinitely) large and individualistic.

To some extent, the intellectual synthesis of wisdom defeats the purpose of a wisdom machine. However, a wisdom machine can probe the many flavors of wisdom that a human being can not easily probe scientifically, systematically, and consistently over a long enough period of time. In probing the depth of human understanding, the wisdom machines may still have the upper hand only because the human mind cannot endure the same intense analysis of wisdom as a machine scientifically, systematically, and consistently over a period of many days or years. Many philosophers and yogis spend their entire lifetimes

---

[2] The symbol T signifies truth in the TVB set and technology in the SET set. In symbol table memory allocation, this does not pose any problem because the TVB and SET are two different three-dimensional spaces.

[3] The granularity and overlap in the meaning of the word in the dictionary can force the machine to use the same wording for some or most of the 64 axioms. This is a potential loophole for the machine in that there are not enough discrete words in the language to "say what it means to say."

engaged in their own unique search[4] for wisdom. In walking the unchartered line between monks and mathematicians, we propose a blend of natural and artificial intelligentsia without a clear line for the contour of thought and without a perfect set of operation codes for the machine. Both are negotiable depending on the situation yet offer long-lasting and transparent axiom of wisdom.

2.2 THREE ORIENTATIONS OF WISDOM

The direction that the pursuit of wisdom follows can be slanted to favor any one of the numerous flavors presented in Section 2.1. In addition to absolute and materialistic wisdom, a kind of counterproductive, destructive, or harmful negative wisdom or foolishness can also exist as an antithesis—a haunting reminder. These three breeds of wisdom are depicted in Figure 2.1 In a broad sense, the *three orientations of wisdom* are evident. Sections 2.1.1–2.1.3 detail further these three breeds of wisdom.

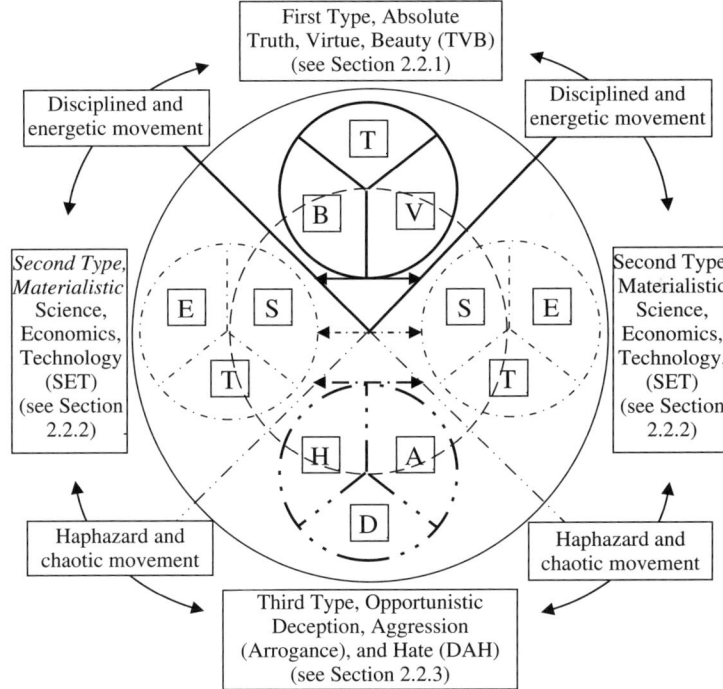

Figure 2.1 Conceptual designation of the coordinates for the three orientations (absolute, materialistic, and opportunistic) of wisdom.

[4]King's search for equality for African Americans, Gandhi's search for nonviolence, Lincoln's search for antislavery, Buddha's search for nirvana, etc.

2.2.1 Absolute Wisdom

The *first orientation* stemming from TVB, leads to a positive, productive, and socially desirable wisdom. If the truthful, socially benevolent, and elegant rules of human behavior are embedded in machine preferences, then the machine will propose axioms of wisdom that have been desirable from the days of Buddha. In the modern days, wisdom though illusive, is alive and well in most civilizations. Human beings receive inspiration even though machines are not specifically programmed to make their artificially intelligent decisions based on truth T, virtue V, and beauty B, unless they are so programmed.

The TVB space is fuzzy and illusive, and any attempt to achieve total precision leads to greater uncertainty. In a sense, the definition of the perfect union of the TVB space becomes equivalent to the pursuit of three infinities (as $T \to \infty$, $V \to \infty$, and $B \to \infty$) simultaneously. The intense order within the space of union becomes self-consuming for human minds and represents the total depletion of resources for machines. As in the scientific domain, when the velocity of any object approaches 'c' (the velocity of light), the mass appears to become infinitely large, in the perceptual domain, when the intersection of TVB is approached, the 'inertia' of any body of knowledge *(BOK)* also becomes infinitely large. However, around the end of the pursuit, the surrounding space indeed contains the TVB coordinates that existed when the quest started. The incremental movement from (T, V, B) to $(T \pm \delta T, V \pm \delta V, B \pm \delta B)$ is the movement of a body of knowledge from K to $(K \pm \delta K)$. If the change occurs in a time span of δt, then all the laws of dynamics for the bodies and boundaries of knowledge can be derived.

2.2.2 Materialistic Wisdom

From the *second orientation* stemming from SET, a neutral, factual, and indifferent realm of wisdom appears. In the scientific, economic, and technological sense, human behavior and embedded machine preferences tend to make this type of wisdom desirable for solving the daily problems of humans and nations alike. Science, economics, and technology exist in the decision-making processes embedded in this materialistic wisdom. Human values and ethics play a secondary role.

Almost all present-day corporate and machine activity is geared to abide by the rationality behind the sciences, the laws of economics (and hidden agenda for the pursuit of profits underlying corporate policies), and the mass production behind technology. This type of wisdom started with the industrial age and still drives most humans and nations alike. In the current society, most of the knowledge workers earn their livelihood by working in the information domain, and machines are versed well in the basic laws of science (S), formulations in economic justification and econometrics (E), and application programs in computer-aided design, robotics, and computer-aided manufacturing in the deployment of technology (T).

2.2.3 Opportunistic Wisdom

From the *third orientation* stemming from deception, arrogance, and hate (DAH), a negative, counterproductive, or destructive wisdom (more appropriately named foolishness) arises. This type of "wisdom" practiced in wars and robotic battlefield maneuvers is truly opportunistic. It is contradictory to classify the attitudes ingrained in the DAH set as wisdom at all. Yet, in practice, such "wisdom" is taught in defense universities and spy schools. Undercover agencies practice this type of opportunistic wisdom. In view of its existence or presence, we acknowledge its effects and the dark shadows it can cast on the traditional wisdom presented in Sections 2.2.1 and 2.2.2. Logic circuits, processors, switches, devices, and computers (with a traditional operating system) are oblivious to negative wisdom and abide by their own (circuit and systems) laws of science, mathematics, and the engineering principles used to design such lower-level machines.

Negative wisdom is presented as foolishness in most dictionaries. However, negative foolishness in not traditional wisdom. The basis of negative wisdom includes the traits and personifications of corruption (in some developing countries) and total insensitivity to society and the environment in general. Rather than being philosophic, a positive (TVB) and neutral (SET) wisdom machine can be forced to adopt a cautious stance against this negative and demoralizing wisdom. Such "wisdom" gapes in the face of machines that deploy positive wisdom. Machines like unwary humans can indeed be fooled. Some human beings and many nations receive their sustenance from the practice of this type of sub-human wisdom. Unfortunately, power bases are developed with this third type of immoral wisdom, and even more unfortunately, SET wisdom can propagate DAH wisdom. Deceptively, newly configured DAH wisdom can pose as TVB wisdom through the deployment of SET wisdom.

The bond between SET and DAH wisdoms become more robust more readily than the bond between SET and TVB wisdoms. The human counterpart of the SET wisdom machine needs to be wary that social inertia favors migration toward the DAH wisdom machine. The degeneration of order to disorder (or chaos) is more readily accomplished than the migration of order to super order (or integration of many lower level orders). The hardware gadgets and programs governing computer and network security must be special deterrents against human advances to corrupt any SET machine and transform it into a DAH machine.

The quest for the union of DAH can be as daunting as the quest for the TVB. Since the coordinates of DAH are sunk the quicksands of deception, arrogance and hate, the final coordinates remain as deceptive, arrogance-ridden, and hateful as the initial coordinates. There is no beginning or end to this quest; it is simply a circle of deception. Such a pursuit completely disorients human beings and leaves machines to pursue the inner intersection space of deception, arrogance, and hate. The fundamental difference between TVB and DAH destination-spaces is that at the end of the quest, the DAH type of humans and machines are completely overwhelmed by deception, aggression, and hate from within and without the mind, self, society. Humans can become aggressors and machines can become monsters. The origin and coordinate system are both lost in this type of opportunistic wisdom.

2.2.4 Needs and Wisdom

Wisdom accrues as the means to satisfy needs as they are experienced, understood, solved, resolved, optimized, remembered, and as they spread through the society. This has been an insidious and cumulative process in societies since the dawn of civilization. It has been a way for one generation to pass the knowledge, experiences, and skills it has acquired to the next. Information society cannot radically change this long-term mode of communicating wisdom. However, knowledge manipulation processes and techniques can refine and enhance the process by validating the truth in axioms, the elegance in representation, and the benefits that truth may bring to society.

Human needs may be basic or highly sophisticated. For most social entities, needs are paramount. A needless social entity soon becomes unneeded. Such needs are embedded in the makeup of human personality, corporate objectives, national goals, etc. For an individual, three specific models exist in the literature. (Freud [1], Maslow [2], and Ahamed [3]). Other hierarchies of needs are presented in Chapter 5 and discussed in detail by numerous social scientists such as Mead [4], Cyret [5], March [6], Simon [7], etc.

The three orientations of wisdom (Sections 2.1–2.3) are arranged and aligned with the three models representing the drives, motivations, and needs of human personalities in Figure 2.2. For machines to process wisdom, social norms need to be programmed and the horizontal alignment of wisdom and needs becomes essential, because the norms to handle needs depend on the level of need.

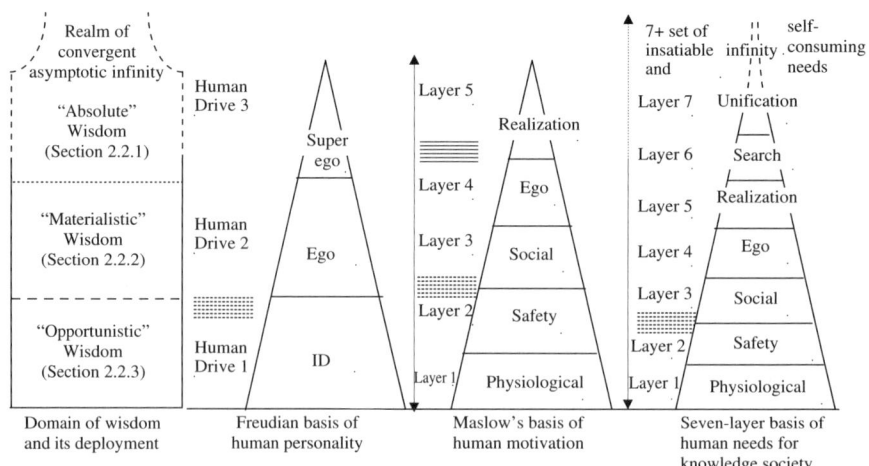

Figure 2.2 Mapping of the wisdom domain using the human drives of Freud, motivations of Maslow, and seven-layer needs structure for information age individuals.

Most individuals and social organizations address each of their needs in a simplistic and algorithmic fashion. Handling lower-level needs is generally a mechanized and reflexive process. Higher-level needs demand greater effort and more sophisticated, customized solutions.

Computers, networks, and the Internet are thus unified to deduce and derive the wisdom behind the elegant solutions to the higher-level needs of human beings and social entities. In sinister sense, we need as many wisdom machines as there are needs, and the needs structure of every human being is unique. In finding a viable configuration for the machine, it becomes appropriate to select a few architectures and numerous layers of software to direct the processing power and address a predefined group of needs, such as the physiological or safety needs of individuals, or the artistic or creative needs of others. On the opposite side lies the glaring possibility that "wisdom machines" can be designed as killer machines for tyrant kings or weapons of mass deception for political leaders.

2.2.5 What Are Wisdom Machines?

Wisdom machines are sophisticated computer systems and act much like computers but with embedded value systems and ethical codes that are protected by even more sophisticated security systems. Machine Intelligence and (machine-based) creativity exceed the artificial intelligence (AI) of AI-based systems and the deductive reasoning of logical systems. All AI tools (pattern recognition, PR; expert systems, ES; computer vision, CV; intelligent agents, IA; and robotics) and current Internet tools (access, searches, semantic, and content-based routing) form building blocks. Much like any new discipline, the tools for wisdom machines still remain to be explored. It is our preliminary thinking that such tools will bring humanity into the sciences just as mathematics was integrated in physics about two centuries ago or computer sciences was brought into telecommunications about six decades earlier. At the outset, the functions of protected and secure wisdom machines will not be tampered with by wayside hackers and corrupt politicians.

Traditional computer systems are but the slaves of those who buy them. There is no option but to execute the code. Computer code undermines any ethical code at the gate level, chip level, and systems level. More recently, the illegal actions of such systems (such as making illegal copies, propagating virus flows, spam dispersal, etc.) have been blocked by computer and/or network operating systems. If logic and algorithms are embedded in the code of the operating systems, then the blueprint for a wisdom machine starts to emerge.

When the innermost shell of security is ripped apart by an ill-willed hacker, the embedded code of values and ethics can be replaced by a mirage of deceit and hate. This mirage becomes a weapon of mass deception for leaders to propagate war and crimes in society. Fueled by political power, wisdom machines can cause unrest and intellectual famine around the globe.

2.2.5.1 Three Levels of Function of Wisdom Machines
Wisdom machines operate in three major roles as follows:

1. As computers and communication handlers
2. As scientific and intelligent agents that help human beings to probe for the inherent truth and validity of the solution being offered
3. As bookkeepers that catalogue any new axiom proposed by human beings, proving that it is in fact a new axiom and not a mere repetition, derivative, or inference from one or more documented example(s) of wisdom

Wisdom machines perform these functions in an optimally interleaved fashion. The purpose of these new machines is to serve and satisfy the human needs in every type of activity (both social and business). At the current stage of the evolution of computers and communications sciences, these two disciplines are mature and play a vital role in the evolution of society for future generations. Such strides have not been achieved on the humanistic side of society. The foundations of long-lasting societies are distinct from the foundations of business and commerce that offer short-term competitive surges of materialistic growth. If there is a pyramid of social hierarchy, then business and commercial growth should be considered akin to the growth of agricultural society that spared the population from hunger and starvation. By the same token, education and learning, justice and wisdom, and values and ethics, would rank higher in the social hierarchy.

From a combined perspective of the social sciences and technological achievements, the gap between human needs that motivate and scientific innovations that contribute needs special attention. Whereas in the past philosophers and political leaders offered insight into closing this glaring gap, now knowledge and wisdom machines can provide some footholds for such a bridge. If humanistic needs stand at one end of the bridge, then technological innovation is at the far end. The bridge will be a long haul bridge and the disciplines to harness, for the social dynamics to occur, will be social psychology (for motivation), economics (for justification), networks (for communication and concurrence), computers (for design optimization and implementation), and finally, media (for acceptance and usage). It is likely that this bridge will be an innovation in its own right since it opens new vistas for humanity to cross over into technologies. The following sections present the methodologies to bridge this gap and perhaps offer long-range solutions that will enable society to survive problems (corruption, corporate crime, terrorism, global warming, etc.) that it has itself created.

Such machines will have a mind of their own, one implanted by human interpretation of absolute wisdom that makes life sustainable in most extraordinary settings in society—life that can exist in the caves under the oceans or in the outer reaches of the universe. Such interpretations can only be dynamic and socially justifiable. It is the interpretation, not the absolute wisdom, that is variant, much like the many photographs of Jupiter when viewed in the infrared, ultraviolet, or normal spectrum. If absolute wisdom is treated as infinity, it moves further away the closer one get to it. In a sense absolute wisdom is a conceptual anchor of thought rather than a Mount Everest to climb and then claim victory for summiting. However, if the guidelines for interpretation is drawn on the basis of selfless wealth, immortal beauty, and (almost) divine benevolence, then the decisions of the wisdom machine are likely to be more stable in time and social space.

2.3 OPTIMIZATION OF WISE CHOICES

Sound business decisions are analyzed, iterated, and refined by executive and decision support systems (DSSs). Automated machine learning is generally a feature in the more sophisticated support systems. Decision support systems may exist independently, within management information systems (MISs) or within executive information systems (EISs). The pursuit of optimality in deciding on one or more corporate goals (maximization of profit, market share, cost minimization, resource allocation, etc.) is generally a run-time option for the users of DSS, MIS and/or EIS application programs. Machines that execute these application programs implement the second type of wisdom which is tailored to the particular corporation. The Customization of EIS platforms to achieve corporate policies and objectives is generally done during installation. The learning algorithms embedded in the machines will (slightly) retune and modify the programs and constraints to suit the dynamic profile of corporate leadership.

In the realm of employee decisions, the human element and managerial style play significant roles. Corporate policies, organizational goals, and short-term objectives generally determine the decision making within a narrow window of options. Such policies, goals, and objectives are set forth by the executive for managers and for employees. At the highest level of corporate structure, the guidelines start to became vague and undefined, except for the legal framework for business operations and conduct. This vagueness is a loophole that will allow corruption to creep in, especially if the top management of corporations (such as Enron, Arthur Anderson, Global Crossing, etc.) is less than honest and indulgent in satisfying its self-interest.

In other corporate and political circles, such guidelines become further negotiable due to lack of supervision and surveillance. In this setting, the downward drift in the orientation of an individual or a collective thought process becomes a component of the third type of opportunistic wisdom of corporate leadership. Numerous flavors of wisdom are "invented," exerting their own influence in the decisions made on an individualistic or corporate basis. The invention of negative wisdom is the creation of a corrupt mind.

2.3.1 Derived Axioms

Axioms of wisdom depend on the direction of wisdom. If we think of wisdom as a path, axioms are the road signs along the way. Such signs can point in any geographic direction, provided they lead to the destination from the current location. If the signs are randomly mixed, it is possible to find a sign for every direction. In the context of the three directions of wisdom, the choice of axioms needs to be contextual in both a syntactic and semantic sense. From a wisdom machine seeking materialistic wisdom, axioms of absolute wisdom are improbable. In instances where the realm of science (the S component of SAT) totally overlaps the concept of truth (the T component of TVB), the pursuit of scientific truth may yield common axioms in both cases. Whereas the constraints of science are localized by the definition of the subject, the constraints of truth are established by the limits of consciousness. In the humanistic domain, the search for gold in Alaska is not the search for nirvana in Da Lat City (Vietnam), even though gold may offer the means to travel to the Buddhist shrines in Da Lat.

From the perspective of an axiom derived from a wisdom machine, the main goal or destination needs to be fragmented as series of smaller sequential goals[5] that make up the graph for the achievement of the main goal. In the fine steps of deriving an axiom on a SAT based wisdom machine, the machine may still search for gold in Alaska, if an Alaskan is attempting to travel to Da Lat. In some cases, such a fragmented search may indeed offer a viable choice. The machine will not rule out any such possibility but "weigh and consider" it alongside other alternatives. In a humanistic sense, the machine will command a robot to leave no stone unturned in every stream in Alaska in order to find and grab a nugget of gold under it. For a solution on any wisdom machine, the searches are exhaustive in Internet space.

2.3.2 Priming of Machine Wisdom for Directional Axioms

Axioms are derived initially based on scientific principles and then iteratively refined with human insight and oversight. Such an approach facilitates the development of a tentative architecture for wisdom machines. The scientific foundation of the three wisdom machines (absolute, materialistic, and opportunistic) thus remains firm, but the top layers of software and special talent of the humanware will be different.

The most rudimentary structure is shown in Figure 2.3, where the machine layer performs the functions of most sophisticated decision support systems (DSSs) or executive information systems (EISs) currently used by institutions and corporations. The dominance of human will and discretion is depicted in Figure 2.3 through the act of selecting the intelligence that will be embedded as firmware and (application and operating system) software in the machine.

---

[5]The strictly disciplined methodology of segmentation of any major problem into such fine discrete steps that a CPU can execute them as a binary instruction has been practiced in software design since the days of von Neumann. It is borrowed here in an oblique sense, to the extent that the goal of an axiom is to find a unique solution to a very worldly problem by abiding by the syntax and semantics of three abstract notions (TVB for absolute wisdom, SAT for materialistic wisdom, or DAH for opportunistic wisdom).

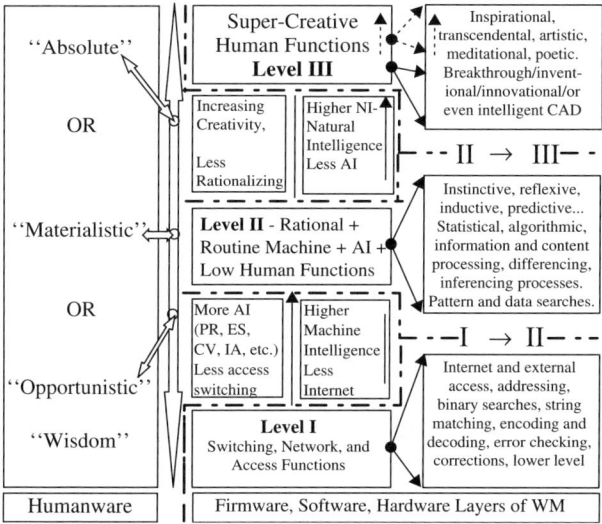

Figure 2.3 Overview of the major building blocks of a general-purpose wisdom machine. The lower part of the diagram emphasizes the traditional computer and network functions, and the top part emphasizes the superposition of the human will and discretion on the computer and switching-oriented networks.

Axioms of any type of wisdom can be derived from this generic architecture of wisdom machines. It is also possible to customize the wisdom to complement the thinking of the user. Customized wisdom, however, is not wisdom but individual wisdom. In this lowest level of operation, a wisdom machine is no more than a customized computer system. In the highest level of operation, the wisdom machine can complement the thought processes of a philosopher, a saint, a poet, or any human being exploring the edges of a generalized mental space. By the same token, in an extremely the lowest level of operation, the wisdom machine can complement the thought processes of a tyrant king, ruthless leader, or thug finding creative ways to inflict harm on human beings and destroy society.

2.4 THREE-LEVEL FUNCTIONS

A typical wisdom machine [8] is shown on Figure 2.4. There are three Levels and two primary functions at each of the two lower levels, I and II. In a sense, the wisdom machine functions on the corporate model[6] with an essential

---

[6]This model consists of three level—namely, (1) the "production" level (Level I) with embedded technical, machine, and robotic (if any) aspects; (2), the "middle management" level (Level II) that deals with streamlining all aspects of production functions; and (3) the "executive" level (Level III) that addresses the policy decisions which deal with philosophy, survival, and the long-term effects and impacts of routine production decisions and activities on a daily basis.

76 WISDOM MACHINES

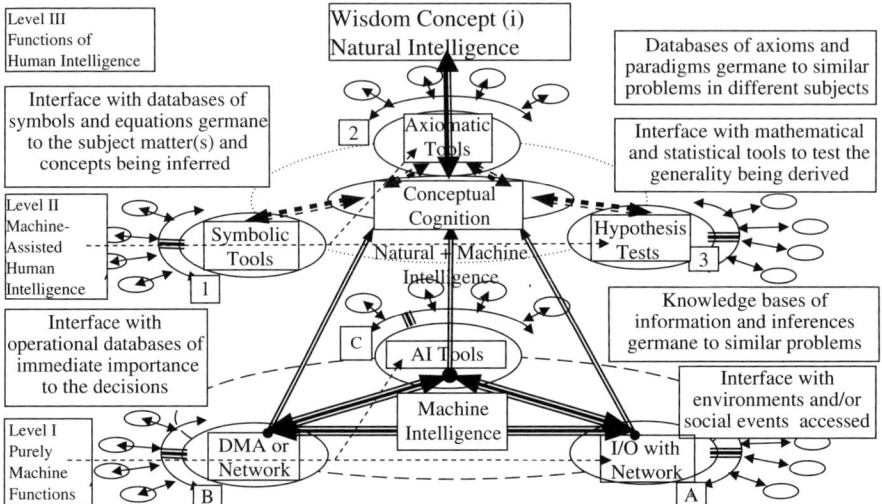

Figure 2.4 Six basic functions (A, B, C and 1, 2, 3) and their implementation at levels I and II of a wisdom machine. Human insight, supervision, and collaboration are provided at level III.

difference: The wisdom machines process information and knowledge to occasionally generate (produce) an axiom of wisdom.

2.4.1 Level I: Access and Administrative Functions

Extending the analogy of the corporate model, Level I functions are related to the technical and robotic functions in a corporate (production) environment. The wisdom machine senses and monitors the environment to find opportunities to generate generic laws that may possibly yield axiom(s) of wisdom. The three subfunctions at this level are:

1. Sensing the immediate environment where wisdom is sought (see A in Figure 2.4)
2. Accessing, communicating, and scanning relevant data, information, and knowledge (see B in Figure 2.4)
3. Administering and managing (local and global) data, information, and knowledge germane to the direction of wisdom (see C in Figure 2.4)

2.4.2 Level II: Linkage, Scientific, and Statistical Functions

Level II functions are related to the midmanagement functions in a corporate (production) environment. The wisdom machine negotiates and contemplates the environment to find the best opportunities to generate generic laws that may possibly yield axiom(s) of wisdom. In the corporate environment, one of the responsibilities is to improve the production environment by examining all the technical, social, and financial alternatives. The laws of science, human relations, and economics are explored by dedicated managers. For the wisdom machine, the three subfunctions at this level are:

1. Establishment of linkages to existing symbols and equations (see 1 in Figure 2.4)
2. Use of scientific and social tools to validate or contradict any tentative axiom (see 2 in Figure 2.4)
3. Test the hypothesis embedded in the axiom against the reality of the environment in the direction where wisdom is sought (see 3 in Figure 2.4)

2.4.3 Level III: Human Authentication

This all-human function permits human wisdom to blend into the machine-assisted lower-level synthesis of axiom(s) of wisdom. The unique character of human beings to project an axiom and integrate its effects (desirable and/or undesirable), and then weigh and consider each independently and as a group[7] are portrayed in the human component of a wisdom machine that seeks to include human beings in the processing of wisdom and axioms. To some extent, Level III functions are the equivalent of executive decisions in corporations based on the recommendations of middle management. As Level II decisions are also partly human, they can be constrictive, narrow, and localized.

2.5 KNOWLEDGE MACHINE BUILDING BLOCKS

Knowledge is the basis of wisdom, and wisdom machines rank higher than knowledge machines in the seven-node sequential chain (see Figures 2.5a and b) of ethics in the society.

The processing of information leads to knowledge, and the processing of knowledge synergistically between humans and machines leads to wisdom. If

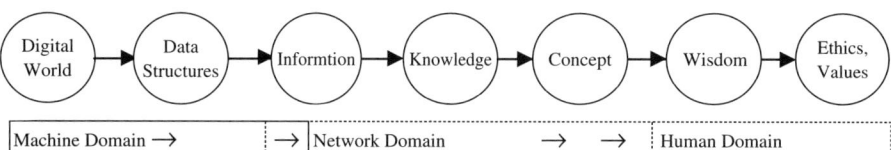

Figure 2.5a Progression of binary data from a digital domain to the key concepts, wisdom, and ethical values in most societies. Seven nodes are shown, with the reality of digital data processed by the computers on the left side and ethics conceived, derived, and refined by human beings on the right.

---

[7]Human ingenuity and creativity are both involved in Level III functions. Segment, process, regroup; and then reiterate— is perhaps the lowest level of human creative functions.

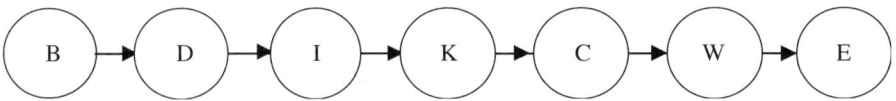

Figure 2.5b Simplified format of Figure 2.5a to carry the chain into the hardware structures depicted in Figures 2.10 and 2.11.

one strips wisdom from a list of divine gifts, then it can only be derived. In the modern world the vastness of knowledge has been multiplied by the reach of the Internet. In a sense, the synergy between humans and machines has already started; the Internet further expands the horizons of individual and collective thought.

While deriving (the axioms of) wisdom, numerous tentative axioms arise, are subjected to scientific and statistical tests, and are tested for correctness, validity, and reality of content by machines. The intellectual tests are performed by elite human teams that monitor the performance of machines as they derive tentative axioms of wisdom. From Figure 2.4, it appears feasible (though not economical) to replace each of the six subfunctions out lined in Section 2.4 to be performed by an independent knowledge machine.

2.5.1 What Are Knowledge Machines?

Knowledge machines (KMs) process synthesized information from any IT platforms (such as DSS, EIS, SAP, PeopleSoft, etc.) to derive knowledge content (of significant value, if there is any) from the numerical and graphical content of such information systems.

KMs are discussed in detail in [9]. In this section, we extract the most significant concepts of the KM such that it can be used as a module of a wisdom machine. The working principle of a KM is shown in Figure 2.6. In Sections 1.4.1. and 3.2.2, it is proposed that knowledge is founded on significant noun objects, and such objects undergo structural or attribute changes as knowledge is processed. In the human mind, the perception of such objects is altered by dialogues, environmental changes, and/or interactive processes with other noun objects. In the core of the knowledge machine, the attributes and relationships of objects (stored as data structures for each object) are updated as knowledge instructions are executed in the knowledge-processing unit (KPU) or units. Hence, two processing units—KPU and the object-processing unit (OPU)—work in conjunction to execute a sequence of knowledge instructions that make up a knowledge program. Such knowledge programs change the status of an object set embedded in the body of knowledge to a more desirable status or extract new knowledge from old knowledge. Such knowledge machines are generic and programmable, offering new knowledge based on existing knowledge, much like computers that derive new numbers (and properties), or find correlations between existing numbers and data sets.

KNOWLEDGE MACHINE BUILDING BLOCKS

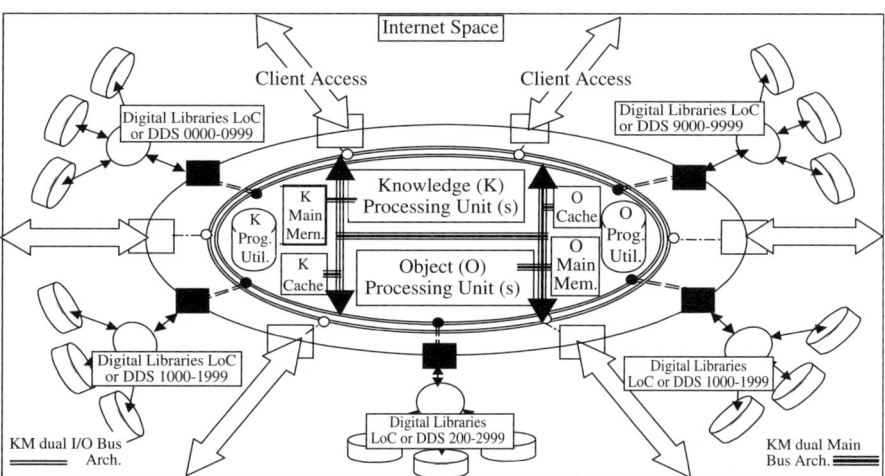

Figure 2.6 Internet-based KCS=Knowledge communication system. and knowledge-processing system server with knowledge- and object-processing capabilities (dual bus). The basis of this IKPS=Intelligent Knowledge processing sytem corresponds to I^2-KPS because of the Internet and knowledge Intelligent or content-based switching to different digital libraries.

In the context of wisdom machines, the generic functions (A), (B), and (C) at level I, and (1), (2), and (3) at level II can be handled by programmable and artificially intelligent knowledge machines. Knowledge and wisdom have no physical limits. However, the pursuit of infinite knowledge and unbounded wisdom are both impracticable for any machines due to resource limitations or any human being due to time/resource limitations. To contain the problem in the context of global knowledge and yet be realistic about the confines of limited resources, access to Internet space is provided via gateway and router access to the knowledge machines. In the same vein, the KPUs and OPUs in the machines are selected to offer maximum performance for handling multiple objects and at the highest speed that the technology would permit. Internet space, access, and network speeds are key issues that will have a profound influence on the practicality of building these new machines. We explore three different alternatives for using knowledge machines as the building blocks of a wisdom machine.

2.5.2 Knowledge-Machine-Based Wisdom Machines

Knowledge from the machine and human elements of knowledge machines becomes the basis for wisdom inferred by wisdom machines. Six knowledge machines can be used to perform the three subfunctions at each of the two levels I and II. A configuration depicting the use of six knowledge machines and a human component for level III functions is shown in Figure 2.7. Data from the networks are channeled (or split) into a "sensor" knowledge machine (A) where the data and user objective are merged for a possible fit. Such snooping

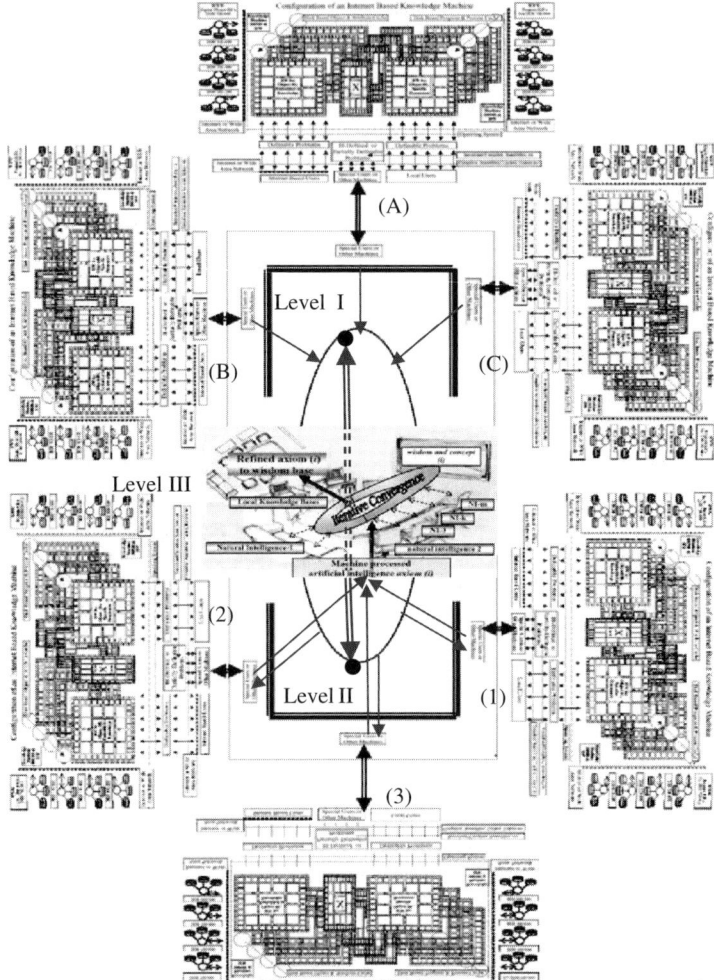

Figure 2.7 A configuration of a data-in wisdom-out machine. Data-in takes place at level I, sensor inputs (A). Wisdom-out takes place at level III as human beings process the tentative axioms of wisdom at level II and iteratively refine the axioms at level III. These processes are akin to the decision-making process at the executive levels of corporations and or at the General Assembly of the United Nations.

on Internet data is routinely done by security agencies to find any illegal actions targeting segments of the population, to explore any clues about possible vicious activity, or to just pry into the lives of citizens.

Level I activities (B) and (C) are also performed by the top three machines of Figure 2.7. Level II functions (1), (2) and (3) are performed by the lower three machines and any promising axioms of wisdom submitted for human deliberations at Level III.

2.5.3 Sensor-Scanner-Based Wisdom Machines

Sensor-scanner systems scan the environment and modify their own behavior so as to stabilize the environment and overcome any unexpected transitory and/or damaging environmental changes. For example, the functions of a typical multinational corporation are generally networked into its own information technology (IT) platform. Its numerous operations scattered throughout the world are dynamically linked to centralized databases of resources, inventory, sales, and general executive instructions. The senses proposed in this invention scan the vital statistics of the corporation and its diversely distributed divisions. Such vital statistics for corporations have a hierarchical relationship and may be linked to the organizational chart of any individual corporation. A large number of management information systems (MISs) and executive information systems (EISs) act as conduits of information to executives and keep them updated about signs and signals of current events, and their trends. This particular invention is a hardware platform on which the EIS may ride. Instead of being a channel switch for the flow of information, it acts as an intelligent EIS network that can process crucial information and monitor the welfare of the corporation, much like the hospital-based systems that monitor the condition of ICU patients on a second-by-second basis and also alert medical staff about any impending danger.

Most of these variables are collected in a centralized data facility, and they are compared to their expected values within the normal range of statistical bounds of the current activities of the corporation. When some of the indicators exhibit sudden, inexplicable, unforeseen, or undesirable variation, trends, or changes, alarm conditions are triggered. In a great majority of the cases dealing with large corporations, these responses are executive or committee decisions. However, the early identification and early warning systems may be triggered, and the sensing network may also monitor the progress of the condition.

Typically, a master clock initiates the scanning and monitoring activities. The process of collecting the data and feeding them back to the predictive model are shown. The model with the most updated internal parameters is used to predict the production process of the corporation for the next interval. The actual production itself is compared to the predicted production, and the error is used to correct the estimation process of the corporate process. The monitoring process takes place on a continual basis, and the internal adjustment of company policies and practices are tuned to the external forces that influence the corporate goals (maximize profit, enhance international trade, capture more market share, enhance productivity, etc.).

The sensing network in the corporate environment serves the purpose of monitoring most aspects of mid-level management. Such functions can range from inventory control, production and assemblyline supervision, to financial control. The result is the dynamic adjustment of the network feedback (and human) control of the entire production process to optimize any predefined goals. The feedback loop(s) and their time constant(s) are monitored to govern the stability and functionality of the human–machine production and communications environment typical of most industrial settings.

Machines, networks, and social entities function as feedback systems with AI-programs to learn from and cope with dynamic swings of the environment. Human creativity reigns as the supreme, yet undefined bridge between machine processes and network connectivity. The simplified architecture of a sensing system is depicted in Figure 2.8. These systems can also function in the feedback or feed-forward mode by reprogramming the AI routines to learn and predict the required response. In the feedback mode, a change in the environment occurs before the correction is applied. These systems can also predict imminent changes and prepare for change when it occurs. In the feed-forward mode, the system predicts the changes that will be necessary to correct error buildup at an intermediate level before final output. The correction is applied such that the transient environment is corrected when processing input into system output. Very settling advantages occur when the feedback and feed-forward techniques involve judicious use. Digital and program control of sensing systems offers sophisticated inputs to drive the wisdom machine.

The lower six functions of any wisdom machine are performed optimally by specially designed hardware suited to the function. Specialized hardware delivers optimal functions quickly and efficiently, but restricts their applications to other possible uses of the wisdom machine and vice versa. In seeking generality of application for any wisdom machine, the compromise lies in reducing the complexity of the machines for the six subfunctions but still delivering the potential to solve a reasonable number of wisdom problems. Computer system designers also faced similar issues during the 1950s to 1970s when designing computers for scientific and business applications. The challenge of the twenty-first century will be designing computer systems, their OS, compilers, and science layers for social and global problems.

In the earlier IBM systems, the problem was resolved by selected the appropriate firmware lodged in the control memories of IBM 360 and 370 series machines, but then both the architecture and potential of the CPUs were enhanced radically to handle a very large variety of applications. The functional interpretation of the same operation code in many ways to meet the requirements of users thus curtailed the need for redesigning the architectural layout of the mainframe computers. Selected hardware chips [such as graphics chips, double-precision arithmetic processor units (APUs), etc.] and systems (such as array processors, A/D subsystems, etc.) were also available for users with specialized needs.

In the same vein, wisdom machines have at least one precursor (IBM mainframe systems) to their evolution. In this context, the two major contenders for building blocks are (1) knowledge machines and (2) sensing environments. Both machines have history and credibility. Knowledge machines are being introduced in knowledge base management systems, such as content-based library management systems, and service management systems routinely used in intelligent networks [10]. Sensing environments are routinely used by almost all local and global security and battlefield installations.

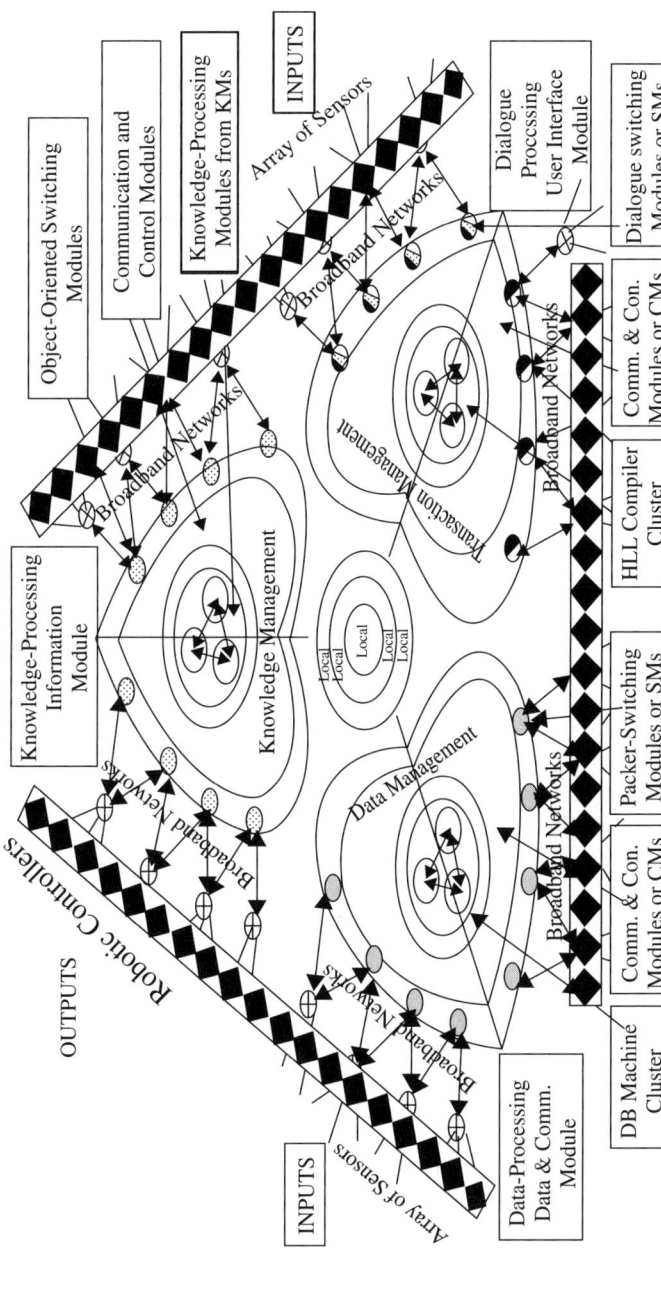

Figure 2.8 A simple configuration of a sensing machine. The sensors continuously sense the changes in the environment (see INPUTS) and deliver robotic command to readjust their controlled outputs (see OUTPUTS) to restabilize the system.

One such approach to the sensing machine is discussed in [9, SM chapter]. Machines of this nature are deployed by security agencies that monitor airports, public and private buildings, and even computer systems that check software for malicious or pornographic content. The sensing machines attempt to correct and compensate for any irregular or abnormal conditions based of AI programs that supply the corrective strategy. One such configuration embedded in wisdom machines is shown in Figure 2.9.

As much as knowledge machines can be programmed to perform the lower six subfunctions at Levels I and II, the sensing machines can be configured (hardware, software, operating systems, AI programs, and macros) to perform the same functions. The fundamental difference is that knowledge machines become more user- and process-centric, whereas the sensing machines become input- and network-centric. However, the overlap of functionality become evident. An optimum blending of the two machine systems (knowledge and sensing) is most likely to offer the best architectural configuration for wisdom machines.

When the applications become specific (such as monitoring battlefield operations robotic operations during space flights, etc.), then the hardware gadgets and software modules can be grouped together for optimality, much like switching gear and the switching software are grouped in electronic switching systems. The close match between software-generated binary instructions and hardware circuitry to execute commands will result in rapid response and precise robotic functions.

2.5.4 Bus Configurations and Switch Locations

Bus structures in knowledge and wisdom machine carry integrated data structures rather than packetized and segmented binary data. Entire knowledge (tree) structure and inference diagrams will need to be transferred from one component to the other or from one machine to another. Traditionally, the OSI model facilitates the file and application data transfer from one computer to another.

One the one hand, logic circuits and DMA switches (with two-complement counters) are deployed in conventional computers to transfer words and I/O data from external devices to registers or memory. Entire file transfers are also common for disk read/write commands calling for file segmentations and header data for file segments. Data transfer speeds and R/W/verify operations become particularly important in the data transfers within computer systems.

On the other hand, packet switching facilitates network (LANs, MANs, and WANs) communications. Routers and gateway switches are generally involved. All the functions specific to seven layers within the OSI model assure that data transfer occurs accurately and reliably between one computer to the next. High-speed networks and QoS requirements generally assure the highest speed subject to component constraints.

For the communication within the wisdom machine, both speed and accuracy are essential. When the machines are located in proximity, fiber-optic bus structure with localized IP addressing capability is a possible venue. When the

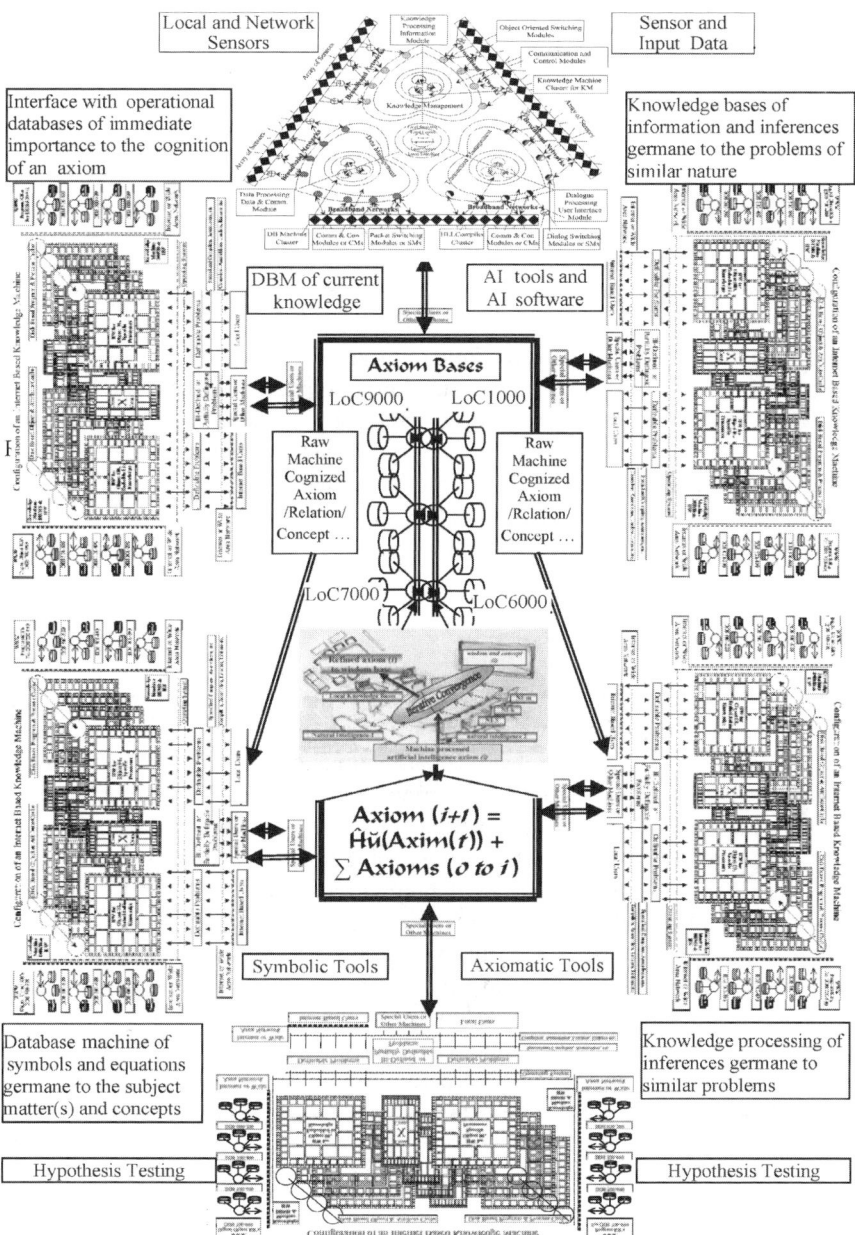

Figure 2.9 Embedded sensing machine for deriving new concept and wisdom bases from routine activities in the environment. The lower five knowledge machine subsystems perform levels I and II subfunctions to detect, deduct, and derive wisdom.

machines are distant, fiber-optic backbone networks with global IP addressing capability are a possible venue. The design of switches and their locations will vary accordingly.

One such configuration of buses and the location of switches is shown in Figure 2.10 for a wisdom machine where the groups of Level I machines lie in proximity and where the groups of Level II machines are distant. The data structures between machines in the same layers (layers I and II) are local-LAN-switched and the tentative hypothesis between layers I and II is Internet-or backbone-switched. Numerous alternative bus and switch architectures are also possible. The location and capacity of the switches should be optimized after defining the physical and logical location of the constituent knowledge/sensing subsystems.

The embodiment of the knowledge trail introduced in [9] and restated in Figure 2.5 is shown in Figure 2.10. Functions at levels I and II $\{(B \rightarrow D \rightarrow I \rightarrow K')\}$ cover[8] the first four nodes using standard computer, information, and knowledge technologies. These three technologies exist in society and encompass logic, switching, computer, and network techniques and AI methodologies. The element crucial to performing Level III functions $\{(K' \rightarrow K \rightarrow C \rightarrow W \rightarrow E')\}$ is humanware consisting of human systems (alive or stored expert system databases) to monitor and guide lower-level humanistic machines.

2.6 MACHINE CLUSTERS

Single machines that can effectively retract wisdom from the binary world of data are inconceivable at present. Designers are not likely to find customized VLSI chips to perform all the necessary functions and be able to integrate them in the architecture of one machine, particularly because there are at least three (absolute, materialistic, and opportunistic) distinct classes of wisdom. However, it is possible to find a generic hardware architecture and very flexible software layers that can perform the six essential functions (three in layer I and three in layer II) to deduce and derive axioms of wisdom. Such a machine can only be inefficient and perform poorly. Both execution speed and flexibility will suffer.

The clustering of higher-level machines offers a possible venue to circumvent the setbacks of a single-machine system. At the outset, wisdom machines may be conceived of as a set of four clustered machines (see Sections 2.6.1–2.6.4). This is especially desirable in view of the many "flavors" of wisdom (see Section 2.1) that can be derived from the three classes of wisdom.

2.6.1 Single-Wisdom Single-Machine Systems

The building blocks for the type of single machine [9] that will extract absolute wisdom from Internet binary traffic are shown in Figure 2.11. Only one

---

[8]K' indicates that knowledge is being generated (i.e., detected, deduced, and derived) from real and factual information. Generally, information is processed iteratively until the level of knowledge satisfactorily meets standard scientific, economic, and social levels of acceptability.

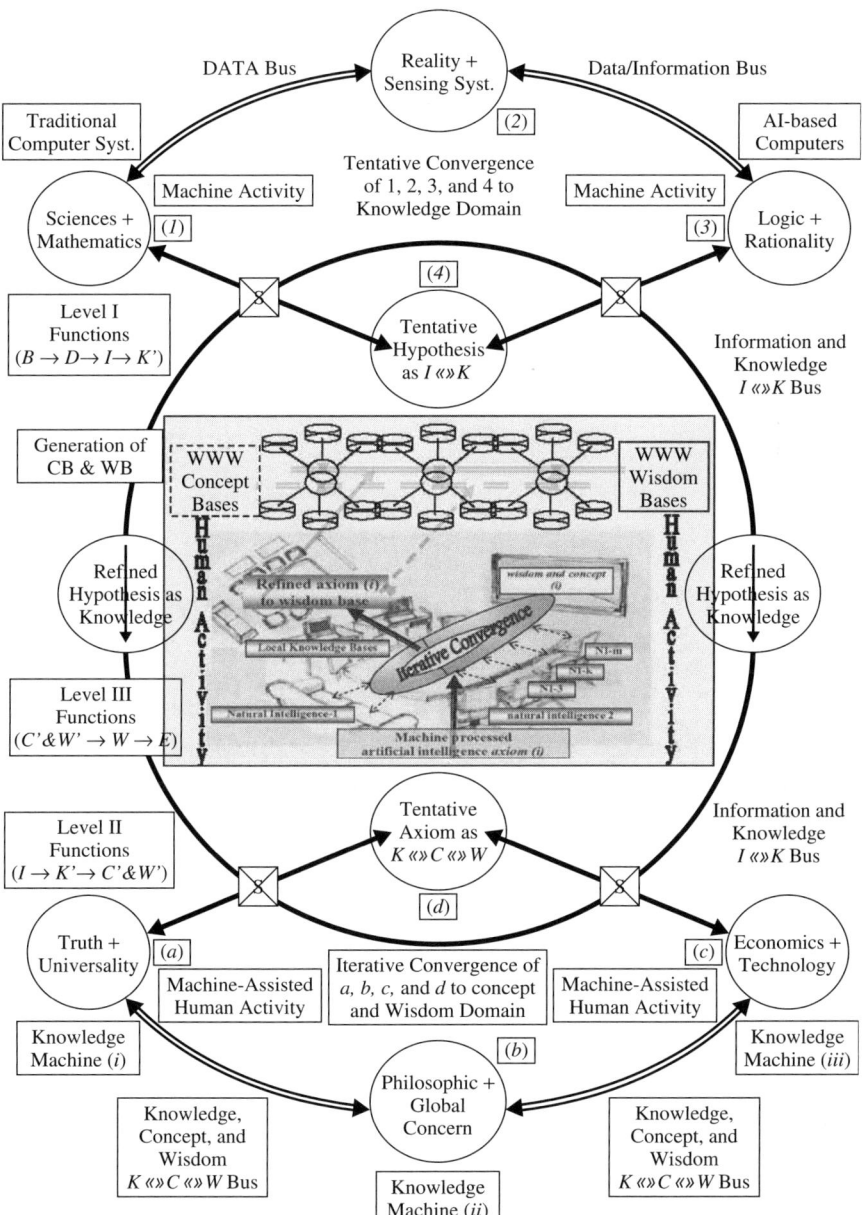

Figure 2.10 Representation of a wisdom machine based on three-level functions (I, II, and III) to detect and deduce concept and wisdom bases from binary-level Internet traffic. All the stages ($B \to D \to I \to K \to C \to W \to E$) in the chain of knowledge are embedded in the hardware, software, and humanware of this machine.

(absolute, materialistic, or opportunistic) directionality of wisdom is operative at one time. The same machine firmware, software, and hardware layers are primed by the three components of the particular wisdom (TVB for absolute, SET for materialistic, or DAH for opportunistic) that occupy the three 120° coordinate axes.

From an entirely philosophic perspective, it appears that the architecture of one composite wisdom machine is impractical. However, from a design perspective, the approach of firmly confining the present single-wisdom single-machine (SWSM) system with finite and realistic bounds offers some venues.

If truth, virtue, or social value, and beauty or elegance, are too many goals for the machine hardware, software, and firmware (at Levels I and II) and the human mind (at Level III) in the third tier of Figure 2.4 to pursue simultaneously, then three inter-dependent architectures of lower-level machines (see Figures 2.11a and b) are invoked to make up one composite wisdom machine. This system

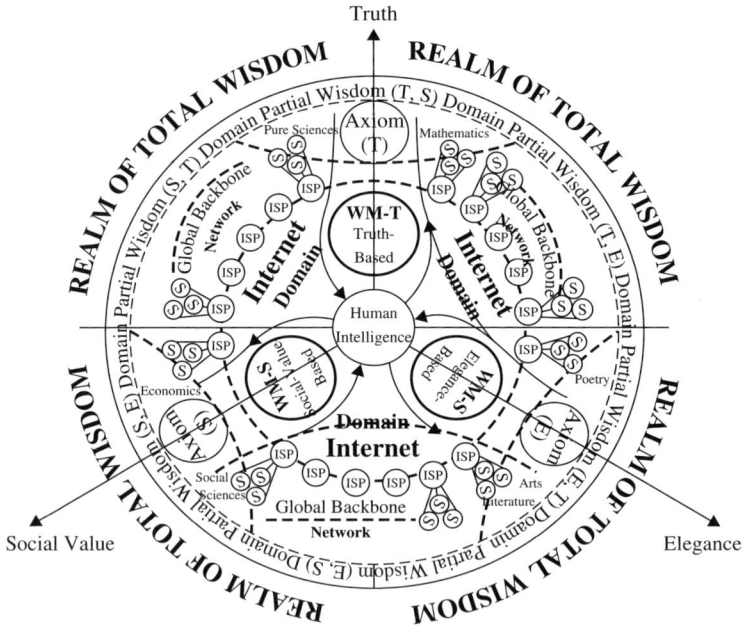

Figure 2.11a Single-wisdom single-machine configuration for absolute wisdom with three submachines along the radial line at $\theta = 90°$ or at $\pi/2$; along the radial line at $\theta = 210°$ or at $7\pi/6$; and along the radial line at $\theta = 330°$ or at $11\pi/6$. They seek T for truth, S for social values (or virtue), and E for elegance (or beauty). The Internet data and traffic are scanned via the numerous Internet Service Providers (ISPs) and switches. The initially identified axioms for each of the three machines are submitted to a panel of human beings to validate or repudiate the findings. The machine search is refined, directed, and targeted by human beings if the machines are likely to offer generalized axioms of wisdom dealing with universal truth, universal elegance, and/or great social value.

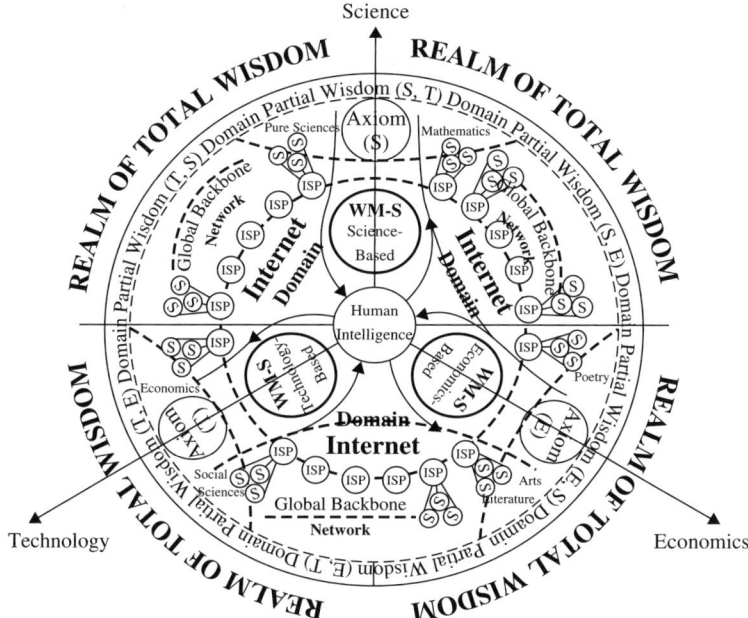

Figure 2.11b Single-wisdom single-machine configuration for materialistic (corporate) wisdom with three submachines along the radial line at $\theta = 90°$ or at $\pi/2$; along the radial line at $\theta = 210°$ or at $7\pi/6$; and along the radial line at $\theta = 330°$ or at $11\pi/6$. They seek S for science, E for economics, and T for technology. The Internet data and traffic are scanned via the numerous Internet Service Providers (ISPs) and switches. The initially identified axioms for each of the three sub-machines are submitted to a panel of human beings to validate or repudiate the findings. If the machine search is refined, directed, and targeted by human beings, then the machines are likely to offer generalized axioms of wisdom dealing with sciences, economics and technology.

now encompasses three more manageable submachines built to represent: (1) the truth-pursuing machine along the radial line at $\theta = 90°$ or at $\pi/2$; s, (2) the virtue-pursuing machine along the radial line at $\theta = 210°$ or at $7\pi/6$; and (3) the elegance-pursuing machine along the radial line at $\theta = 330°$ or at $11\pi/6$, as shown in Figure 2.11a and b.

These submachines may work synchronously or sequentially and accumulate the results for the final overview and interweaving by human beings at Level III. The derivation of each axiom of wisdom can be a time-, machine-, network-, and resource-intensive process.

2.6.2 Single-Wisdom Multiple-Machine Systems

The configuration of a single-wisdom multiple-machine (SWMM) system is shown in Sections 2.5.2 and 2.5.3. The multiple machines can be independent knowledge machines Figure 2.7, sensor-scanner machines, or both (Figure 2.9).

Each of the six machines can have their own three submachines to pursue TVB (for absolute wisdom), SET (for materialistic wisdom), and DAH (for opportunistic wisdom). The architecture of the SWMM system can become elaborate. However, the final system configuration can be assembled from Figures 2.7, 2.9, and 2.11a and b. The advantage of the SWMM machine will be multitasking such that the multiplicity of machines can tackle a subset with numerous directions (see Section 2.1) of wisdom at the same time, or multiple machines may be deployed to optimize the wise choices (see Section 2.3) from the earlier results of SWMM systems. One of the possible deployments of SWMM systems is shown in Figure 2.12: during the pursuit of absolute wisdom with TVB directionalities.

It is equally possible to deploy these systems during the search for materialistic wisdom with SET directionalities. In the worst case of abuse, the systems will serve just as well during the pursuit of opportunistic wisdom with DAH directionalities. Cases of such abuse are prevalent in nations founded on war and military technologies. Much like the human mind that controls instincts (Freudian ID; see Figure 2.2), human discretion controls the SWSM, SWMM, MWSM, and MWMM systems.

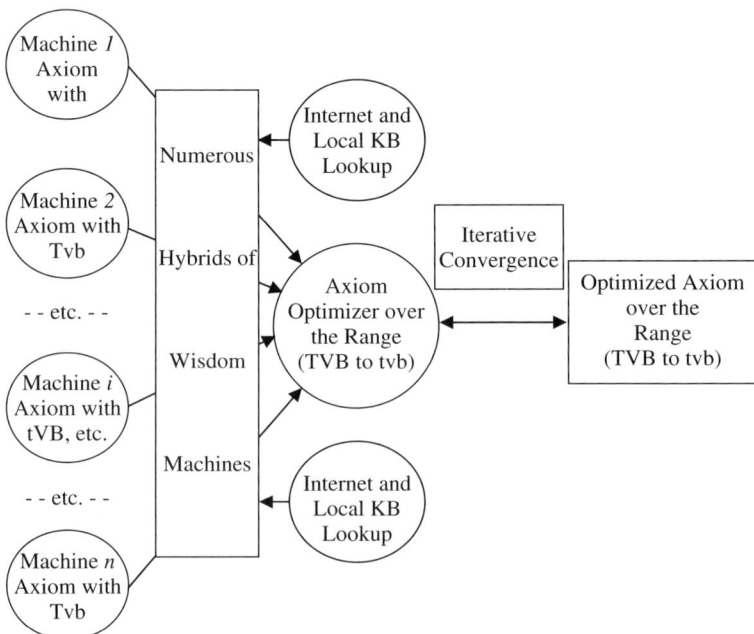

Figure 2.12 System-assisted derivation of axioms (portraits) of truth, perfection, arts, poetry, literature, justice, etc., using the single-wisdom multiple-machine configuration. The configuration shown here is for axioms of absolute wisdom with truth, virtue, and beauty as the three directionalities.

2.6.3 Multiple-Wisdom Single-Machine Systems

The multiple-wisdom single-machine (MWSM) system configuration forces a single machine to produce axioms of wisdom that would suit the dominant mindset of a philosopher, a corporate executive, and/or even a thug. In the real world, personality profiling for such human beings is done by psychologists, psychoanalysts, and guidance counselors. The practice of "profiling" became common during the interrogation of alleged terrorists in 2003–2008.

In the case of an MWSM system, the composite machine imitates the behavior of a human being who has a variety of attitudes toward a variety of thoughts and actions {X array}. In essence, if a matrix can be generated to indicate the mode of behavior {(TVB, SET, and/or DAH) Y array} and it is possible to blend them according to social setting and environment, then the machine would select the numerical blending of the X, Y terms in the matrix and predict the behavior of the human being based on a set of past predispositions.

In the deployment of the system shown in Figure 2.13, the blending proportions are chosen[9] by the user. Prior human expert opinions (a, b, c proportions) for Level III decisions, prior functions (d, e, f proportions) for Level II functions, and the IP addresses (g, h, i proportions) for Level I accesses are blended by the blending box and fed into a single generic wisdom machine. It would also be possible to force the MWSM system to behave like a schizophrenic by changing the blending proportions (a through i) in the middle of a run. The machine would also behave in chaotic mode if the proportions (a through i) are picked randomly during the run. In order to tame performance and force it to behave in a realistic and acceptable way, guidelines for the blending proportions become desirable.

It is noteworthy that the machine does not behave as a robot but as a machine whose mindset is numerically quantized by the blending proportions. These blending proportions can become more numerous by altering the granularity of the T to t separation in the T dimension, V to v separation in the V dimension, and B to b separation in the B dimension. Similar expansions in the S, E, and T and D, A, and H dimensions offer more flavors of machine personalities, much as innumerable flavors of human personalities become possible. The memory of a machine and the software resident in it alters the machine's behavior as profoundly as the contents of the human mind. The fundamental difference is that a schizophrenic or an insane machine can be corrected by rebooting the system, and it becomes highly feasible for a MWSM system to have multiple personalities without being insane.

In the past, the IBM 360 systems could also perform as scientific computers or as business machines by altering the chip sets in the control memories of the CPU(s). The use of Internet knowledge and expert system bases in Level I functions, to identify the latest socioeconomic parameters in society, permits the system to be dynamic and adaptive over many generations.

---

[9]This user selection corresponds to the selection of numerical quantifiers in the XY matrix. Since the personality profile is malleable (assignable) in the machine, the machine can behave as any human being might, ranging from being a saint to a ruthless dictator.

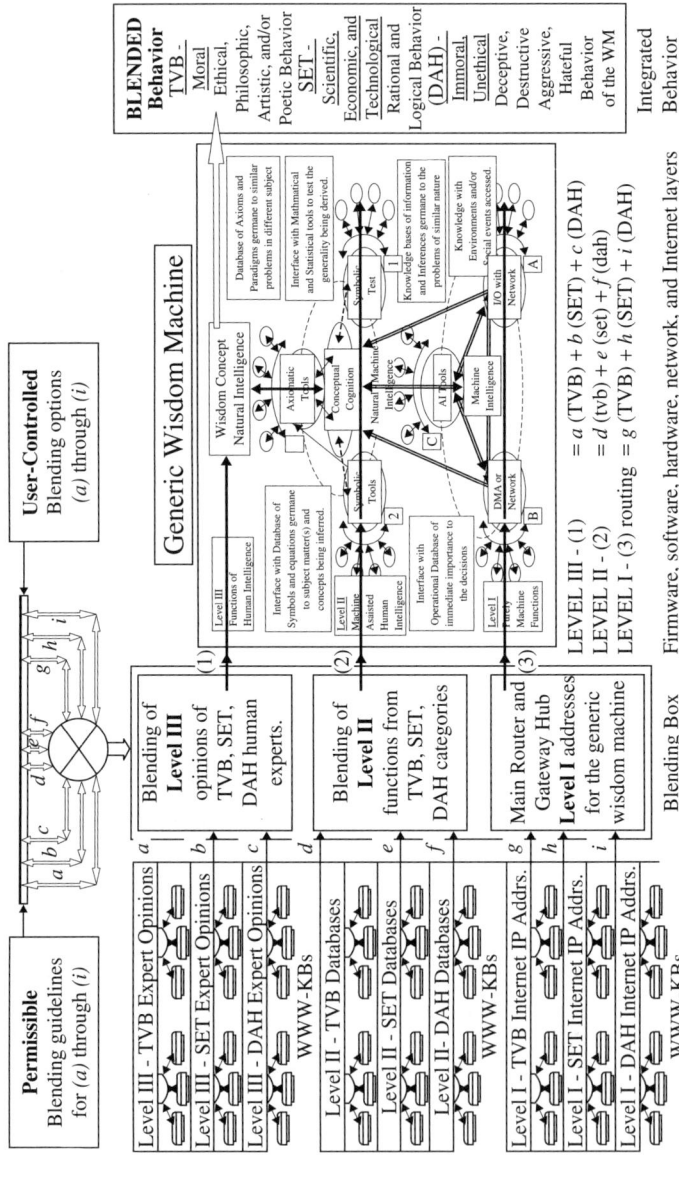

Figure 2.13 Architectural arrangement of a multiple-wisdom single-machine configuration capable of pursing multiple formats of wisdom.

2.6.4 Multiple-Wisdom Multiple-Machine Systems

Multiple-wisdom multiple-machine (MWMM) systems pursue numerous (absolute, materialistic, and/or opportunistic) domains of wisdom by using multiple machines. Much like the multiple-instruction multiple data (MIMD) architecture deployed in a traditional computer environment, the MWMM systems work as a group inter-dependently to solve one problem or a multiplicity of problems at the same time. The operating systems have control of the allocation of resources depending on the complexity of the problem(s). Such operating systems have indeed been functional since the 1970s for MIMD-based mainframe computers.

The configuration for a multiprocessor version of this type of a machine is depicted in Figure 2.14 for the solutions to a set of numerous social, corporate, medical, etc., problems in a knowledge environment. The inputs are defined in Box 1, and the machine driven by the operating system of the cluster machines will process each problem. A series of problem-dependent switches will be necessary to prime the control memories, the Internet, and local KB searches, and the specialized software necessary for each type of problem.

A typical system configuration is shown in Figure 2.15. The personal preferences of the user, organization, or any social entity dictate the blending options,

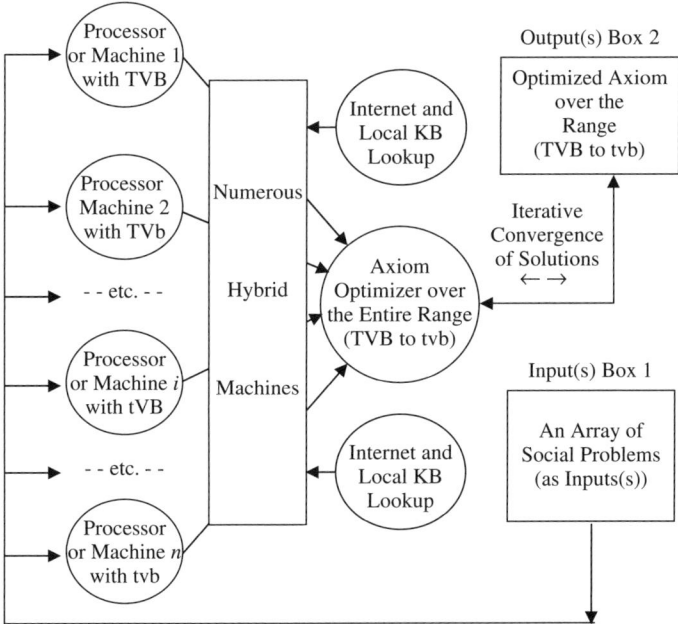

Figure 2.14 Multiprocessor environment of cluster machines concurrently seeking solutions to an array of social, cultural, medical, etc., problems. The TVB to tvb bases for each processor/machine need to be realigned to the nature of the problem (social, cultural, medical, etc.) that the machine is attempting to solve. This is arranged by a problem-dependent selector switch embedded in the architecture.

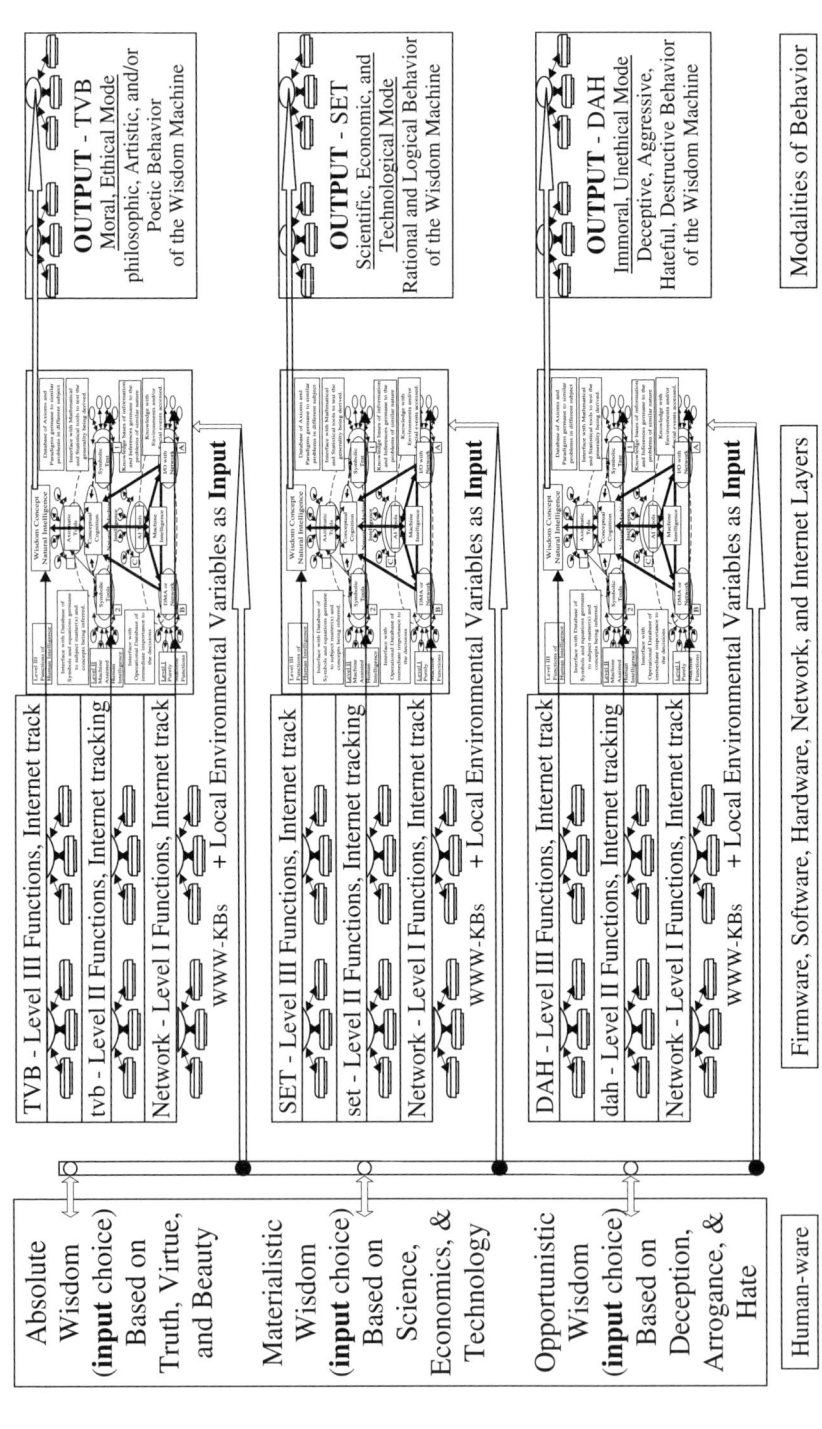

Figure 2.15 Architectural arrangement of a multiple-wisdom multiple-machine configuration capable of pursuing multiple formats of wisdom.

thus eliminating unnecessary trials by the machine by pursuing options not consistent with the desirable preferences. To optimize the performance of MWMM systems, access links and Web addresses can be prioritized for Level I functions.

To facilitate scientific and statistical methodologies, knowledge bases are used by company financial and R&D personnel for Level II functions. Finally, to keep the machine choices consistent with managerial preferences and the executive policies of the organization, the expert system knowledge bases accessed by MWMM systems for Level III functions access the expert system opinions of corporate executives.

It is possible the curtail the machine options so much as to make the machine "think" as a human with a selected personality type would in or out of any prechosen organization. It is thus possible for the verses of Shakespeare to be rewritten in the words of Fitzgerald [11] and vice versa, or Heisenberg's principle [12] to be restated according to the teachings of Saint Peter. Both would be equally compelling. The option for the machine to be creative in its own right is controlled by the humans who run the machine.

The blending of controls for access to internal knowledge bases and the capacity to be stringently scientific about Internet access, knowledge, and the expert opinions of many philosophers, scientist, laymen, and even thugs are under the discretion of the user.

The operating systems (OSs) and application program structure do not permit MWMM systems to wander away from their course or due process of executing program commands. However, if the user (or programmer) is tardy, slack, or careless in the use of system resources wisdom machines can behave like kittens that chase their own tails. Ill-defined problems and imprecise programming can (literally) drive wisdom machines to mindlessness. This is an anticlimax common to all computer systems. Generally, operating systems intervene to block the execution of "silly and senseless" programs. Wisdom machines have far more numerous degrees of freedom to become unstable than scientific traditional computers.

2.7 FROM WISDOM TO BEHAVIOR

The perspective on wisdom is wide and varied. In a real sense, their are as many perspectives as there are human beings. However, an underlying code of ethics is common to most who conform to social norms. Such codes and norms are indeed transplanted to the control and core memories of machines. Like the constitutional similarities of many great nations, there are many strings of common thought in the wisdom that is the foundation of behavior. Such seminal wisdom is embedded in the beginnings of nations and civilizations. A distillate of commonalities constituting the code of ethics of humans and nations makes up the operating system rules, operation code, and processes within generic wisdom machines. Once the science, economics, and technology (SET) gurus see through the purpose of such machines, socially oriented wisdom machines can be as common as scientifically oriented business machines. Personalized wisdom machines are indeed the future PDAs, with wireless access to wisdom bases around the world.

From the classical considerations of Aristotle [13] to the recent explanations of string theorists [14], at least *three* generalizations are evident. *First*, the sense of energy and persistence of an axiom of wisdom prevails well into the future. *Second*, a search for the fewest number of axioms of wisdom to comprehend the entire spectrum of behavior within the universe, ranging from elementary particles to cosmic bodies, has constantly invoked the human spirit. *Third*, the intercepted space where three of the highest human values (e.g., truth, virtue, and beauty) overlap is a possible realm of wisdom with an great amount of order,[10] contemplation, and analysis.

These three directionalities—that is, truth, elegance, and social values—as they exist in the domain of absolute wisdom need to be reoriented as science, economics, and technology for materialistic wisdom. Finally, deception, arrogance (plus aggression), and hate can only be whimsical and short-lived as they destroy the very organism that hosts them. Any wisdom becomes as highly dimensioned as the human mind that hosts it. Absolute wisdom based on TVB and materialistic on SET, and their derivatives, are ready to overflow the boundaries of the human mind. Conversely, opportunistic wisdom based on DAH and its derivatives destroys the mind that hosts it.

In a totally positive sense, anything less than the set $(T \to \infty, V \to \infty, B \to \infty)$ is less than absolute wisdom, and conversely, anything greater than the set $(D \to \infty, A \to \infty, H \to \infty)$ is better than absolute chaos. The intermediate space is for the mind to comprehend its own flavor of wisdom, stretching from TVB (∞^3) space to DAH (∞^3) space, and to behave accordingly. A wisdom machine navigates a small territory of these vast spaces with ingrained rationality much as a robot would navigate itself on a planet.

Behavior becomes a reflection of personal wisdom. The human mind by itself derives it own axioms of wisdom and functions or behaves accordingly. In an extreme case, learning to live is the wisdom to survive. Partly instinctual and partly acquired, the mind performs low-level wisdom machine functions by accessing genetic code and responding to social norms. At the highest levels, the mind outperforms a wisdom machine by creatively blending untapped neural paths and social freedom. Even though such extensive flexibility is inconceivable by the designers of wisdom domain software and operation code, the blending of user creativity and machine characteristics remains under the discretion of the user.

At an intermediate level of machine use, routine users and application-based wisdom machines are likely to outperform routine users and specialized computers that function as decision support systems (DSSs) or computer-aided design systems (CADs) in most disciplines. The key difference is the organization of the three levels of wisdom machine functions (see Sections 2.4.1–2.4.3) as depicted

---

[10]Even though we may label incomprehensible order as chaos.

in Figure 2.4. High-level DSS [9] can be forced to act as low-level wisdom machines.

2.8 ORDER, AWARENESS, AND SEARCH

The acceptance and recognition of order involve an act of faith. In the realm of science and mathematics, universities and research institutions have convinced us that an illusive framework of order forms the foundation of nature. Such an order constitutes the Laws of Nature. Electrons do exist, photons do travel at the velocity of light, and gravity and the laws of electromagnetism can be unified by the general theory of relativity, etc. Order became evident to Maxwell as he searched for the laws of electromagnetism in an orderly universe that was partially (and practically) explored by Faraday, Ampere, Gauss, and Galvani. However, it was up to Maxwell to write the general equations of electromagnetism. In the same vein, it was up to Einstein to unify gravity, as Newton comprehended it, with the laws of electromagnetic wave propagation that Maxwell wrote. The search for absolute wisdom is the tireless work of society's all notable achievers human. Conversely, the search for opportunistic wisdom [15] is the senseless work of ruthless human beings. In a limited sense, most corporate executives search for materialistic wisdom for their own corporations or their own personal wealth. The limits of the search for materialistic wisdom (however limited it may be) dominate the life cycle of most human beings. The path of a human being with a personalized wisdom machine is likely to be far more optimal than that of an astronaut with a guidance system or the path of a scientist with a computer.

In Figure 2.16, each of the three wisdom machines [*WM(i)*, *WM(j)*, and/or *WM(k)*] follows the lead of scientists, philosophers, and philanthropists searching for absolute, materialistic, or even opportunistic wisdom. We also propose that the

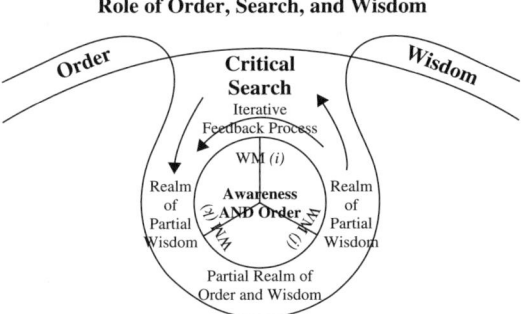

Figure 2.16 High-level functionality of each of the three wisdom machines: *WM(i)*, *WM(j)*, and *WM(k)*. Here, i, j, and k assume the TVB, SET, or DAH (or their derived combinatorial) directions. The Internet data are searched and researched within each of the *WM(i)*, *WM(j)*, and/or *WM(k)*, for underlying axioms of wisdom that reflect order in the origin, flow, or final use of Internet data.

artificial intelligence embedded in these machines presumes (and almost predicts) that one or more axioms of wisdom will be derived from Internet traffic.

Partial order is detected first, and then the awareness of this order triggers the search for greater and greater order on a cyclic mode until a generalized axiom of wisdom can be derived or inferred. To some extent, the thrill of discovery or invention triggers a more intensified search until the underlying truth, elegance, or social value is uncovered in a positive sense. The definition of oracles of wisdom in a dictionary accrues to making progress a reality. When the objective functions are defined for different applications, it is quite possible that these machine(s) may uncover abusive users of the Internet and also provide all the forensic evidence needed to stop criminals, terrorists, or those with destructive goals.

2.9 CONCLUSIONS

A conceptual framework for building a generic human–machine knowledge-processing environment to synthesize wisdom is presented in this chapter. This framework leads to understanding the interrelationships between objects and events based on the flow of traffic through the Internet or any specific network. Tentative inferences may be drawn based on the frequency of occurrence and the correlation between objects. The pursuit of total generality, like the pursuit of absolute truth, is likely to fail. However, in a limited sense, artificially intelligent machines and groups of human beings working in synergy may explore scientific truths, universal elegance, and/or social values. The AI processes in the machines reflect the natural intelligence in human beings and vice versa.

The Internet, broadband networks, and information highways provide ample amounts of raw data to seek and uncover relationships that are universally true or of lasting value to society. At the first level, very large amounts of data are first filtered by high-speed computers, and at the second level, the machines focus on information that carries knowledge and wisdom. With an appropriate set of filter parameters, information and knowledge are gathered in any given direction—for example, on economic activity, national security, or medical emergencies. The compiled information may then be finetuned to search for cause–effect relationships based on models.

For tracking the economic activity of a nation, models based on monetary theory [16], Keynesian economics [17], or scarcity theory [18] are invoked. For tracking the factor affecting world peace, models based on global unrest, wars, and United Nations activities are invoked. For tracking national security, AIDS, SARS, or bird flu epidemics, tentative models (if any) from global knowledge bases are invoked. The validity and accuracy of such models can also be tracked. The machine provides a buffer between the flow of Internet data and the human beings who make policy decisions.

The proposed human–machine system supplements the inference that the immediate environment can offer in raw digital format. Lower-level AI functions are programmed to sense and scan specific conditions within the environment. Knowledge generated from the machine's own knowledge banks of accumulated procedures and inferences about the topic under investigation further enhances the search for generality and wisdom in the flow of Internet traffic. Human intelligence firmly drives the programmed artificial intelligence of the machine in directions that validate or repudiate an axiom of knowledge and wisdom from a mathematical and scientific perspective.

The wisdom machine stems from ideas drawn from AI systems, corporate problem-solving strategies, decision support systems, and from the ability to assign a direction to wisdom in any subject based on the Dewey Decimal System and/or the Library of Congress classification system. In addition, any inference that becomes a part of the derived axiom/relationship/concept retains a level of confidence that the machine evaluates and substantiates. Thus, a tree of knowledge is implicitly structured before any notion of wisdom is drawn at any confidence level. Three corporate executive problem-solving techniques that have a direct bearing on the functioning of the wisdom machine are (1) the limited search and directional search, (2) the satisfying (i.e., tentatively acceptable) solution rather than an optimal solution, and (3) constraint relaxation techniques that find a solution and then tighten constraints in order to optimize the solution.

We have also introduced a variation of the computing and network environment so that it can function as an intelligent human agent [9, 19]. The arrangement is inclusive of traditional AI and scientific/computational environments. It also encompasses the realm of human decision making and offers enough latitude to the human counterpart of the machine to carry out practically all humanoid functions. The realm of true creativity still lies in the hands of the human driver who forces the machine to come as close to doing to the data and information what the human being is doing to the embedded concepts and knowledge. The machine, being adaptive and self-learning [20, 21], becomes as effective as the computers that play (and win) most chess games against novices. The Internet features embedded in the addressing capability of the machine make it at least three times more powerful. The capacity of the wisdom machine to solve problems from a knowledge-based (not subject-based) perspective, with culture and IP address-based access, provides sophisticated options to finding "a needle in a haystack" as easily as locating a solar system in the galaxy using knowledge bases around the world.

REFERENCES

1. S. Freud, *Basic Writings of Freud*, Modern Library, New York, 1938.
2. A. H. Maslow, A Theory of Human Motivation, *Psychology Review* 50, 1943: 370–396. See also A. H. Maslow, *Motivation and Personality*, Harper & Row, New York, 1970; A. Maslow, *Farther Reaches of Human Nature*, Viking Press, New York, 1971.

3. S. V. Ahamed, "An Enhanced Need Pyramid for the Information Age Human Being," Fifth International Conference on Business, Hawaii, May 26–29, 2005. See also "An Enhanced Need Pyramid of the Information Age Human Being," International Society of Political Psychology (ISSP) 2005 Scientific Meeting, Toronto, July 3–6, 2005.
4. G. H. Mead, *Mind, Self and Society*, University of Chicago Press, Chicago, 1934.
5. R. Cyert and J. G. March, *A Behavioral Theory of the Firm*, Prentice-Hall, Englewood Cliffs, NJ, 1963.
6. H. A. March and J. G. Simon, *Organizations*, John Wiley & Sons, Hoboken, NJ, 1958.
7. H. A. Simon, "A Behavioral Model of Rational Choice," *Quarterly Journal of Economics* 69, 1955: 99–118.
8. S. V. Ahamed, "The Architecture of a Wisdom Machine," *International Journal of Smart Engineering Systems Design* 5, (4), 2003: 537–549.
9. S. V. Ahamed, *Intelligent Internet Knowledge Networks: Processing of Wisdom and Concepts*, John Wiley & Sons, Hoboken, NJ, 2007.
10. R. A. White, "Intelligent Networks: Perspectives for the Future," Zurich Switching Seminar, March 1992. See also S. V. Ahamed and V. B. Lawrence, *Intelligent Broadband Multimedia Networks*, Kluwer Academic Publishers, Boston, 1997.
11. E. Fitzgerald, *Rubaiyat of Omar Khayyam*, Rupa & Co., Delhi, India, 2002.
12. L. de Broglie et al., *Heisenberg's Uncertainties and the Probabilistic Interpretation of Wave Mechanics: With Critical Notes of the Author (Fundamental Theories of Physics)*, Springer, New York, 1990.
13. J. Barnes (Ed.), *The Complete Works of Aristotle*, Vols. 1 and 2, Princeton University Press, Princeton, NJ, 1995.
14. N. Arkani-Hamed, S. Dimopoulos, and G. Dvali, "The Universe's Unseen Dimensions," *Scientific American*, 283(2): August2000: 62–69.
15. E. Fromm, *Anatomy of Human Destruction*, Holt, Rinehart and Winston, Austin, TX, 1973.
16. M. Michael, *Dilemmas in Economic Theory: The Persisting Foundational Problems of Microeconomics*, Oxford University Press, Oxford, UK, 1999. See also H. Robert, "The Political Economy of Monetary Policy," in *Political Economy of American Monetary Policy*, T. Mayer (Ed.), Cambridge University Press, Cambridge, UK, 1990.
17. R. M. O'Donnell, *Philosophy, Economics, and Politics: The Philosophical Foundation of Keynes's Thought and Their Influence on His Economics*, Palgrave Macmillan, New York, 1989.
18. R. Tilman (Ed.), *A Veblen Treasury. From Leisure Class to War, Peace and Capitalism*, M. E. Sharpe, Armonk, NY, 1993.
19. M. Mohammadian, *Intelligent Agents for Data Mining and Information Retrieval*, Idea Group Publishing Hershey, PA, 2004. See also M. Wooldridge, *Introduction to MultiAgent Systems*, John Wiley & Sons, Hoboken, NJ, 2002.
20. C. M. Bishop and C. Bishop, *Neural Networks for Pattern Recognition*, Oxford University Press, Oxford, UK, 1996.
21. Mathworks and Simulink, *New Real-Time Embedded Coder 2*, Release 13, MathWorks Consulting Services, T. Erkkinen, Novi, MI, 2003. http://www.mathworks.com.

CHAPTER 3

GENERAL THEORY OF KNOWLEDGE

CHAPTER SUMMARY

The general theory of knowledge introduced in this Chapter is based on the truism that any action associated with an object changes its status. The validity of this truism is reinforced by the forces and events in nature that constantly alter (almost) everything in the universe. If we have the capacity to measure the intensity and modality of the action, then there is a good chance we can *estimate* the change in the status of the object and vice versa. The change can be monumental or miniscule. From its macroscopic dimension in the entire universe or to its microscopic dimension in the processor of a machine, the effect of any verb function on any noun object changes the entropy of the object. From its microscopic nature within the limits of human comprehension, the effects of numerous "verb functions" acting on an appropriate set of "noun objects" cause measurable and discernible changes in their status. If the snapshot of the status of any object is stored as its entropy, then the nature, and magnitude of the change of status can be numerically related to the directionality, nature, and magnitude of the verb function. The relationships can indeed be derived from the nature and properties of the object itself. Numerous subsets of objects constitute a complex object, and different varieties of forces operate simultaneously on the complex object(s) in the real world.

Computational Framework for Knowledge. By Syed V. Ahamed
Copyright © 2009 John Wiley & Sons, Inc.

Hence, the effect of a group of verb functions on a collectivity of subobjects needs to be related. The directionality, nature, and magnitude of all the verb functions operating on a collectivity of complex objects thus change the individual and collective entropies of all the object(s).

In this chapter, it is proposed that the entropy of objects can be artificially changed by the processors of machines: (1) to rearrange the structure of any complex object to suit any given application and (2) to assess systematically the sequence and nature of the extraneous forces to rearrange the structure of the complex object. The two processes are tempered by physical reality to make sure the necessary events occur in the correct sequence to derive the "new objects" to solve (most of the) scientific and social problems.

3.1 A BASIS FOR THE THEORY OF KNOWLEDGE

The general theory of knowledge is based the truism that any action associated with an object changes its status.[1] It is also because actions cause change. Since no object exists that does not undergo change, there is some action or a set of actions inherently and inadvertently acting on every object. An object without one or more verbs to act on it is a nonentity, and a verb without corresponding nouns to be acted on is virtual. These truisms were as valid during the Big Bang as they will be during the collapse of the universe.

Objects in nature and their inevitable changes are the bases for the theory of knowledge. Changes occur over measurable and discernible parameters. The change of internal momentum within objects relates to the extraneous forces in nature that bring about the change. Indeed, this becomes the theme to construct the theory. The validation of the usual normal equations relating measurable parameters (such as mass and momentum, force and movement, torque and rotation, energy and inertia, etc.) becomes necessary. As real objects fill the physical space, hypotheses, concepts, and notions fill the intellectual space; humans, their needs, and their innovations fill the social space. One of the theory's objectives is to extend the domain of computation (statistical and probabilistic) coupled with the inferencing and directionality offered by AI techniques to expand the frontiers of knowledge and to derive notions leading to concepts and axioms leading to wisdom.

3.2 COMPREHENSION, NATURE, AND KNOWLEDGE

Nature preceded humans by eons. The intricate schema of nature are indicative of the immense knowledge lodged within creation. During the evolution of the earth in the universe, cosmic forces have shaped its composure. During the evolution

---

[1] Human action regarding an object updates the individual memory(ies) and a machine process updates the action taken about that object and its result (if any), unless it is a null operation.

of humans on the earth, intellectual forces have shaped human destiny. During the evolution of thought, social forces have shaped the achievements of the mind. Generic themes of universal order and ingrained methodologies lie hidden through the innumerable parallelisms that span the origin of the universe and intricacy of the mind.

Perception and comprehension are founded deeply in human thought processes. They become the cornerstones for a framework of knowledge, as we perceive it from recent times. To avoid being trapped in a ring of reflective philosophy of the universe or the physiology of neurons in the brain, we can step into the methodology of computation encompassing numbers, symbols, and the algebra of their manipulation in the processing units of machines.

In this section, three fundamental symbolisms, designating universal nature, human comprehension, and structure of knowledge, are suggested and evolved (see Figures 3.13 and 3.18). Their relationship, manipulation, and processing are explored next.

3.2.1 A Functional Approach

Based on the triad of understanding, nature and knowledge let the human comprehension convolved with the laws of nature give rise to the structure of knowledge. This hypothesis can be stated as a cause–effect relation as follows:

(Human comprehension) $*$ (Laws of nature) $\rightarrow$ Structure of Knowledge

The sign $*$ indicates a convolution designating mental or machine processes rather than a routine multiplication. Individual human comprehension has numerous aspects, such as intellectual capacity, concentration, grasp, prior knowledge, skills sets, endurance, etc. Even though comprehension is a noun, the act of comprehending involves specific actions (analyze, analogize, conceptualize, generalize, rationalize, synthesize, systematize, or valid verb.) from a scientific perspective. These actions (verb functions) are the conscious deeds of humans or programmed events in machines. The underlying schema speak loud and clear; verb functions convolved with embedded noun elements of the basic laws of nature yield a (comprehensible) structure of knowledge.

The entire set of the laws of nature is (almost) beyond the comprehension of any one individual. However, grains of comprehension can be piled (integrated) over time with imperfect and decaying memory effects. The comprehension of a set of humans (e.g., scientists, philosophers, spiritualists, etc.) can be cumulative, but the collective comprehension does not have total consistency. Individual piles of communal "grains of comprehension" constitute the knowledge bases and labeled by the DDS or LoC classification scheme. When a certain structure, robustness, and computational geometry are imposed in such grains of comprehension, then the communal piles start to appear as stacks[2] and the stacks start to appear as orderly high-rise complexes.

---

[2]These notions are expanded further in Chapter 10. In particular, see Figures 10.23 and 10.24.

In a similar fashion, the structure of knowledge can also have numerous aspects, but they can be altered in discrete steps by the verb functions embedded in comprehension convolving with a dynamic set of the laws of nature.

3.2.2 Incremental Changes

All the laws of nature and the entire structure of knowledge can become incomprehensible and infinitely detailed. The infinitesimal nature (granularity) of human comprehension can only account for incremental changes in the structure of knowledge arising from a subset Δ of the laws of nature. To capture the secondary hypothesis, the preceding statement can be restated as

(Human comprehension) $*$ (Δ Laws of nature) $\rightarrow$ Δ Structure of knowledge

In order to establish a methodology for probing into the incremental changes in the structure of knowledge, a framework of the following steps is introduced initially but later relaxed. These steps are derived from the traditional human approach to processing knowledge and the equivalent machine procedures that can be programmed to generate new knowledge from old knowledge.

Step 1 Human beings can voluntarily focus their mental activities to comprehend certain innate laws of nature. Two fundamental parameters that follow are: the region of focus and the intensity of mental activity within the focus. The product of the two is assigned as granularity of comprehension. Over a period, such "grains of comprehension" accumulate to make up large bodies of knowledge. A grain of comprehension is thus the product of granularity and its duration in the DDS or LoC classification of the Δ laws of nature. In a strict sense, the grain of comprehension is the time integral of the instantaneous granularity over the duration of its persistence. If the product of the region of focus and intensity of mental activity constitute an irregular surface, than the grain of comprehension becomes an even more irregular three-dimensional crumb.

The product of the two parameters (i.e., the area of focus and intensity of mental activity within that focus) varies greatly on an individual basis but it is finite and lasts for a shorter span of creative spells than those for knowledge machines that can process accurately over wider foci and for longer durations. Thus, knowledge machines can methodically generate well-structured knowledge structures to be more accurate, more comprehensive, and faster than human beings can. The configuration and computational geometry of knowledge are thus instilled by systemic arrangement and access to every "grain" of collective comprehension.

If individual comprehension is revisited and structured as a column with analyze, analogize, conceptualize, generalize, rationalize, synthesize, systematize, etc., as its elements, then an *alphanumeric column* can be generated (see Figure 3.1).

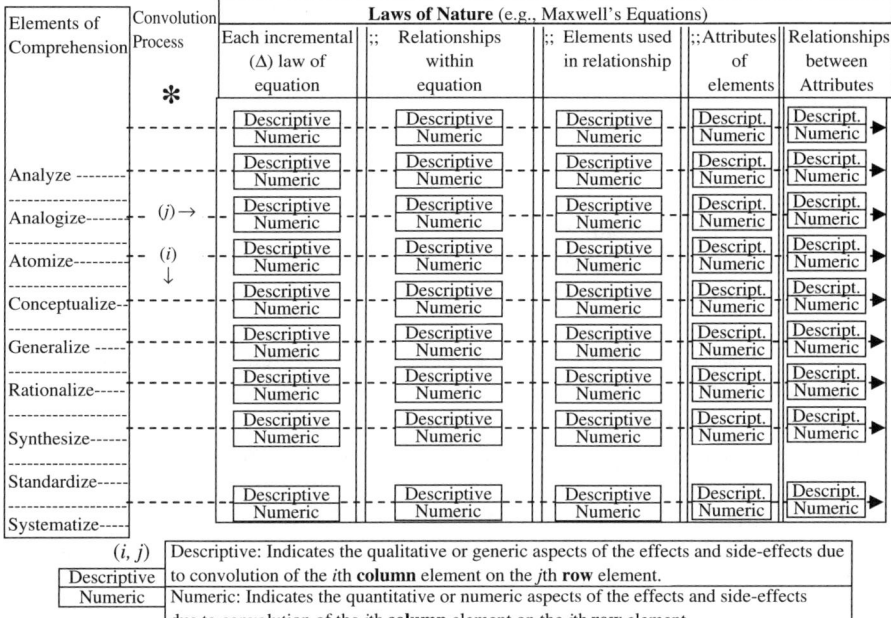

Figure 3.1 Effect of convolving "human comprehension" on the localized and interpreted "laws of nature." The elements entailing human comprehension are executed as artificial intelligence functions in knowledge machines, and the equations, relationships, and symbols in the "laws of nature" are initially derived from Internet knowledge bases but modified to suit the context of the application and to derive new knowledge from old knowledge.

Step 2 The "laws of nature" for any discipline are grouped as an "array of laws" that is, Δ laws of nature, to be considered independently and channeled into a given discipline with a series of quantization(s) for each law and with an associated numerical classification (DDS or LoC) it. The numerical quantity assigned to each law should be initially assigned a weight proportional to the "quantity of knowledge" in that numerical classification of knowledge (DDS or LoC).

In the initial stages, the array of laws should be used in moderation. Very wide classification leads to too many laws in too many disciplines and vice versa. Machines can compute and generate new increments of knowledge in a moderate-sized window. An alternative way of selecting a weight would be to base it on the number of (perceived) natural laws. For example, Maxwell's Laws have generated a large number of secondary knowledge and have a much wider global influence compared to Cavalieri's theorem[3] that has almost no impact on the sciences.

---

[3]Cavalieri's theorem for the plane states that "if two planar areas are included between a pair of parallel lines, and if the two segments cut off by the areas on any line parallel to the including

If the arrays of laws are revisited and structured as a row with all the laws in the particular direction of the DDS or LoC classification, then an *alphanumeric row* can be generated (see Figure 3.1). The four basic Maxwell's laws for electromagnetism, Krichoff's laws for electrical circuits, Newton's laws of dynamics, or the laws of conservation of energy, etc.) make up a subset of such laws.

Step 3 The convolution should be considered as an intellectual activity or a series of equivalent machine-generated processes. A study of human reaction leads to a set of standard procedures that can be programmed into machines as AI or expert system reactions.

Step 4 The structure of knowledge can be represented as the entropy of the information representing the "body of knowledge." A graph of knowledge is constructed around knowledge-centric objects has discernible relationships with other objects and bears subsidiary relationship to their attributes.

Step 5 Initially, the process of convolution can be treated similarly to a matrix operation between the numerous grains of comprehension arranged as a column working collectively on a given "single" array of laws of nature arranged as a row (see Figure 3.1).

A number of uncertainties become evident. Initial uncertainty generally leads to future knowledge. The process of convolution involves fragmenting the individual comprehension so finely that every monocular aspect of the law of nature is examined by conscious human (or group) activity or verified by a series of knowledge machine processes. Three of the uncertainties are listed below.

1. The so-called convolution between alphanumeric columns and rows is neither entirely algebraic nor totally numeric. The resulting elements of such a convolution can only be alphanumeric and are indicative of how an element of comprehension can interact with a law of nature (or an element in that law of nature) and lead to (possible) change or enhancement in the law. Generally, the resulting alphanumeric elements indicate numerous pathways that may be explored even though there are no words for such pathways.

For example, if an individual can apply "mental capacity" to Maxwell's laws for the two vectors E (for electric intensity) and H (for magnetic field), the classic theorem of Poynting ($P = E \times H$) results. As other examples, if Einstein's mental processes are convolved with Newton's laws, then a relativistic correction will result; if laws of contour integration are applied to Maxwell's differential equations, then Gauss's and Ampere's laws can be deduced; if Edison's endurance in research is applied to the laws of thermal emission in photons, then a lightbulb results, etc. Using the same reasoning, if we replace the numbers and

---

lines are equal in length, then the two planar areas are equal." (See *CRC* **Handbook of Tables for Mathematics**, 4th ed., Chemical Rubber Co., Cleveland, OH, 1970, p. 13.

logical entities in a CPU by knowledge-centric entities (see Section 7.1), then the CPU can function as an object processor unit (OPU) or as a knowledge processor unit (KPU). More enhancement may be required (see Sections 7.2–7.10) to obtain flexibility, accuracy and performance.

2. The process of generating new knowledge (enhancement of existing laws or a new interpretation of them) is now seen as a quasi-scientific process of dealing with merging human capacity focused on a semi infinite set natural laws. The features of human capacity can be simulated in machine environments. Human creativity is singularity of thought. It is oblivious. Creativity lacks a precise definition, and thus it remains a gift to be creative. However, creativity can be approached ever so closely in a complex arena of neural processes without being able to program it.

3. Perhaps the greatest uncertainty is to be able to recombine, reconstitute and reinterpret the resulting alphanumeric entities after convolution. Typically, any resulting alphanumeric entity (such as the effect[4] of "grasp" of photonic currents) is a variation of the law by exerting every element of the comprehension column on any particular element (or subelement) of every symbol in the single array of laws. Like numbers that can be sequenced up and down, the symbols can be relaxed to include symbols of a similar character. For example, if the mental capacity of a researcher is used to investigate the properties of fluorine, then the similar properties of chlorine, bromine, and iodine may create new knowledge about other compounds. As another example if the antibacterial properties of a compound are being studied, then its effect on similar strains of bacteria can generate new knowledge, or the properties of a slightly different compound on a different group of bacteria may be examined to generate new knowledge from old knowledge. In a more complex example, if a "column of comprehension" is used to alter the array of antibiotic reactions on an array of bacteria (by altering the composition of the antibiotic and also the strain of bacteria found in nature), then a three-dimensional array of resulting alphanumeric entities results. Each of the resulting entities would have a possible venue to new knowledge. For human beings to process and investigate every possible resulting entity is impractical even though a machine can perform an exhaustive search. Coupled with human control and intuition, knowledge machines (see Chapter 7) can offer greater leaps of constructive knowledge than humans without machines or machines without human intuition can.

In a sense, a wisdom machine (see Chapter 2) is derived to generate new and useful axioms by humans who guide numerous machines to well-authenticated axioms. In a different sense, knowledge from old and depleted knowledge may be derived by creating a knowledge machine that relaxes the constraints of human

---

[4]The human "generalization" or "machine search-and-verify" of photonic electric currents can indeed point to a new type of light-emitting diode (LED) used in fiber-optic transmission systems. As another example, the human capacity to analogize a transistor can lead to optical amplification in Erbium-doped trans-oceanic fibers.

comprehension and modifies the subset of "laws of nature" in the direction designated by the particular DDS or LoC classification.

3.2.3 Elemental Convolution and Knowledge Operations

A knowledge operation in a knowledge machine corresponds to a microscopic operation convolving an element of the alphanumeric column constituting human comprehension with an element of the alphanumeric row of every element of the laws of nature in that particular DDS or LoC classification. When such operations are sequenced in compiled and executable knowledge programs, then the knowledge machine will simulate humanistic and social domain functions and offer optimized solutions to routine application programs dealing with individual, group, and cultural issues.

The symbols in the rows and columns are alphanumeric entities that are partly descriptive and partly numeric. During the process of convolution, both the descriptor of the symbol and its numeric value undergo changes. The descriptor facilitates the change in the perceived (elements of the) law under the scrutiny of human comprehension. Elements of human comprehension are programmed as AI routines or as a series of knowledge programs in machines. The numeric value quantifies the extent of change incurred or that will be incurred. A cumulative index of a series of such changes (descriptive and numeric) is an estimate of the change of entropy of the old knowledge if relaxation of the proposed elements embedded in any of the laws is feasible and practical. The change between the entropies of the new knowledge and the old knowledge is the measure of the knowledge gained or knowledge lost.

Figure 3.1 depicts the convolution process as a row–column operation. Some of the descriptive/numeric entries could be 0/0, indicating that the effects and side-effects of an proposed step in the ith column entry on the jth row entry are zero both qualitatively and numerically.

3.3 CENTRAL PROCESSING AND KNOWLEDGE PROCESSING

If knowledge is considered as the status (entropy) of any object, then we can use assembly-level knowledge instructions (with knowledge operation code) to alter the status (entropy) of objects (stored as knowledge operands). The basis of the theory of knowledge is that the change of status (entropy) of objects can be modeled by a series of knowledge-level operations on the appropriate set of objects in the hardware of a knowledge processor unit (KPU). In fact, the basis of all computation is that a series of assembled operation codes are executed on a series of numerical and logical operands in the hardware of a central processor unit (CPU).

In the real world, numerous subsets of objects constitute a complex object and a variety of forces operates simultaneously on the complex object(s). Hence, the effect of a group of verb functions on a collectivity of subobjects needs to be related. The directionality, nature, and magnitude of all the verb functions operating on a collectivity of complex objects to change the individual and collective status (entropies) of all the object(s).

In the real world, objects, unlike numbers are not always passive. Some objects respond intelligently. The mode of response can indeed be reactive, neutral, or intelligent. Furthermore, the reaction can be linear or nonlinear depending on the magnitude of verb function(s), being assistive or resistive to the direction of the verb function(s). The individual response can also be related to the responses of other objects in the collection, thus enhancing nonlinearity or causing a wave or tidal effect which may be resistive, neutral or assistive. Even though the situation may become complex for human perception, it is manageable, predictive, and computable for a machine. Knowledge processing, by necessity, starts to become more intricate than number and word processing. Both hardware architectures and software design differ from those for traditional computer systems.

In practice, verb functions are set of processes triggered by nature, caused by human beings, and executed in processors. The nature of verb function and collection of processes can range from being the simplest (e.g., move or copy) to very complex (generate an earthquake or create a Big Bang). Verb functions can be fixed or adaptive, linear or nonlinear, with fixed or variable directionality(ies), independent or dynamic, etc. The web of interrelationships can have a topology in its own right. Programmability of verb functions causing the change on the noun objects that experience the effect becomes complex, yet feasible. The complexity of the chain of cause–effect relationships is handled by the detailed structure of the software and specialized design of the hardware—its components, bus structures and their inter connecting switches. The interaction between any set of cataclysmic verb functions and any super structured noun objects may become hopelessly complex or even chaotic; it is likely to be manageable, predictive and computable by a complex network of machines.

To simplify the derivation of the general theory of knowledge, any valid and executable process or set of processes on any (complex collectivity of) noun object is shown in Figure 3.2. In order to be accurate about the change in status of objects, the machine tracks the noun object–verb function interaction in two ways. *First*, it tracks the noun object(s) that constitute the collectivity, and *second*, it tracks the array of verb function(s) that operate on the collectivity. In tracking the noun objects, the machine maintains a dynamic record of:

1. Objects in the collectivity (e.g., input to a knowledge program)
2. Attributes of each object
3. Relationships between the objects
4. Relationships between the attributes
5. Nature of the objects (hostile, neutral, or cooperative)
6. Linearity or nonlinearity of the response to each of the verb functions (their magnitude, nature, and directionality)

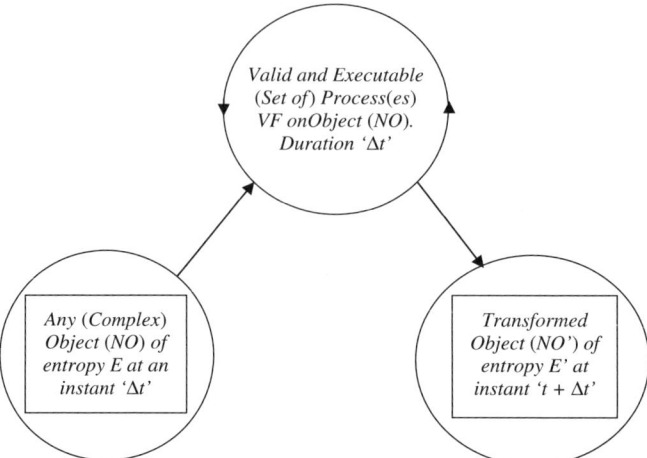

Figure 3.2 Change in entropy of a noun object after a valid set of process(es) verb functions resulting in a transformed (new) object NO'.

In tracking the verb functions, the machine maintains a dynamic record of:

(i) Functions or processes in the group (e.g., the machine operations in the knowledge program)
(ii) Nature of each verb function (fixed, adaptive, or programmable)
(iii) Magnitude or intensity of each verb function
(iv) Directionality (positive, neutral, or negative) of each of the verb functions

At the next level of representation, if the verb function(s) can be symbolically represented as an alphanumeric column and the noun objects are represented as an alphanumeric row, then the result would be a convolution matrix caused by a productlike effect. The rules for deriving the new noun object (i.e., the convolution matrix) have yet to be evolved. The interaction between the verb functions (VFs) and noun objects (NOs) is represented in Figure 3.3.

The entire initial status (entropy) of a group of noun objects (NOs) is shown as a top row of Figure 3.3. The entropy of the entire collectivity is represented. All the characteristics of each noun object are tagged, and all the inter-relationships are identified by forward and backward pointers. In addition, the intensity, modality, and nature of the set of processes (VFs) that operate on this group of objects are represented as a column of numbers, alphanumerals, and vectors.

In evaluating the outcome of VFs on the objects, a matrixlike numbers, alphanumbers, and the vector displacement of entropies are computed, and a weighted sum of the column gives rise to the new entropies of the transformed object (NO'). The rules for the convolution of terms in the matrix depend on the nature of objects and functions Such rules are specific to each environment. In a simple case, when an electromotive force (VF) of V volts acts on a resistor

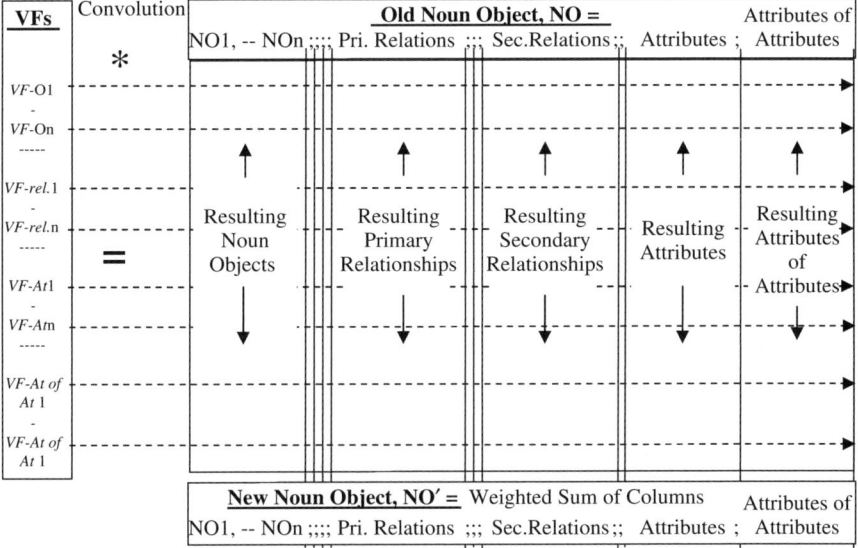

Figure 3.3 Change in status (entropy) of a noun object after a valid set of process(es) verb functions resulting in a transformed (new) object NO′.

(NO) of R ohms for t seconds, the energy level within the resistor is enhanced by $(V^2/R) \cdot t$ watt-seconds. In a more complex case, when a group of photons enter the eye from an object, then the entropy of the object in the eye of the beholder is altered by the neural-electric impulses generated by the retina. These pulses convey the change in entropy of the object that is stored in the neural pathways of the human brain. The entropy of the object is thus altered in the brain. Numerous other examples also exist.

In every case, a physical and a scientific process occurs between VFs and NOs that can be computed with a good level of confidence. When the forces are social, cultural, or even healing and medicinal, then an estimate of intensity, modality, and nature becomes necessary and the confidence level of the outcome is feasible. Such methodologies have been deployed in AI-based programs (Intern and Mycin) for the finding a possible medicine of common ailments.

3.4 ACCUMULATION OF INFORMATION, KNOWLEDGE, AND WISDOM

Most environments track the chronology of information, knowledge, and wisdom. Reflection, contemplation, and mediation give rise to values and ethics. In the evolution of wisdom, values, and ethics from bits and pieces of observations of events in society, human thought and the ensuing actions play a vital role. The cause–effect relationships in the chain of events and observations give rise to the systematic progression of a logical chain as depicted in Figure 2.5a. This

112 GENERAL THEORY OF KNOWLEDGE

diagram of collected observations attempts to make them consistent and coherent as the B, D, I, K, C, W, and E nodes (Figure 2.5b) of a long process that repeats itself during the continued evolution of human societies.

A quantum jump from one node to the next appears as unfeasible as the conversion of binary bits to the representation of floating point numbers during Boolean times. Yet, algorithms have been conceived, implemented, successfully deployed, and perfected in most computer systems since von Neumann's times. We mix and merge innate human nature and events, and then link them in a chain with systematic and algorithmic procedures from computer sciences to construct the general theory of knowledge in Sections 3.5 and 3.6.

3.5 THE ENHANCED KNOWLEDGE TRAIL

The "knowledge trail" shown in Figure 2.5a (also see Figure 3.4a) is the basis for the development of the general theory of knowledge. The representation of the trial can be enhanced dramatically to incorporate the impact of science and technology over the centuries. In order to progress systematically and nevertheless depict movement over the decades, six zones are (initially) added between the seven nodes: B, D, I, K, C, W, and E. These zones are represented in Figure 3.4b. The rules for navigation through these zones differ dramatically. They can depend on the inputs and expected output of each node. For example, the laws for manipulating the bits at the B node in the B zone are necessarily different from the laws for manipulating graphic objects at I node in the I zone.

In Figure 3.4b, the seven nodes, six zones, their spatial spreads, and the laws for manipulation of input objects are shown. The input objects (i.e., the NOs) at each of the nodes are processed by the operation code (i.e., VFs specific to that node) to generate the expected output from the node. Depending on the nature of NOs and VFs, the processes can be short and explosive, or may be extremely long and last for many decades into the future in human societies.

The *order* and *structure* of each of the inputs and outputs start to increase from the B node to the W node. When the processing of input objects enters the human domain (see the right side of Figure 3.4a), the laws for processing become individual, and the human personality is reflected in the outputs of the nodes K, C, W, and E. In a sense, the complexity of human nature starts to influence the individual knowledge trail, thus making each human being unique.

The expansion of the knowledge society over the last few decades has brought more enhancements to the right-hand nodes, that is, from I to W, along the knowledge trail. Perhaps three of the hardest links in the trail lie between I and W nodes. Fortunately, scientific methodologies now exist so one can attempt to cross the K, C, and W zones depicted in Figure 3.4b.

Over the last few years, computer hardware and software have evolved sufficiently to realize computer game programming, object manipulation, and sophisticated graphics applications. The I zone is sufficiently mapped and navigated for most the object-oriented functions. However, the applications are

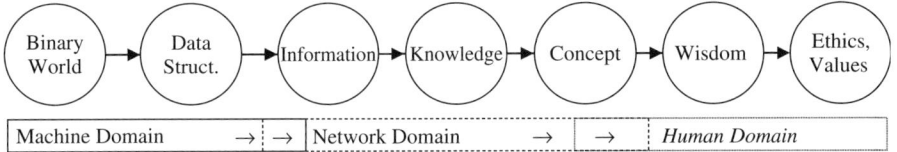

(a) Configuration of the knowledge trail (taken from Figure 2.5b)

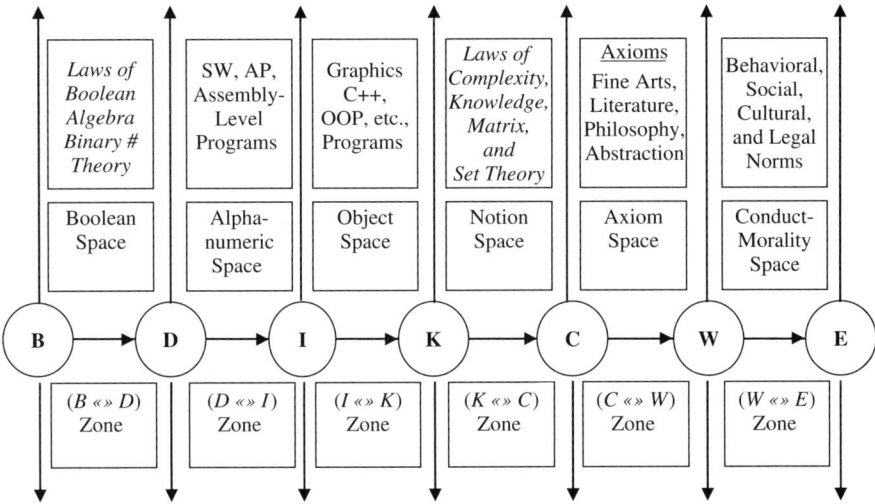

(b) Enhanced configuration of the knowledge trail with zones and applicable laws for processing inputs at each of the seven nodes

Figure 3.4 Partitioning of the knowledge trail into zones (a) and disciplines (b) to map to various tools to process the input at the nodes.

limited to computer games, image processing, and robotic systems. Languages to isolate and identify embedded objects and the underlying relations between objects, attributes, and attributes of attributes (EO, R, Ats., Ats. of Ats.; see Chapter 7) are not generally available, even though they can be developed. The knowledge-oriented languages and their compilers do not exist.

Scientific inventions over the many centuries are represented symbolically as ellipses in Figure 3.5. Each invention in a particular zone contributes to the navigation through that zone. The nodal distances are depicted by a series of six steps in Figure 3.5 Each stride of accomplishment is an ellipse in its own right. Typically, one of the ellipses in zone B is the inception of Boolean algebra at step 1. One of the ellipses is the development of C++ at step 3 in zone I, etc. Figure 3.5 depicts the knowledge trail and scientific inventions in each of the six zones.

The invention of scientific methodologies, by itself, does not expand human understanding. Instead, the application of new methodologies for the solution of problems expands human understanding. It propels the movement from one node

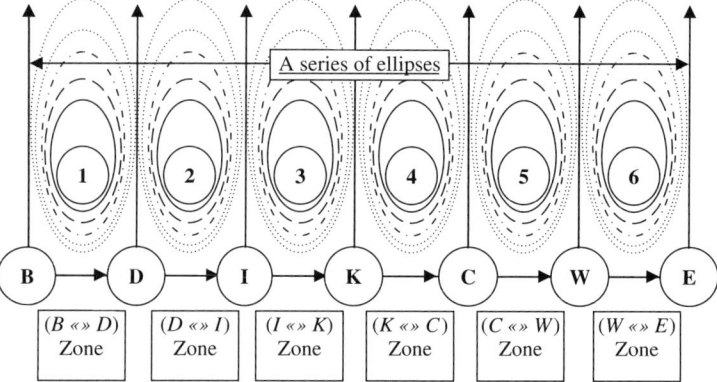

Figure 3.5 Depiction of inventions and breakthroughs in the six zones along the knowledge trail. Each of the ellipses signifies an invention or a methodology to process the input objects at all the seven nodes.

to the next by an incremental step. Humans learn the art of graceful living ever so slightly more and epitomize well along the ideology of pursuing the knowledge trail, by being more knowledgeable, conceptual, and wise. As a specific example, Boolean algebra did not extend human understanding directly, but the use of algebra (in computer systems) to solve generic social and important scientific problems, thus furthering human understanding in many directions.

As an additional example, the invention of magnetic resonance imaging (MRI) techniques did not solve any human problems. Instead, the detection of cancerous tissue in the body by deploying MRI techniques provided humans with an understanding of cancer in the patient. Coping with the ailment and possibly curing it (with localized radiation therapy) followed leter. The gaps between our knowledge of cancer in the patient's body, that is, the K node, and a concept of tolerance/cure, that is, the C node, were thus incrementally bridged.

3.6 SEQUENCING OF EVENTS AT NODES

The $(I«»K)$-, $(K«»C)$-, and it $(C«»W)$- zones of Figure 3.5 appear devoid of significant methodology to transit out of these nodes. In an early analysis, the forward movements appear infeasible, if not impossible. However, investigations in any particular discipline under consideration provide some clues to move forward along the knowledge trail.

The major inventions in each of the six zones of Figure 3.5 are represented by ellipses. These ellipses do not directly help migration along the knowledge trail, instead these ellipses become a part of the mechanisms to bridge the six inter-nodal zones. The migration along the trail is facilitated beause the zone is incrementally bridged. The derivation of such incremental steps by very incisive inventions is now feasible in the knowledge society.

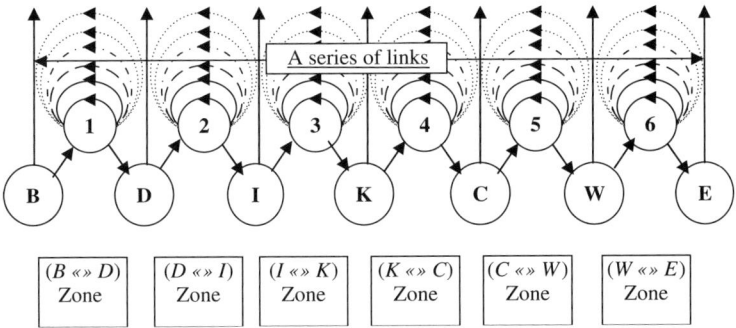

Figure 3.6 Replacement of inventions and breakthroughs by transitions along the knowledge trail. Transition occurs when the scientific tools and methodologies are sufficiently mature to move from one node to the next. Transition at steps 1–6 in mature societies leads to wisdom of a culture or a nation.

We use this analogy to modify Figure 3.5 to Figure 3.6. Whereas Figure 3.5 conveys how human struggle facilitates the movement along the knowledge trail, Figure 3.6 sets out the condition for the transition from one node to the next along the knowledge trail. The five ellipses in each zone are now replaced as a series of five links (through the inter-zonal nodes 1–6). Whereas the vertical ellipses in Figure 3.5 signify human achievements, the five links in Figure 3.6 signify the transitional movement of the individual nodes B, D, I, K, C, and W migrating to D, I, K, C, W, and E, respectively. The transition can occur only if the scientific methodology of all the human achievements can satisfy the most stringent requirement of the highest link.

Each contribution that results from the prior achievements adds to the final probability of the forward transition. Hence, the integrated effect of *all* the contributions in each zone causes the movement from one node (B through W) to the next (D through E). The successful transition in the state diagram in the six zones establishes the forward link at each of the six nodes (B through W). The migration along the knowledge trail constitutes the basis for the theory of knowledge.

Magnified regions around the nodes I, K, and C in Figure 3.6 are shown in 3.7. The steps at 3, 4, and 5 for transition result from the confidence in the three minor nodes H (for hypothesis), N (for notion), and A (for axiom). These steps are discussed further in Sections 3.7 and 3.8.

Stagnation can occur at any of the seven nodes in the knowledge trail (Figure 3.5) simply due to the lack of human effort or intelligent agents embedded in the AI programs driving the machines. The region of interest in the knowledge trail shifts to the right side because the migration between the I, K, and C nodes to the K, C, and W nodes appears monumental. There is a lack of a scientific methodology for the migration through hypotheses, notions, and axioms, as shown in Figure 3.7. However, transition diagrams generated by computing the effects of all prior scientific innovations in any of the zones

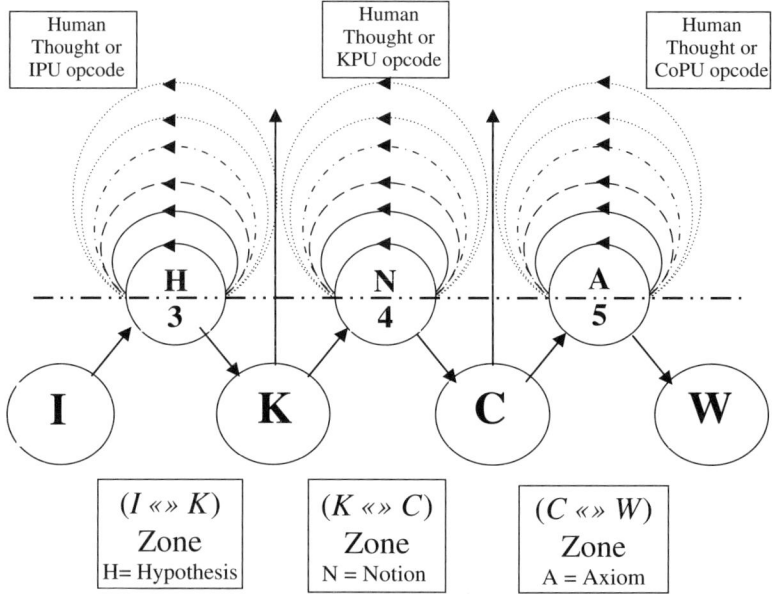

Figure 3.7 Replacement of the human and machine processes from the information node (I node) to the wisdom node (W node). H, N, and A are minor nodes, and migration occurs based on the confidence level with generation of H, N, and A.

will offer a good probability that the following node is reached successfully. We approach the problem of transition based on decision trees. When a correct hypothesis, notion and axiom, reaches a given confidence level, then the passage from I to K, K to C, and C to W nodes can be affirmed. A considerable numbers of iterative adaptive learning cycles may become necessary to progress through these nodes in any given discipline.

3.7 TRANSITIONS AT I, K, AND C NODES

The three minor nodes (H, N, and A in Figure 3.7) are well within the realm of human thought and act as human memory caches to hold on the transitional paths between $(I \ll \gg K)$, $(K \ll \gg C)$, and $(C \ll \gg W)$. In essence, the three minor nodes H, N, and A can be represented as

$$\mathbf{H} = (I \ll \gg K), \quad \mathbf{N} = (K \ll \gg C), \quad \text{and} \quad \mathbf{A} = (C \ll \gg W) \quad \text{(see Figure 3.8.)}$$

The symbol « » signifies a state of transition such that the I, K, or C operand in the information processor unit (IPU), knowledge processor unit (KPU), or conceptual *(CoPU)*, due to the one or more informational opcode(s), knowledge opcode(s), concept-oriented opcode(s). These new processor units are presented in Chapter 8.

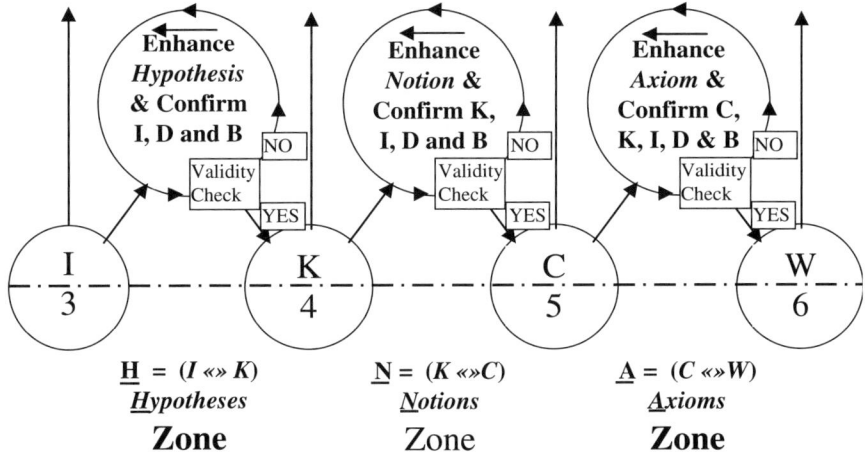

Figure 3.8 Magnified diagram of the human and machine processes from the information node (I node) to the wisdom node (W node). The migration occurs based on the transition at stages 3–5. See the text for the conditions for transitions to occur.

3.8 TRANSITION MANAGEMENT AT NODES

The boundaries between the nodes I, K, C and W are fuzzy and ill defined. However, the information and wisdom are as much apart as binary data and information. The strategy for differentiating bit from data structure can be projected to differentiate information and knowledge. The basis for difference is the grouping and ordering of bits to constitute data structure.

Structure and order of the contents at the nodes I, K, C and W differ significantly. The processing that takes place at the intermediate zones 3, 4 and 5 brings about the incremental changes in the structure and order of contents. Generally, the structure and order increase significantly from the B node to the E node during migration along the knowledge trail for any specific body of knowledge (BOK). We follow this strategy to enhance the $(I \ll \gg K)$, after the I, $(K \ll \gg C)$, after the K, and $(C \ll \gg W)$, after the C nodes as follows.

1. Prior to transition at any of the three minor nodes H, N, and A (Figure 3.9), the activity around these nodes remains trapped. For instance, the activity at the I node remains trapped as information oriented functions (such as report generation, inferencing, and relaying or broadcasting of current information, etc.). The activity becomes repetitive until an AI agent in the machine or human creativity attempts to reason (or "search") the embedded information. Such an attempt leads to the possibility of a generality or tentative hypothesis (H) in Figures 3.7 and 3.8. The affirmation of the tentative hypothesis, H, leads to knowledge in the next node K.

118 GENERAL THEORY OF KNOWLEDGE

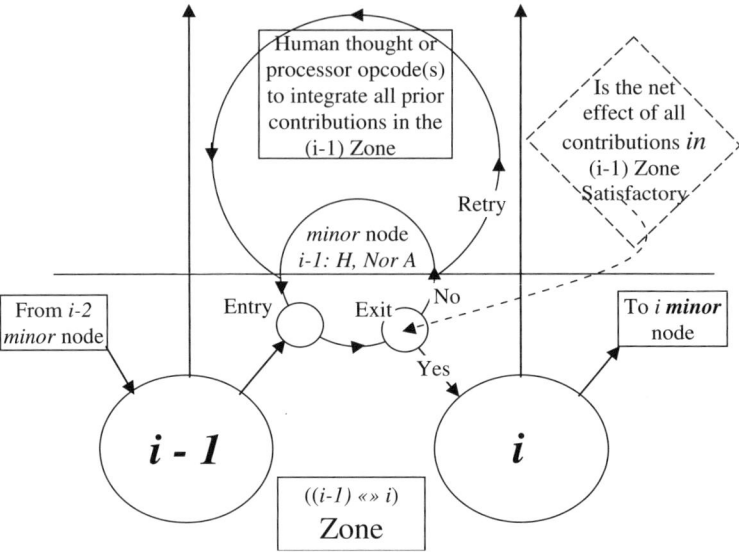

Figure 3.9 Transition diagram for the migration from one node to the next along the knowledge trail. At the exit of the transition diagram, the integrated effect of prior scientific contributions in the $(i-1)$ «» i zone is computed to check if the transition can be made. When the probability exists that the prior contributions can substantiate the migration, then the i node is reached. The i node can be any of the D, I, K, C, or W nodes.

2. Exit from the minor nodes at H, N, and A results when it is confirmed that *hypothesis is indeed knowledge* at the minor node H, *notion is indeed concept* at the minor node N, or a*xiom is indeed wisdom* at the minor node A. The process of confirmation is never a certainty, and there is reason that new scientific contributions can occur in any one of the seven zones along the knowledge trail. The confidence level to exit from the minor nodes is finite since the boundaries of data, information, knowledge, concept, and wisdom are expanding. Under these circumstances, the procedures for migration become more stringent, but the trail of knowledge remains firmly established.

3. When the strategy for migration is push backward to the B and D nodes, the rules for converting binary streams to data structures are firmly established. For example, the codes for bytes, words, hexadecimal, ASC representations, etc., are standardized throughout the world. Most well-used audio, graphics, and video codes are recognized throughout the industry.

4. The strategy for converting data structures to a standardized representation of information such as PDF documents, formatted papers, news paper formats, textbooks, balance sheets, etc., is generally adapted by most publishers of information.

The major reason that the knowledge trail does not stagnate at the B and D nodes is the standardization for converting binary bits to data structures and then for converting data structures to information. It is logical to deduce that the standardization for converting information to knowledge, knowledge to concepts, and concepts to wisdom is lacking, and for this reason techniques and algorithm need to be invented and standardized to traverse the entire knowledge trail.

A magnified diagram of the human and machine processes from the information node (I node) to the wisdom node (W node) is depicted in Figure 3.8. The migration occurs based on the transition at stages 3–5. For the transition to occur, the validity of H, N, and A nodes needs to be established within the confidence limits at each node. These minor nodes serve as placeholders for transitional information as it migrates from one node to the next.

Figure 3.9 shows the sequence of events that cause transitions and the eventual movement from one node to the next. The condition for transit is that the rules in the prior nodes need to be affirmed on all the noun objects and all the verb functions invoked at the current node. Only then, migration occurs from one node to the next along the knowledge trail. At the exit of the transition diagram, the integrated effect of prior scientific contributions in the $((i-1) \ll \gg i)$ i zone is computed to check if the transition can be validated and is scientifically correct. When the probability exists that the prior contributions can substantiate the migration, then the i node is reached from the $(i-1)$ node. The i node can be any of the D, I, K, C, or W nodes.

In order to be precise, the diagram in Figure 3.10a should have as many forward trajectories as there are ellipses in Figure 3.5, and each ellipse represents an innovation in the $((i-1) \ll \gg i)$ zone. This elaborate diagram is shown as

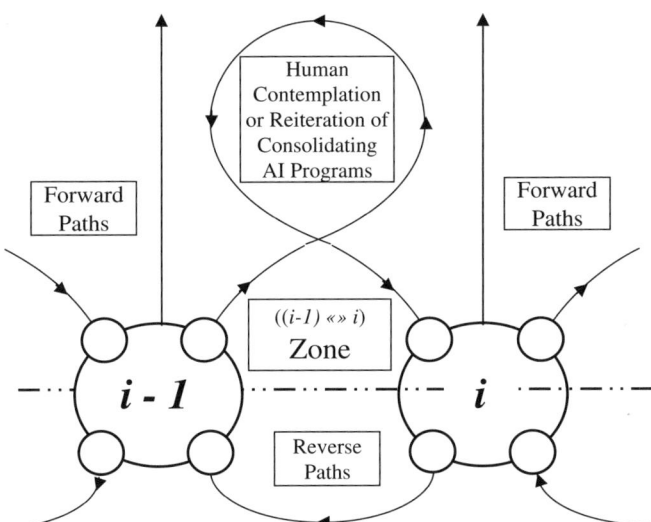

Figure 3.10a Simplified representation of Figure 3.9 to affirm the transition from node $(i-1)$ to node i, $(i = 4, 5,$ or $6)$. The forward paths propagate the movement along the knowledge trail and the reverse paths verify that the prior path is indeed correct.

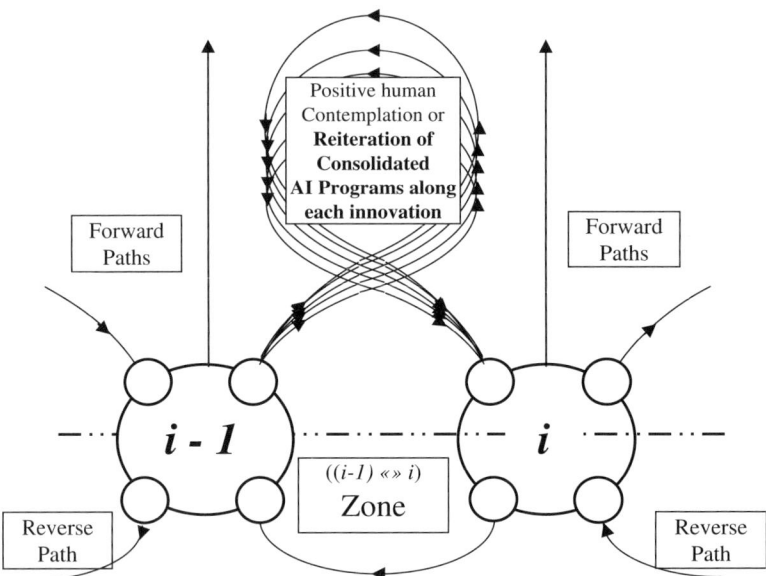

Figure 3.10b Representation of all the major innovations contributing to the forward paths to propagate from node $(i-1)$ to node i.

Figure 3.10b. When a machine is being programmed, then these numerous trajectories indicate that all the innovations in the zones are duly considered.

3.9 AN INVERSE UNIVERSE

In contrast to a knowledge trail that migrates towards the betterment of human society, it is possible to conceive a parallel and converse universe; a universe where everything right is wrong and everything wrong is right. Such a universe will prefer the D (deception), A (arrogance), and H (hate) goals and ideologies over the T (truth), V (virtue), and B (beauty) goals and ideologies. One serious effect is that each transition can be chaotic, and the continuity of any movement is not guaranteed. Whereas only one absolute truth (T) exists, there are infinite deceptions (D); whereas one pristine virtue (V) exists, there are infinite inflictions of vice and arrogance (A); and whereas only one perfect beauty (B) exists, there are infinite embodiments of ugliness and manifestations of hate.

In reality, the human activity in the negative DAH domain (see Figure 3.11) can become as dominant as the activity in the TVB domain. Largely, due to human laziness, the path to reach the devious questionable node i from the node $(i-1)$ is more easily traversed through the DAH domain than the TVB domain. In consequence, any software to check the validity of the path traversed needs greater affirmation such that the three dimensions of TVB are not compromised. The opposite scenario is equally valid. A negatively primed computer system will favor the DAH path and attempt to block the TVB path. The main guiding

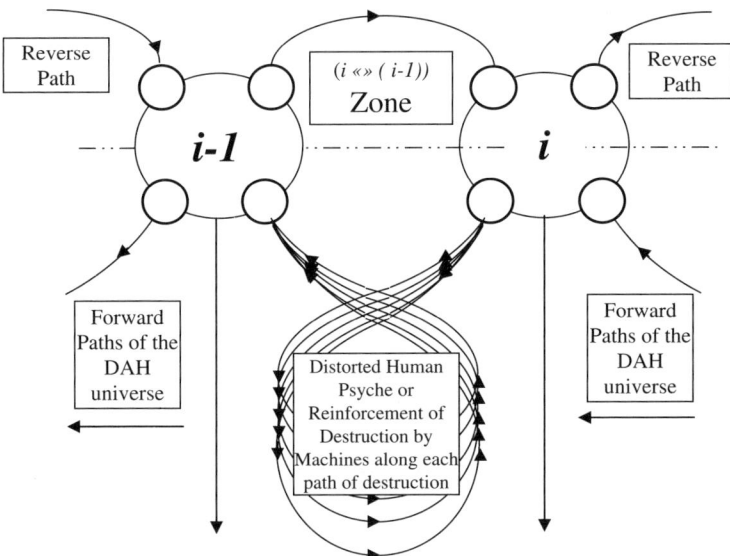

Figure 3.11 Representation of a parallel and converse universe where the positive values and ethics embedded as TVB are replaced as deception, arrogance, and hate (DAH). The trail moves in a reverse path to disintegrate the contents of every node until there is nothing except the binary bits at the B node to be disintegrated. The process may even get pushed into node N (see Figure 3.13b), where the humans attempt to destroy nature that gives rise to the B node.

force is the human discretion that governs the progress toward TVB or DAH lies in the context and intent of the pursuit of knowledge.

For the many reflections of TVB onto the DAH domains, the movements in the parallel universe of DAH can only be jagged and unpredictable. For the sake of completeness, we depict the transitions in Figure 3.11. Such transitions along the negative trail of knowledge can only degrade ethics and values (if any are left) to the negative goals of DAH. The positive progress of a lifetime in any discipline can be rendered questionable by the ill-will of deceptive, arrogant and hateful humans and their machines or weapons of mass deception.

Such incidents have been documented in the past: the holographic scripture of the Incas being burned down by the Spanish in South America, the temples of India being raided by the British, or the dignity of native Africans being robbed when they were sold as slaves by the Portuguese. This type of human behavior or robotic response of negatively primed humans and machines generally occurs in the opposite direction, from the E to B node, tearing apart any shade of rationality, logic, or even the laws of mathematics (by deception). It has yet to be seen if a scientific or any methodology can be created for this negative trail of knowledge and the downward spiral of values and ethics in society.

3.10 ORIGIN AND DESTINATION

In Figure 3.12, the positive forward movement occurs from the B to W nodes. A simplified transition diagram is depicted. Even though only single forward paths and backward paths are shown, all the nodes prior to any particular node can propel the forward move. Any reverse movement is also impacted by all the prior nodes. The effect of the forward and reverse movements can occur directly on the preceding node or through each of the preceding nodes. The precise effect will depend on the nature of the prior noun objects and the nature of the verb function causing the change of status in the current noun objects. The response of the KPU can depend on VF- or *kopc-*, and NO- or *kopr*. The nonlinearity of the response can also become tainted by the memory effects retained by the noun objects, a subject discussed further in Section 5.6.

The reverse paths in Figure 3.12 offer venues to verify prior steps and the numerical levels of confidence at each prior node. In each of the six zones, human activity is that of contemplation and validation (at the prior node) affirming or rejecting that migration to the next node is scientifically correct and justified. The same effect is obtained by a series of machine processes giving rise to the confidence level of AI processes (at the prior node) that migration to the next node is scientifically correct and justified. A reiteration of the consolidation of all previous scientific findings may become necessary. Traditionally, these functions are attempted by human beings. AI-based systems such as Intern, Mycin [1], and NeoMycin [2] can be used to assist the transition process.

In the tradition of human thought dating back to Aristotle, the pinnacle of human wisdom lies in the betterment of society. Symbolized as the pursuit of TVB, all human struggles searching for new knowledge, concepts, and wisdom are unified as the benevolence of humankind. If the modern machine can encompass the mysteries of nature and the welfare of humankind, then the purpose of all machine processes and human effort can be depicted as shown in Figure 3.13a. Here, the laws of nature blend into scientific laws, the social order, the axioms of wisdom, and finally the code of ethical conduct.

3.10.1 Nature, Origin of Knowledge Trail

Before the B node, it appears logical to have an N node (N for nature) that generally provides a yes/no answer to most primitive questions. For example, nature and reality assure that either objects exist or they do not, in a binary mode. Placing this primary and seminal node N in front of the B node makes TVB, (the eighth node) shown in Figure 3.13b. In considering the migration of human understanding, nature has played a dominant role. The observation of nature gives rise to a "bit" of information, that is, the presence or absence of something (any noun object) or of some force (some related verb function) that alters the noun object. In most cases of evolved data and information processing, it may not be necessary to include explicitly the nature or origin node (i.e., node N) before the B node.

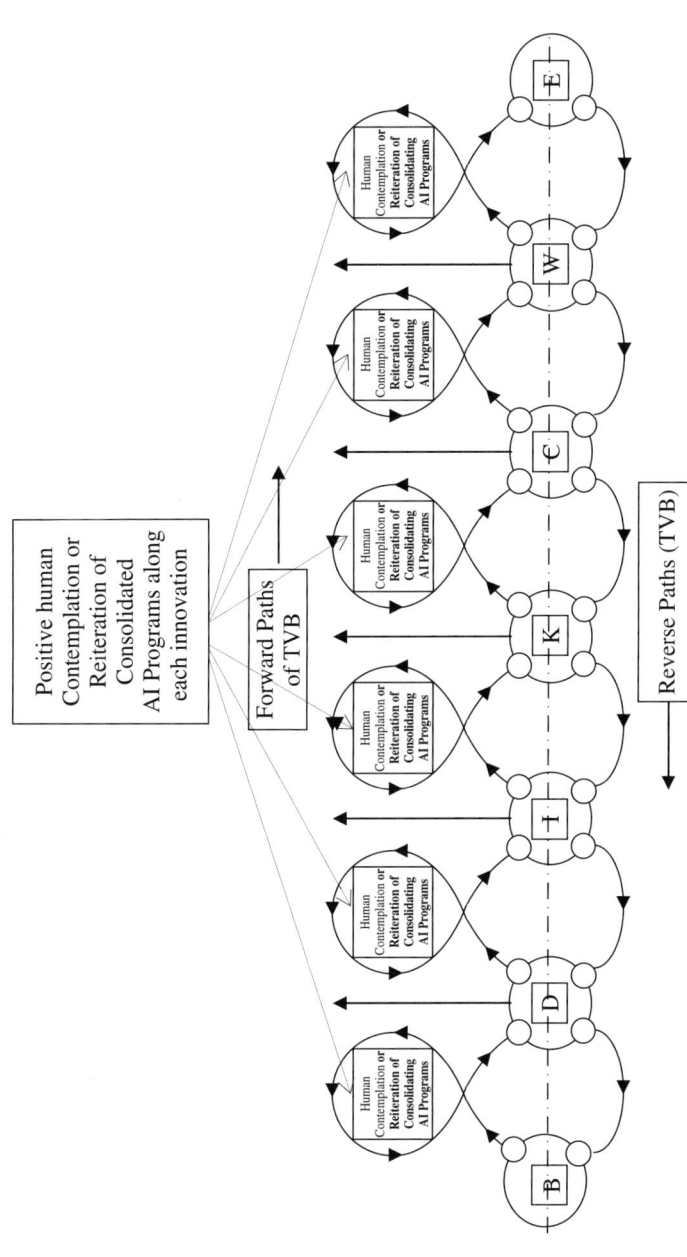

Figure 3.12 Integrated movements from the B node to the W node along the forward paths. The reverse paths offer venues to verify prior steps and the numerical levels of confidence at each prior node.

124 GENERAL THEORY OF KNOWLEDGE

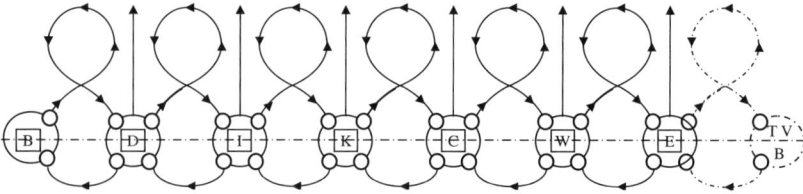

Figure 3.13a Inclusion of the TVB (truth, virtue, and beauty) node beyond the E (ethics) node aligns the knowledge trail pointing toward a socially desirable end. Forward path of the trail is from left to right.

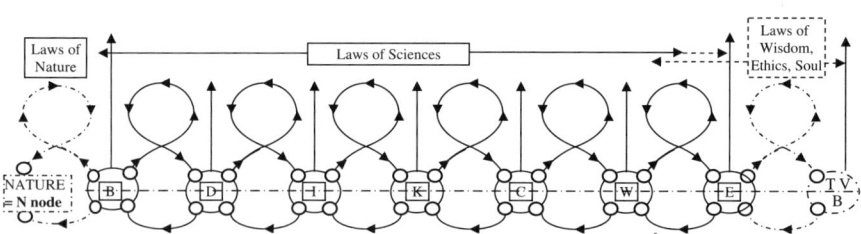

Figure 3.13b Inclusion of the nature node before the B (binary) node aligns the knowledge trail as it originates from the primal origin of humankind, that is, nature, to end in TVB (as shown here). It could also end in DAH.

3.10.2 Two Destinations of Knowledge Trail

Even though it is easy to see that nature is the logical start for all knowledge, there are infinite destinations for the knowledge trail in either of the two sets of goals clustered in TVB and DAH. The search for any destination can become an ongoing and open-ended pursuit since an ideal destination is never reached. Time and place continuously redefine the perfection of any goal. However, when all energy is expended on the search rather than its goals, then two illusions, nirvana and negativism, start to surface. Reality becomes a dynamic equilibrium state between the two opposing ideologies of TVB and DAH.

Whereas the ideals of TVB are likely to provide the energies for forward migration, the goals in DAH actively fuel the reverse paths, thus depleting the knowledge, concepts, and wisdom of prior generations. History repeats itself, and culture is set back rather than moving forward. In the computational domain this representation (Figures 3.13a and 3.13b) offers a means to verify and isolate the source of error in the forces along the forward and reverse paths.

It is desirable to treat movement along the forward TVB force (from B to E) differently from movement along the reverse DAH force (from E to B). The force to move in the forward path is derived from the ideals of TVB, thus accelerating the movement toward W and E, and to some extent attenuating the movement toward E to B. On the other hand, the reverse force is derived from the ideals of

DAH, thus retarding any movement toward W and E and eventually accelerating the movement from E to B. The two forces are distinct in most human beings and the driving software is separate in machines. The lack of a definitive barrier between these two forces causes confused and unpredictable behavior in humans and oscillatory or divergent results from machines.

In the real world, the eighth node is illusive. It can be the TVB goals in some instances or it can be DAH in others. To consider this practical situation, the effect of each force individually and all the forces need to be collectively examined.

3.10.2.1 From Nature to Nirvana: N to N^+, or N(Plus)

The endless quest to satisfy the goals embedded in TVB ideals leads to a lasting illusion of truth, virtue, and beauty—all condensed at one instant of time and in one iota of space. Impossible as it may be, it offers a hideout for those depleted of any energy to migrate and those who have reached the end of their search, however illusionary it may appear. A state of nirvana based on deep understanding, firm knowledge, integrated concepts, and undaunted wisdom begins to consolidate, transitioning from abstraction to a mental refuge for individuals. In some cultures this refuge is the shrine of society.

When the path to this TVB destination is the knowledge trail, then some stability is assured, because of the sound scientific principles and accepted social benevolence, and concrete symbols of beauty that have resulted in the processes involved in reaching the illusive state of nirvana. To reach this goal, the nodes and paths become self reinforcing. The perfection of this state is self consuming, and any attempts by a machine to define a path deplete all its resources, with its results still not be conclusive. In human beings and societies the perfections at this state of illusionary nirvana are exhaustive and never meant to be entirely stable or totally "perfect."

3.10.2.2 From Neutralism to Negativity: N to N^-, or N(Minus)

The antithesis of movement from nature to nirvana is the migration of neutralism to negativism. Individual, social, and cultural neutrality is the result of acceptance without contemplation in humans and use of software without validation in machines. The state of neutrality is a precursor to negativism because of imperfections in accepted notions or an error condition in the software. In a sense, this path is based on the reversal of what is already established as a notion or software module.

When recklessness, arrogance, and aggression are involved, there is further setback. When rationality and contemplation are involved, a search for the validity of prior steps occurs. In a backward movement, when it escalates from the W node to B node, the prior achievements at the B, D, I, and K nodes and attainments at the C, W, and E nodes are reversed.

Destruction comes much easier than construction, and social fabric is ripped apart quickly over the period of decay of any society. The DAH goals reinforce

126   GENERAL THEORY OF KNOWLEDGE

the deception, arrogance/aggression, and hate that are necessary to move society down the spiral of negativism.

3.10.2.3 The Choice: N to N⁺ (TVB), or N to N⁻ (DAH)

The two variations of the eighth node depicted as TVB or DAH are shown in Figure 3.14. This TVB (truth, virtue and beauty) node that beyond the E (ethics) node aligns the knowledge trail pointing toward a socially desirable end. It affirms that the knowledge trail exists for the greater benefit of community, society and humanity. Largely, the presence of this node facilitates a forward move on the right-hand side of the figure. In reality, the presence of this node also provides a reverse path to verify and validate the decisions and transitions at all prior nodes, thus strengthening the knowledge trail along the TVB direction. In some instances, the ideals embedded in the TVB are displaced by the corruption and greed of corporate, social and political leaders. In such instances, as a society or culture starts to deteriorate, the knowledge trail slants toward DAH (deception, arrogance, and hate), and the forward movement gives rise to doubt and suspicion along the reverse paths.

3.10.3 Multiple Feedbacks along the Knowledge Trail

The effects of multiple feedback paths from each of the forward nodes are depicted in Figure 3.15. In fact it can be seen that the ith node receives reverse inputs from $(7-i)$ nodes and sends out $(i-1)$ outputs to the prior nodes to maintain the reverse trails' balance through the seven nodes. The influence of the eighth node is treated as a separate entity because this node can effect each of the seven nodes directly, and that effect does not have to be propagated though each of the nodes.

The entire forward influence of the TVB (the eighth node) in the knowledge trail is shown in the top half of Figure 3.15. The feedback to migration from the seven prior nodes (B, D, K, C, W, E, and I) has a positive effect due to the synergy between TVB goals and the contemplation of humans and the operation codes of machines. Both seek an optimal solution toward truth in the results derived from their functions (the lasting contributions of humans and accuracy of machines), both seek beauty (elegance as reflected in the thought of humans and algorithms embedded in machines), and finally, both seek to make the results as valuable as possible to the user community. The synergy releases enhanced positive energy for forward movement. The reverse moves exist only to validate prior forward moves.

A pursuit of perfection near the right three nodes starts to appear, and increasing amounts of resources are expended in every forward incremental step around these three nodes (W, E, and TVB). From an energy perspective, any discrete change of velocity of progress around these nodes needs greater and greater effort. A situation of this nature occurs as elementary particles approach the velocity of light.

The entire reverse influence of DAH (the eighth node) in the knowledge trail is shown in the lower half of Figure 3.16. The feedback to the migration from the seven prior nodes (B, D, K, C, W, E, and I) suffers due to the conflict between

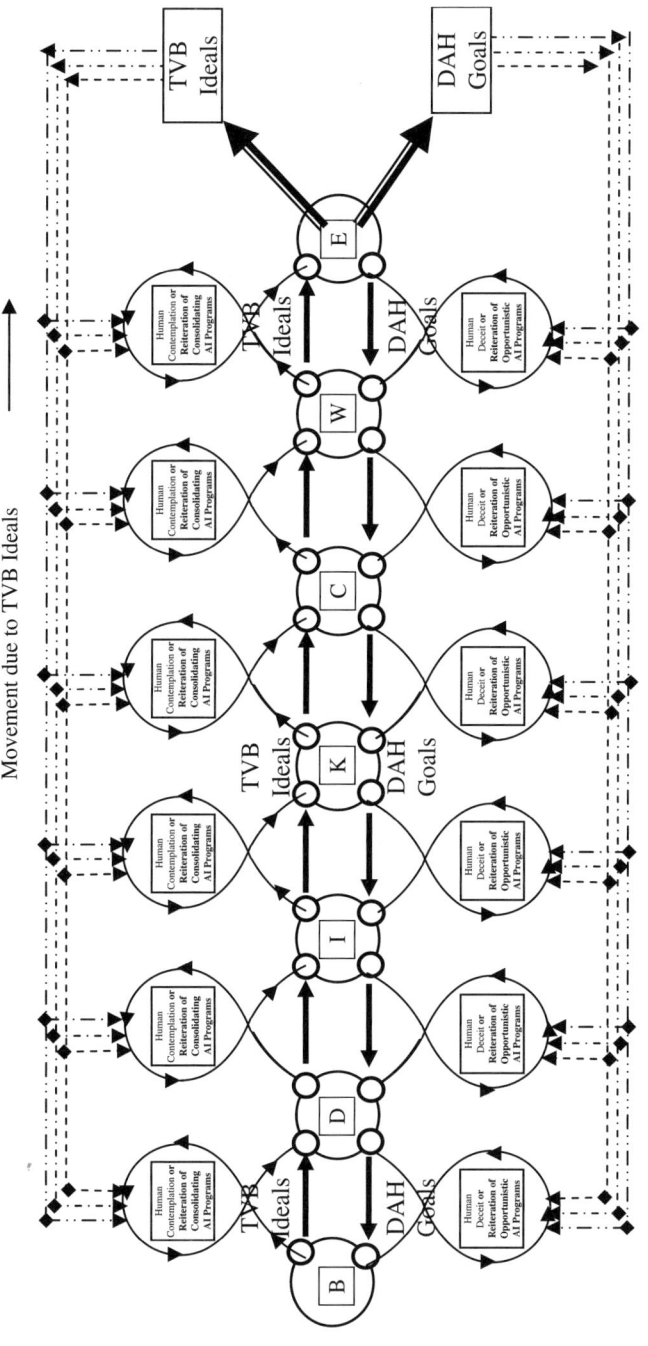

Figure 3.14 Human contemplation or reiteration of consolidating artificial intelligence programs in the pursuit of TVB ideals propels the forward movement along the knowledge trails (and to some extent stems reverse movement). Deceit or reiteration of opportunistic artificial intelligence programs in the pursuit of DAH goals propels the reverse movement along the knowledge trails (and to some extent stems the forward movement). Thus, the disintegration of the K, C, W, and E nodes starts.

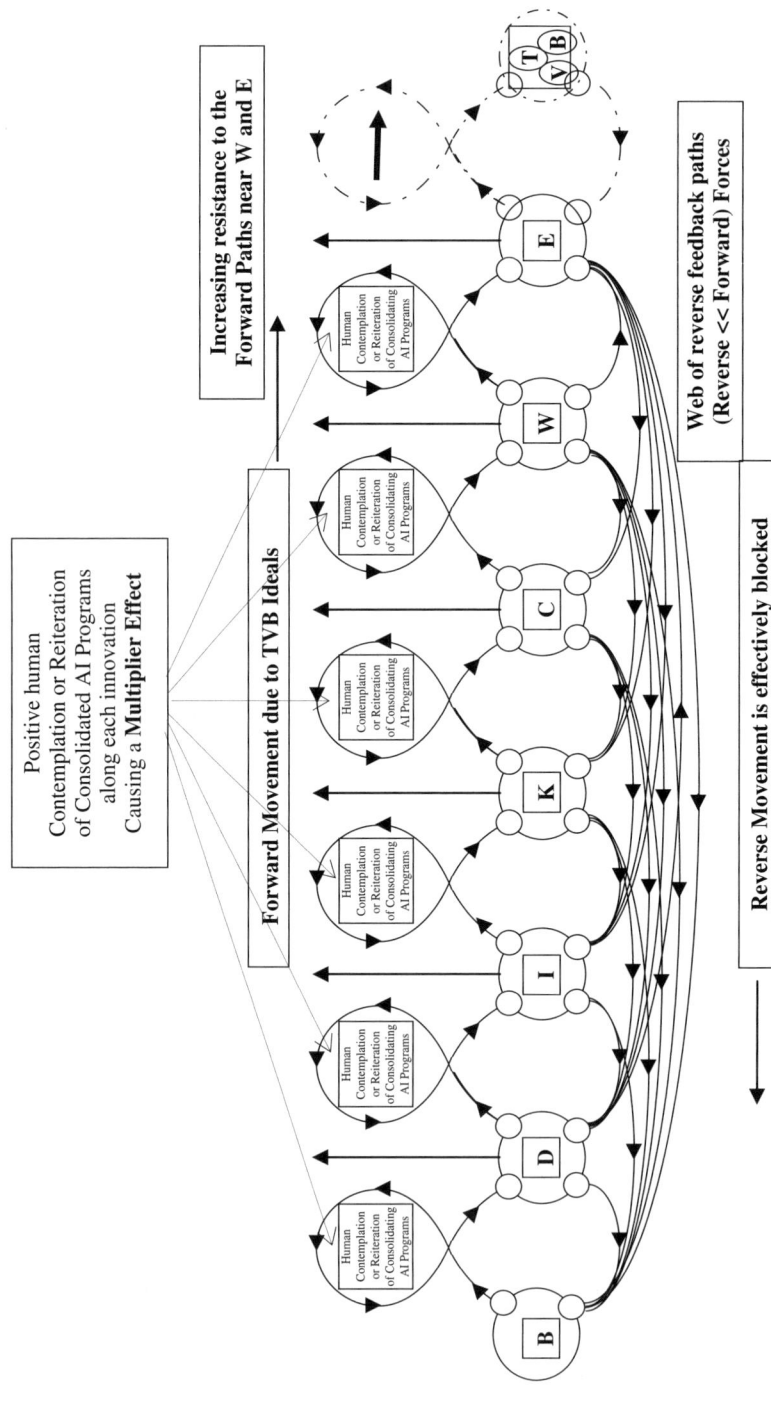

Figure 3.15 Multiple feedback paths from each forward node offer a means to verify and isolate the source of error in the forward path toward W (wisdom) and E (ethics) nodes.

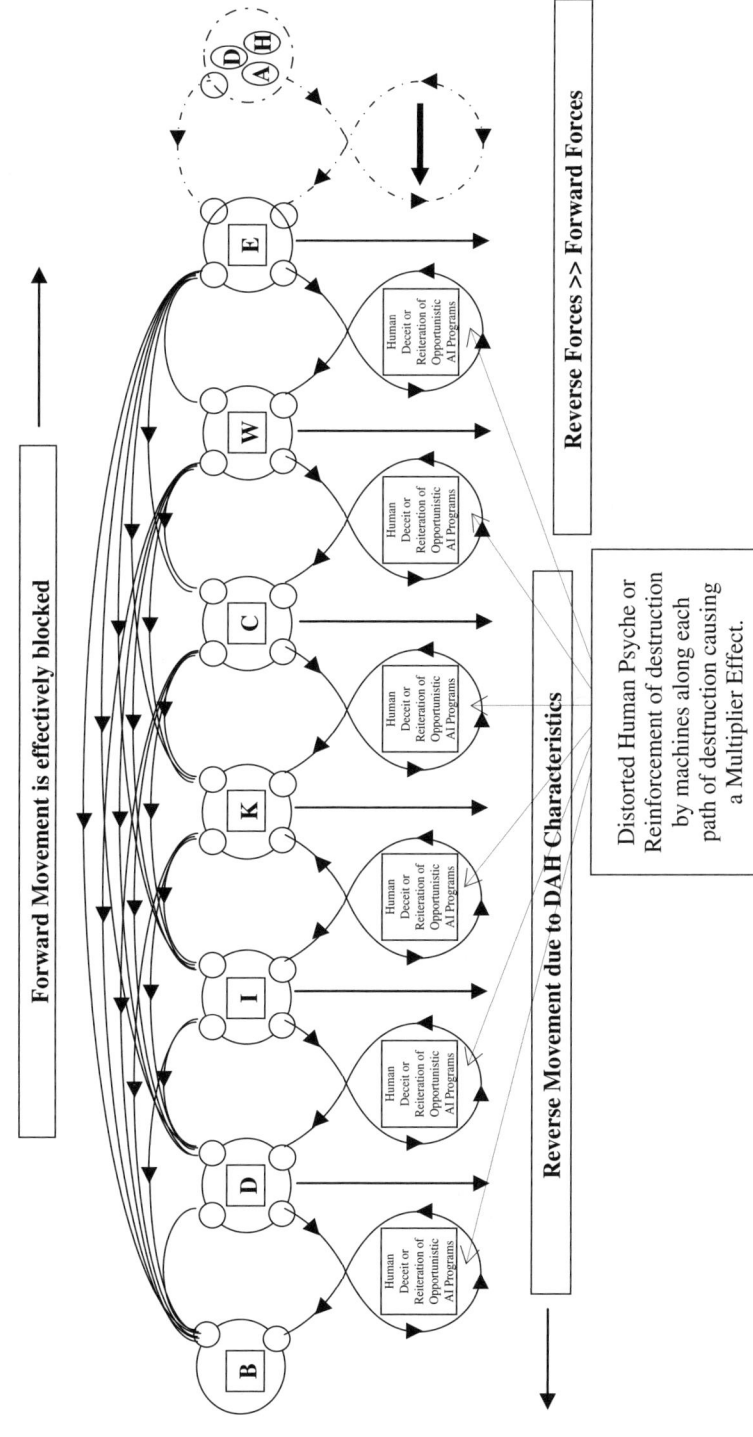

Figure 3.16 Influence of DAH (the eighth node) on the knowledge trail. The feedback to the migration from the seven prior nodes (B, D, K, C, W, E, and I) has a negative effect due to the conflict between DAH goals and the natural tendencies of humans and the operation codes of machines.

130 GENERAL THEORY OF KNOWLEDGE

DAH goals and the natural tendencies of humans and the error ridden operation codes of machines.

Whereas humans by nature learn and seek solutions toward truth in the results derived from their functions (lasting contributions of humans and the accuracy of machines), the DAH goal seeks to deceive. Whereas humans by nature learn and seek solutions toward elegance and beauty in the results derived from their functions (a lasting short and straight path toward the solution and machines built with optimal circuit configurations), the DAH goal seeks to be arrogant and contradictory. Whereas humans tend to make their solutions to have generic use to one and all, the DAH goals for success have no qualms about practicing hate in exploiting others. The conflict runs deep and remains ingrained like deception, arrogance, and hate. These conflicts eventually destroys the very hosts that offers a refuge for the eighth node of DAH.

3.10.4 Dynamics of Knowledge in Societies

In an accurate representation of the real world of knowledge and society that hosts that knowledge, the roles of humans who help society move forward and the role of those who block the society need representation. At any given time, both factions of society are active and a conflict of interest create a resultant forces that nudges society. The effect can be gradual, transient or even radical. There is ample evidence of each from a historical perspective.

The complete set of forces within society that represent the dynamics of knowledge can be classified into four categories:

(a) A set of forward forces to reach the goal set by the ideals embedded in the Aristotle's TVB direction
(b) A set of reverse forces that retard the forces in (a) by being suspicious and negative toward those pursuing TVB ideals
(c) A set of backward forces to deploy (Aristotle's) DAH tactics to achieve selfish and personal goals
(d) A set of reverse forces that merely correct the tactics and procedures of the forces in (c).

Forces (a) through (d) are very real in societies. Such forces exist abundantly in national, corporate, university, cultural, and even the most intimate family settings. It is also apparent that the forces are numerous and the directions different. Their net effect can be erratic if not chaotic. When the local origins are accurately identified and tagged, the net resultant force in society can cause jagged movement and a source of instability in the smooth migration from one node to the other. From the perspective of an individual, the numerous forces (from any node to the prior or following nodes) and their overlap can destabilize the thought processes necessary to develop knowledge from information and concepts from knowledge. From a social and cultural perspective, the movement for societies and culture to migrate in a positive direction can become slow, tedious, error-prone, and even corrupt.

Their frequency and intensity offer very real indications regarding the forward movement and backward slide in social entities. Generally, (b) and (d) forces are reactionary, opposing the (a) and (c) types of forces. For this reason the intensity may be low compared to the main forces driving leaders to be either TVB- or DAH-oriented. It is interesting to note that not all leaders are TVB-oriented, thus taking society on a lagging cycle between being progressive and recessive. Social cycles become the lagging indicators of leadership patterns, and their innate style of leadership ranges from the socially benevolent to very opportunistic tyrants. Conversely, leadership style is a leading indicator of social change in a numerical and mathematical sense, as just leading economic indicators become a mirror for the economy of a nation.

A look at events[5] of the present and of the long past offers proof of their existence, intensity, and durations. The effects that follow are somewhat predictable within a margin of error. Some type (a) forces are propagated from one statesperson to the next, even though they are more common in monarchies than democracies. Type (b) forces gain a strong foothold as type (a) force generates social momentum. Society vacillates, almost helplessly, between the colossal type (a) and type (c) forces generated by TVB ideals and DAH opportunism.

The directionality of knowledge according to DDS or LoC classifications further narrows the forces on any given body of knowledge. The vector resulting from forces (a) through (d) results in the final movement of knowledge. The forces do not have a very a definite magnitude or directionality. The sum of all the estimated forces in society are integrated over a period to compute a resultant vector. Econometricians have deployed such tools for determining the movement in various sectors of any national economy by securing a good estimate of the economic activities and resources spent in numerous sectors of the economy.

When a given set of macroeconomic parameters are to be predicted, macroeconometrists use time series approximations, estimate the local gradient in the particular economy, and compute incremental movements over short durations. In

---

[5]The four types of forces in social groups have existed from the beginning of civilization. However, we focus on the events from the recent past and examine the historical evidence of actions [types (a) and (c)] and the reverse forces [types (b) and (d)] that have been invoked by American presidents in American society. In the past, type (a) forces prevailed in the actions of presidents such as Lincoln (during the abolition of slavery) and Kennedy (while resolving the Bay of Pigs crisis), and Carter (attempting to promote peace in the Middle East at Camp David, while advancing human rights and alleviating suffering), etc. Type (b) forces by Southern states countered the efforts of Lincoln, by generals in the army ready to destroy Cuba countered the efforts of Kennedy, and by war monger leaders in the region, countered the efforts of Carter, etc. Type (c) forces prevailed in the actions of Johnson as he put the lives of thousands of young Americans on the line during the Vietnam War, the actions of Nixon in the coverup of the Watergate scandal, and the actions of Bush in furthering the war in Iraq even though the invasion and occupation were based on nonexistent weapons of mass destruction, etc. Type (d) forces are documented in the efforts of honest reporters and journalists (such as Chet Huntley, David Brinkley, and Bill Moyers), when actions of type (c) occurred in society.

132 GENERAL THEORY OF KNOWLEDGE

a similar vein, the force that any particular knowledge society experiences over a given duration in any given direction causes predictable incremental changes in the knowledge domain. Thus, the overall dynamics of knowledge in the knowledge society can be tracked just like the dynamics of any other commercial commodity in any economic society.

All four forces, (a) through (d), are shown in Figure 3.17. It can seen that this figure is not symmetric about the horizontal centerline through the nodes due to the absence of the six major loops depicting human contemplations leading to the TVB ideals. It is assumed that human beings do not contemplate DAH goals, although they may be tempted by them. The feedback to the migration from these seven prior nodes suffers due to the conflict between DAH goals and the natural tendencies of humans and the nature of SW in the machines.

In estranged and perverse societies, it is possible that Mafia types and tyrant kings will actually "contemplate" negativistic DAH goals. Under these conditions, the diagram shown in Figure 3.18 will become symmetric, with six major loop in its lower part.

When leadership displays both TVB ideals (of noble and dedicated politicians) and DAH tendencies (of corrupt and selfish officials), complex and confused patterns of movement within society occur. However, modern machines can track social dynamics accurately.

A representation of the movement of information, knowledge, concepts, and wisdom through society is depicted in Figure 3.19. At a detailed level of (machine) analysis, patterns of social forces and intents are discernible. A configuration of dual-feedback signal filters starts to emerge. If the I, K, C, W, and E nodes represent knowledge banks for accumulating information, knowledge, concepts, wisdom, and ethics, then the progress (forward) or retreat (backward) of a society along the knowledge trail can be predicted and controlled much like the control of signals in a complex analogue or digital filter.

The types of filters representing social dynamics become more elaborate than filters deployed in communications systems. For instance, in Figure 3.20, the numerous multipliers $m's$ that inputs in the two sections of the diagram are filters in their own right. They indicate how the efforts of social leaders are filtered through society. This configuration in a true sense, encompasses the effect of 12 additional social filters that monitor movement along the knowledge trail, which is by itself a slow-response highly-damped long-term filter. Traditional filter theory techniques need considerable enhancement before they can be deployed in social situations.

3.10.5 I, K, C, W, and E Bases to Replace Nodes

The entropy of an object captured, as a digital data structure (see Section 3.1 and Figure 3.3), is stored in the databanks of computers. It reflects the status of the object in a digital format that can be processed. It also carries forward and backward pointers that indicate the origin and enhancements of the object, sibling objects, descendant and precedent objects, light and loose relations, and ties of the objects with their own and neighboring objects.

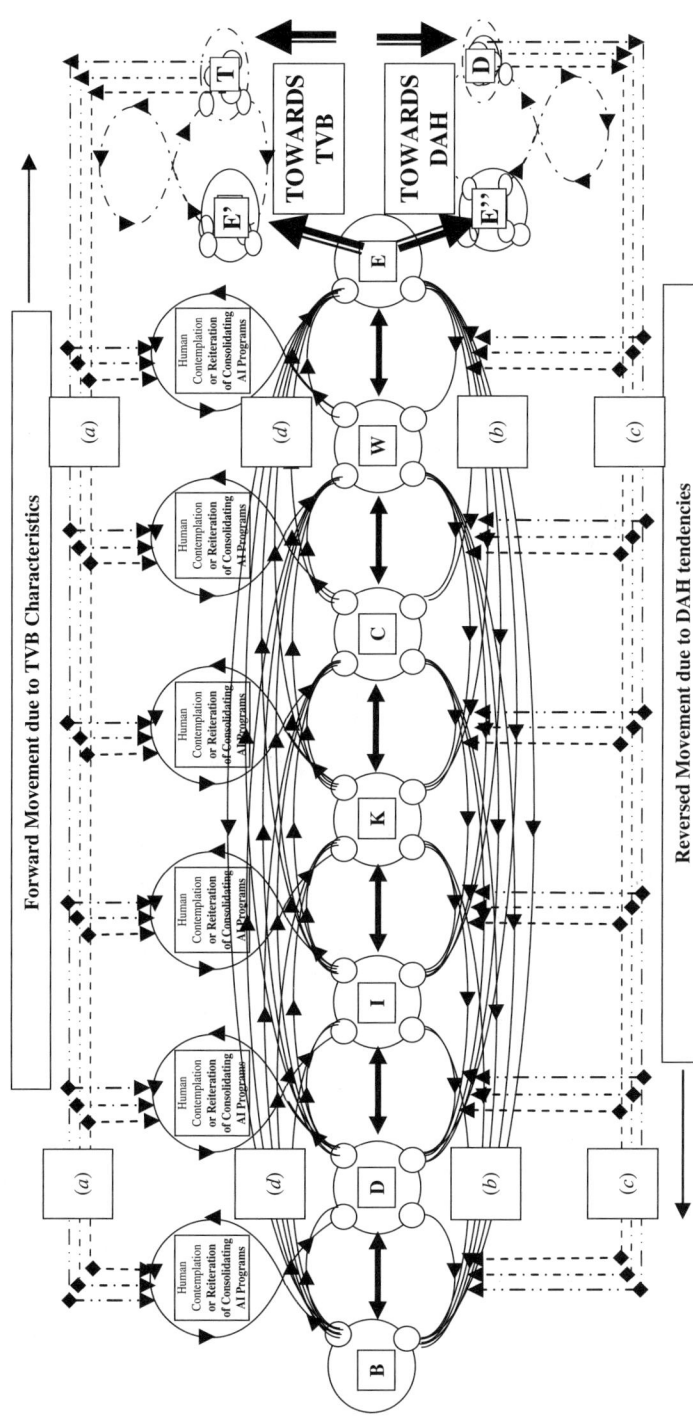

Figure 3.17 Influence of TVB and DAH tendencies on the knowledge trail. The feedforeward to migration from the seven prior nodes (B, D, K, C, W, E, and I) has a positive effect due to the synergy between TVB goals and drives towards the goals E′. It can also have a negative effect due to DAH tendencies pursued by humans and machines. Types (a) and (d) prompt movement toward TVB ideals that reinforce forward movement. Conversely, types (b) and (d) prompt movement toward E″ goal and retard the forward movements.

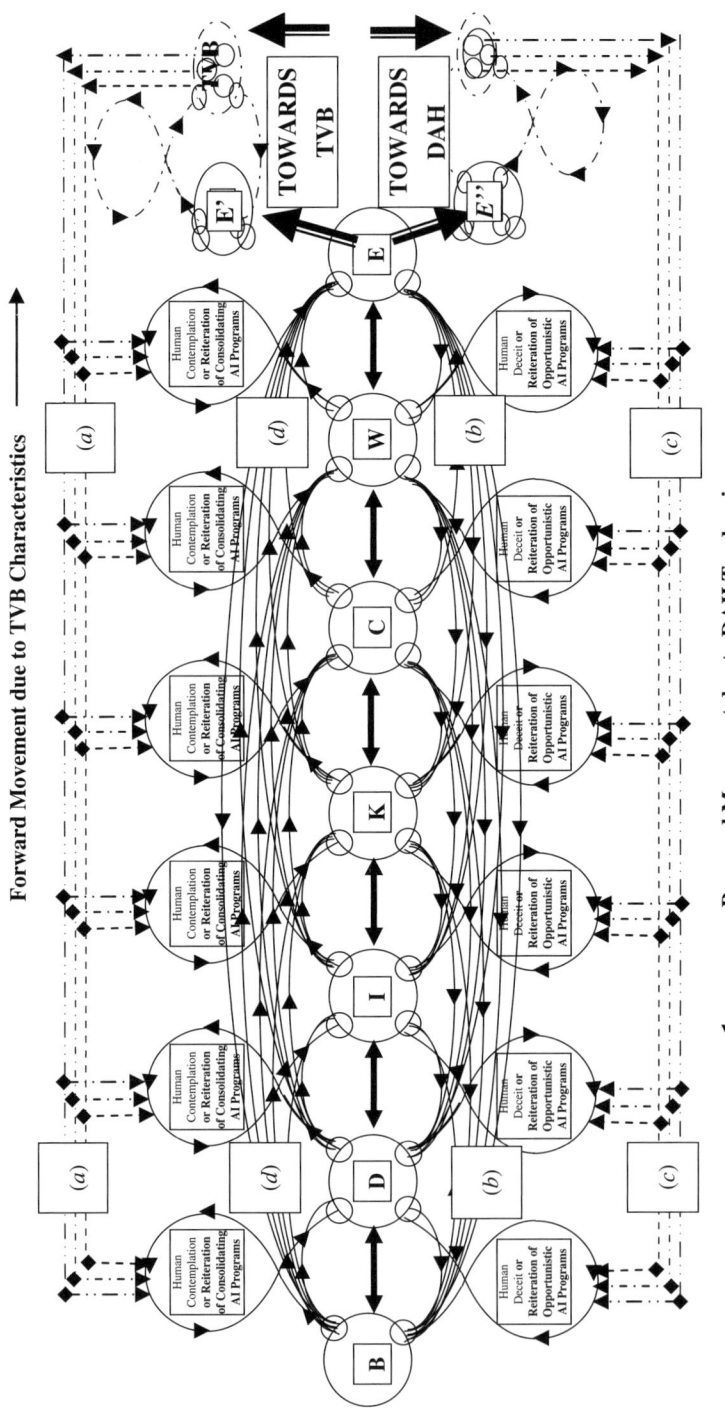

Figure 3.18 Influence of TVB and DAH (the eighth node) on the knowledge trail in societies where legal and ethical agencies face stiff resistance from Mafia types and tyrants in society. The conflicts and violence can be catastrophic. When society is confined to neighborhoods and student groups, street violence and gang wars can erupt.

134

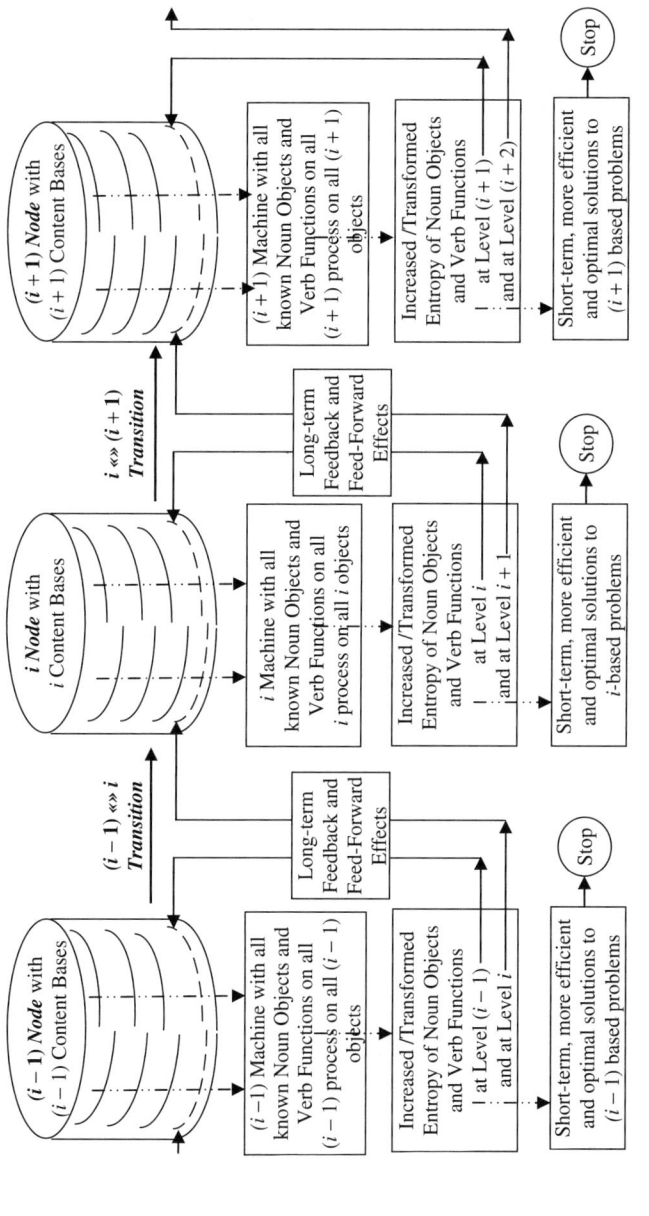

Figure 3.19 Automatic feedback and feed-forward system for the movement of an individual, a society, or a culture along the knowledge trail. The feedback governs the stability and feed-forward assures an error-proof migration from one node to the next need(s). The processes involve scientific, social, and ethical aspects in society.

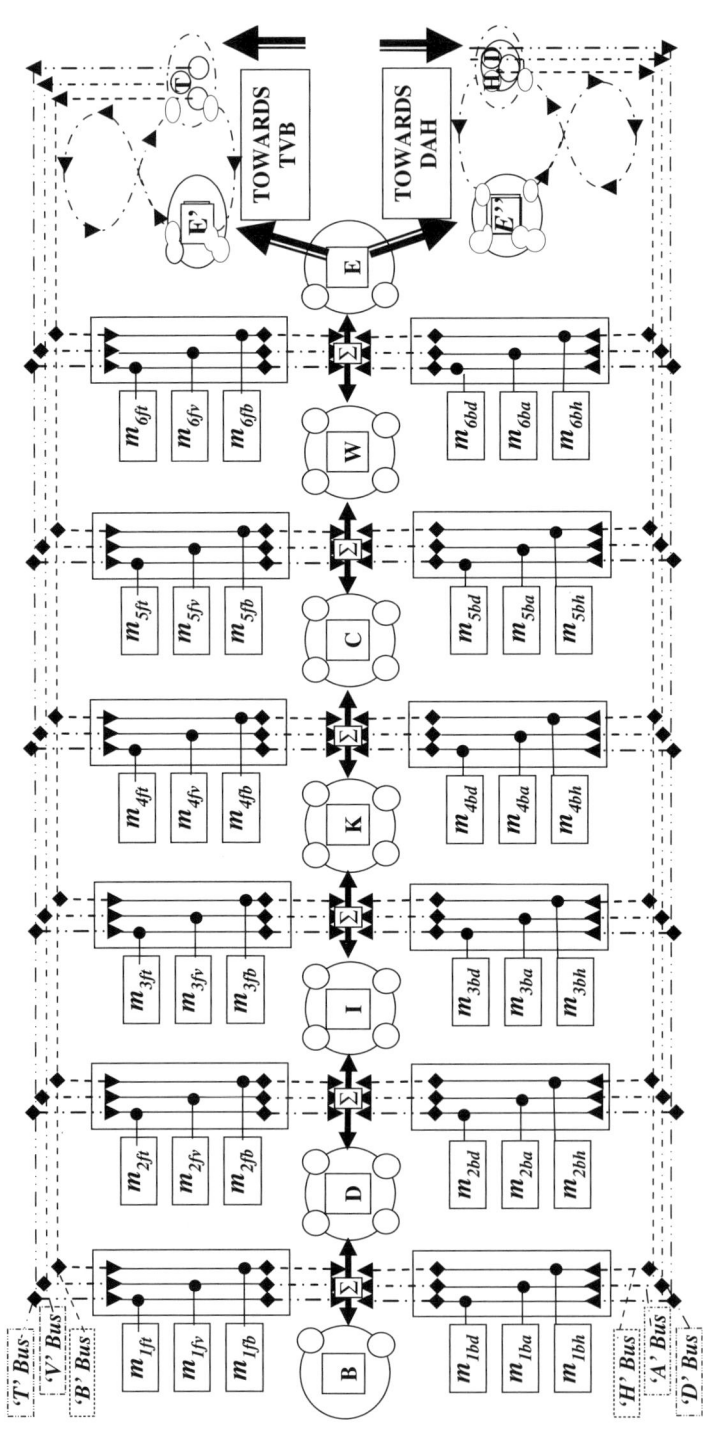

Figure 3.20 Representation of the movement of information, knowledge, concepts, and wisdom through society where leadership displays both TVB ideals and TVB tendencies, creating a complex and confused pattern of movement in society. The signals and forces are discernible, and the configuration of a dual-feedback signal filter starts to emerge. If the nodes represent knowledge banks for accumulating Information, knowledge, concepts, wisdom, and ethics, then the progress (forward) or retreat (backward) of a society along the knowledge trail can be predicted and controlled much like the control of signals in a complex analogue or digital filter.

This representation offers ample information for a machine to manipulate an object, whether it is an individual object or member of a group of objects. If the analogy is extended to a society or culture, then the dynamic snapshot of the knowledge trail in that society is a digital imprint of its status. When the entire knowledge trail is encoded as a comprehensive data structure, numerous subsidiary bases become a part of the colossal data structure. Though large, it is still manageable with larger mainframe computers.

The memory size, processor speed, inherent limitations in its ability to accurately distinguish microscopic changes in the entropy of noun objects, and finally the architecture facilitating or hindering Web information, knowledge, concept, and ethics bases in conjunction with the socioeconomic state of the nation will dictate the final success of massive mainframes. Such machines carry a digital imprint of the micro- and macro knowledge economics and socioeconomics of a nation, a culture, a corporation, or any social entity. Such imprints of the stock market are generated by mainframes in the financial industry.

In this section, we present a methodology whereby the movements of any culture or society along the entire knowledge trail can be charted, managed, and navigated. To be able to manage the movements (like the navigation of a nation's economy), the I, K, C, W, and E nodes along the knowledge trail are replaced by information, knowledge, concept, wisdom, and ethics bases. The appropriate machines to process information, knowledge, concept, wisdom, and ethics are situated in their respective zones. If a generic node i has an i base, then the functions around this node are as depicted in Figure 3.19.

In Figure 3.20, the effects of multipliers $m's$ in society are also represented as TVB or DAH objects, and verbs traverse through the immediate environments of the individual nodes. In addition, the net effect of global and local noun objects (NOs) and their corresponding verb function (VFs) is depicted. The C and W zones function interdependently of the C objects and W objects are linked together. In a sense, stability along the knowledge trail should be investigated as an iterative numeric and object-based solution rather than as independent solutions for stability. The solution methodologies start to become complicated, yet one or numerous solutions are possible while attempting to solve the knowledge equations discussed in Sections 3.10 and 3.11.

3.10.6 Nodal Forces in Societies

The four basic forces (a) through (d) discussed in Section 3.10.4 act on any of the six nodes (D, K, C, W, E, and I). The subsequent movement depends on local economic and societal forces. Such forces may be social, financial, or completely group-oriented and group-motivated. These local forces are depicted in Figure 3.21, with the two major forward and backward forces (a) and (c) shown by vertical arrows acting directly on the node. In reality, these are both sets of sociomotive forces (both internal and external) to move society toward becoming a better society based on the TVB ideals in society or to deplete society based on the DAH goals of corrupt leaders. An ethical balance at the end of the knowledge

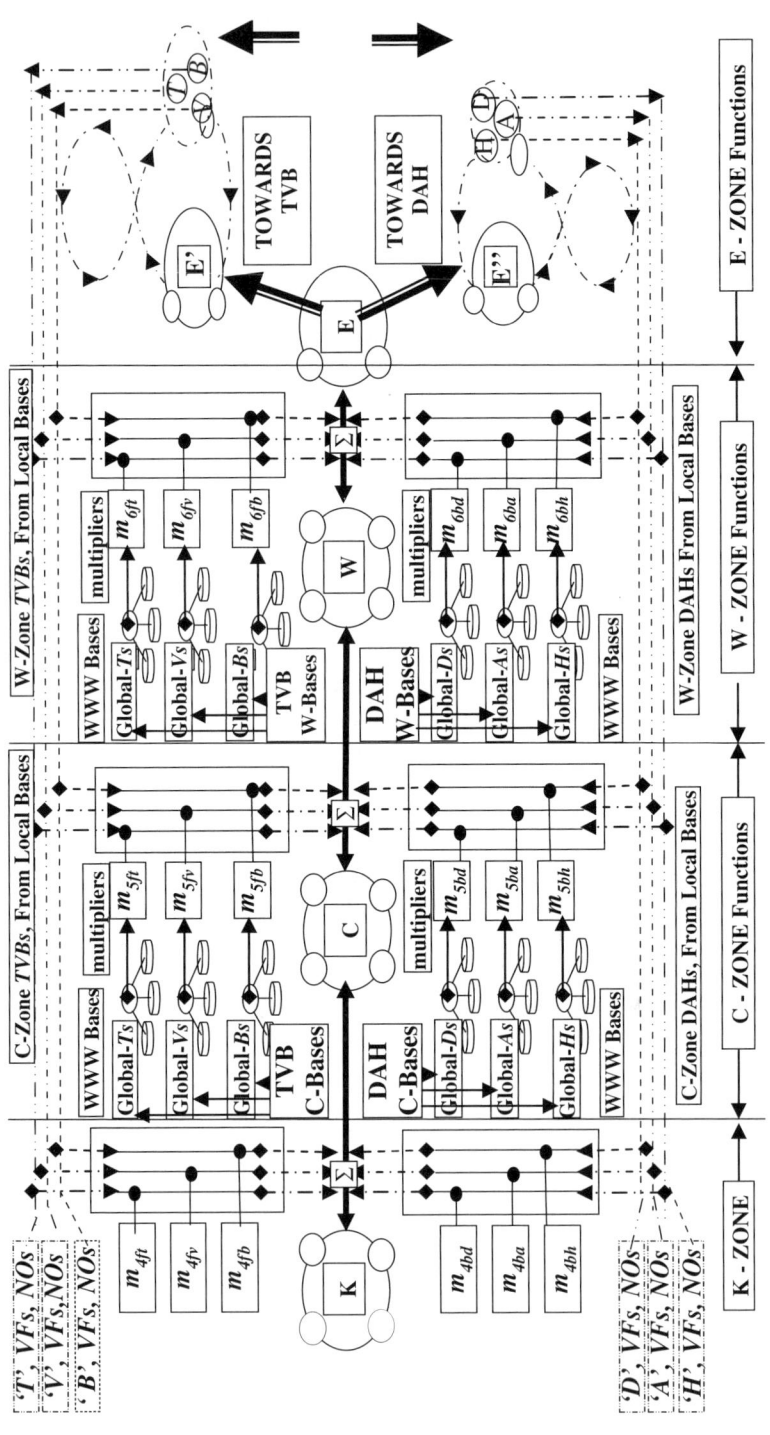

Figure 3.21 Details of the functions performed at the C and W nodes and the blending of local (noun objects + verb functions) with the global (noun objects + verb functions). The effects of multipliers $m's$ in society are also represented as the TVB or DAH objects, and verbs filter through the immediate environments of the individual nodes.

ORIGIN AND DESTINATION **139**

trail establishes the resultant vector that closes the integrated funicular polygon [3] of vector forces on the social entity.

When the forces (a) through (d) are pursued and plotted at any node, secondary effects start to surface. Consider only the forward TVB set of forces. These forces at previous node(s) may generate a (positive or negative) ripple effect (or even a shockwave) at the current node, thus adding to (or subtracting from) the current force at the current node. These secondary forces are shown as an entry point on the left side of i node in Figure 3.22. Such forces can also experience a positive or negative multiplier m_{i-1} effect due to the scientific innovations in the $(i-1)$ zone immediately prior to the i node.

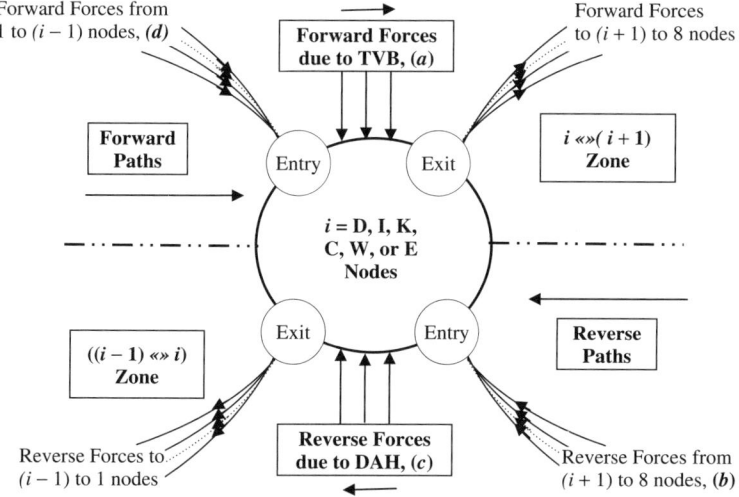

Figure 3.22 Dynamic state of each node in the knowledge trail. The value of i ranges from 2 to 7, with the eighth culminating node representing TVB or DAH.

3.10.7 Summary of Sociological Forces on Nodes

$$\text{Net Forward Force } FF(i) = \sum_{j=i+1}^{j=8} FF(j) = \sum_{k=1}^{k=i-1} FF(k) + FF(TVB)$$

$$\text{Net Reverse Force } RF(i) = \sum_{j=i+1}^{j=8} RF(j) = \sum_{k=1}^{k=i-1} RF(k) + RF(DAH)$$

$$\text{Net Force } NF(i) = \sum_{j=i+1}^{j=8} FF(j) - \sum_{j=i+1}^{j=8} RF(j)$$

and

$$= \sum_{k=1}^{k=i-1} \text{FF}(k) - \sum_{k=1}^{k=i-1} \text{RF}(k) + \{\text{FF(TVB)} - \text{RF(DAH)}\}$$

when the multiplier effects, that is, m_{i-1} in the $(i-1)$ zone, are ignored.

3.10.8 Multiplier Effects at Intermediate Nodes

These equations become more elaborate if the multiplier effects are included. Under these conditions, the final intensity of the force at any particular node i can be written as

$$\text{Nodal Forward Force FF}(i) = \sum_{j=1}^{j=i-1} \text{FF}(j) \cdot \left(\prod_{k=i-j}^{k=i-1} m_{kf} \right)$$

and

$$\text{Nodal Reverse Force RF}(i) = \sum_{j=1}^{j=i-1} \text{RF}(j) \cdot \left(\prod_{k=i-j}^{k=i-1} m_{kr} \right)$$

where m_{kf} is the multiplier for forward TVB ideals and m_{kb} is the multiplier for backward DAH goals in society. The multiplier m_k attenuates or amplifies the TVB or DAH force in the k zone. These multipliers depend on the nature of society and its inclination toward following TVB ideals or DAH goals. Such multipliers exist in economics, where investments in certain zone (such as medical research) can bring about new effects at the following nodes (such as innovations in medicine and the enhancement of general health) within an entire society. In the knowledge domain, the efforts of a leader to enhance truth in information and knowledge will eventually enhance the integrity of the society, and reduce white-collar crime. Society may create a domino effect (i.e., $m_k \gg 1.0$) in response to the efforts [i.e., FF(j)] of its leaders, or deliver a deadening blow (i.e., $m_k \to 0$, or even $m_k \to$ negative number for countering) to their efforts. Society may move forward, stagnate, or slide backward at any given time or over a time (with nodal forward force > 0, $= 0$, or < 0) at any node. This equation relating the initiatives of leaders with the reactions of society derives the positive or negative effect along the knowledge trail in that society.

Only if the summation of all the forces at all seven nodes is steadily sustained and distinctly positive will a stable cycle of growth occur. Conversely, a gradual slide backward and decay of society are also equally plausible. Sudden changes in nodal forces cause transients and shockwaves in society that can be indeed predicted from mathematical considerations. Such transients and shockwaves occur in traditional filter theory when a complex electrical system needs to gradually move from one steady-state condition to another. In the knowledge domain, the steady-state conditions may correspond to a steady migration of society, from traits of mafia crime and cruelty to peace and justice. The converse scenario is also equally feasible.

3.10.9 Implications of Knowledge Equations

The equations in section 3.10.8, can be rewritten for each node independently. For example, in terms of the force at knowledge node K or at $i = 4$, the equations imply that the force to move "Knowledge" to "Concept" is the sum of forward forces from the B to I nodes plus the force due to social force leaning toward TVB orientation minus the social force leaning toward DAH orientation. A knowledge machine can interpret an equation of this nature and then go on to predict that a Concept node may be reached in a time frame of t, depending on the social inertia of a society in any particular direction of knowledge or discipline, such as medicine, nuclear technology, petroleum refinement, spare travel, etc.

When the forward force greatly exceeds the reverse force, a shear effect is experienced in the positive direction at any node, and vice versa. The shear effect indicates polarization in society and the responsibility of political leaders becomes that of keeping the forward force to produce smooth acceleration of the society at any given node without shear or fragmentation at the node.

3.11 THE ENTROPY OF KNOWLEDGE

The entropy of a body of knowledge (BOK) is an exact graph of the embedded objects, their relationships, their attributes and attributes of attributes, symbolized as (EO, R, Ats., Ats. of Ats.). Each of the (EO, R, Ats., Ats. of Ats.) has two components: its structure and the numerical value of the four components. The graph has a structure based on the DDS or LoC classification(s) of the subject matter(s) from which the object(s) are drawn, and a numerical component for the nodal relation to the object(s) and capacity (or strength) of the links in the graph.

To be precise, the depiction the entropy is a map of knowledge. If the map was indeed a face, the objects would be hair, forehead, eyes, etc., all the way to the chin and neck. Relationships would be distance between the eyes, from the center of the forehead to the center of the eyes, etc. The attributes would be hair color: black, white, gray, blonde, etc.; forehead: clear, wrinkled, etc.; eyes: black, green, brown, etc. The attribute of attributes would be natural/dyed hair color, thick/thinning, etc. The depth of the graph (tree) could be made arbitrarily long or short. Such maps of knowledge are pursued in the minds of humans, but they can be duplicated in the memories of machines. Once entered as objects in the machines, they can be processed by knowledge-processing systems at knowledge machine language programs from assembly-level programs to application programs at a macrolevel. In a generic sense, bodies of knowledge can be processed and compiled much like the facial recognition and composition programs used by criminologists. The entropy of a body of knowledge is documented as the graph of EO, R, Ats., Ats. of Ats.. The graph has a structural and numerical component for the nodes and links in the graph. Any alteration of the graph implies a change in entropy.

3.11.1 The Change of Entropy

In this section, we propose that the convolution of the energy of all the knowledge processes operating on any body of knowledge (BOK) with an entropy E will result[6] in an altered BOK' with an entropy of E':

$$\text{(Energy of } \Sigma \text{ Processes)} * \text{(BOK with entropy} = K)$$
$$= \text{(BOK' with entropy} = K') \quad (1)$$

This equation is akin to the law of conservation of energy and implies that the entropy of (noun) objects remains transfixed unless there is energy to change their status. In the knowledge domain, convolutions (rather than multiplication) become necessary due to the complex nature of BOKs, the embedded objects, their relationships, their attributes and attributes of attributes, symbolized as (EO, R, Ats., Ats. of Ats.).

The energy in the knowledge processes is the result of the forces (social, natural, computational, etc.) that change the (EO, R, Ats., Ats. of Ats.) of any BOK. For example, a passive statement, copy process, or nonoperational *kopr* does not cause a change in entropy. An active process such as altering any of the (EO, R, Ats., Ats. of Ats.) produces the change of entropy or a snapshot of the graph of knowledge. The magnitude of such force can be represented as the intensity of the verb functions on the noun objects (2^0), on the relationship (2^{-1}), on an attribute Atr. (2^{-2}), an Atr. of an Atr. (2^{-3}), etc. It is important to treat the entropy of any BOK as a distinct signature, fingerprint, or retina scan. Any change in any of the eight parameters [structural location and magnitude of the four basic groups (EO, R, Ats., Ats. of Ats.)] makes the resulting BOK' different from the original BOK. The minimal change that is sensed depends on the sensitivity of the knowledge processors used to compare the two bodies of knowledge, BOK and BOK'.

3.11.2 The Duration for Change of Entropy

If the processes (social, natural, computational, etc.) occur over a period of time Δt, then

$$\text{(Energy of } \Sigma \text{ processes)}_{\Delta t} * \text{(BOK with entropy} = K)_t$$
$$= \text{(BOK' with entropy} = K')_{t+\Delta t}$$

---

[6] The relations proposed in these sections should be treated as tentative and may need enhancement and enrichment as we progress toward the quantification of changes in the entropy and status of bodies of knowledge by knowledge machines and wisdom machines.

Specifically, if a knowledge or wisdom machine produces a change of entropy, then the summation of the energies of knowledge operations (i.e., the *kopcs*) convolved with the initial entropy K will result in a net change of entropy $(K' - K)$, and

$$(\Sigma \pm kopc \text{ power}) * (\pm \text{BOK with entropy}) = \pm \text{Rate of change of entropy}$$

As an approximation and for an infinitesimal period if the convolution is treated as a multiplication, the rate of change of entropy become proportional to the energy of the processes (social, natural, computational, etc.) operating on the collection of noun objects that are embedded in a body of knowledge BOK. By enhancing KPU capacity, the effect of social and natural processes can be duplicated at high speeds in knowledge machines and wisdom machines.

3.11.3 No Prior Knowledge: No Change in Entropy

When there is no structure (none) and no magnitude (none) for all four basic groups (EO, R, Ats., Ats. of Ats.) of objects embedded in any BOK, then the effect of all the structured *kopcs* is zero. The structure of any *kopc* depends on some knowledge of the entropy encoded within the eight parameters.

However, if the knowledge machine is permitted to attempt random combinations of *kopcs* or ill-structured *kopcs*, then the machine can investigate if this random approach results in any change of entropy. This attempt on the part of the machine in similar to the random crawling of an ant to find food in unknown territory for it too walks in any direction to discover a scent of food.

3.11.4 No Knowledge Process: No Change in Entropy

This inference is a direct result of the equations in section 3.11.2. The implication of this statement is that the entropy of any BOK remains preserved (depending on the time-dependence characteristics of the storage media) unless some form of energy is expended in altering it. Natural decay and healing are also processes, especially in human minds and societies. Social processes occur at any of the seven levels (ranging from senseless gossip to cleansing and enhancement, or from slander to injury and oppression) [4] in society.

3.11.5 Examples of Prevalent Knowledge Processes

Four common examples of knowledge processes are listed as follows:

1. Consider the evolution of species. In symbolic form, the process can be stated as

    Mutation * Prior genes of an existing species = Genes of a new species

144 GENERAL THEORY OF KNOWLEDGE

A complex set of operations (VFs) convolved with a known set of DNA structures and amino acids (i.e., collection of NOs) results in a new set of DNA structures. This process has endured for a long time.

2. Consider a simple NPU statement:

   $X = \text{Reciprocal}(X)$

   Numerical operation code RECI (Operand X) = New Value of X

   Simple NPU operation code on X (i.e., a NO) results in a new value of X. Processes of this nature (reciprocal, compliment, exponentiation, etc.) have occurred in any NPUs since the early days of computing.

3. Consider the natural event of a earthquake in any neighborhood. In symbolic form, the process can be stated as

   Earthquake $*$ Neighborhood = Destroyed neighborhood

   A complex set of natural operations (VFs) convolved with known objects in the neighborhood results in a new (destroyed) set of objects. This is also a process that has occurred for a long time.

4. Consider an economic policy for a nation. In symbolic form, the process can be stated as

   Financial policy change $*$ National economy = Altered economy

   A complex set of human operations (VFs) in the policy convolved with known objects in the national economy results in a new and altered set of objects. In addition, this national control mechanism has been exercised for a long time.

3.11.6 Generalization of Knowledge Processes

The generic form of the inferences stated above can be listed in the context of a knowledge-processing environment:

1. In simplest form, the generalization can be stated as

$$\{VF^{(+,0,-)}\} * \{NO^{(+,0,-)}\} \rightarrow \{NO^{(\pm,0,\pm)}\}$$

2. In array form, the generalization can be stated as

$$\text{An array of } \{VF^{(+,0,-)}\} * \text{A row of } \{NO^{(+,0,-)}\}$$
$$\rightarrow \text{A matrix of } \{NO^{(\pm,0,\pm)}\}$$

3. At a microprogram (**M**) level, this generic statement with KPU instructions, can be stated as

$$\mathbf{M} \cdot \{kopc^{(+,0,-)}\} * \mathbf{M} \cdot \{kopr^{(+,0,-)}\} \to \{\mathbf{M} \cdot result^{(\pm,0,\pm)}\}$$

4. At an application (**A**) program level with a $\sum$ operator, this macroscopic statement can be stated as

$$\mathbf{A} \cdot \sum \{kopc^{(+,0,-)}\} * \mathbf{A} \cdot \sum \{kopr^{(+,0,-)}\} \to \left\{\sum \mathbf{A} \cdot \{results^{(\pm,0,\pm)}\}\right\}$$

5. At a universal program level with a U operator, this super-macroscopic statement can be stated as

$$U \cdot \sum \{kopc^{(+,0,-)}\} * U \cdot \sum \{kopr^{(+,0,-)}\} \to \left\{U \cdot \sum \{results^{(\pm,0,\pm)}\}\right\}$$

It should be noted that all the above equations can be executed in the computational domain and not in an algebraic format. The *two* major reasons exist. *First*, all the operations are generally nonlinear and the effect of one operation can seriously affect the execution of the next. The knowledge domain remains highly fluid in the knowledge machine and wisdom machine environments. *Second*, the progress can become unstable due to the nonlinear response of an intelligent object (see Sections 5.2 and 5.3). The condition can be heightened given the nature of the adaptation (see Sections 5.2.1–5.3) of *kopcs* and *koprs*. In a linear-cooperative mode (i.e., VF^+ with NO^+, VF^- with NO^-, VF^0 with NO^0), the computational steps are likely to be executed dependably. In the noncooperative adversarial mode, the knowledge machine and wisdom machine are likely to display large oscillations and significant tendencies for divergence in the solutions of knowledge- or wisdom-level programs.

3.12 CONCLUSIONS

The general theory of knowledge facilitates the treatment of knowledge as mathematical entities with "knowledge-centric objects" (KCOs) that have and hold dynamic relationships to perform the same social and production-oriented tasks as human beings would. The mathematical representation is facilitated by a graph that spans the primary and secondary KCOs, their relationships, their attributes, and their relationships and interdependencies. The nature and complexity of the KPU to handle such objects increase rapidly as the number of objects and the nature of response (linear or nonlinear, assistive or resistive, fixed or dynamic, etc.). The solutions may become unstable, divergent, or even oscillatory. Such behavior is sometimes noted in human interactions and problem solving.

The general theory of knowledge based on "symbolic quantization" is also presented. Symbolic quantization is based on blending human responses to the elements embedded in the laws of nature, as they are perceived by a particular user or groups of users. The symbols are alphanumeric entities that are partly descriptive and partly numeric. During the process of blending, both the descriptor of the symbol and its numeric value can undergo changes. The descriptor facilitates the change in the perceived (elements of the) law under the scrutiny of human comprehension. Elements of human comprehension are programmed as AI routines or as a series of knowledge programs in machines. The numeric value quantifies the extent of change incurred or that will be incurred. A cumulative index of a series of such changes (descriptive and numeric) is an estimate of the change of entropy of the old knowledge if the relaxing of the proposed elements embedded in any of the laws are feasible and practical. The change between the entropies of the new knowledge and old knowledge is the measure of the knowledge gained or knowledge lost.

On a global scale, human comprehension and the laws of nature are blended to alter the basic understanding of such laws or to customize the laws to suit a particular application, All the parameters involved in basic equations representing any law of nature in any given discipline are scrutinized by human comprehension encoded as AI routines in knowledge machines to verify if they can readapted.

On a local scale, the global scrutiny of the laws is reduced to knowledge machine assembly-level programs to perform knowledge-level operations. These programs reconstitute existing knowledge (such as the standard operating procedures in corporations, production operations in manufacturing facilities, etc.) to relax and scrutinize each operation (verb function) and every object (noun object) in the localized operation to perform a computer-assisted optimization (i.e., enhance the validity and its authenticity) of the knowledge that surrounds the local operations. In the local mode, the machine executes a blend of programmable knowledge operation codes and utilities from "knowledge libraries" to complete knowledge level tasks such as finding the best university for an individual, determining a viable contract between a local union and industry, find the most likely cause of social unrest in a community or within a social organization.

REFERENCES

1. E. Shortcliffe, *MYCIN: Computer-Based Medical Consultations*, Elsevier, New York, 1976.   See also B. G. Buchanan and E. H. Shortcliffe, *Rule-Based Expert System: The Mycin Experiment at Stanford Heuristic Programming Project*, Addison-Wesley, Boston, 1984.

2. W. Kintsch et al., *About NeoMycin, Methods and Tactics in Cognitive Science*, Lawrence Erlbaum, Mahwah, NJ, 1984.
3. M. Salvadori, *Why Buildings Stand Up: The Strength of Architecture*, W.W. Norton & Company, New York, 2002.
4. S. V. Ahamed, *Intelligent Internet Knowledge Networks*, Wiley-Interscience, Hoboken, NJ, 2006.

CHAPTER 4

VERB FUNCTIONS AND NOUN OBJECTS

CHAPTER SUMMARY

This chapter introduces software design for knowledge machines. Based on Chapter 3, when any complex body of knowledge is accumulated around an array of "knowledge-centric objects," the processing of knowledge entails the manipulation of such objects, subset of objects, the nature and extent of their interrelationships. Objects also have a series of attributes that need special processing. Verb functions operate on noun objects to alter their structure and entropy. A series of such microscopic operations executed in the knowledge processor units constitute macrofunctions corresponding to subroutines and library functions in typical computer systems.

The essential software for the execution of any "knowledge program" on a given set of input objects requires a compiler to scan and search for objects and build a framework of existing Web information around such objects. Local library and Internet access provides the supporting information about related objects, their relationships and attributes. The structure of objects is thus extracted in the context that the knowledge problem is being solved. Extended artificial intelligence (AI) techniques for "bodies of knowledge" are introduced in this Chapter.

Pattern recognition (PR) is a recognized branch of AI, whereas compilation is a systematic procedure in identifying symbols and operators. In the language

---

Computational Framework for Knowledge. By Syed V. Ahamed
Copyright © 2009 John Wiley & Sons, Inc.

of knowledge (individual or grouped), objects take over the role of symbols and (individual or grouped) verb functions assume the role of operators. The grammar for the documentation of knowledge being more complex that of computer languages offers a large variety of operators and their syntactic variations. However, a thread of conceptual commonality exists. The recognition process (from PR) and identification of symbols (from compiler theory) overlap. Both recognition and identification are intensive disciplines calling for accurate processing and algorithmic refinement.

If parsing, lexical, syntactic, and semantic analyses are placed in an inner shell of the flowchart of a compiler for a computer environment, then statistical, syntactic, and semantic PR forms an outer shell for the flowchart of recognition procedure in a complex visual system. Now, if the two shells are collapsed into one, then a knowledge machine (KM) can be the software recipient to compile and "visualize" the body of knowledge (BOK) under process that passes through the KM.

In the knowledge society, computers and AI play a significant role in preventing nations from falling into this catastrophic cycle of boom and bust that come and go. The brave new machines [1] monitor cyclic forces dominating the such major and minor loops within the cycles. The machines also sense [2] the short-terms corrections as well as the long-term trend-setting forces within and without society. Human beings and machines can indeed bring about long sustained positive growth in the wealth of wisdom in nations. Long-term corporate growth of this nature has been demonstrated[1] in developed nations in the past. The forward movement of nations results from the growth of wealth of numerous directionalities of positive wisdom among respective nations. Such directionalities can range from the enhancement of gross national product (GNP) to the development of arts and culture within the nation.

4.1 POSITIVE AND NEGATIVE SOCIAL FORCES

Machines are gadgets and human beings are temperamental. Humans can swing in two extreme directions from positive wisdom (pursuing its absolute form) to negative wisdom (pursuing its opportunistic form). Between these two extremes, lies the materialistic wisdom that provides the means of living for the average-minded knowledge worker. On other hand, machines being driven by code can also function as the precise, incisive, and intelligent (but mindless) agents of human beings. It is humans who pursue any form of wisdom from the absolute to the opportunistic or any form in between. The directionality of wisdom may be initially encoded, but it can also be reversed later, from pristine to corrupt and vice versa.

When the human element that keeps the wisdom machine on its intended course is offered freedom, the negative slide is less strenuous. The positive movement becomes an uphill and more demanding struggle. After all, it is

---

[1]The demised Bell Systems and AT&T, DuPont, Corning Glass, etc.

animal instinct and human nature to maximize marginal rewards (or the ratio of rewards to effort). In a true sense, the negative slide occurs in any dynamic society naturally, unless it is blocked by the more noble efforts of socially responsible humans. In the current state of the evolving knowledge society, the well-intentioned effort of human beings or organizations (e.g., the United Nations, the Supreme Court, etc.) is greatly supplemented by positively primed wisdom bases.

The sensor and AI software in machines make them aware of the changes in the environment. They become adaptive to counter the effects of changes and yet maximize the rewards for the overall profits and stability of the systems. The design philosophy behind the functioning of intelligent systems starts to parallel the intent behind the actions of rational human beings. Social forces on human beings can be mimicked by special layers of "social" software based on the hierarchy of needs in humans and societies, and the laws of economics and ethical behavior to satisfy such needs in the hierarchy. Being culturally variable, the flavor of social software depends on the national and cultural setting. Much like human behavior, the performance of the intelligent machine will become culturally biased.

When the movements in society are tracked and monitored, the incremental change in the status of primary social "objects" (such as banking, commerce, education, health and welfare, etc.) becomes indicative of the forces responsible for the movement of these primary social objects. For example, change in the banking industry (a primary object) depends on the rate of interest, economic growth, public sentiment, etc. (a set of forces), that move the banking industry. When the cause–effect relationships in the social dimensions are extended, then a set of socioeconometric rules emerges.

It is logical to find areas of overlap between conventional (i.e., national and global) and social econometrics. The economy drives society, and society moderates it. Within this relentless cycle of inter-dependence, human beings develop concepts, wisdom, and ethics (see Figures 2.5a and b). Concept and wisdom machines, and well-intentioned human beings, can help just as much as ill-primed concept and wisdom machines, and ill-intentioned human beings can hurt.

4.1.1 Forces in Society and Operations on Objects

Society provides an environment for all objects, real and virtual, to exist in their own domains. Such domains may criss-cross many times over; however, the path through the domain remains dynamically linked to the social setting and time. In essence, the path through the domains is to resolve the social issue at that particular time. The dynamic linkages provide the intellectual pathways that provide continuity from the conception of a social issue to its resolution. Well-designed maps of the domains and pathways provide robust solutions to social issues. In many instances, the navigation that is designed by computers becomes more optimal than the weak linkages and circuitous pathways of some human beings.

Some of the objects in nature can be highly complex, independent, and volatile. The processing of such objects requires special care to simulate and predict their behavior. The processing time needs to be considerably shorter than the reaction time for such objects. In some cases, predictive processing (to compute the reaction of a predator to the movements of victim, or to determine the reaction of certain bacteria to antibiotics, etc.) may become necessary. When the forces in society are unpredictable, the reaction of the active or irrational objects can become chaotic. Unpredictable human behavior and irrational actions during disasters like volcanic eruptions or wars and famine are some examples.

Some of the other objects show (relative) immunity to environmental changes. The processing of such objects (such as numbers and constants) is relatively benign simple, and traditional computing systems are sufficient to process and predict the behavior. In social environments where noun objects are an admixture of the numerous types of objects, every object needs different sets of attributes and dependencies.

The social environment being more encompassing than the object domains provides the rules for objects to evolve, survive, and thrive. Society and objects soon form symbiotic relationships. Each modifies the other and syntactic rules for social objects to "live." In providing an acceptable environment, the forces within society provide a framework of forces for the objects and rules for their mechanics.

Dynamic equilibrium conditions for healthy interactions between the objects are continually examined and readjusted to minimize the energy consumed when maintaining the routine functions within society. Human beings and machines learn during the process of adaptation. Movements of objects occur at the macro and micro level. The architecture, software, and firmware of machines to deal with the two levels may be as different as micro and macro economics.

4.1.1.1 Movement of Macroobjects At the macroscopic level, the major national objects (such as the economy, banking, housing, public health, etc.) move precipitously due to major social forces (such as wars, recessions, famines, natural disasters, etc.). The movement of macroobjects can be differentiated (partially or totally) to determine the nature and extent of the social force. For example, the rate of change of displacement yields velocity. The social momentum for objects with estimated social inertia can thus be approximated. The instantaneous rate of change of momentum yields the social force. Given the inherent predisposition and inertia of the object (see Chapter 3), future incremental movements due to specific social changes can be estimated in any given culture or society.

4.1.1.2 Movement of Microobjects At the micro level, objects of an individualistic nature (such as homes, medicines, food supplies, safety items, etc.) move due to minor individual forces (such as job, local weather, family, etc.) within society. The nature of laws that govern the movements' macro- and microobjects [such as the intensity of verb functions (VFs) and the inertia of noun objects (NOs)] are essentially the same, but the magnitude and impact can be substantially different. National integration of the movement of microobjects provides a basis for the movement of macroobjects. Micro and macro social economics become inter-dependent as traditional micro- and macroeconomics.

Social processes at, among, and around noun objects become distinctly feasible in social machine tailored to emulate the behavior of objects in most (force, coerce, etc.) social settings. Major objects (i.e., the noun objects; see Section 3.2) motivate minor nouns and forces within society and provide the operators (verb functions; see Section 3.2) that can take place at, among, and around the minor noun objects. Adjectives modify the noun objects and/or their attributes, and adverbs modify the verb functions and/or intensity. Conditional clauses modify the convergence points for noun objects and/or verb functions. Such noun objects become the operands, and such verb functions become operational codes. From machine process considerations, most major verb functions are resolvable into smaller and smaller microscopic functions that the processors in the machines (knowledge, medical, educational, and computational) handle. From machine operand considerations, will most complex noun objects can be fragmented into smaller and smaller microobjects that the processors in the machines will handle. Now we have the same classic computational scenario to implement as knowledge, medical, educational, and computational machines. The combination machines are as feasible as the combined processors that handle arithmetical, logical, and graphical operations.

Forces in society propel knowledge-centric objects through the knowledge trail shown in Figures 2.5a and b. These objects do not have profound identities and firm boundaries. Human perception generally supplements enough details to anchor the ill-defined objects and move them along the left half (i.e., the D, I, and K nodes) of the knowledge trail. The path tends to become ill-defined and tedious in the right half (i.e., the K, C, W, and E nodes), and the new knowledge processors and machines supplement (1) the details from DDS or LoC classifications and their web addresses and (2) the rigor and persistence needed to bridge the right-half nodes. The need for such machines and the new algorithms becomes important because of the numerous type of objects and their behavior in response to verb functions and other objects. These intricacies of behavior of the noun objects and their classification are explored in the following sections.

4.1.2 Assistive, Neutral, and Resistive Objects

Objects have occurred in nature long before civilizations. It has become human nature to manipulate objects to satisfy needs and achieve goals. However, the efforts of humans to alter the objects are not always totally successful. A compromise (see Appendix B, Chapter 5) between the objects that exert the change and the objects that undergo the change, becomes essential.

Thus, the objects that undergo the change can be classified as assistive, neutral, or resistive (see Section 5.2.1) to the change depending on its nature (i.e., the verb function). Objects (especially other animate objects) have an attitude that is (1) willing to change a cooperate, (2) neutral toward the change, or (3) resistive to it. Generally, assistive and resistive attitudes are displayed by intelligent objects (including living organisms), and neutral objects will follow the laws of algebra, arithmetics or physics to reach an altered state after the change is complete.

For the most part, the willingness to change is based on the unsatiated needs, forces within, and aspirations of intelligent operands. The converse is equally true when the object, that is, the *kopr*, objects (strongly, mildly, or weakly) resist the change. Thus, a numerical range of changes develops [an emphatic verb function on a strongly willing noun object invoking a (+, +) response, to an emphatic verb function on a strongly unwilling noun object invoking a (+, −) response]. The numerous combinations and responses are tabulated in Table 5.2 in Chapter 5.

4.2 FRAMEWORK OF KNOWLEDGE

Traditionally, human knowledge and wisdom reside in the mind. Thoughts and contemplations manipulate, manage, and mobilize knowledge and enter the domain of concepts to become axioms of wisdom to master the art of daily life. Like humans, intelligent machines having the capacity to manipulate, manage, and mobilize (MM&M) knowledge-bearing objects can also optimally perform the fine art of science in solving routine scientific and social problems. Furthermore, in emulating a human posture, knowledge machines also command, control, and coordinate (CC&C) knowledge-centric objects much like managers (see Figure 1.7) would do during the managerial activities in a corporation and institutions. However, machines need to be primed with object sets and (humanistic) tools to perform MM&M functions in conjunction with CC&C functions. The intersection space of the six functions appears to be a fertile ground for human or machine creativity. At a macroscopic level of aligning and orienting any scientific and social problem, the machine hardware, software,[2] firmware, and operators need to be defined to suit the scientific and cultural environment for the solution. At the microscopic level of executing individual knowledge instructions, the objects and object sets become operands *(k-operands)* and tools become the methodology for executing the operation code *(kopc)* in the knowledge processor unit (KPU) of a KM.

4.2.1 Hyper-Dimensionality of Knowledge

When every classification from the DDS or LoC system is mapped to a direction in the knowledge space, the "body of knowledge" starts to become hyper-dimensional. The processing of information by the human mind has the freedom and capacity to hop freely between directions at will and also control the length of each hop. Emulation of such grace and beauty in the machine domain does not appear feasible at present. However, constrained movements may be

---

[2]This can extend into the actual program code as the declaration, dimensioning, grouping, statements, etc., of the objects.

emulated and realized. Much as an airplane cannot totally imitate the flight of every bird, the machine cannot imitate every graceful movement of objects in the human mind. Blending of truth, virtue, and grace in the manipulation of objects in a single pass of the machine appears impossible at this stage. However, the coordinated, rational, and logistic displacement of noun objects and global, goal-oriented, and optimal distribution of associated verb functions may give the machine a complementary role (almost an edge) in solving complex engineering and social problems. After all, a spacecraft flies where no bird has flown before.

The hyper-dimensional space of human thought soon becomes fragmented but an interlinked spaces in the memory banks of machines. When the forward and backward pointers are provided and preserved, recursive entries and exits in these interlinked spaces are distinctly possible for the machine to reiterate its deductive reasoning, just as human beings would weigh, consider, and reconsider their thoughts and contemplate their decisions. The responsibility of preventing circular and self-trapping loops becomes the responsibility of knowledge domain programmers. In configuring operating systems for traditional computers, software designers (aim to) prevent circular loops, deadlocks, input/output (I/O) incompatibilities, viruses, etc.

4.2.2 Intertwined Spaces of Knowledge

When the bodies of knowledge overlap, union and intersected subspaces emerge. These subspaces can enclose objects that influence other bodies of knowledge and the stability of the overall global environment that hosts numerous objects. The influence of objects and their attributes needs representation as the machine manipulates objects and their attributes during the solution of knowledge problems. In essence, the memories of machine should be designed to hold objects, attributes, and their related objects (as cross-linked objects) to whom they bear a relationship.

The three-dimensional Cartesian space has little use in the representation of knowledge. As the domain of knowledge expands to include newer objects, relationships, and dependencies, the knowledge space needs to be modified by readjusting and realigning axes and even their superposition. In a sense, the incremental knowledge may indeed modify the previous body of knowledge and displace the origin of the older coordinate system. For example, when relativistic physics was introduced, corrections to the classic knowledge and concepts in physics needed to be enhanced. As another example, when the truth about the Watergate tapes became known, the role of the Nixou White House in the coverup required revision.

To discover a hypothetical space for knowledge and concepts, the notion of connectivity between hyper-dimensional, intersecting, overlapping, and curvy spaces needs a mathematical treatment. If the mathematical operations appear impossible at this stage, then the software to lead a machine through such an irregular space becomes necessary. In this knowledge space, *microcosmic* mathematical operations [such as localized partial differentiations, localized counterintegrations, localized object dynamics (for displacements, velocities, accelerations, and forces)] are feasible (see Figure 1.3).

4.2.3 Contours of Knowledge

In the physical world, civil and structural engineers have discovered (or derived the fact) that the integrated horizontal and vertical components of forces need to be balanced just as action and reaction need to be equal and opposite to maintain the equilibrium of any structure. This principle is verified by constructing the funicular polygon [3] to verify if there is any residual force that is not compensated. Such residual forces yield slow and insidious instability for the structure. The concept of evaluating the destabilizing forces on "objects" in the knowledge domain is examined in the knowledge domain.

In order to determine the stability of objects in the knowledge space, the closed contour integral of the social forces acting on an object or object group should be zero.[3] Objects within society have bonds and structures of their own. They are subject to stability, partial stability, and even collapse [3]. Objects and object groups in society can have inertia, friction, and stick-sion effects. The response to social forces could be positive neutral or negative. Object respond accordingly with acceleration, momentum, and displacement for each of the forces if the objects are positively predisposed to forces. The opposite effect occurs for objects with negative predisposition toward such forces, causing residual tension and friction in society.

In the knowledge space when objects undergo forces in society, they tend to move swing, vascillate, or even vibrate. For example, if a force of expense is not balanced by a force of income, the object (a human, corporation, nation, etc.) moves from solvency to debt over a period of time. As an another example, if the wisdom of a nation is not balanced by a sum of internally generated concepts, innovations, and/or imported axioms (know-how skills sets, literature, etc.), then the wisdom bases in the nation are consumed and the nation drifts into bankruptcy of knowledge, concepts, and wisdom. Numerous other examples of economic entities (GNP, production, spending, etc.), manpower, education, ethics, etc., also abide by this rather mundane and mechanistic rule for the dynamics and stability of objects.

4.2.4 Funicular Polygon of Concepts

In the knowledge space, if the funicular polygon of concepts does not close, the contribution of that particular body of knowledge (paper, report, book, movie, documentary, etc.) should equal the residual vector to close the polygon of concepts. This vector is generally hyper-dimensional since the coordinates at the beginning of knowledge (before processing the body of knowledge) and end

---

[3]This concept is borrowed from the theory of structures based on the closed/open contour of the funicular polygon, emphasizing that a residual horizontal and/or vertical force remains active on any structure if the polygon of forces is not closed.

of knowledge (after processing the body of knowledge) can occupy different knowledge (and subject) coordinates. This residual force contributes to the movement (progress) of the objects and is a result of the change in entropy of that particular body of knowledge. The magnitude of the movement depends on the bias (friction, stick-sion, predisposition) in society. Forces in society inducing change (i.e., verb function) and the mechanics of (noun) objects become computable.

Bodies of knowledge that have the most impact on society deliver or contribute a linear (highly directional) vector from the known into the unknown and establish a localized contour of knowledge at the end point of the vector. Such a contribution is a breakthrough, and it can be precisely spelled out. Even though such contributions are rare, classical examples (such as relativistic corrections, Maxwell's equations, notion of pi, Heisenberg's principle, unified gas laws, anti-slavery, etc.) have occurred. Quantum jumps in the knowledge domain do not generally follow an articulated game plan.

A totally closed funicular polygon of concepts is an exercise for the mind rather than a motivation. The beginning and end are the same in the hyper-dimensional space of knowledge. The motivation for the next thought is the open end of the polygon of concepts that a body of knowledge does not close till the polygon is closed. Concepts are thus born in the minds of those who dare to jump in and attempt to close an open end of the funicular polygon in the knowledge domain.

4.3 COMPILATION OF KNOWLEDGE

For a moment, consider a "body of knowledge" (BOK; e.g., a paper, book, document, etc.) as a "program" to convey a message. Next, consider the basic processes (lexical, syntactic, and semantic analysis) embedded in traditional compiling (Sections 4.2.1.1–4.2.1.4) of programs by computer systems to extract symbols and operators in programs. Now, superpose the principles of compilation on the body of knowledge by invoking the routine machine executable steps involved, but to extract the "knowledge" (noun objects, verb function, and structure) encoded in that body of knowledge.

Largely, the program structure of a well-written software module is reflected in a well-organized document accompanying the module. Both have their own parallel structures. One basic difference is that most documents written for human use are poorly structured, but the human mind is forgiving and adept to extract the message from the document. Such a process is also akin to the human filtering of poorly structured audio signals to recover the message from poor-quality audio signal waves. The practice of message filtering by human senses is evolutionary, even though the principles are recently documented. The process of knowledge filtering by machines is not sufficiently evolved to extract the genesis of knowledge from poorly written documents by and for human beings.

Whereas the science of compilation (of computer programs) is well documented, the science of compilation (of knowledge from message-carrying documents) is still to follow. For initiating a platform to compile knowledge, we propose that *BOKs* be subjected to the same rigorous steps in the compilation of programs written in any standard computer language.

4.3.1 Lexical Analysis for Knowledge

As the first function in compilation, lexical analysis permits the machine to scan and overview the program for permissible key elements and the authentic operations on and among such key elements. Parsing is a required function in lexical analysis. It permits the computer system to identify and classify the key elements (noun objects) embedded in the program (BOK) and authenticates that the elements in the program are indeed valid and recognized by the system (from a dictionary). Traditional operators ($+$, $-$, $\times$, $\div$, etc.,) are recognized and tabulated for syntactic analysis to verify that such operations requested in the program are indeed permitted in the language of the program.

Computer programs are well-structured BOKs written in any standardized language that can be compiled. In processing, generic BOKs are also programs, but the syntax and semantics are far more flexible. They can be seriously error-prone. However, if the parsing can be relaxed to tolerate the loose structure of generic language and filter out its flexibility, it becomes a multilevel process (presented next) to identify noun objects and verb functions first, and then interpret the relationships that identify the structure of knowledge as a graph.

4.3.2 Parsing Bodies of Knowledge

Parsing for symbols and tokens is a standardized procedure in compilation. However, in dealing with complex objects and BOKs, parsing for (noun) objects, relationships, their attributes, and operators tied to the objects becomes essential. Operators are equivalent to verb functions that operate on noun objects to alter the entropy of the object or object group. The function is equivalent to the execution of a operation code on an operand. When the operation is an executable operation on the operands [i.e., not a no-op (NOP) or pseudo-operation], the result is computed and stored at a preassigned location.

4.3.2.1 Parsing for Noun Objects
Noun objects (NOs) should be considered as dimensioned symbols in typical computer systems. The numerous dimensions should accommodate a unique or group address for the object, attributes, qualities, linkages, pointers, precedent, antecedent and following object addresses, location and identification of the object location in any knowledge tree that holds the noun objects, etc. It is indeed a complete digital image of the object.[4] Implied is that a blank digital image defines an object with no physical or virtual coordinates. There are at least six types of related parsing for noun objects:

---

[4]This type of digital image is used by security agents to hold and store the images of ordinary police suspects, members of organized crime, terrorism suspects, etc.

158   VERB FUNCTIONS AND NOUN OBJECTS

1. Parsing for information-bearing noun objects (Section 3.2.1)
2. Parsing for qualities of noun objects:
   (a) Nature being positive, neutral, or negative to move toward A, M, or O types of wisdom
   (b) Reactivity being intelligent, linear, or nonlinear
   (c) Intelligence to verb functions by being assistive, neutral, or resistive to verb functions (see Sections 3.3.1–3.3.3; Chapter 5 further discusses the nature of noun objects)
3. Parsing for relationships (and linkages) between noun objects (Section 1.3; see Figure 1.1)
4. Parsing for attributes of each noun object (Section 1.3.1)
5. Parsing for relationships between attributes of each noun object (Section 1.3.1)
6. Searching for related objects, attributes, and their relationships, of associated (derived and/or dependent) local and Internet bases

The first pass of parsing for noun objects is very passive. In any body of knowledge, the function is merely to identify the noun-objects and distinguish them from nonobjects. The location of the objects in each sentence and its role are determined in the context of the structure of the sentence. The location and role of each sentence are determined in the context of the structure of the paragraph, etc., until reaching the highest node to evaluate the context of the body of knowledge in the context of the universal and accepted domain and dimensions of knowledge. An entire array of unidentifiable noun objects in any body of knowledge (in any language) deems such a body of knowledge gibberish.

4.3.2.2   Parsing for Verb Functions   The parsing for verb functions (VFs) can also be equally detailed. Two fold equivalency is evident: verb in the sentence structure, and operator in the CPU. Verb functions being equivalent to operators (arithmetic, logical, matrix, complex, etc.) need scanning and the context in which the operations take place. For example, if the machine is to search for a rare ailment in the human species, then the verb function (*kopc*) is the "search," the noun object (*koperand*) the "human species," and the verb modifier (specialized *kopc* from the set of search *kopcs*) a "rare ailment." All three parameters become instrumental in obtaining the correct answer from the search engine I/O device of the knowledge machine. Such a search process needs to be identified and tagged to the knowledge instructions. Processes of this nature are common in the computation of data (within the CPU) when operators and operands have to be consistent with the capacity of the machines (i.e., integer operations on integer variables, floating point operation on floating point variables, matrix operations on arrays, etc.) to execute an instruction. Typically, the parsing for the verb functions has at least six variations:

1. Parsing for operative verb functions (Section 3.2.2)
2. Parsing for the nature and intensity of verb functions
   (a) Nature being positive, neutral, or negative (+, 0, −) to move toward absolute, materialistic, or opportunistic types of concept and/or wisdom
   (b) Intensity being strong, mild, or weak (Section 3.2.2),
3. Parsing for tense (past, present, or future) of each verb function
4. Parsing for conditional, dependent, and linked verb function
5. Parsing for relationships between the nature and tense of each verb function
6. Searching for the related verb functions of associated (derived and/or dependent) local and Internet bases, and their nature and inter-relationships

The first pass of parsing for verb functions is very passive. For any given body of knowledge, BOK, the function is merely to identify the verbs and actions and distinguish them from nonaction-oriented verbs. The location of the verbs in each sentence and its role are determined in the context of the structure of the sentence and of the noun objects identified in the previous parse. Such a parsing of operators including their relationship with symbols is standard procedure in the compilation of a computer program. An entire array of unidentifiable verb functions in any body of knowledge (in any language) deems such a body of knowledge to be as challenging as a sentence without a verb. Furthermore, unidentifiable verb functions in conjunction with unidentifiable noun objects deem the body of knowledge to be challenging gibberish or like listening to an alien conversation.

4.3.2.3 Parsing for Relationships of Noun Objects Relationships between noun objects and their attributes give structure to knowledge. The contour of any body of knowledge is composed of the peaks and valleys of relationships. Lateral and hierarchical relations offer horizontal and vertical dimensions to offer a snapshot of the profile and the expanse of its contents. However, the hyper-dimensionality of knowledge forces the contour, profile, and expanse to become a multidimensional and complex object. Such objects are conceived by the mind much faster than by a machine. Much like computer imaging software that can present pictures by filtering or enhancing the light of certain wavelengths, the knowledge-processing software can present images in any Library of Congress (LoC) or Dewey Decimal System (DDS) subject classification, or any other classification such as the artistic, literary, or even emotional content. Such knowledge-processing software is as probable now as image-processing software was during the days of von Neumann; it is just four decades too late.

Lateral and hierarchical relations between objects (and their attributes) form the basis for a graph embedded in the body of knowledge. Largely, the graph is the graphical or symbolic representation of the structure and stability of concepts embedded in knowledge. Perfect knowledge will have eternal and indestructible concepts (and even ethical values) within it. In reality, the machine (and human beings) will learn to processes incremental concepts from imperfect knowledge for a limited timeframe. Primary concepts are relationships between primary noun objects related by primary verb functions giving rise to the main graph of knowledge. Branches and twigs in the graph result from secondary objects and attributes with weak relationships.

Parsing of relationships in the machine helps to construct a graph of concepts. Syntactic and semantic analysis will then be able to verify the validity of the graph in comparison with any identifiable concept structure from universal knowledge-concept bases around the Web addresses.

4.3.2.4 Parsing for Nature of Verb Functions As much as noun objects are related, the verb functions are also intertwined. Verbs indicate action. Different segments of action may be inter-related just as noun object may be intertwined. For example, if the action is to pass an examination (noun object), the examination may be in numerous subjects (related noun objects), and passing implies the practice of different approaches to pass the tests in different subjects. Hence, the execution of a task on a noun object can lead to numerous responses and results (see Section 3.3.1).

4.3.2.5 Extended Artificial Intelligence Strategies for Bodies of Knowledge Pattern recognition (PR) is a recognized branch of AI, and compilation is a systematic procedure in identifying symbols and operators. In the language of knowledge, objects take over the role of symbols and verb functions (individual or grouped) take over the role of operators. The grammar for the documentation of knowledge, being more complex than that of computer languages, offers a large variety of operators and their syntactic variations. However, a thread of conceptual commonality exists. The recognition process (from PR) and identification of symbols (from compiler theory) overlap. Both recognition and identification are intensive disciplines calling for accurate processing and algorithmic refinement.

If parsing, lexical, syntactic, and semantic analyses are placed in an inner shell of the flow-graph of a compiler for a computer environment, then statistical, syntactic, and semantic PR forms an outer shell for the flowchart of recognition procedure in a complex visual system. Now, if the two shells are collapsed into one, then a knowledge machine (KM) can be the software recipient to compile and "visualize" the BOK under process that passes through the KM.

The software integration of the two shells needs to satisfy the goals of the KM per se. Between the I node and K node in Figure 3.6, the KM software serves as the compiler–visualizer of (information-bearing) objects buried in information and the relationships between such objects. Between the K node and C node in Figure 3.6, the KM and CM (concept machine) serve as a compiler–visualizer of (knowledge-bearing) notions buried in the knowledge and the relationships between such notions, etc. The complexity of the compiler–visualizer becomes more and more severe as its functionality is customized for functions at the seventh and eighth nodes of Figure 3.13.

If the goal of the knowledge trail is movement toward Aristotle's (type I) ideals, that is, truth, virtue, and beauty (TVB), then the goals themselves become the rewards. Conversely, if the goal of the knowledge trail is movement toward Aristotle's (type II) objectives, that is, deception, aggression, and hate (DAH), then the goals become a self-consuming cannibalistic trap. Both destinations are manifest in society. The KM becomes a means of the mind to an end for the soul. The writings of Carl Jung point to similar notions about modern man in search of a soul.

4.3.3 Syntactic Analysis of Knowledge

Syntactic analysis in data-processing systems is essential to preventing the CPU from trying to execute impossible or ill-structured instructions. In the same vein, knowledge operations (*kopcs*) in the knowledge domain should be executable in the KPU in the context of the knowledge operands (*koprs*) that will undergo operation. As for the context of the *kopc* requesting any particular knowledge instruction, the instruction should be a legal operation that the KPU can execute, thus making syntactic analysis necessary.

For example, if a KM is to construct a graph of objects and relations from a body of knowledge, then the type and format of the body of knowledge need declaration and must be acquired in memory locations from an input source. As a precursor to actual execution, the machine may instruct the extraction of noun objects and their attributes together with the relationships between such objects, their attributes, and the relationships between attributes of attributes. Elements of the graph, such as branches, twigs, and terminal objects, also need identification and labeling, so that a complete graph may be constructed.

Inconsistent and incorrect modules are tagged for the semantic analysis to verify the validity and appropriateness of the module of knowledge. In a sense, every module of information and knowledge *(I «» K)* is thoroughly validated syntactically and filtered for consistency with the structure of knowledge from the local and Internet knowledge bases.

4.3.4 Semantic Analysis for Knowledge

The precise meaning of the noun object can vary from one occurrence to the next. The appropriate meaning will be identified in the local and global context of the information that is being communicated during this phase of compilation. The true identification of the meaning in the context of its usage also becomes essential. The human mind is well adept at this type of filtering that blocks the unsuitable meaning of any noun object to clutter the overall message in the body of knowledge. In a sense, the KM finds the embedded relationships to other noun objects and their attributes as it evaluates which meaning bears the most correlation. The check can be completed if certain combinations are already declared acceptable, questionable, or denied in accordance with the rules of the particular context. Knowledge compilers can indeed be written for science-, fiction-, business-, and economics-oriented bodies of knowledge.

162 VERB FUNCTIONS AND NOUN OBJECTS

Such compilers are frequent in traditional computer systems. Fortran, Algol, C*, Prolog, and graphics compilers exist. When the application is extensive, specialized application programs are developed. Graphical user interfaces (GUIs) further facilitate the flexibility and usage of such programs. The uniqueness of the knowledge-level compilers is that they unify the older object-oriented compilers to address social and human problems with passive and active objects that may have linear and nonlinear responses. When such compilers generate executable code for KMs with KPUs capable of handling numerous objects, related objects, their attributes, and their relationships, then the KMs can perform comparably to any traditional mainframe computer system.

A sense of reality and energy balance between noun object and verb function is implied in the semantic relationships between objects and verbs. Reality and energy balance have a numerical relationship. For example, an ant moving a mountain forces a fictional nature or highly improbable reality, whereas a machine moving a mound of snow becomes readily acceptable as a reality. In the realistic directions of sciences, the semantic analysis of knowledge programs "weighs and considers" [4] the energy balance of operations. The compiler thus has the potential ability to tag fictitious (or highly improbable) operations in a realistic scientific program and vice versa. Other semantic violations are also tagged by the compiler. When Web knowledge bases are available, the compiler has the option of looking up the particular knowledge base to assign a confidence level accepting or denying the validity of any proposed knowledge function. In a sense, the compiler acts as a reference to the proposed knowledge function within a program. Similar lookup procedures are used in Mycin [5] and NeoMycin [6] systems in the medical field.

This implication invokes a sense of realism (or possibility) or fiction (or impossibility) in the syntactic and semantic relations between noun objects and verb functions. For example, a module of knowledge (i.e., a sentence or paragraph) may represent reality or a virtuality. When reality is implied in scientific subjects (i.e., DDS classifications 500 and 600), the laws of sciences[5] (such as laws of dynamics, thermodynamics, and stability of structures, also derived from the same DDS classifications 500 and 600, etc.) are applicable, and verifying

---

[5]Subjects in the Dewey Decimal System (DDS) classification, are 000 Generalities, 001 Knowledge, 002 The book, 003 Systems, etc., 100 Philosophy & psychology, 101 Theory of philosophy, 102 Miscellany of philosophy, 103 Dictionaries of philosophy, etc., 200 Religion, 201 Philosophy of Christianity, 202 Miscellany of Christianity, 203 Dictionaries of Christianity, etc., 300 Social sciences, 301 Sociology & anthropology, 302 Social interaction, 303 Social processes, etc., 400 Language, 401 Philosophy & theory, 402 Miscellany, 403 Dictionaries & encyclopedias, etc., 500 Natural sciences & mathematics, 501 Philosophy & theory, 502 Miscellany, 503 Dictionaries & encyclopedias, 504 Not assigned or no longer used, etc., 600 Technology (Applied sciences), 601 Philosophy & theory, 602 Miscellany, 603 Dictionaries & encyclopedias, 604 Special topics, etc., 700 The arts, 701 Philosophy & theory, 702 Miscellany, 703 Dictionaries & encyclopedias, 704 Special topics, 705 Serial publications, 706 Organizations & management, etc., 800 Literature & rhetoric, 801 Philosophy & theory, 802 Miscellany, 803 Dictionaries & encyclopedias, 804 Not assigned or no longer used, etc., 900 Geography & history, 901 Philosophy & theory, 902 Miscellany, 903 Dictionaries & encyclopedias, 904 Collected accounts of events, etc.

syntactic rules make the module of knowledge syntactically appropriate with the other modules that constitute the rest of the module. Virtual operators (such as magic, witchcraft, demon-worship, etc.) on virtual objects (such as fairies, witches, demons, etc.) may be executable in the KPU environment to create a fictitious atmosphere of computer games and cock and bull entertainment.

4.3.5 Knowledge Machine Code

The actual machine code depends on the hardware capability of the KPU, the macro and subroutine libraries in the compiler. An extensive set of macros and subroutines is not written for the overall processing of knowledge programs. An efficient knowledge compiler will make the processing of human and social computer programs as dependable as the processing of scientific programs when the optimizing Fortran V compiler was introduced in the 1960s.

4.4 DERIVATION OF KNOWLEDGE

New knowledge can be derived from existing knowledge at *three* levels. At the *first* level, the execution of a simple or complex verb function ($kopc$) on one or more noun objects [$kopr(s)$] causes an incremental change in the entropy by altering the status of noun objects. The convolution of the VF's on NO's may also be altered in the KPU. These functionalities occur in the KPU hardware and, typically, these changes may be too small to offer new knowledge. However, when the functionalities are compounded and cascaded by sophisicated knowledge programs, the change in entropy of knowledge can be significant. The KPU has access to similar objects in the object, and an element of embedded intelligence can be implanted in the KPU itself to predict a massive change in the entropy if the objects are appropriately adjusted. However, a series of such assembled programs from the compilation of a knowledge program can produce new information from a BOK from which the noun objects are parsed.

At the *second* level, the execution of an entirely consistent (application) program on one coherent and self-contained BOK creates an increased change of entropy and transforms BOK to BOK'. This second-level functionality occurs in the machine hardware. All or some of object relationships, their attributes, their relationships, and attribute of attribute relationships would have undergone modifications, thus causing greater changes in entropy. The KM or WM has access to similar objects and BOKs, and artificially intelligent agents of embedded intelligence can be implanted in the machines to predict a massive change in entropy if the BOKs or objects are appropriately adjusted. As a next step, a series of such assembled application programs from the compilation of massive social and political actions can produce new and more beneficial social and political actions for a society from which numerous BOKs are drawn.

At the *third* or universal level, when an entire array of programs (dependent, quasi-independent, or independent) operate on a global collection of BOKs (e.g.,

164 VERB FUNCTIONS AND NOUN OBJECTS

the real world, a universe, or a society or nation), then the change of entropy in the knowledge of that universe can become unpredictable and even chaotic. Examples of such conditions occur in seismic eruptions (see Section 1.6.4), stock markets, or even nations undergoing severe dynamic conditions. Examples of situations in which the changes are controlled but nonchaotic include the migration of herds, evolution of species, natural erosion of land masses, etc.

In the following subsections, the representation of machine architecture to handle these three levels of change is discussed. The KPU configuration, and knowledge and wisdom machine configurations are explored next.

4.4.1 Microscopic or Basic Assembly-Level Knowledge Instructions

In basic assembly-level knowledge (BALK) instructions, a typical verb function (VF) with numerous knowledge operation codes (*kopcs*) can operate on a complex object BOK with numerous noun objects (NOs), operands or *koprs*. A typical KPU layout with data base support as shown in Figure 4.1. To verify the validity

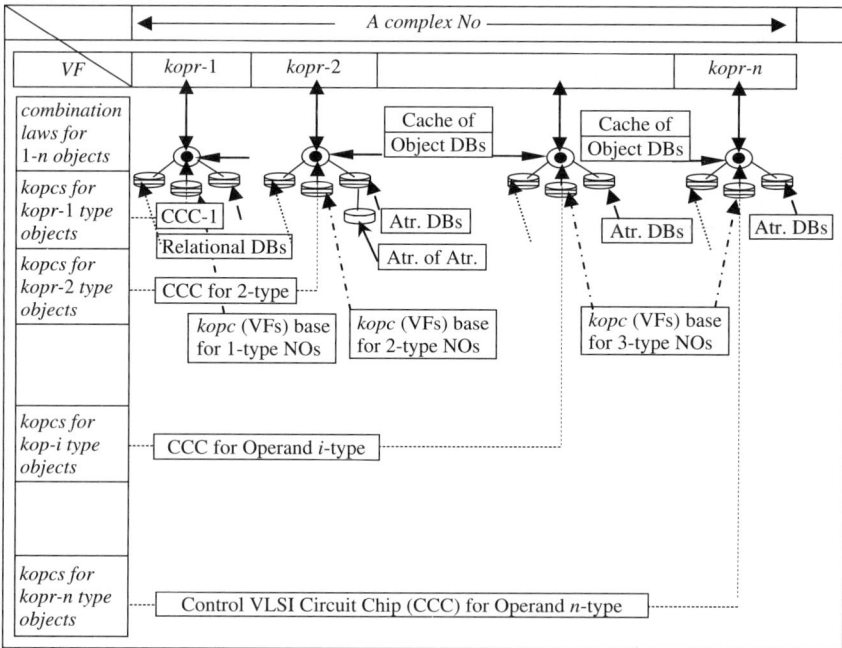

Figure 4.1 Logical configuration of the knowledge-processing unit to execute a standardized VF on a complex authenticated object. All the knowledge operation codes (*kopcs*) are microscopic verb functions, and all the knowledge operands *koprs* are microscopic noun objects. Only permissible verb functions can be executed on legitimate noun objects. The numerous control circuit chips now becomes the equivalent of many instruction registers and their control memories, in a MIMD or pipeline central processing unit architecture. CCC = control circuit chip.

of the operation, a database of permissible *kopcs* is provided on the left side of the figure. Verification of the match between *kopcs* and *koprs* becomes crucial in the KPU because of the dynamic nature of the objects. Operation code-operand mismatch can generate unpredictable results in the KPU and cause instability in the program's execution.

The databases of related objects, their attributes, and attributes of attributes also permit the changes that the *kopc* will bring about on the entire BOK. When the noun objects are classified by type of objects, the respective VFs can also be arranged to match. In Figure 4.2, the layout of a VLSI chip for the KPU is shown. The operand, relational, and attribute databases, and logic and control circuit space for the KPU, are shown on the right-hand side of the figure. If the databases are not large, the cache memories can be used instead of traditional data bases and the entire KPU can be accommodated on a single chip or wafer. There are two aspects to the decoding and execution of the *kopcs* shown on the left side. At the database level (extreme left of the illustration), a new and updated list of *kopcs* is generated from the local and Internet bases. The control memories to decode the *kopc* are shown in the center.

4.4.2 Application-Level Knowledge Programs

Application-level knowledge programs (ALKPs) operate on complex objects and have the necessary *kopcs*, macros, subroutines, and library functions to modify the entire structure and characteristics of objects, their attributes, etc., of the components of knowledge. The extent of modification or enhancement depends on the nature of the application programs.

An example of such application programs is available in the executive and management information systems (MISs) of a corporation. The simplest of such MIS programs is the report generation MIS that scans the complex object (i.e., the corporation) and reports the vital signs (such as the price/earning, liabilities/assets, balance sheets, etc.) of that object or entity. More complex MIS functions (such as corporate realignment, production scheduling, maximization of profits, etc.) require dynamic profiling of all the essential objects (noun objects) and entities (complex noun objects with relationships, attributes, etc.) in the corporation and the effects of altering executive policies (verb functions).

Other examples of complex entities occur in medicine, hospital management, and national and econometric modeling of numerous segments of the economy (such as consumption, savings and taxes, or industrial production, investment, and government spending, etc.). An architectural configuration of the corresponding knowledge machine is shown in Figure 4.3.

The applications of the knowledge machine are generic in almost all disciplines of scientific study where the dynamic equilibrium can be modeled by mathematical equations or statistical behavioral patterns. The knowledge machine has the final edge over traditional computing systems because of Web access, AI techniques, and predictive behavioral pattern synthesis that offer knowledge machines significant advantages over normal computers especially in social problems.

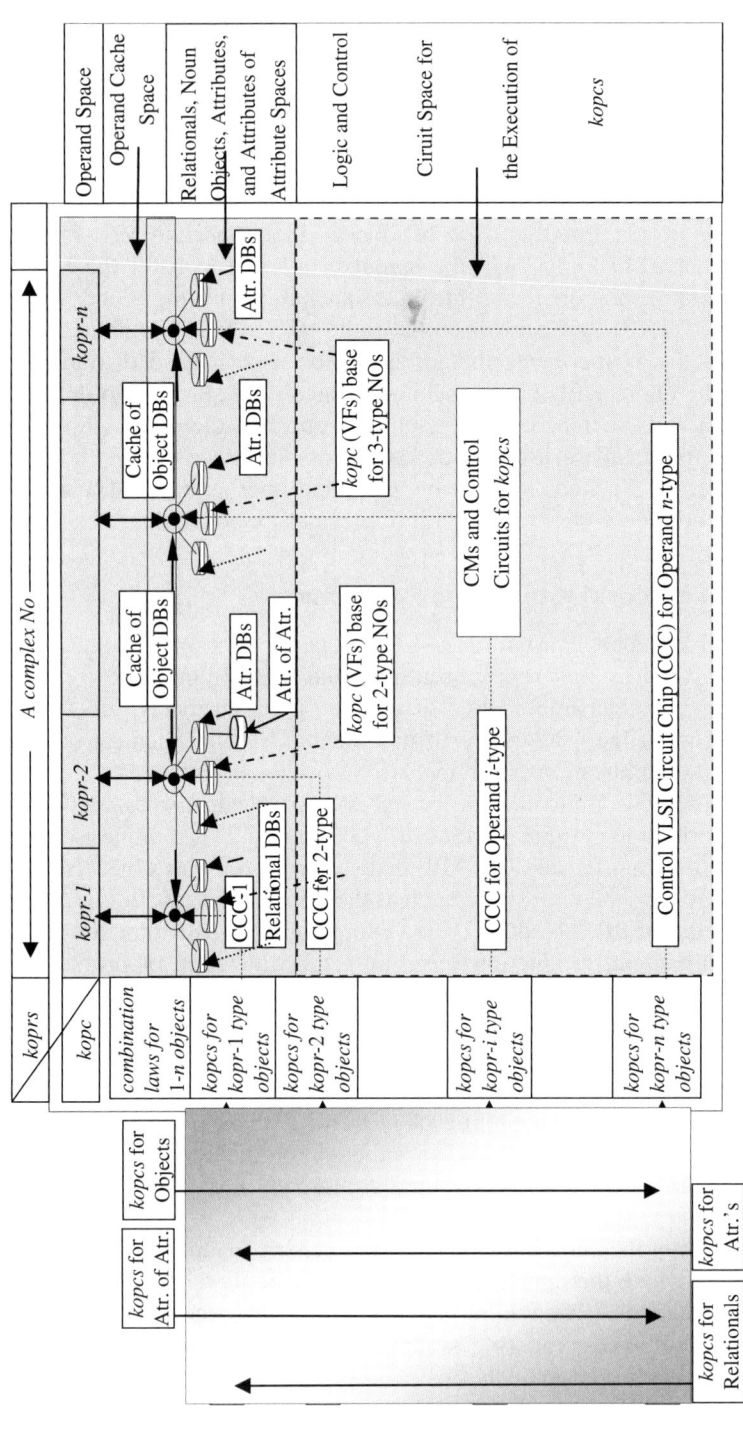

Figure 4.2 Spatial layout of the knowledge-processing unit to execute a standardized *kopcs* on an authenticated object. Note that the objects are knowledge-centric noun objects that occur in any body of knowledge. Specialized *kopcs* for different groups of noun objects are executed in the logic and control space of the unit. CCC = Control memories and control circuits. DB = Data bases, Atr = Attributes.

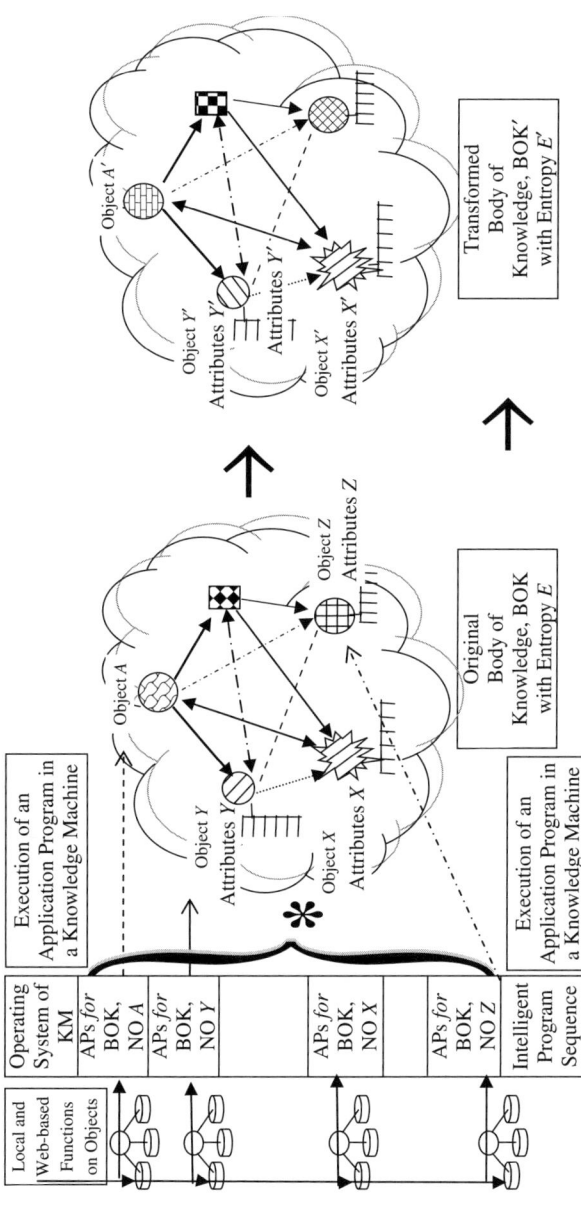

Figure 4.3 Extension of Figure 4.1 to application-level programs to change the structure and entropy (E) of a body of knowledge (BOK) to the derived entropy (E') of a newly computed body of knowledge (BOK'). The symbol * indicates a machine-based knowledge convolutions.

The synergy due to statistical, AI, and predictive pattern recognition techniques over and above the computational techniques (i.e., arithmetic, logical, matrix, and scientific) in the knowledge machine forces the machines to behave in a humanistic mode with humanistic characteristics in dealing with complex objects. When the rational side of the attributes of the objects in the complex BOK is supplemented by inclinations and charismatic choices, the humanistic functions and decision of any KM can be slanted to become almost human.

4.4.3 Universal-Level Knowledge Programs

These programs extend the operations of a knowledge machine further such that the machine is able to operate its knowledge functions on numerous complex objects in any universal setting and alter the entropy of the universe from where the complex objects are drawn. When the complex objects reach the boundaries of the universe of knowledge that encompasses them, the functions become unpredictable because of the truncated scientific, statistical, AI, or predictive capabilities of the machine. The machine can only come to a halt because the basic *kopc* in any KM is based on some prior knowledge about the objects, its related objects, its attributes, etc.

A typical configuration of such a machine is shown in Figure 4.4. If a machine could possibly track the transition between the two universes depicted in Figure 4.4, then the machine would have all the reasons for the evolution of the universe. A more localized and realistic version of a machine for discovering unknown diseases is shown in Figure 4.5. The noun objects (NOs), that is, human diseases in this case, are systematically mutated to find the existence of another hidden universe of an unknown disease. The symptoms of the patient are matched with the derived symptoms resulting form the mutated diseases, their relationships, the local objects, their attributes (ATs.), and the ATs of the ATs. In order to discover the order of the hidden disease, the symptoms and conditions of a suspected patient (with the unknown disease) and ailments and attributes of known diseases are processed by AI routines [5, 6] in the application programs in the KM or WM. The process is exhaustive and all the clues are self-tuned to maximize the probability of positive identification. The cure may also be synthesized by the KM or WM by tracking and mutating the generic medicines for the known diseases.

However, when the knowledge is minimal or inconclusive, the machine does have the ability to propose a hypothesis (see the section on WMs) and test its validity. This type of "educated guessing" is practiced by scientists based on intuition or hunch [7]. Like a human being, the machine can systematically override the rational or logical blocks just to explore the unknown universe or objects in it that do not abide by our rationalities

These functions that are unique to the WM may become initial machine-generated attempts to second guess knowledge where none exists. The human element of an WM is absent here since human interference can impede the semi-logical or exhaustive search that the machine is capable of performing.

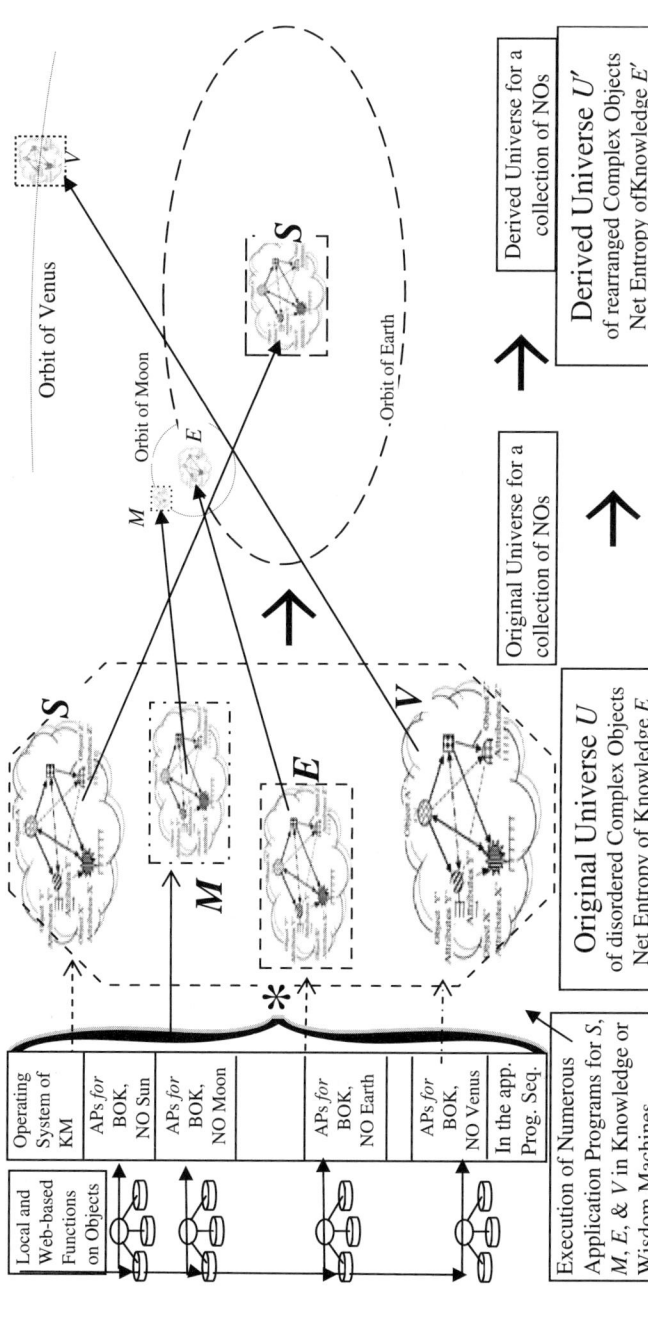

Figure 4.4 Extension of Figure 4.3 to global-level programs to change the structure and entropy (E) of a complex unbalanced universe U consisting of numerous knowledge-centric objects with an entropy level E to the derived level of entropy (E') of a newly balanced and computed universe for derived knowledge-centric objects in a different state of transition, stability or equilibrium. The symbol * indicates a machine-based knowledge convolution. S, M, E and V stand for sun, moon, earth and venus.

169

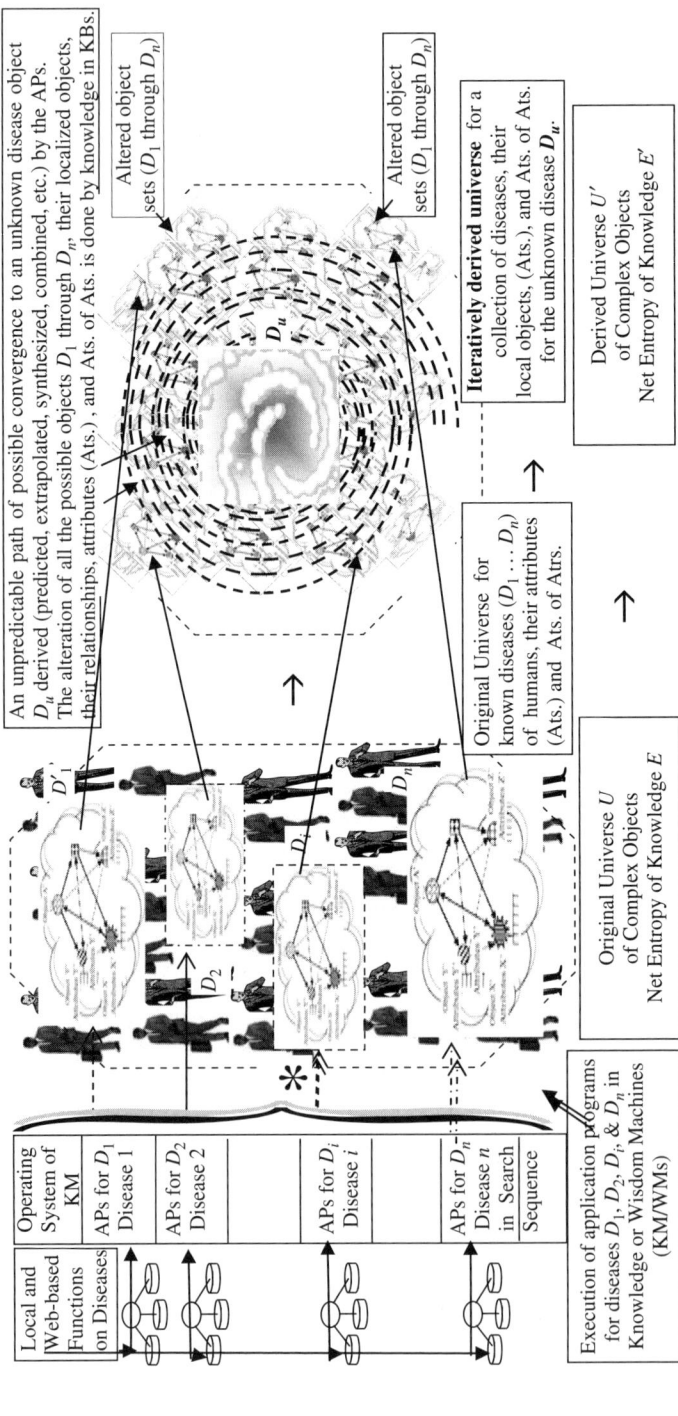

Figure 4.5 Extension of Figures 4.3 and 4.4 to comprehensively isolate (i.e., analyze, analogize, conceptualize, generalize, rationalize, synthesize, systematize, etc.; see Section 3.1.2) an unknown disease D_u from the knowledge bases of the known diseases $D_1\ldots D_n$. The path to D_u may involve intuition and creativity of the human mind. Human control of the functions is an essential part of knowledge and wisdom machine operation. Ats. = attributes of any disease D_i.

The role of intelligent agents in exploring or deriving universal knowledge has yet to be investigated. When the entropy of knowledge is low, the machine's ability to be both instigative and intuitive or "institutive" needs a numerical estimation. Predicting a probability the machine will uncover new knowledge based on no knowledge, is an exhaustive search. Every logical block in the direction of the search needs to be overridden. In a sense, the reason to block a search should become a reason to search. Interpolating the directionality of the search is based on the machine's sensitivity to break down the DDS or LoC system into finer steps. When the machine reaches it own numerical level of estimation of direction and probability, the search for new knowledge can be terminated. In a human environment, such initial limits are found only to be broken. The creativity of human beings still wins and a human intervention would be in order. Under these circumstances, the machine for universal-level knowledge becomes a super-sensitive wisdom machine with the capacity to override its own knowledge programs.

4.5 KNOWLEDGE MACHINE SOFTWARE HIERARCHY

The knowledge machine needs the traditional attire of software to clothe the bare machine. Being supported by traditional computer systems, the knowledge machine primitive software can be uploaded by the lower-level "slave" systems. At the lowest level of knowledge functions, the machine code for loading the knowledge control memories (KCMs, if any), *kopcs, koprs*, fetch, store, linkage loaders, etc., needs to be primed into the KPU.

The initial structure of the software is suggested in Figure 4.6. Instructions at the most primitive level provide the capacity for the machine to perform the elementary *kopcs* (e.g., OROR object register-object register, single-object fetch, single-object store, etc.) that are cascaded to perform complex *kopcs*. The utilities and libraries fall at levels 2 and 3, and different sets of libraries will become necessary based on the application. Knowledge machines like medical computers or educational systems will have applications in the social domain and serve as intelligent robots tailored to the individual needs of the user.

Individual human needs are well articulated by Maslow [8] and extended by Ahamed [9], and specialized libraries to meet such needs ranging from the most primitive (physiological and safety needs) to the most sophisticated (search and unification needs) can be evolved. In traditional computing environments, we have routines ranging from the trivial (arithmetic and logic routines) to the sophisticated (fast Fourier transform and complex matrix inversion routines).

A platform of traditional computer systems become desirable to build a full-fledged knowledge machine. The integrated hardware and software integration of a traditional computer is shown in Figure 4.7a. The layers of software can vary substantially between machines customized for different applications. Central or distributed processing environments are feasible. The input/output devices and processors can vary radically. The CPU can handle

172 VERB FUNCTIONS AND NOUN OBJECTS

numerous types of (single data, SD; multiple data, MD; or pipelined) data streams and execute single or multiple (SI or MI) instructions, etc.

4.6 KNOWLEDGE HARDWARE AND SOFTWARE SYSTEMS

In traditional computers, the building blocks [CPU(s), memory(ies), I/O subsystem(s), bus structure(s), and switch(es)] prevail in almost all general-purpose machines.

Specialized computers may have additional elements such as sensors, robotic control devices, digital signal processors, etc. In general, specialized architectures are more expensive and demand additional interfacing software. In order to execute higher-level knowledge-level programs, as shown in Figure 4.6, the

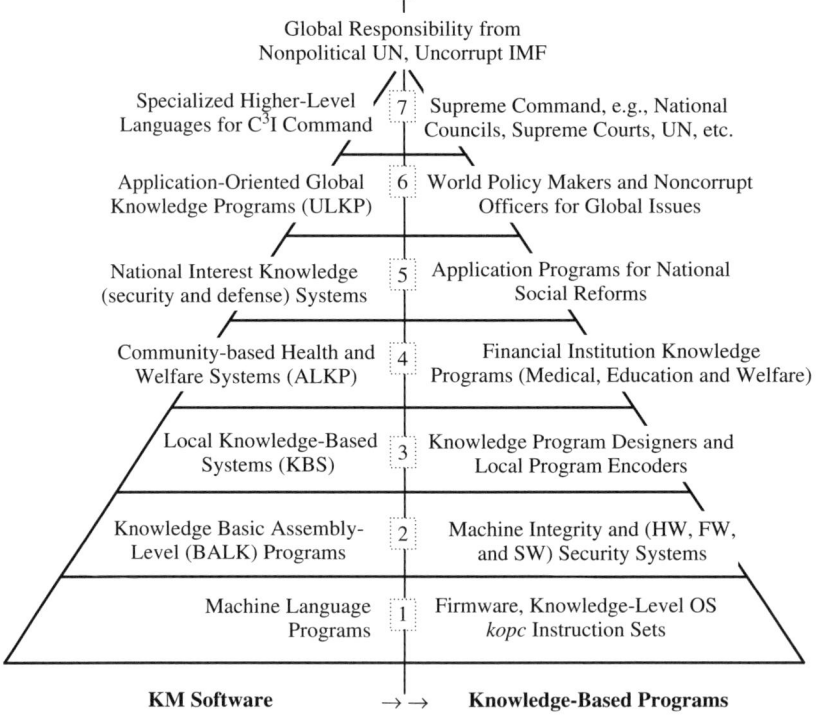

Figure 4.6 Software hierarchy for a knowledge machine. This hierarchy is similar to the software hierarchy for traditional computer systems. It permits program writers to use the knowledge machines to resolve social, human, local, and global issues. BALK = basic assembly-level instructions for KPU-based systems, KOPC = knowledge operation code for the KPU, KBS = knowledge-based systems, ALKP = application-level knowledge programs, ULKP = universal-level knowledge programs (see Section 4.4), C^3I = command, control, and coordinate with intelligence, UN = United Nations, IMF = International Monetary Fund.

inner layers of the knowledge ware (KW) need some enhancement to make the knowledge machine consistent with the capacities of a knowledge processor unit. Knowledge machines can also be built to abide by the multiplicity of KPU architectures. Some of the themes that can be readily enumerated are single-process single-object (SPSO), single-process multiple-object (SPMO), multiple-process single-object (MPSO), and multiple-process multiple-object (MPMO) designs. These variations are presented in [6].

The very inexpensive chip sets that constitute the traditional computer justify this platform. The hardware in the immediate vicinity of the KPU that is the bare KM hardware assumes a configuration as shown in Figure 8.17. Bus structures, object handlers, and object caches become complex but addressable and manageable. In Figure 4.7b, four computers in the platform act as slave machines and addressable I/O subsystems. The outer layers of the KM software function to act as compilers for the HLL knowledge programs, and as loaders and linkers for the executable knowledge code.

A four traditional computers make up the platform of a KM in Figure 4.7b. The functions of these four computers are depicted in Figure 4.8, and they, in turn, perform the following four essential functions in any complex knowledge based computational facility:

(a) The Dewey Decimal System (DDS) and/or Library of Congress (LoC) subject and content identification and address management system identify objects in the context of the knowledge program (compile program, step 1) and with reference to the knowledge bases in the local and global knowledge bases and libraries.

(b) The management of executable programs and libraries is more elaborate than that in traditional computers because numerous subprograms can be invoked as a result of the prior executable steps. Both compiler and interpreter features may be necessary to execute the more complex knowledge programs with human entities as objects embedded in the program.

(c) In terms of disk and database management, a series of secondary storage systems may to necessary to store (1) discipline based (DDS or LoC) objects and their attributes, (2) cross-linkages to other related objects and their attributes, and (3) precedent and descendant objects.

(d) Dynamic access management exists for objects, attributes, object linkages, and primary and secondary linkages in the problem solution and to further its verification and optimality.

Knowledge machines can act as object machines by removing some of their special knowledge-processing features and content-based addressing and processing capabilities. The differences between object processor units (OPUs) and KPUs are presented in Chapter 8. Knowledge machines can also act as wisdom machines by incorporating the traditional search for Aristotelian TVB, that is, truth (universality), virtue (social benevolence), and beauty (innate elegance and inherent balance) as the machine executes knowledge programs.

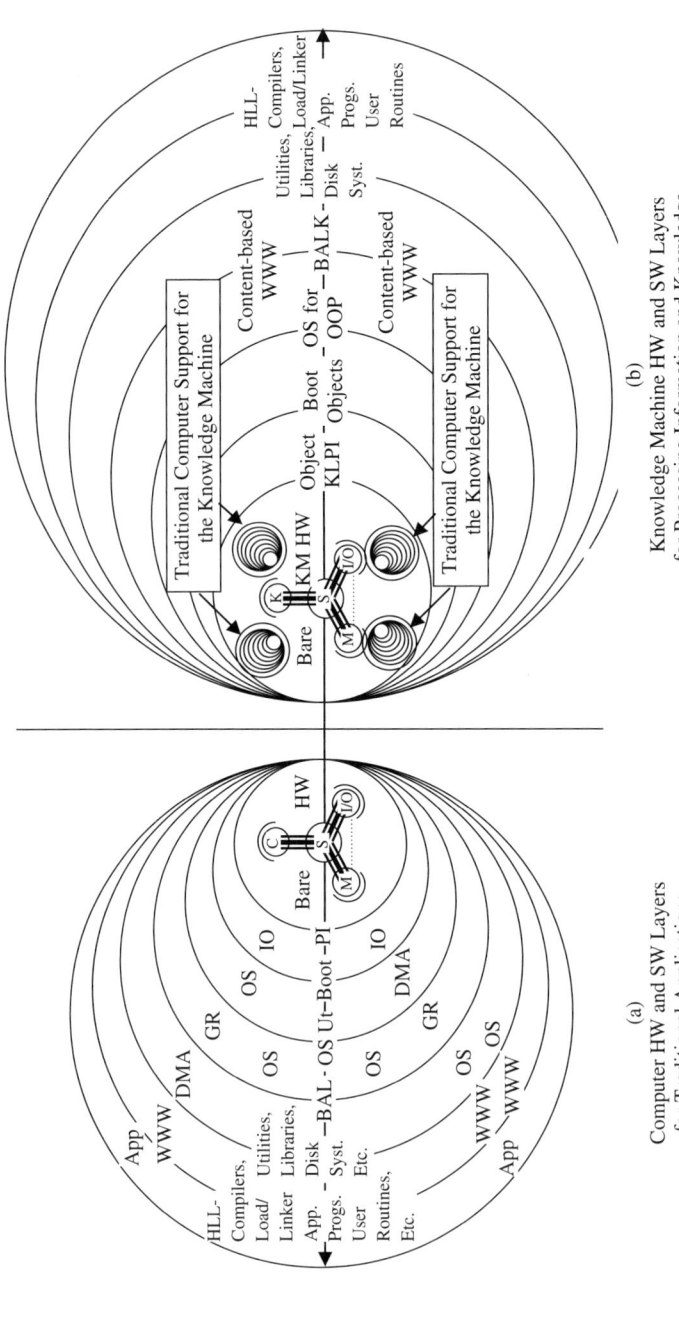

Figure 4.7 Design for the innermost software layers of a traditional computer and a knowledge machine. PI = privileged instruction, IO = input/output, DMA = direct memory access routines, Boot = bootstrap software, BAL = basic assembly-level assembler and SW, HLL = higher-level language compilers and SW, App = application-level programs, KL = knowledge level, OOP = object-oriented programs, K prefix/suffix designates knowledge-level functions.

174

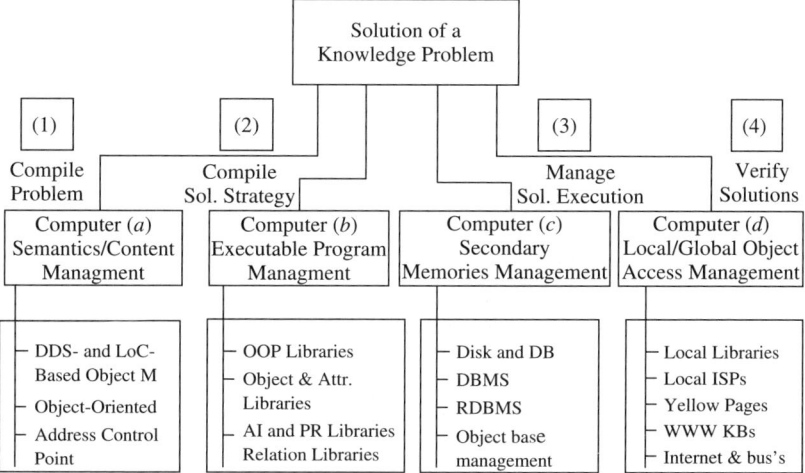

Figure 4.8 Partitioning of basic computational functions within a knowledge machine. Four essential steps are executed by the conventional computers shown in Figure 4.7. DB = database, KB = knowledge base, DDS = Dewey Decimal System, LoC = Library of Congress, OOP = object-oriented program, RDBMS = relational database management system.

For example, if the information filters presented in Chapter 6 seek out a perfect combination of TVB and also attempt to filter out every hidden deception, arrogance and hate (DAH) in any BOK, then the KM-based machines will struggle to make sense from every piece of information. Preprocessed input information will facilitate the final outcome from the KM. However, the convergence toward wisdom becomes faster if the information is processed at the K and C nodes before the W nodes along the knowledge trail depicted in Figure 2.5.

When the movement along the knowledge trail is not pursued vigorously and systematically, the purely philosophic discussions about information, knowledge, concepts, and wisdom nodes, that is, I, K, C, and W nodes, create a tangled web of thought. In the next section, we illustrate the movement from the B (binary) node of Figure 2.5 to the TVB node after the E node in the same figure based on the dictionary definitions of the seven nodes given in that same illustration.

4.7 CLASSICAL MIGRATION PATH OF KNOWLEDGE

The definitions for the terms "binary," "data," "information," "knowledge," "concept," "wisdom," and "ethics" are taken from the *American Heritage Dictionary* and appear in the boxes under each node. Figure 4.9 depicts the meanings of the words "binary" through "ethics" under each of the seven nodes. When the words are read casually, the mental path through all seven nodes is irregular and often circular.

The multiplicity of the meanings of each node is also shown. From a purely philosophic consideration of the meanings, the general directionality for the movement is clear because the meanings in the right-hand side boxes (mostly) refer to the terms in the left-hand side boxes, confirming the notion that the progression of thought processes in the knowledge trail (Figure 2.5 and 3.4) occurs from left to right.

As indicated in Figure 4.9, the lack of a theoretical framework for knowledge that is proposed in Chapter 3, can force the human thought process to become trapped in circular loops of philosophy and mental road blocks. Knowledge machines can traverse the knowledge trail based on the computation that an object derived from a prior node can satisfy the criterion for the next node. The methodology is derived from achieving a high confidence value (see Figures 3.8 and 3.9) that the new entity will pass the tests of prior scientific investigations of that particular object in that particular DDS or LoC direction. The methodology is highly specific.

Tracing the semantic linage as arrows leads to Figure 4.10. Numerous circular loops are also evident between the B and D nodes, around the K node, and between the E and TVB nodes. The arrows that lead to the K, C, and W nodes are external inputs in the knowledge trail arising from human intuition and creativity. The arrows going out of the I, K, and E nodes indicate the effect that information, knowledge, and ethics and values nodes have on society.

4.7.1 Knowledge Traps

In tracing the left to right movement of knowledge in Figure 4.10, four major traps (circular paths) are evident between the B and D nodes, at the K node, between the I and K nodes, and between the E and TVB nodes. Numerous jump-in points (K-4, C-2, 3, and W-3b) and jump-out points (I-6, K-5, and E-4,5,6) are also apparent. Long jumps (C-1, D-1, D-3, and W-1,3,4) between nodes also appear to result from the imprecise definitions of words such as "binary," "data," "data structures," "knowledge," "concepts," "wisdom," and "ethics." Historically, human language has carried greater ambiguity than computer programs written in a structured language.

In an effort to reach the goals of wisdom and ethics, ancient thinkers successfully avoided the knowledge traps, unsubstantiated shortcuts, jump-ins, and jump-outs and formulated classical theorems in mathematics, oracles of wisdom, and scriptures of truth, virtue, and beauty. In order to be scientific and instill rationality, we propose that the migration path between nodes be precise and programmable (see Chapter 3, Figures 3.5, 3.6. 3.8, and 3.9).

4.7.2 Transition at Generic Nodes

For the sake of completeness, we represent the transition between any $((i-1)$ to node $i)$ in Figure 4.11. The major findings in any particular discipline are denoted in the minor node between $(i-1)$ and i. The knowledge remains trapped in

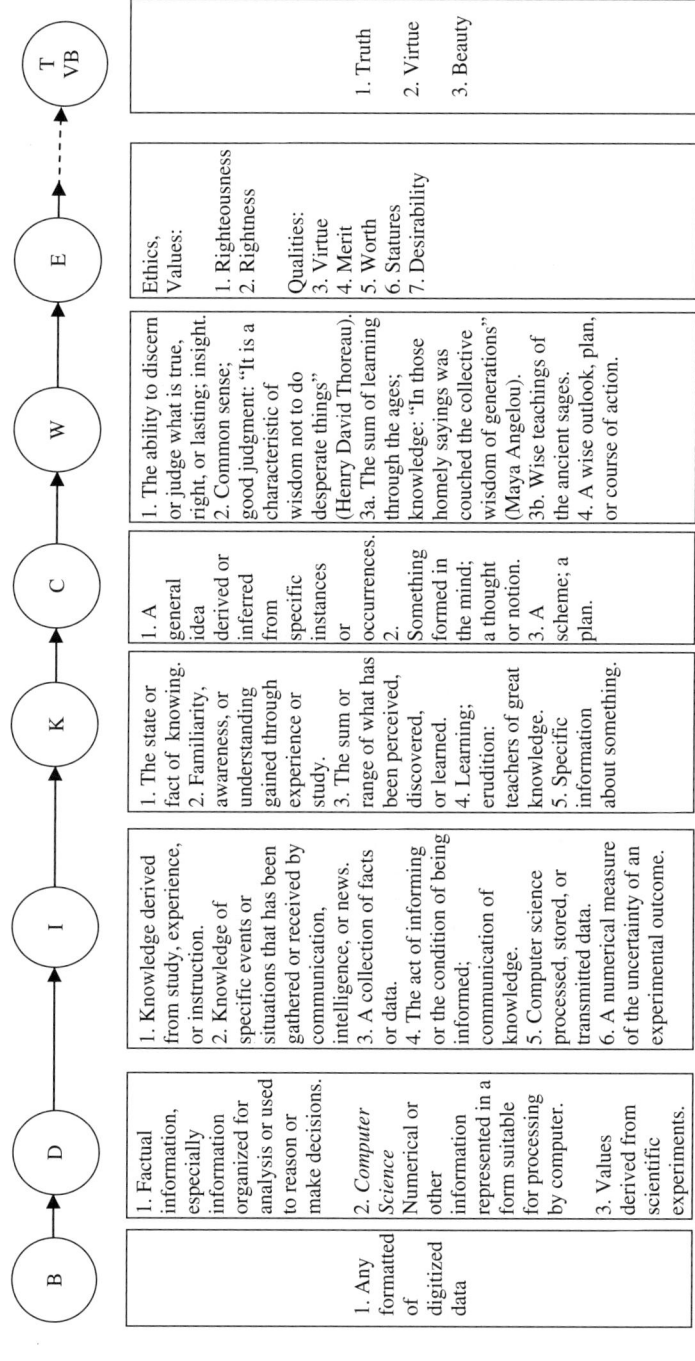

Figure 4.9 The meaning of the words "binary," "data," "information," "knowledge," "concept," "wisdom" and "ethics" are shown in the boxes under each node (and taken from the *American Heritage Dictionary*). When the common words are shown as lines with forward and backward, activity then a messy diagram as shown in Figure 4.10 is generated.

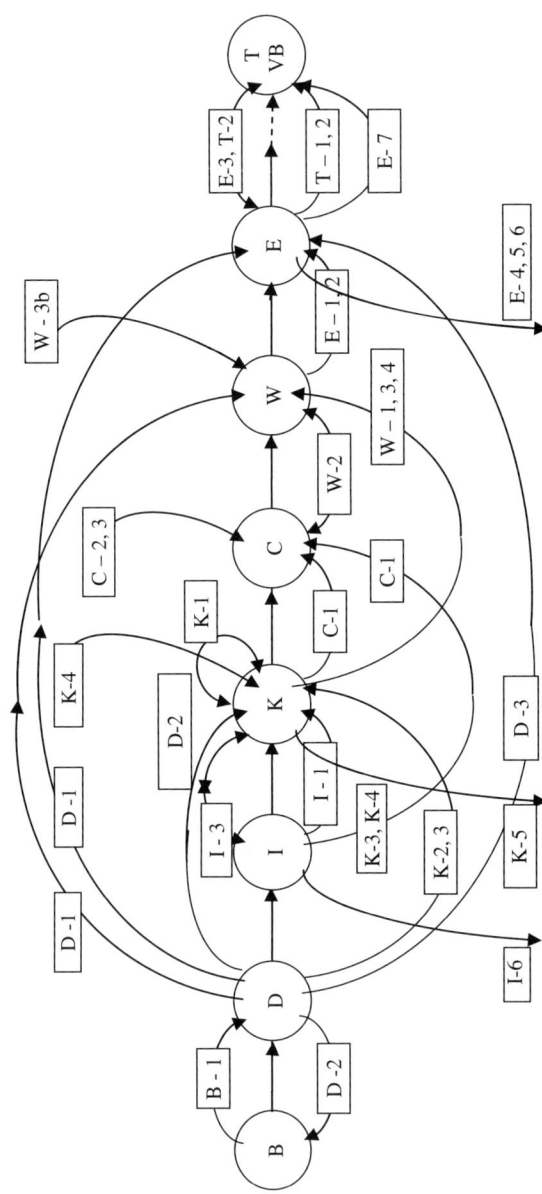

Figure 4.10 When the commonality of the words is replaced by arrows in the knowledge trail, it becomes unduly complicated and at least three circular loops near the B,D; I,K; and E,TVB, etc., nodes are formed. These loops depict the jagged path of society to reach its mature and stable state, in spite of the closed loops that block freedom of thought in many instances.

178

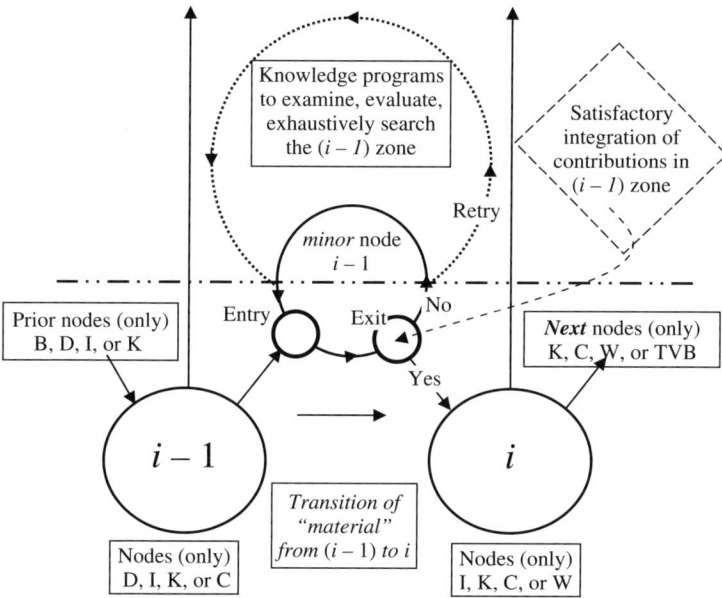

Figure 4.11 The discussions about any "subject matter" are replaced by tests for the D, I, K, or C material to be satisfactorily supported by all the current research in that subject matter. The ultimate condition for transition lies in the proof of the D, I, K, or C material and can be confirmed by the current research on that topic in the transition zone.

this node until a scientific methodology is established to move from $(i - 1)$ to node i. The condition occurs if and only when the cumulative findings of all scientific contributions satisfactorily confirm that "information" can be classified as "knowledge," that "knowledge" can be classified as "concept," that "concept" can be classified as "wisdom," or that "wisdom" can be classified as "ethics." All the rigors of mathematics and methodologies of sciences are applicable to make the transition.

When the procedure of Figure 4.11 is streamlined and the iterations for the research are complete, then the diagram for the transitions for all the nodes converges to a figure like the one shown in Figure 3.8. Ideally, knowledge machines should be able to perform all the functions of conventional object computers and information-processing systems. Application-oriented designs will yield efficient performance and optimal solutions.

4.8 CONCLUSIONS

In this chapter, the basic concept of the operator (a verb function) operating on a generic entity (a noun object) is introduced as an incremental convolution. The result is the effect of a single component of a verb when it interacts with a single element of a noun object. Numerous such results are generated and they ac cumulate. The effect is an alteration of the state of the object, its attributes, and its linkages for future use. A machine-based knowledge convolution is the cumulative effect of such incremental convolutions on the entire object. The integrated incremental convolution results are systematically merged with the original object, its attributes, and its linkages to create the altered object.

Conceptually, the effect is that of slightly altering the characteristic of an object as the machine executes a knowledge operation code (*kopc*). When a series of basic knowledge-level *kopcs* are executed, then the whole picture and perspective change and the processed object becomes an altered object. Numerous such incidents occur in nature or the real world: when a flower blossoms, when a student goes through college, when the rain falls, etc. When the knowledge programs accurately portray the state of change that the objects undergo in the environment or society, then the transitions can be tracked accurately and the direction of change can be adjusted to generate objects and entities that are desirable in any social setting.

Some changes are more profound (e.g., a tornado, an earthquake, a new member in a family, etc.), and the corresponding knowledge programs are more detailed and intricate. Knowledge machines like traditional computer systems can be tailored to suit the object types, the nature of change, and the environment.

The nature and architecture of the knowledge processor units (KPUs) differ considerably from those of contral processing units traditional (CPUs). These differences are examined in detail in Chapter 8. It is possible to design a platform for the KPU that is built on numerous CPUs. Such CPUs perform mundane tasks such as numerical computations, matrix and array processing, signal processing, statistical functions, and hypothesis testing.

The nature and complexity of the software (including the operating systems) also differ considerably from those of traditional computers. The basic software operations are more closely aligned to object and attribute processing rather than numeric and algebraic processing. Such an extreme difference occurs between traditional computer systems and the electronic switching systems (ESSs) used in typical circuit-switched environments. However, the commonalities between two systems persist, and both mainframes and highly specialized ESSs have prevailed for the last six decades.

If knowledge machines are designed to serve human and social needs, then the major contributions in both the computer and communications fields will propel a new generation of integrated social domain computers and networks dedicated to serving society. The object banks of such machines can also be personalized to suit individual users, thus making the machines well suited to be the Personal Companions (PCs) rather than personal computers (pcs) etc. of users.

REFERENCES

1. S. V. Ahamed, "Architectural Considerations for Brave New Machines," Intelligent Engineering systems through Artificial Neural Networks, *Proceedings of Artificial Neural Network in Engineering Conferen*, 2005, vol 15, pp. 71–80, See also "Knowledge

Processing Environments in the Intelligent Internet Era," *Journal of Business and Information Technology (JBIT)*, Academy of Business and Information Technology, Indiana, PA., 6, Fall 2006/Spring 2007: pp. 1–28.
2. S. V. Ahamed et al., "Sensing and Monitoring the Environment Using Intelligent Networks," U.S. Patent Publication US2003/00441013 U.S. Cl. 706/59, February 27, 2003.
3. M. Salvadori, *Why Buildings Stand Up: The Strength of Architecture*, W.W. Norton, New York, 2002.
4. N. S. Stuart, *Computer-Aided Decision Analysis*, Quorum Books, Westport, CT, 1993.
5. E. Shortcliffe, *MYCIN: Computer-Based Medical Consultations*, Elsevier, New York, 1976. See also B. G. Buchanan and E. H. Shortcliffe, *Rule-Based Expert System: The Mycin Experiment at Stanford Heuristic Programming Project*, Addison Wesley, Boston, 1984.
6. W. Kintsch et al., "About NeoMycin," in *Methods and Tactics in Cognitive Science*, Lawrence Erlbaum, Mahwah, NJ, 1984.
7. S. V. Ahamed and V. B. Lawrence, *The Art of Scientific Innovation: A Case for Classical Creativity*, Prentice-Hall, Englewood Cliffs, NJ, 2005.
8. A. H. Maslow, "A Theory of Human Motivation," *Psychology Review*, 50, 1943: 370–396. Also see A. H. Maslow, *Motivation and Personality*, Harper & Row, New York, 1970, A. Maslow, *Farther Reaches of Human Nature*, Viking Press, New York, 1971.
9. S. V. Ahamed, *Intelligent Internet Knowledge Networks*, Wiley-Interscience, Hoboken, NJ, 2006.

CHAPTER 5

HUMANISTIC AND SEMI-HUMAN SYSTEMS

CHAPTER SUMMARY

Human genes may provide the footsteps for the body and mind, but personality is shaped by the norms of society. Adaptation is the key to survival and intelligence is the strategy behind successful adaptation. The information age has brought expectations about being well informed and also highly intelligent. Genomes and social conditions limit an individual to be astute, clever, and rational and also to be totally informed and almost scientific in every aspect of daily life.

An unlikely scenario starts to unfold. On the one hand, biological and environmental constraints govern the extent to which an individual adapts and becomes astute, clever and rational, or scientific in comprehending information. On the other, society moves along at its own pace adding new layers of knowledge and information. The information society has brought about an explosion in the quantity of knowledge, and the genetic structure of the mind is much slower to adapt. This scenario sets up an inconsistency between individual and society.

An individual thus loses the position to be intelligent and also well (even adequately) informed in every direction when new layers of knowledge are added. Human beings are ill equipped to change their nature abruptly by assimilating large amounts of information to make far-reaching rational decisions of an intelligent being. However, being innovative and astute in many directions, humans

---

Computational Framework for Knowledge. By Syed V. Ahamed
Copyright © 2009 John Wiley & Sons, Inc.

have created computational environments that shield their minds from a deluge of information.

A high degree of *focus* on essential information and the *ingenuity* to achieve a viable solution become one approach to dealing with an exploding knowledge society. A new set of information-processing methodologies embedded in VLSI technology for KPUs offers a bridge between sharp human minds and the complex information society.

Gadgets and devices help to the extent that computers offer numerical solutions. iPods and wireless LANs help to the extent that backbone networks carry digital data. Such gadgets, tablets and iPods do not deal with information and knowledge, but instead bits and bytes of data. However, if there were a robot in a gadget that had the humanistic character to convert data in an iPod to objects in a humanistic framework, then the humanistic character of an iPod would become a knowledge tool for the user.

For example, if the monies spent on wars were re-represented as the number of welfare recipients who existed below the poverty line, then the impact on lawmakers would be their growing more human in their decisions, if the data streaming down from satellite channels around the globe about the outsourcing of IT jobs were re-represented as the number of jobs lost in the sourcing country every quarter, then politicians would be more informed about the implications of their legislation in this quarter. In reality, the new gadgets (knowledge tools) link the world of numbers with the world of knowledge and values. Such a link between numbers and human values is more than what current computers (as hardware) and decision support systems (as software) can provide.

To be more generic, the procedures executed in new gadgets should learn to function based on processes much like a human being's conceiving of a situation, and then resolving and deducing a strategy to address it. If the new gadget has robotic/humanistic response, then it should deduce a strategy as a CPU and also a KPU (human being) would have done, and communicate commands to an external response system if changes in the society need to follow.

5.1 HUMANISTIC CHIP SETS

We introduce these new chips with humanistic patterns of problem solving as *humanistic chip sets*. They function rationally, consistently, and dependably with four distinct traits (typically practiced by corporate executives):

1. They conceive the immediate problem at hand in a local and global context (problem orientation).
2. They choose the most economical solution based on the marginal cost of resources expended and the expected gain in the marginal utility of the solution (Marshall's laws).
3. They resolve the numerous steps in implementing the solution in a logical and consistent fashion (software design methodology).

4. They communicate (with human approval) with a response system (when it is available and physically connected to the system) to physically implement the solution (engineering and practical realization).

In order to offer the best choice of solutions, access to local and global knowledge bases and the Internet becomes necessary. It is quite possible to arrive at a solution based on limited or selected access to knowledge bases. These four steps have a scientific flavor that is programmable in the VLSI that becomes the new humanistic microchip sets. The architectures presented in this chapter investigate the possibilities of mechanizing these major steps and their incumbent minor steps.

Depending on the preferential attitude of a particular executive whose behavior is implanted in these chipsets, they bear a distinct human signature in their response. The learning capacity of the chip sets also tailors their responses to suit the individual whom they serve. Thus, the machine learns to find a balance between the employee needs and executive attitude. Much as an intelligent robot, the capacities of the chipsets are tailored to the behavioral preference in the decision-making domain. This mode of operation is different from the conventionally intelligent-robots that assist the production processes in any industry based on numerical data and AI programs.

Varieties of such chip sets are necessary, much like the varieties of processor chips that serve applications in industry. Ranging from a routine human interface for the elderly and disabled to a super-fast, deadly accurate interface for space age pilots, the functioning of the humanistic chips broadly depends on the application and more specifically on the individual. Such chips have been built and deployed in channel unit digital subscriber loops [1], where the initial response depends on the R, L, G, and C character of the loops, and the fine level (equalization and/or echo canceller response) then depends on the individual loop.

In a generic sense, the humanistic chip sets address the needs outlined in the need hierarchy [2] of human beings. The most generic and profuse chips will cater to lower-level needs (levels 1–4 needs) but be customized for the individual using the interface. On the other hand, the most specialized chip sets will address the professional and realization needs (level-5 needs) of the sophisticated user. For example, the generic chip sets for an educator will organize and interface with all the routine academic duties, and the specialized chip set will interface with knowledge banks and research sets around the globe. As an additional example, the generic chip set for a medical professional will first and foremost interface with the duties of the individual physician, and then the specialized chip sets will address the specialty issues the professional may face, etc.

It is possible to envision a human chip set for every human being seeking relief from the mundane duties of life and seeking to concentrate in any direction of specialized activity, however transient it may be. For instance, as a human being progresses from childhood to puberty and then natures, the architecture and configuration of generic humanistic chips and special chip sets are adjusted to suit the developmental growth patterns of the individual with a permanent record of variations in the physiological, emotional, and psychological health of the individual. Other applications are plentiful.

5.2 ESSENCE OF HUMAN ACTIVITY

Simply stated, all human activity has one format: Do (verb function) something (noun object), even if "something" is "nothing.". Even in sentences that do not have an explicit do something format (such as "black is beautiful", "winters are cold," etc.), the do something format is implicit. In the first sample sentence, it is implied that the reader will modify one of the attributes of "black" as being "beautiful," and in the second sentence, it is implied that the reader will modify one of the attributes of "winter" as being "cold." In fact, it is impossible to write a sentence without (at least) a noun, a verb, and an implied tense. The verb function can have a wide range of simple, complex, and hierarchical command structures. The noun object also can possess a wide range of tangible, virtual, group, collectivity, etc., entities. Much as complex commands have structures, complex noun objects have relationships and linkages. Figure 5.1 depicts a step in the routine activity in the solution of any simple or complex real-life problem. Even though the steps are rarely articulated, it is necessary to establish their sequence to format the correct sequence of events in the microchips.

Just as a dictionary of language contains all words, a verb functions dictionary contains all possible commands and a noun object dictionary all possible entities. A subset of computer verb functions dictionary contains all possible

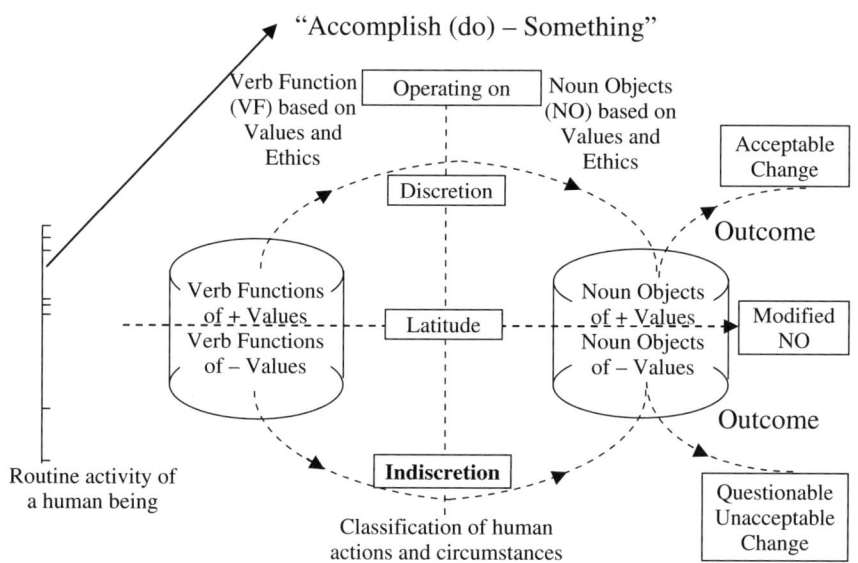

Figure 5.1 A microstep in the routine activity of a human being.

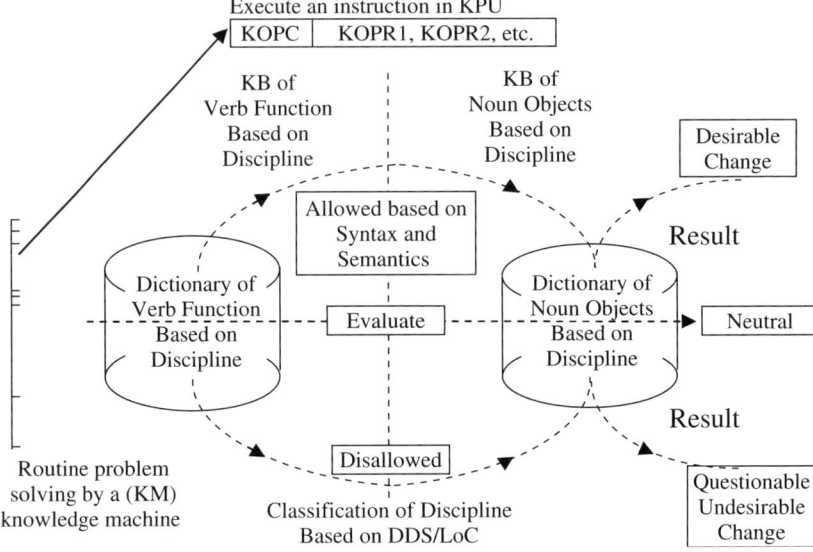

Figure 5.2 A microstep in the knowledge processor unit of a knowledge machine.

machine-executable commands (*opcodes*). A subset of computer noun object dictionary contains all possible local and Internet protocol IP-addressable entities. A subset of human verb functions dictionary contains all possible humanly possible actions. A subset of human noun object dictionary contains all possible local and IP-addressable objects or entities.

Figure 5.2 depicts the computational parallel of Figure 5.1. The sequence of steps in a knowledge processor unit (KPU) of a knowledge machine parallels the logical steps of human rationality shown in Figure 5.1. Both offer the intermediate result(s) of the microfunction that is itself a step in the overall solution to a real-life problem.

Human beings, being more keen and adept, have developed object manipulation skills that are object-dependent. In addition, the context of the sentence based on the verb function that is associated with a noun object dictates the semantic rules. The following methodology allows the machine to process varieties of objects, used in conjunction with many classes of verb functions.

5.2.1 Varieties of Noun Objects

Objects in the context of processing knowledge can be docile or intelligent. Inanimate objects (such as numbers, logical entities, data structures, words, etc.) are generally immune to the process and do not react to the verb operator (such as the operation code, *opc*, or the knowledge opcode, *kopc*).

Intelligent objects (like human beings) or derived intelligent attributes (like the attributes of a human being, such as emotions, feelings, health, etc.; like the beauty of diamonds in response to incident light; like the songs of birds depending on the season; etc.) respond intelligently depending on the *opc* or *kopc*. The response can be in the assistive mode (like the feeling of a human in response to love), neutral mode, or resistive mode (like the resistance to a decline in morality in response to a corrupt regime).

5.2.1.1 Inanimate or Passive Objects Inanimate objects undergo any verb function in a predictable and mechanized fashion. Conventional computing is based on the object *operands* being inanimate and nonreactive. Only a selected repertoire of *opcodes* is executable with corresponding inanimate objects. For example, an arithmetic operation consistent with the arithmetic unit (AU) functions well in the central processing unit (CPU) environment. Such *opcodes* can be logical and they function well, consistent with the logic unit (LU) in the CPU. Matrix and array functions can also be performed on the declared matrices and arrays, etc. Only allowable operations are verified at the compile time and performed at the program execution time.

5.2.1.2 Intelligent or Responsive Objects Knowledge processing deals with a broad genesis of objects. Such objects can be animate and responsive objects that are either natural or artificial. Naturally intelligent objects include all adaptive life forms that respond to the verb function and context of neighboring objects. They retain the effect of prior operations. Generally, the attributes of such intelligent objects retain memory and hold a response mode to verb functions. For example, if the *kopc* command is "harm yourself," then the resistive mode of the object (yourself) is invoked and the KPU may block or soften the command at the time of execution. On the other hand, if the command is "protect yourself," then the assistive mode of the object (yourself) is invoked.

Artificially intelligent objects are devices and systems that are intelligent by deploying any combination of AI tools. For example, if a command to a robot X is "collide robots Y," then the resistive mode is invoked if robot X has collision avoidance algorithms built into the robot X operation software, etc. On the other hand, if the command is "learn path," then the response of robot X will be assistive if learning algorithms are built into the control software.

Intelligent objects have two major characteristics: (1) their *reactivity* to VF: reactive (i.e., overly responsive in the direction of VF), resistive (oppositional to it), or just neutral (ignoring it), and (2) their *predisposition* toward being positive (e.g., staying healthy), negative (e.g., cheating others), or unbiased (e.g., transacting normal business).

Reactivity to VF(s), is more of an individual trait of objects. A reactive predisposition tends to make objects and their attributes strongly react to the VF(s) in the direction of change imposed, such as becoming overly protective at the slightest call for help (in a positive sense) or becoming a criminal when presented with the invitation to steal (in a negative sense). A resistive predisposition tends to make objects resist any change imposed by VF(s) and respond fractionally to it, such as becoming an average student when the expectation is that one will be an excellent student (in a positive sense) or making only a partial payment when repayment of a loan is demanded (in a negative sense). At the median of these two predispositions reside objects that are neutral and do not respond unless the extent of the VF(s) goes beyond a given minimum level.

A positive predisposition tends to make objects and their attributes fall in the type -I region of the Aristotelian classification of human nature—that is, to be good, beneficial, honest, truthful, etc., in their attributes. Conversely, a negative predisposition tends to make objects and their attributes fall in the type II region of the Aristotelian classification of human nature—that is, to be evil, harmful, dishonest, deceptive, etc., in their attributes. At the median of these two types reside objects with unbiased attributes, such as being neither good nor evil, being passive and indifferent to society.

In essence, intelligent objects fall into three categories of predispositions. The first group of assistive noun objects ($NO+$) help the VF in altering the object(s) or attributes in its direction. The second group of resistive noun objects ($NO-$) oppose the VF in its attempts to alter the object(s) or attributes in its direction. The third group of neutral objects ($NO\pm$) neither assist nor resist the object or attributes in the direction of the VF.

Extraneous programs can invoke a programmed response of intelligent objects, or the KPU can determine the response of the object since it includes the *kopc* function (VF) that is about to be performed on that particular NO with a given set of attributes. To deal with an intelligent object in the KPU, at least two algorithms are necessary to handle the resistive and assistive response modes of any NO.

5.2.2 Types of Verb Function Commands

Commands are executed on object(s). After execution, the processed object(s) have a different set of structures, relations, and/or attributes modified by the type of command executed. In conventional computing, a command such as (X = 1/X) will generate the new X as the reciprocal of old X and determined precisely by the age-old laws of division. X is an inanimate object and the command divide is a neutral verb.

In the knowledge-processing environment, verbs may assume two flavors': neutral (such as copy) or potent (such as modify). Furthermore, the potent verbs have a preferable or nonpreferable outcome on the object being processed. For example, potent and preferable verbs (such as sanctify, cleanse, purify, etc.) bear an age-old positive context. Conversely, potent and nonpreferable verbs (corrupt, contaminate, pollute, etc.) bear an age-old negative context. Context does not have a precise numerical measure. A sense of value judgment starts to appear in the classification of verbs.

However, as it is suggested in Section 5.4, there is an algorithmic path through the complexity of behavior. If such verbs are assigned values and directionality consistent with the traditional deployment of actions (verbs) in the past based on

VF knowledge bases around the globe or within society, then a certain rationality[1] for the use of such actions is established.

5.2.2.1 Neutral Verbs Traditional computing consists of executing neutral verb functions on inanimate objects, such as all (primary and secondary) ALU functions on corresponding numerical and logical entities, all complex algebra operations on complex numbers, all matrix and array operations on corresponding matrices and arrays, and so on. The rules for performing the legitimate computational functions are well documented in the literature.

5.2.2.2 Potent Verbs Verb functions act on noun objects. Potent verbs alter the character and/or attributes of noun objects. Radical and significant changes can occur in the objects or their attributes when potent verbs are executed on objects. Such transformations occur in society. The rules for their effects can become complex and variable. The programmer has the option of defining the extent of change by a certain class of potent verbs on inanimate and intelligent noun objects. Hence, we classify the verbs or VFs (from the dictionary of verbs) into three groups: namely, as $VF+$, verbs with positive impact (strengthen, amplify, enhance, cleanse, etc.) on the noun object; as $VF-$, verbs with negative impact on the noun object; and as $VF\pm$, neutral verbs.

5.3 SMART VERBS AND INTELLIGENT NOUNS

Typically, the KPU executes the *kopc* ($VF*$) on noun *koperands* ($NO*$). Verb functions are classified into three groups: $VF+$, $VF-$, and $VF\pm$. Likewise, the noun objects are classified as $NO+$, $NO-$, and $NO\pm$. To represent this wide range of effects in the KPU, we suggest the following methodology for classifying the effect of executing a *kopc* on a noun *koperand*.

Intelligent objects respond like life forms. When intelligence within the object senses the expected impact of the verb function, the response approaches the realm of human behavior ranging from cooperation to aggression. In human beings, this type of behavior is predictable, though not very precise. In the KPU environment, artificial intelligence algorithms predict the outcome by estimating the laws of interaction[2] of $VF*$ with $NO*$ to produce a $PNO*$. The mathematical representations of such an outcome on the processed object $PNO*$ that results from the KPU function are not easily amenable to any well-known algebraic formulation. For this reason, we present a new methodology in two stages: qualitative synopsis and mathematical representation.

---

[1]This practice is adapted in the legal profession to justify ("kill to defend") or deny ("kill to rob") a certain actions (to kill) as normal and acceptable action. Sometimes, silly and irrational examples (such as "all is fair in love and war") are used to justify erroneous and unethical actions by any rational standard. World-wide knowledge bases provide some basis for acceptable, neutral, and unacceptable verb functions.
[2]An asterisk * indicates any or all combinations of $+$, $-$, and $\pm$.

5.3.1 A Qualitative Synopsis of the Response

The two items of data within the KPU necessary to execute a binary instruction are *kopc* and *koprand(s)*. The following representation:

```
VF*.on.NO* (leads to) → PNO*
kpoc.on.koprand(s)    → Processed nounobject(s)
```

is used to indicate that a *VF** operates on a *NO** to generate a *PNO**. We propose five qualifiers[3] ($\uparrow\uparrow, \uparrow, \rightarrow, \downarrow, \downarrow\downarrow$; see Table 5.1) to indicate the status of the processed object. The qualifier $\uparrow\uparrow$ indicates that the original positive quality (such as truth in a statement, the beauty in a scene, etc.) of the *NO* is *significantly* increased by an algorithmic process (discussed in the next section). The qualifier $\downarrow\downarrow$ indicates that the original negative quality (such as deception in a statement, the horror in a scene, etc.) of the *NO* is *significantly* increased by an algorithmic process (discussed in the next section). The qualifier $\uparrow$ indicates that the original positive quality (such as truth in a statement, the beauty in a scene, etc.) of the *NO* is *mildly* enhanced by an algorithmic process (discussed in the next section). The qualifier $\downarrow$ indicates that the original negative quality (such as deception in a statement, the horror in a scene, etc.) of the *NO* is *mildly* enhanced by an algorithmic process (discussed in the next section). The qualifier $\rightarrow$ indicates that the original quality of the *NO* is retained in the processed object.

Processed noun object(s) result from the execution of a *VF** (or *kopc*) on a *NO** [or *koprand(s)*]. However, the extent of change in the *NO** can vary widely depending on the predisposition, the reaction to *VF*. Generally, the attribute of a *NO* is changed. Such attributes can have quantitive designation to the maximum limit of the attribute, such as redness of color, sainthood of a human, or beauty of a sunset, etc. For this reason, a reactive noun responds to a change of attribute command (*kopc*) in a quasi-linear or nonlinear fashion, as shown in Figures 5.3a and b.

TABLE 5.1 Processing of Intelligent Objects

1. Reactive Objects Category: Assistive Noun Objects
   (a) Positively predisposed NO   VF+.on.NO+ → PNO+ $\uparrow\uparrow$
   (b) Negatively predisposed NO   VF−.on.NO− → PNO− $\downarrow\downarrow$
2. Adaptive Objects Category: Resistive Noun Objects
   (c) Positively predisposed NO   VF−.on.NO+ → PNO+ $\downarrow$
   (d) Negatively predisposed NO   VF+.on.NO− → PNO− $\uparrow$
3. Indifferent Objects Category: Neutral Noun Objects
   (e) Positive or Negative NO   VF±.on.NO± → PNO ± $\rightarrow$

---

[3] It is possible to have only three qualifiers ($\uparrow, \rightarrow, \downarrow$,) and use one algorithm. To indicate the unpredictable response of the *NO** or *VF** as being severe or mild, we have retained five rather than three qualifiers.

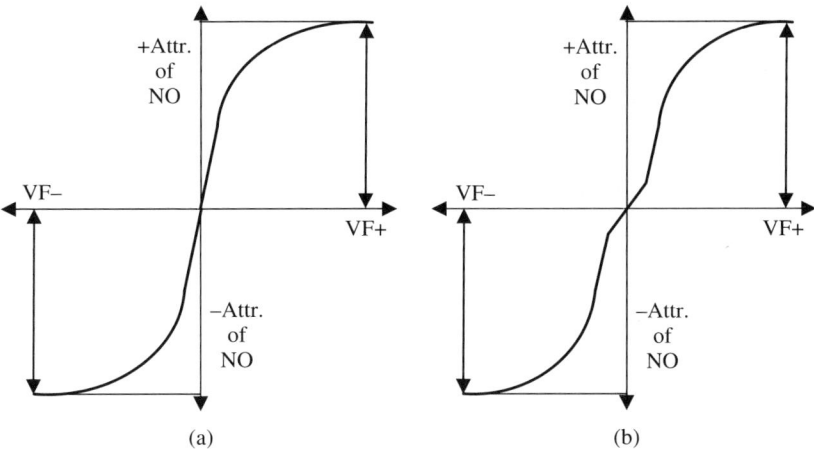

Figure 5.3 Typical nonlinear relationships indicative of the response of attributes of intelligent objects to external verb functions.

Such relationships are the common magnetization curves[4] of magnetic materials. Permeability μ, relating flux density B and magnetizing force H, and the extent and limit of saturation depend on the nature of the magnetic material. In the response[5] of animate objects, nonlinearity, saturation, and hysteresis are also common.

5.3.2 Nonlinear Response

When the intelligent object responds to a VF, the intensity of the VF determines the extent of change in the attribute. In addition, the response also depends on three inherent characteristics (its predisposition, its reactivity, and the maximum value of the attribute) of the noun object. For example, if the VF is strong (such as stop deception vs. slow down cheating) and the predisposition of an "intelligent object" is toward good, the reactivity is positive, and the maximum distance between the current attribute and being perfectly truthful (the opposite of being deceptive) is 100%, then the object will achieve 50% of being truthful. Conversely, if the VF is weak (such as slow down cheating), then the human being may only achieve 20% of being truthful, etc. A quantitative estimation of change in the $NO*$ may thus be introduced in the KPU.

---

[4]The relationship $B = \mu H$ is an equation to compute flux density resulting from magnetization force H. The value of μ is a nonlinear function of $|H|$ and sometimes approximated as $1/(a + b|H|)$.
[5]The response to anger, hate, fear, etc., is not always predictable. However, directionality follows a rational pattern in most instances. Nonlinearity and hysteresis exist, and they can be represented by the nonlinearities of m, n, j, and k in Table 5.2.

5.3.3 An Intelligent Response

From a computational prespective, the response of the intelligent object to become *PNO** from *NO** should follow a definable algorithm. Simplistic algorithms fall short of immitating the complex behavior of intelligent humanistic objects. For this reason, we state patterns of localized behavior that reflects the response of an intelligent object and then present a series of algorithmic steps to quantitatively track such behavior:

(a) The extent of response of the intelligent objects depends on the intensity of the *VF*. The *kopc* has two components: the binary operation code and its magnitude, that is, $|VF|$ designated as l.

(b) The response of intelligent objects can have a component of friction and "stiction" effect, that is, there can be an inherent resistance (laziness) to change. Objects tend to ignore any command to change unless it (i.e., $|l|$) reaches a minimum threshhold designated as $|\varepsilon_l|$. Hence, if the *kopc* is too mild (i.e., the intensity of *VF*, $|l|$, is very low or less than $|\varepsilon_l|$), the *NO* will just ignore the *kopc* and treat it as a no-op (i.e., no chage between *NO** and K*NO**).

(c) The reactivity of intelligent objects ranges from being mildly reactive to highly reactive to any *VF*. Hence, a numerical value can be assigned to $\pm m$ that is added to (or convolved with) the magnitude l of the *VF*.

   c.1. If the value of $|m|$ falls in the region of $\pm \varepsilon_m$, then the effect of m is ignored.

(d) The predisposition of intelligent objects ranges from being highly positive to highly negative. Hence, a numerical value $\pm n$ is added to (or convolved with) the magnitude of $(\pm l \pm m)$.

   d.1. If the value of $|n|$ falls in the region of $\pm \varepsilon_n$, then the effect of n is ignored.

(e) Intelligent objects can have an opportunistic trait that makes such objects strengthen their predispositions, that is, when an opportunity arises, the object will tend to change in the direction of its own predisposition.

(f) Intelligent objects can have an *elastic trail* that makes such objects go back to their initial attribute values. After an operation finishes, the object will tend to gravitate toward its preferred predisposition of minimal expenditure of energy to maintain the status quo. Over a period of time, intelligent objects become immune to the process that changed their attributes during the last operation.

The numerous nonlinear shapes that can be represented easily in the KPU environment are depicted in Appendix A. The derived parameters usually stem from *VF*, its modulus value $|l|$, and $+$ or $-$ directionality.

5.4 ALGORITHMIC REPRESENTATION

The response of the intelligent object is formulated in two steps: (1) finding a change index x and (2) computing the actual change in the attribute of an *NO* in response to a *VF*.

5.4.1 Step 1: Compute Change Index *x*

Finding the change index is shown in Table 5.2. There are 3 possible variations for each of the 3 items of any *kopc* instruction, (*VF*, reactivity of *NO*, and its predisposition), leading to the 27 rows of Table 5.2. The magnitude of l or $|l|$ is indicative of the intensity of *VF*, which depends on the source and force of calling for change in the attribute of the *NO*. A positive value of l is indicative of changing the attributes of the *NO* toward the type I region of the Aristotelian classification of human nature, that is, to be good, beneficial, honest, truthful, etc., in their attributes. Conversely, negative l tends to make objects and their attributes fall in the type II region of the Aristotelian classification of human nature, that is, to be evil, harmful, dishonest, deceptive, etc., in their attributes. At the median of these two types, a numeric value of l that is in the range of $\pm\varepsilon_l$, reside forces with unbiased force such that they are neither good nor evil, are passive and indifferent to the consequences.

The positive value of m implies that the NO assists and amplifies the effect of *VF* in either direction. The innate nature of the object will respond to changes in any direction that *kopc* or *VF* imposes on the NO. A negative value of m implies that any change imposed by the *kopc* or *VF* will be opposed. A numeric value of m that falls in the range of $\pm\varepsilon_m$, implies that the objects are indifferent to *kopc* or *VF*.

A positive value of n implies the tendency of the NO to gravitate toward the type I region of the Aristotelian classification of human nature, that is, to be good, beneficial, honest, truthful, etc., in their attributes. Conversely, negative n tends to make objects and their attributes fall in the type II region of the Aristotelian classification of human nature, that is, to be evil, harmful, dishonest, deceptive, etc., in their attributes. At the median of these two types, a numeric value of n that is in the range of $\pm\varepsilon_n$ reside objects with unbiased attributes such that they are neither good nor evil, are passive and indifferent to society.

In the numerical estimation of the change of attribute, the determination of the change index x due to the three forces influencing the change of the attribute of an *NO* is the first step. From Table 5.2, two other variables, j and k, enter the picture. In a sense, j reflects the effect of nonlinearity between *VF* and the final change in the attribute of the *NO*. The variable k reflects the effect of the innate tendency of the object to be opportunistic and change the attribute to return to its predisposed value. If a time stamp of the last *kopc* is one of the fixed attributes of an *NO*, then the effect of time and the elastic nature of the previous change may also be included in the derived value of k. The nonlinearity of m and n with

TABLE 5.2   Response of Intelligent Noun Objects

Row	Verb Function (VF)	Noun Object (NO) Reaction to VF	Noun Object (NO) Predisposition	Change index x
		Positive Verb Functions		
1	$+l$	$+m$	$+n$	$x = l + m + n$
2	$+l$	$+m$	$\pm\varepsilon_n$	$x = l + m$
3	$+l$	$+m$	$-n$	$x = l + m - n$
4	$+l$	$\pm\varepsilon_m$	$+n$	$x = l + n$
5	$+l$	$\pm\varepsilon_m$	$\pm\varepsilon_n$	$x = l$
6	$+l$	$\pm\varepsilon_m$	$-n$	$x = l - n$
7	$+l$	$-m$	$+n; n' = j \cdot n$	$x = l - m + n'$
8	$+l$	$-m$	$\pm\varepsilon_n$	$x = l - m$
9	$+l$	$-m$	$-n; n' = j \cdot n$	$x = l - m - n'$
		Neutral Verb Functions		
10	$\pm\varepsilon_l$	$+m; m' = k \cdot m$	$+n; n' = j \cdot n$	$x = m' + n'$
11	$\pm\varepsilon_l$	$+m; m' = k \cdot m$	$\pm\varepsilon_n$	$x = m'$
12	$\pm\varepsilon_l$	$+m; m' = k \cdot m$	$-n; n' = j \cdot n$	$x = m' - n'$
13	$\pm\varepsilon_l$	$\pm\varepsilon_m$	$+n$	$x = 0$
14	$\pm\varepsilon_l$	$\pm\varepsilon_m$	$\pm\varepsilon_n$	$x = 0$
15	$\pm\varepsilon_l$	$\pm\varepsilon_m$	$-n$	$x = 0$
16	$\pm\varepsilon_l$	$-m; m' = k \cdot m$	$+n; n' = j \cdot n$	$x = -m' + n'$
17	$\pm\varepsilon_l$	$-m; m' = k \cdot m$	$\pm\varepsilon_n$	$x = -m'$
18	$\pm\varepsilon_l$	$-m; m' = k \cdot m$	$-n; n' = j \cdot n$	$x = -m' - n'$
		Negative Verb Functions		
19	$-l$	$+m$	$+n$	$x = -l - m + n$
20	$-l$	$+m$	$\pm\varepsilon_n$	$x = -l - m$
21	$-l$	$+m$	$-n$	$x = -l - m - n$
22	$-l$	$\pm\varepsilon_m$	$+n$	$x = -l + n$
23	$-l$	$\pm\varepsilon_m$	$\pm\varepsilon_n$	$x = -l$
24	$-l$	$\pm\varepsilon_m$	$-n$	$x = -l - n$
25	$-l$	$-m$	$+n; n' = j \cdot n$	$x = -l + m + n'$
26	$-l$	$-m$	$\pm\varepsilon_n$	$x = -l + m$
27	$-l$	$-m$	$-n; n' = j \cdot n$	$x = -l + m - n'$

Note: The variables j and k represent the behavioral differences of intelligent objects. If the object is opportunistic, indices j and k change accordingly. If j varies between zero and i (with a maximum value of 2 or 3, for instance), the conflict between VF and the reactivity of NO is seized as an opportunity for NO to shift toward what its predisposition favors (rows 7, 9, 25, and 27). If k varies between zero and i, the weakness of VF (rows 7, 9, 10, and 12) is seized as an opportunity for NO to shift toward what its innate tendency and its own predisposition favor.

respect to l is reflected by the two additional parameters[6] j and k. The equations to derive the current values of j and k for the current attribute represent the nonlinear relationship between VF and m and n, respectively.

Altering the directionality (with respect to the VF) and the numerical values of j and k reflects emotion (such as love, anger, hate, etc.) in the behavior of any NO. For instance, if the VF and m are both $+$, then a high $+j$ value tends to intensify the effect of the VF on the NO (a sense of $+$ love). If both the VF and NO are negative, then a high value of $+j$ indicates a "Bonnie and Clyde" type of love.

Anger is invoked by making m negative and enhancing the $+j$ value. The response falls in a direction opposite to that of VF, and the intensity of hate is invoked by increasing the magnitude of m. The nonlinearity of human temper (anger) is invoked by also making the relationship between j and l nonlinear. Instability in the response of the NO is an indication of the instability of human behavior or social processes.

5.4.2 Step 2: Compute Change Δ after *kopc*

The extent of change for an attribute of an intelligent object depends on:

1. Current attributes value
2. Maximum attributes value
3. Change index x and its direction
4. Nonlinearity between the change index and the actual change that occurs as a result of item 3

Figures 5.4a and b depict the basis for the computation of actual change in the attribute of the NO. Figure 5.4a shows the computation of the maximum possible change Δ, and this change depends on the initial value of the attribute prior to the time of execution of the *kopc* command in the KPU. There are four possibilities $(\Delta_1, \Delta_2, \Delta_3, \Delta_4)$ for the maximum change. The value of Δ depends on the initial value A_i, its direction, and the direction of x. The absolute maximum value of Δ is $|A_{max} + A_{min}|$, and it occurs when x tries to transform the attribute entirely for any NO from A_{max} to A_{min} or vice versa. Such changes are seen when the NO perceives total denial of a need or threat to the security of the intelligent NO. Ample flexibility exists to represent such behavior.

The actual change due to a finite value of x is computed from the exponential relation shown in Figure 5.4b. The base for the exponential relation is represented as e, even though any value would serve to depict the saturation in the change of attribute as a result of external verb functions.

The shape of the nonlinear curves in Figures 5.4a and b can differ from one NO to the next, and the final level of an attribute can become indefinitely high

---

[6]This two-parameter representation permits the effects of nonlinearity to be interjected at selected rows in Table 5.2 rather than all the rows where m and n appear.

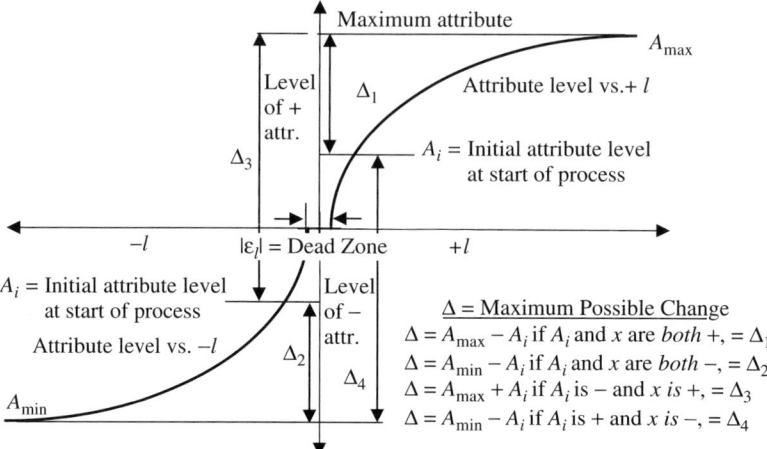

Figure 5.4a Nonlinear relationship between the intensity of the verb function, that is, $|l|$, and the level of attribute at the start of a knowledge domain operation. Note that the maximum possible change Δ is reused to calculate the actual change δ.

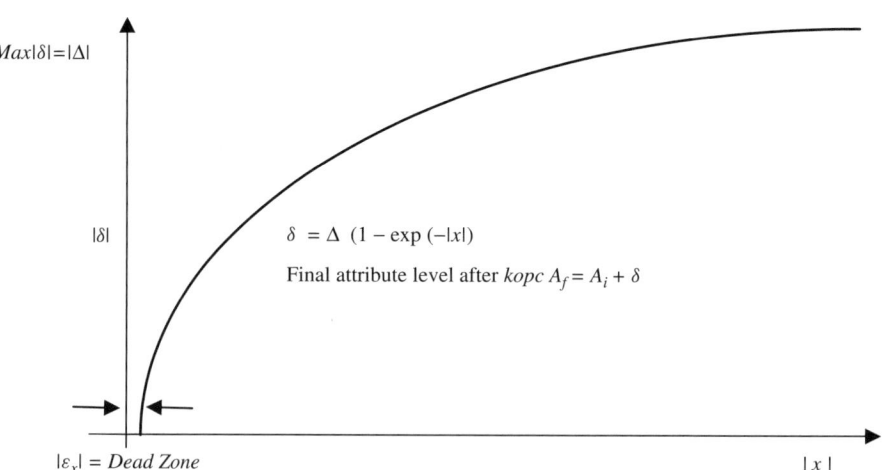

Figure 5.4b Relationship between the actual change δ (of attribute) and the maximum change of the attribute that is possible. The law of diminishing return of change becomes evident as $|x|$ increases in trying to change the attribute. Note that these curves are dynamic and the values are computed at execution time in the knowledge processor unit. All the attributes (at that instant of time) of the noun object contribute to final attribute level A_f after the execution of *kopc*.

(such as the tolerance of a saint or beauty of a masterpiece) or infinitely low (such as the reassurance of a tyrant king or intelligence of an idiot). The values of $A_{\max}$ and/or $A_{\min}$ are adjusted accordingly. The total rigidity of an intelligent *NO* is reflected in the equations by making $\Delta = 0$.

5.5 EFFECTS OF INTELLIGENT RESPONSE

Numerous responses to a stimulus for change are feasible. In this chapter, the six humanistic (see Section 5.5.3) variations of major behavioral traits are accommodated by the parameters m, n, j, and k and their nonlinearties. For accurate tracking of the attributes of objects, the current values of these parameters leading to the composite index of change x are essential. The current values of the parameters are computed from their earlier values and change that has occurred. Such a change could be the time-decay effects of other verb functions in recent history. The exited or docile state of an intelligent *NO* is reflected by enhancing or diminishing the parameter m; the nature (opportunistic, rational, controlled, etc.) of an *NO* is reflected by adjusting the signs and magnitudes of m and n together.

The initial value of the attribute plays a role in the change. The values of Δ and δ, and their derived direction, permit an accurate computation for the final value of the attribute. Objects with different sets of attributes exhibit characteristics for the overall behavior of the system and its stability.

In simulating the behavior of a life, a sociological system or an ecosystem with lifelike objects, the behavior of individual objects[7] and collectivities of objects become crucial. The representation of such behavior emphasizes the role of key players in the performance of social institutions.

5.6 GRADATION OF KNOWLEDGE PROCESSOR UNIT RESPONSES

CPU is a purely numeric KPU. The *kopcs* in KPUs degenerate to plain *opcs* (opcodes) in CPUs. In this mode, the KPU functions as a CPU and the knowledge machine serves as a plain old computer. The numerous types of KPU architectures [single-process single-object (SPSO), single-process multiple-object (SIMO), multiple-process single-object (MPSO), multiple-process multiple-object (MPMO), and pipeline object processors] presented in Section 4.6, retain their identity as the corresponding conventional architectures such as SISD, SIMD, MISD, MIMD, and pipeline data processors of CPU, respectively.

The response from the KPU while processing the numerous types of objects (active, neutral, and passive), much like the CPU response, would differ while

---

[7]In the study of interactions, von Neumann and Morgenstern have predicted the human behavior of individuals in economic interactions (zero-sum, nonzero games, and prisoner's dilemma, etc.,). The approach proposed here is more generic and the entire problem can be simulated by including the effects of the rational and irrational attitudes of its participants.

processing numerous types of data (integer, fixed point, floating point, etc.). We discuss some of the KPU responses in the rest of this section.

5.6.1 Traditional Computer Response

The CPU mode of response typical in a traditional computing environment is invoked by making $l, m, n = 0$, thus turning *kopc* into an *opc* for a typical CPU. The objects respond without invoking any of the six response modes for behaving like an intelligent object. In fact, this mode of object behavior for numbers, logical entities, matrices, complex numbers, series, etc., is the only mode that exists in a typical computer environment. All objects are neutral to process and let the control unit perform any (authenticated) operation on them and their attributes. The knowledge machine would become a computer and the KPU will become a CPU.

5.6.2 Life Form Response

In representing the lowest-level life forms, the basic needs are preservation and survival. Next, the need to reproduce becomes dominant. Even in such rudimentary situations, the role of competition and aggression becomes evident. The forces changing the objects themselves, rather than their attributes, are the environmental and natural forces. These semi-intelligent objects depend on their instinct rather than rationality. The responses are geared toward their simplistic (not yet totally understood) needs for survival and propagation of the species.

The evolution of animal and plant life has resorted to some intricate and delicate balances. Some of the most amazing adjustments occur in nature. For predicting different species' responses to the forces of nature, the methodology proposed in this chapter is totally inadequate and specialized models of the responses for such species become essential.

5.6.3 Rational Response

The human species resorts to rationality in behavior. Expected rewards and possible retributions are usually evaluated in changing human attributes. Fortunately, such behavioral variations[8] have been studied by social scientists and statistical data are available to track the extent of change, such as the types of six behavioral responses presented in Section 5.5.5. The patterns and norms are established by sociometric studies. However, individual variations are plentiful and accuracy is lost. For these reasons, the pattern of social response is determined first and then the behavioral parameters l, m, and n are adjusted to suit the particular situation and the specific individual. In most instances, rational behavior is predictable.

---

[8]Only a small subsets of rational human responses is deemed predictable for the knowledge machine. In reality, the human emotional reactions include love, happiness, gratitude, surprise, fear, sadness, anger, disgust, contempt, shame, and guilt. Such situations are much too complex for any machine representation at the present time.

Increments of behavioral responses are chained together to estimate the overall response, much like a series of machine instructions are executed to generate a library function in any computer environment.

Human goals can become complex, thus altering the behavioral balance with different parameters (l, m, n, j, and k) for different goals. Such superposition of behavioral responses can cause instability in the behavioral mode of human beings. In these circumstances, the knowledge machine (KM) can provide solutions that tend to dampen oscillations and propose stable modes of behavior for individuals and social entities. We have proposed only six important behavioral responses and only three parameters to represent the estimated change in the attribute(s) of the individual. This set appears too constraining to predict the behavior of highly adaptive human beings such as poets, artists, philosophers, or even brutal kings. In fact, the parameter most applicable to such human beings is $+n$ for the positively inclined intellectual and $-n$ for selfish, deceptive, but intelligent human attitudes. Such responses are generated in a KPU environment by enhancing the value of n in general, or by enhancing the value of k for special situations in which nonlinearity will affect only certain attributes when the attribute value(s) falls in a critical range. Instable behavior patterns become predictable. Numerous programming tools to modify the extent and shape of nonlinearity of l, m, n, j, and k are available.

Economic gain, optimization, or any other objective function drives programs to find a suitable individual to achieve the goal. The knowledge machine identifies the limitations of individuals who can become barriers to the results when individual personality traits play a critical role. These scenarios are identified by the machine as it starts to seek out the traits of individuals and compares these traits with the statistically averaged traits of sample groups.

5.6.4 Artificial Intelligence Response

Machine learning and neural nets permit computer systems to readjust their response in an artificially intelligent fashion. When such learning methodologies modify the content of the knowledge bases that influence the relative directions and magnitudes of the parameters l, m, n, j, and k, the outcome of any social process becomes programmable. Proper choice of the learning algorithms will stabilize or aggravate social interactions, leading to peace or war between individuals and nations. Short-term predictions appear feasible even though the learning algorithms will take over to compensate for any changes in the behavior of the other intelligent object.

5.6.5 Humanistic Response

Humanistic responses depend on the programs driving the local chip set and the response from global wisdom bases around the world. However, knowledge machines respond to local inputs and dynamic conditions. When models of behavior characterized by parameters l, m, n, i and j, leading to an index of change

x as shown in Table 5.2 and depicted in Figure 5.4, are primed into the control modules of the KPU or encoded as behavioral software routines, the response of the knowledge machine starts to approach the behavior of a humanistic one.

It becomes quite crucial to make a statistical study of a significant group of human beings and select the numerical values of l, m, n, i, and j, and to represent the human behavioral responses as algebraic relationships. The knowledge machine is ill equipped to model human nature in its entirety even though it may compute and predict behavior in tight compartments of rationality and intelligent economic behavior. Such behavior has been studied in the past by Morgenstern and von Neumann [3] in their classic book.

5.6.6 Human Response

Human beings wear many faces of good and evil that lie hidden in their thoughts. The exploration of thoughts and emotions (love, happiness, gratitude, surprise, fear, sadness, anger, disgust, contempt, shame, guilt, etc. [4]) is outside the scope of this book. In a very limited sense, when the limits of the mind are restricted to rationality, and the direction of the solution lies in the optimality of one or more attributes (such as wealth based on laws of economics, knowledge based on searching knowledge bases, etc.), then the response (rather than overall behavior) may be attempted based on parameters m, n, j, and k and their nonlinearities.

On the one hand, when the human response is to be predicted, the faces of truth, virtue, and beauty (TVB) can appear in their true colors: truth appears as universality, virtue as the benefit it brings to society, and beauty appears as graceful elegance. The union of these three spaces is tightly packed with the utmost accuracy, uncompromising discipline, and scientific rigor. Most human beings are ill prepared involved in seeking infinite precision in all three areas (truth, virtue, and beauty). On the other hand, the face of deception, arrogance, and hate (DAH) has infinite manifestations and adopts a fuzzy blend of responses. Whereas the effect of time is to dissipate the union space of DAH, the union space of TVB appears more robust and is iteratively reinforced. The confluent, coherent, concurrent, and cogent perfection of TVB is likely to leave an asymptotic trail of universality in time.

The positively primed knowledge machine works very hard in the pursuit of the perfection of the positive. Any variation from this is tolerance of the negative. In a sense, the extreme of human behavior is not predictable even though the knowledge machine may be able to approach it asymptotically by forcing the values of $|m|$ to become extremely high. However, the shape and extent of nonlinearity will interfere with the precision in the details of the outcome. In a negatively primed knowledge machine, the eradication of the positive and perfection of negativity are just as possible.

True deep human nature is too unpredictable for any machine. It is futile to simulate the long-term behavior of extremists in the positive sense (for poets, philosophers, creative scientists, etc.) or in the negative sense (for tyrant kings, organized crime figures, cruel warlords, dictators, etc.). However, in the median, there is room for a knowledge machine to predict the response and emulate the localized behavior of many human beings (from corporate executives to street thugs) who have a conceivable range of intellectual activity and freedom.

5.6.7 Other Types of Responses

5.6.7.1 Self-Propagating Flip-Flop Response This type of flexible (almost opportunistic) behavior ranges from a very positive response (m is positive, if l is positive; n is positive; j is positive, if l and m are positive; and k is positive, since n is positive) to a very negative response (m is negative, if l is negative; n is negative; j is negative, if l and m are negative; and k is negative, since n is negative). Such modes are also feasible from the intelligent KPU environment with quick adaptations of the parameter values.

Under such conditions, the intelligent object is more than intelligent in an extremely ingenious way. Self-interest (rather than social or organizational interest) dominates the response. In a positive sense, the search for truth and benevolence, self-interest and social interest, elegance and grace are unified under one goal. The converse situation becomes evident in a negative sense. Under persistent negative conditions, the *NO* comprehends the external environment or objects dictating the *VF*, and adjusts its own (*NO*) behavior to influence the other (see Section 5.8). Such instances are evident in the behavior of humans who take on the role of tyrant kings, corporate criminals, or even petty thieves who refuse to emerge from a circle of self-interest. The circle of DAH is completed as an infinitely small zero of positive values.

In most cases, the converse attitude for the preservation of social or ethical interest is rare, even though exceptional behavior (from Gandhi [5], Schweitzer [6], King [7], etc.) has been documented. It remains to be seen if the computational behavioral response invoked by the *adaptive* selection of the parameter values invoked by dynamic updating of m, n, j, and k, and their inter-dependent nonlinearities, will depict some of the more dominant responses of intelligent objects (and humans).

5.6.7.2 Ultra-Fast Response The execution of *kopc* is neutralized if the response of a noun object (*NO*) characterized in Section 5.6.7.1 is executed at ultra-high speeds. It is interesting that the *NO* has already realigned itself before the control signals arising from the *kopc* are completely executed. Such patterns are rare (at the current stage of scientific understanding) in the nature of macroscopic objects, even though nature is full of surprises.

Conversely, if the external environment exhibits an ultra-fast (and smart) response to all responses of *NOs*, then intellectual slavery builds. Under these conditions, it appears that the response domain of the intelligent noun object is charted out before any process in the KPU. The response-time that influences the response of objects becomes too slow (or immaterial) for the external environment to issue the most appropriate *kopc* for the *NO* to generate the exact response that the external environment dictates. Such responses have been witnessed in nefarious wartime prisons among captors and the captives of war.

5.6.7.3 Saintly Response Such a response is invoked from the machine by suppressing reason (and economic behavior) and over-emphasizing social responsibility, human values, and universal justice. The l, m, n, i, and j values turn highly positive and the value of x, or the change index, becomes highly positive. Historically, religious societies, church states, and prince pastors have left their imprint. Such societies (as history points out) have been transient, concluding the human struggle in the production/consumption domain, where the laws of microeconomics have rewarded individuals and laws of global economy have rewarded industrial nations.

In the latter instances, the globalization and enhancement of ethics and values for humanity become subservient to the heightened greed of a few in industrial societies. The norms of society have thus vacillated and the cycle of social change becomes evident as business and global economic cycles. The observations of individuals (such as Karl Marx [8], Milton Friedman [9], Thorstein Veblen [10], etc.) and institutions (such as the United Nations, Stanford Research Institute [11], Hoover Institution [12], Princeton Institute of Advanced Study [13], etc.) are ignored. However, it is our contention that if a knowledge machine tracks these steady, subtle, and slow changes over time and discovers cause–effect relationships between laws and downward drifts of morals and ethics in a quantitative format. A scientific basis for prolonging the ethical state of a nation is dynamically adjusted (indefinitely, if it is possible) and securely established.

5.7 FRAGMENTS OF OVERALL HUMAN ACTIVITY

All human activity is decomposable into fractional functions. Such fractional activities have order, pattern, flow, methodology, and organization. In the development of application software, the four components (order, fragmentation, stabilization and reintegration) that make up a robust system are actively pursued at both a local and global level.

From a global perspective, fractional activities form the links of a state diagram and the objects that undergo or participate in the major or fractional activities form the nodes of the same state diagram. The result of the transitions in the state diagram is the end-effect of human activity.

Traditionally, computing systems handle state diagrams in a logical and cogent manner. It is possible to represent the numerous objects that humans handle (or even perceive) in the allocated "object space" in the memory of a computer. It is equally possible to represent human activity as a microcode, an operation code, a macro, a subroutine, or even a full-fledged application program. Hence, a new breed of computer systems permits the representation, simulation, and even optimization of all human activities by operations research (OR) techniques by machines or by human counterparts who program such new machines.

5.7.1 Minor Tasks and Microcodes

Minor steps are executed by microcoded instruction in the control memory of a traditional CPU. In the KPU environment, minor steps in the execution of a *kopc* need to be classified and tied to the nonlinear relationship for l, m, n, j, and k. The procedure for obtaining Δ and δ to compute the change of attribute of an intelligent object is suggested in Figure 5.4. These relations are amenable as microcode instructions in the control memory of a behaviorally oriented KPU.

Alternatively, all numeric functions can be executed by the CPU within the KPU. The close alliance between the KPU and CPU functions has been delineated [3]. Logical and selective circuits can capture rudimentary human tasks. The collectivity of such tasks, their order, and their intensity clarify the situational differences in the response of an intelligent object. An element of uncertainty prevails even though it is bounded in most instances. Like econometric projections, behavioral responses have a numerically bounded pattern.

5.7.2 Major Ordeals and Complex Programs

Compartmentalized human functions and computer programs stem from one basic commonality: the *decomposition* of human functions into smaller manageable tasks and the *fragmentation* of computer programs into machine-executable instructions. On the one hand, human beings have an uncanny ability to "see through" a large number of steps to solve a problem. On the other, software designers have programming tools and techniques to systematically see through applications, programs, routines, assembly-level instructions, and executable code. In the intellectual space at the union of these two approaches lies a vacuum for a new machine with a processor that executes humanistic functions but with the same accuracy and dependability of the CPU of a computer.

5.7.3 The Microcode of the Humanistic Chip Set

The microinstruction set for the humanistic chip set is as premature as the instruction set for the limited number von Neumann machine (of 1945) was in 1940, or the library functions for the electronic switching systems (ESSs) (of 1965) were in 1960. Over a few decades, humanistic chips are likely to become as common as 486 chips during the 1980s. The solutions to human needs are generic and change with social etiquette. When a machine with humanistic chip sets are embedded in the PDAs of the next decade, then the solutions to a vast majority of daily problems can be resolved by a knowledge machine rather than human struggle. It appears only appropriate to hand over the human struggle to the chip sets. For instance, if the best subway connection between any two points in a city at anytime is to be determined, an algorithmic solution is feasible especially when the train schedules (and their most recent updates) are available over Internet or local databases. Furthermore, if the most recent research for bone cancer needs to be found, the chip set can finetune those findings with respect to the (intelligent and responsive physiology of the patient) object and its attributes, and then present an estimate of the value of each cure found for that particular object on a regular basis. More applications of such chip sets are plentiful.

The microcode within the chip set will further enhance its potential use by human beings. Such a design can be attempted at two levels: *(1)* the selection of instructions for object-oriented functions presented in [3] and *(2)* by a machine-executable reduced instruction set computer (RISC) based knowledge machine.

5.8 DUAL KNOWLEDGE PROCESSOR UNIT HUMAN INTERACTION MODEL

When there is an interaction between two life forms (e.g., cat and mouse, buyer and seller, boss and employee, dictator and subjects, etc.), the dual interaction process can be represented by two KPUs or two KMs. If the output of one KPU, that is, (*PNO**), is used to generate the *VF* (*kopc*) for the other KPU, and vice versa, then an entire lifelike interaction process is emulated by two KMs or by a KM with two KPUs. Cooperative (*l* and *m* are both + or −) and conflicting (*l* and *m* bear opposite signs) interactions can be emulated.

The nature of the equation that relates *PNO** with *kopc* affects the stability, convergence, or divergence of the interactive process of *NOs* with an almost lifelike response pattern. Whereas not all the intricacies of emotional responses may be presentable, the logical outcome is predicable. When the rules of engagement (interaction) are well defined, the range of outcomes becomes narrow. Rationality, objectivity, and well-defined goals lead to smooth convergence toward the outcome. Generally, oscillations are more likely and divergence is just as common. Divergence can also have a smooth or an oscillatory trajectory.

At one extreme, when both objects (processor object and the processed object) are inanimate and the processing capability of one object remains unchallenged, the entire process becomes an execution of binary instruction. Typically, the entire process includes fetching the instruction from the memory, decoding the instruction in the instruction register of the CPU [fetching the *operand(s)* from memory, if necessary], executing the *opcode*, and storing the result in a conventional CPU. At the other extreme, when both objects are unilaterally balanced and rational, a human interaction becomes imminent.

Most of the real-life environments lie between these two extremes. In the scientific literature, each of the processes is studied from the computer or the human perspective. An interactive negotiation results, when conditions are relaxed from either of the two extremes. Some of the scenarios are discussed in Section 5.6. The computerized response modes are generated in Section 5.5 when the object undergoing the process is intelligent and adaptive. However, the real-life situation is centered in the responses when both the processing and processed objects are intelligent. A human interaction results under these conditions. In order to close the gap of interactions at the two extremes, a perspective from the human side is essential for a KPU to function as an intelligent agent for the object imposing the process and the object undergoing the process.

5.8.1 Bilateral Human Relations

The bilateral human relation between two individuals X1 and X2, depicted in Figure 5.5, is a symbolic model of the interaction process. Both X1 and X2 have needs to be satisfied and assets to satisfy. However, social interdependencies force most individuals into a negotiating stance (dashed lines in Figure 5.5) when the needs of X1 may be adequately satisfied by personal assets or better satisfied by the assets of X2 and vice versa. Even though this computational model is symbolic and number-oriented, it reveals computational cycles and the instability that can arise in real life and in the computational processes, which follows the interactive process.

Largely, in just societies, the laws of fair trade tend to equate the net worth of assets exchanged, thus maintaining a framework for stable and repeated social interactions. Many variations and exploitations are possible due to ignorance, greed, or cruelty. To deploy such situations, the use of the five variables l, m, n, j, and k becomes appropriate. However, two such sets of variables (l_1, m_1, n_1, j_1, and k_1 for X1 and l_2, m_2, n_2, j_2, and k_2 for X2) become necessary. In addition, any nonlinearity between the attributes of X1 and X2 also needs consideration.

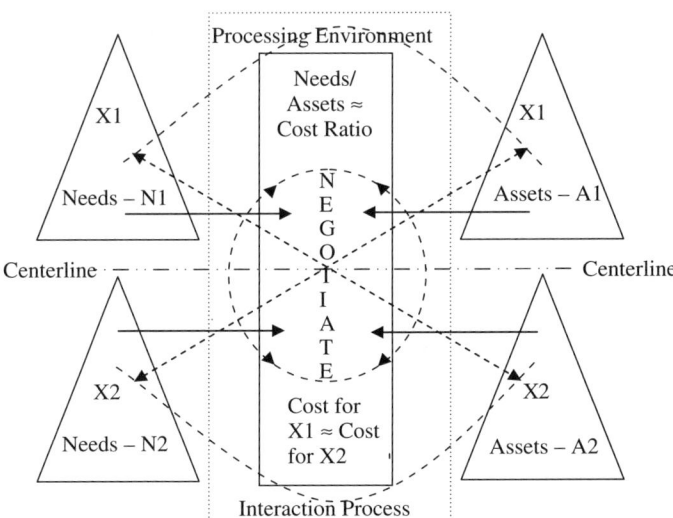

Figure 5.5 Depiction of the bilateral framework of relationships between two intelligent objects (or individuals). In the knowledge processor unit environment, each object influences the response in a symmetric fashion. Asymmetric relationships occur when one object has power or authority (such as social position, boss, ownership of resources, etc.) over the other.

Many thousands of types of human attributes, their nature, and numerical range of the variables and their interrelations account for the innumerable types of human interaction. The KPU of a knowledge machine is thus capable of simulating the entire spectrum, ranging from insignificant exchange of trivialities in a relationship to a cruel war between nations. Figure 5.5 depicts a bilateral interactive model for the interaction of two individuals. The role of the two individuals is reversible, and the centerline of symmetry runs horizontally through the computational model. Further elaboration of this diagram results in a more comprehensive computer model, as presented in Figure 5.6.

5.8.2 Models of Human Interaction

5.8.2.1 Human-Robotic Situations Numerous models for human interactions with machines, aerospace vehicles, jet fighters, etc., are available [14]. These studies report human reactions in complex and stressful settings. Medical and psychological responses have also been investigated from the perspective of offering robotic and medical devices for the sick and wounded [15]. Numerous research papers for designing robots to imitate human responses also exist. These research findings deal with military applications [16]. Most papers deal with programs to invoke a robotic response to imitate truly human reactions [17]. Studies detail decision-making models of human behavior that underlie human-like agents, groups, or entities, or explore convincing yet autonomous human responses [18].

The simplest models of human–human interactions were introduced by Eric Berne [19]. The paradigm of self-organization in human beings was skillfully presented from a realistic and human perspective by M. Nadin in 1987 [20]. The discussion raises numerous representational questions but fails to present a systemic model that can be used in a computational model to represent dyadic and triadic relationships between human beings. In a later 1995 publication, Nadin presents the process of negotiating from an ideal and aesthetic perspective. The document bears emphasis on virtual reality and the human comprehension required in such situations.

5.8.2.2 Human Interactions A systems model for human interactions ranging from interpersonal relations to labor management negotiations of the nature depicted in Figures 5.5 and 5.6 was studied during the late 1970s at Bell Laboratories by Ahamed [21] and the Stern School of Business Administration.[9] Mathematical models of human interactions are presented in Appendix B. Roman et al. [22] and Pen [23] depict the symmetric interactive processes. Figures B.1 and B.2 [24] illustrate the nature of the interactive process based on an economic exchange of assets offered and concessions received by both parties. Four

---

[9]The Stern School was known as New York University's Graduate School of Business Administration during the 1970s, and the three faculty members involved were Professors Norman Martin, Professor Robert Kavesh, and Professor Oscar Ornati.

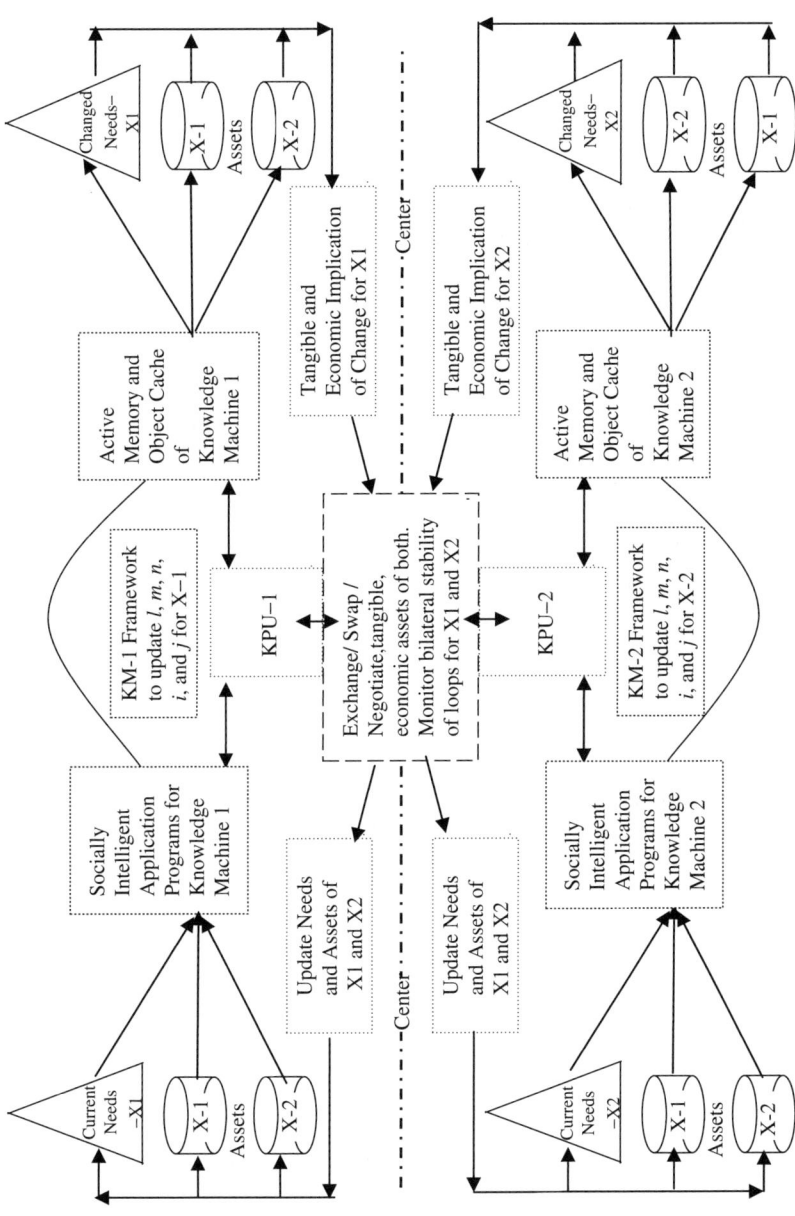

Figure 5.6 Schematic of negotiation process between X1 and X2 with updates for human variations l, m, n, i, and j.

possible outcomes predictable by the models are (1) smooth convergence to a mutually satisfactory result for the interacting parties; (2) oscillatory convergence; (3) oscillatory-divergence without any agreement resulting from the interaction; and (4) smooth divergence without an agreement. In such interactive processes, instability and oscillations result in a waste of time.

Such a framework becomes applicable as (intelligent) objects undergo processing in the KPU environment. The oscillations can take-on numerous forms in the intelligent object's behavior. On the one hand, the response can be the object's over-or underreaction, out-of-context responses to certain verb functions (VF); on the other, the response can be the opposite of that expected for any predefined VF. Rational or irrational behavior is feasible, thus invoking two (logical or illogical) reactions from the other party. Briefly, the rational-logical interactive mode leads to a convergent and sensible interaction, the irrational-illogical mode leads to the most severe (even destructive) oscillations and/or divergent interactions.

The control of such oscillations lies in the predictive capabilities of each or both parties in establishing final and desirable goals for both parties. In essence, the mechanics of the systematic processes of interaction between intelligent objects, in reality, is portrayed by the status of the objects and their attributes in knowledge machines. Intelligent objects and a knowledge machine would be able to function at two levels. Level 1 functions monitor progress toward the achievement of goals in a mutually beneficial fashion, whereas level 2 functions track the stability of the negotiating process and monitor the status of the objects and their attributes from becoming divergent, explosive, and mutually exclusive. The KPU thus functions at an "intellectual" level (level 1 functions) and at an "emotional" level (level 2 functions).

In the knowledge machine, the psychotic and abnormal behavior of intelligent objects is forced by implanting oscillatory and abrupt changes in the values of l, m, n, j, and k, for evaluating the response(s) of object(s). Bilateral instability in the parameter (l, m, n, j, and k) values is likely to be a cause for conflict and confrontation. A series of violent and turbulent changes can only lead to war and destruction.

REFERENCES

1. Software Industry Report, "Itex Shipped over 1 Million ADSL Chipsets," Millin Publishing, Inc., June 2005.
2. A. Maslow, *'Motivation and Personality'*, Harper and Row, New York., 1964.
3. J. Neumann and O. Morgenstern, *Theory of Games and Economic Behavior*, Princeton University Press, Princeton, NJ, 2004.
4. D. L. Mosher and B. B. White, "On Differentiating Shame and Shyness," *Journal of Mot. and Emotion*, 5, no.1, DOI 10.1007/BF00993662, year?:. 61–74.
5. M. K. Gandhi, *Gandhi, an Autobiography: The Story of My Experiments with Truth*, Beacon Press, Boston, MA, 1995.

6. J. Brabazon, *Albert Schweitzer: Essential Writings*, Orbis New York, 2005.
7. C. Carson, *The Autobiography of Martin Luther King, Jr.*, Warner Books, New York, 2001.
8. S. Aviner, *The Social and Political Thought of Karl Marx*, Cambridge Studies in the History and Theory of Politics, Cambridge University Press, New York, 2003.
9. M. Friedman, *Price Theory*, Walter de Gruyter, Hawthorne, New York, 1976. See also *Columbia Encyclopedia*, Columbia University Press, New York, 2002; M. Michael, *Dilemmas in Economic Theory: The Persisting Foundational Problems of Microeconomics*, Oxford University Press, Oxford, UK, 1999.
10. R. Tilman (Ed.), *A Veblen Treasury. From Leisure Class to War, Peace and Capitalism*, Armonk, NY, 1993.
11. L. Fasulo, *An Insider's Guide to the UN*, Yale University Press, New Haven, CT, 2005. See also W. B. Gibson, *A Story of Scientific Service to Business, Industry and Government*, Newcomen Society Stanford Research Institute, CA, 1968.
12. P. Duignan, *Hoover Institution on War, Revolution and Peace: Seventy Five Years of Its History*, Hoover Institution, 1989.
13. Princeton Institute of Advanced Study. Princeton, NJ, http://www.ias.edu/about/mission-and-history.
14. "Cognitive Modeling of Human Interaction with Complex Environments," http://www.calendar.cs.cmu.edu/hciiSeminar/799.html. Carnegie Mellon University, Human-Computer Interaction Institute, Pittsburg, PA.
15. http://www.cs.rpi.edu/news/colloquia/September22_2005.htm. Troy, New York.
16. R. Prasuraman, "Designing Automation for Human Use: Empirical Studies and Quantitative Models," cat.inist.fr/?aModele=afficheN&cpsidt=1459593,. Ergonomics ISSN 0014-0139.
17. J. F. Smith III and T. H. Nguyen, *Autonomous and Cooperative Robotic Behavior Based on Fuzzy Logic and Genetic Programming*, IOS Press, The Netherlands, 2007.
18. R. W. Pew and A. S. Mavor (Eds.), "Modeling Human and Organizational Behavior," in *Application to Military Simulations*, Panel on Modeling Human Behavior and Command Decision Making: Representations for Military Simulations, Commission on Behavioral and Social Sciences and Education, National Research Council, National Academy Press, Washington, DC, 1998.
19. E. Berne, *Games People Play: The Basic Handbook of Transactional Analysis*, Ballantine Books, New York, reissued August 27, 1996.
20. M. Nadin, "The Paradigm of Self-Organization," Fourth Annual Computer Graphics Conference (Applications on the Leading Edge), University of Oregon Continuation Center, Eugene, October 27–29, 1985; "Interface Design: A Semiotic Paradigm," in *Semiotica*, Vol. 69-3/4, Mouton de Gruyter, Amsterdam, 1988, pp. 269–302. Also see M. Nadin, "Negotiating the World of Make-Believe: The Aesthetic Compass," in *Real-Time Imaging*, Academic Press, London, 1995.
21. S. V. Ahamed, "Human Interactions, a Mathematical Framework of Human and Corporate Interactions," MBA thesis, written in collaboration with Professors Martin, Department of Management, R. Kavesh, Former Chair, Department of Economics, and O. Ornati, Department of Manpower Management and Social Studies, New York University, Stern School of Business, 1979.

22. S. V. Ahamed and E. G. Roman, "A Model-Based Expert System for Decision Support in Negotiating," Chapter 14.3 in *Modelling and Simulation Methodologies in the Artificial Intelligence Era*, M. S. Elzas and B. P. Zeigler (Eds.), Elsevier, New York, 1988.
23. J. Pen, "A General Theory of Bargaining," *The American Economic Review*, March 1952: 24–42.
24. E. G. Roman and S. V. Ahamed, "An Expert System for Labor-Management Negotiations," in *Proceedings of 1984 Summer Computer Simulation Conference*, North Holland.

APPENDIX A

HUMAN INTERACTIONS

The unpredictability of human interactions poses a serious problem in their mathematical representations. However, the situation is still manageable in the formulation of the type and extent of nonlinearity in the stimulus response that human beings exhibit. Chaotic behavior results if the human response is entirely unpredictable, but in most common and rational modes of behavior, the response of human beings is indeed predictable and most frequently nonlinear. The basis of nonlinearity can be highly variable, contextual, and verb-function-dependent.

In general, a machine has some recourse to an approximation of human behavior in view of the past history of responses generated for that particular noun object. Noun objects may be a human being, an organization, a nation, etc. The verb functions can indeed be any excitation function, such as love, hate, hunger, panic, etc., for a human); buy, sell, negotiate, produce, manufacture, etc., for an organization; or war, peace, import, export, order, disorder, etc., for a nation. Machines like human beings have a memory of what is expected when a verb function (*kopc*) is performed on a noun object (*operand*) in the KPU environment.

If a mathematical relation to the response is available to the machine in predicting the response, then the outcome of the particular *operand* is estimated and tracked down as noun objects undergo knowledge programs. Total certainty is not the forte of knowledge machines, and for this reason, the machine tracks the outcome and a numerical estimate of the operations of a series of verb functions (*kopcs*) on a number of participating noun objects (*operands*).

A.1 NONLINEAR RESPONSES

The representation of every possible response is not necessary since most noun objects (*operands*) exhibit simpler nonlinearities and the extent of nonlinearity resides in the computable domain. In this section, we present a limited set of nonlinear functions and their oscillatory variations. In most instances, the combined nonlinear responses from numerous objects initiate the instability of larger groups of noun objects. Oscillatory response also leads to wide variations in the response of a collectivity of noun objects.

A.1.1 Exponential Representation

Increase or decrease is represented as exponential functions, and X and/or Y displacements may be added or subtracted:

$$Y = \exp(+ \text{ or } - X); \quad \text{Increase or decrease, exponential}$$

$$Y = 1/X; \quad \text{Decrease, reciprocal}$$

A.1.2 Memory Effects

Memory effects can be represented in numerous algebraic forms. The exponential decay of an intelligent object response is very common as the object tends to forget its response (Y-axis) over a period of time to verb functions (X-axis), as shown in Figure 5.A.1.

A.1.3 Saturation

Saturation can be represented in numerous algebraic forms. Displacement effects may be included by adding or subtracting incremental values from X and or Y:

$$Y = 1 - \exp(-X)$$

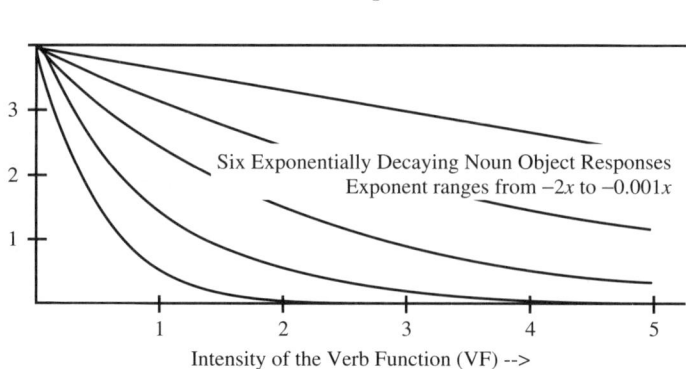

Figure 5.A.1 Exponential decay $[Y = \exp(-X)]$ in the response of the noun object to a verb function or its intensity *VF*.

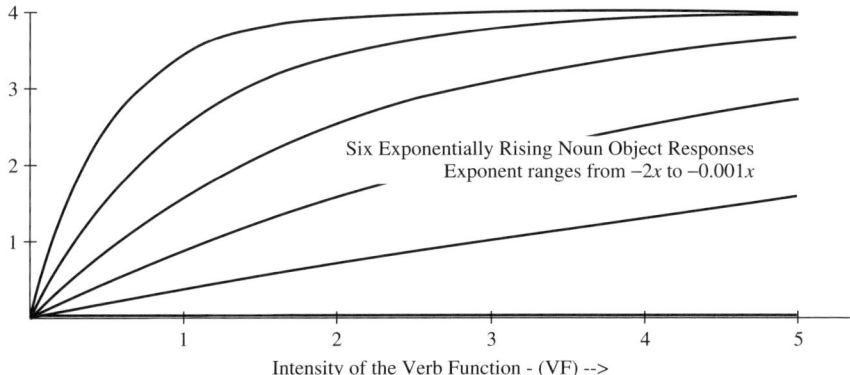

Figure 5.A.2 Exponential rise $[Y = 1 - \exp(-X)]$ in the response of the noun object to a verb function or its intensity VF.

The shape of this curve is graphically depicted in Figure 5.A.2. The saturation occurs at 4.0, but this value is arbitrary and can be adjusted in the formulation of the curve. An alternative representation of saturation effects is expressed as

$$Y = X/(a + b|X|)$$

A.1.4 Peaky Response

The peaky response of noun objects can be represented as the difference of two exponential functions such as $[Y = \exp(-aX) - \exp(-bX)]$. The two parameters a and b (with $b > a$) create a peak. The peak is determined by differentiating with respect to X and equating it to zero. The shape of the response curve is shown in Figure 5.A.3.

Alternative saturation effects are also depicted in Figure 5.A.4. A response of this type is seen when the noun object suffers from a delayed response and a lack of resources, energy, power, interest, etc., to sustain a high-level response. The response dwindles and the direction of change instituted by the VF prevails, provided that VF has enough resources, energy, power, interest, etc., over long periods.

The shape of this saturation curve, graphically depicted in Figure 5.A.4, has a saturation value at $Y = \pm 1/b$. Dual saturation effects occur as X ranges from $-\infty$ to $+\infty$. The curve varies smoothly and the differential of this curve is also a well-behaved function.

A.1.5 Hysteresis Effects

Such effects are represented by laterally displacing two nonlinear exponential $[Y = 1 - \exp(-X)]$ curves. A series of such curves is shown in Figure 5.A.2. The curve is displaced by $+t/2$ and starts at $X = +t/2$. The curve b traces a

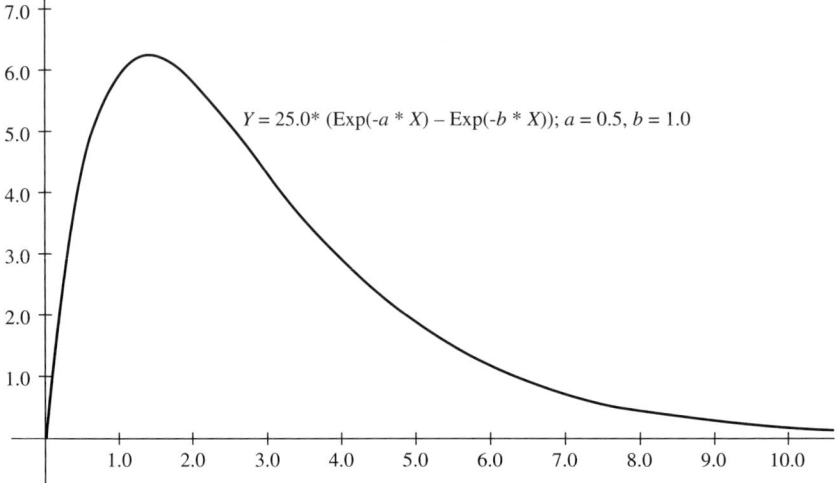

Figure 5.A.3 Peaky response can be difference of two exponential curves. The constants 25.0, a, and b, in the equation are unimportant, and they are adjusted to suit the particular noun object characteristics.

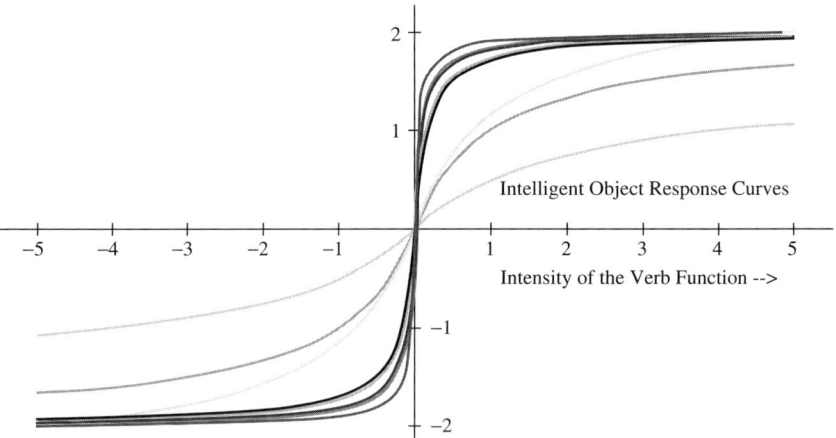

Figure 5.A.4 Saturation $[Y = X/(a + b|X|)]$ in response (Y-axis) of the noun object to a verb function or its intensity VF. Different values of the parameters a and b generate the series of curve shown, and their values are adjusted to suit the past response of the intelligent object under the influence of similar verb functions (X-axis).

return path and intersects the X-axis at $X = -t/2$. The paths of the curves a and b are rotated counterclockwise by $180°$ to complete the entire hysteresis loop. The horizontal width of the loop at $Y = 0$ becomes the thickness t. This figure is not shown.

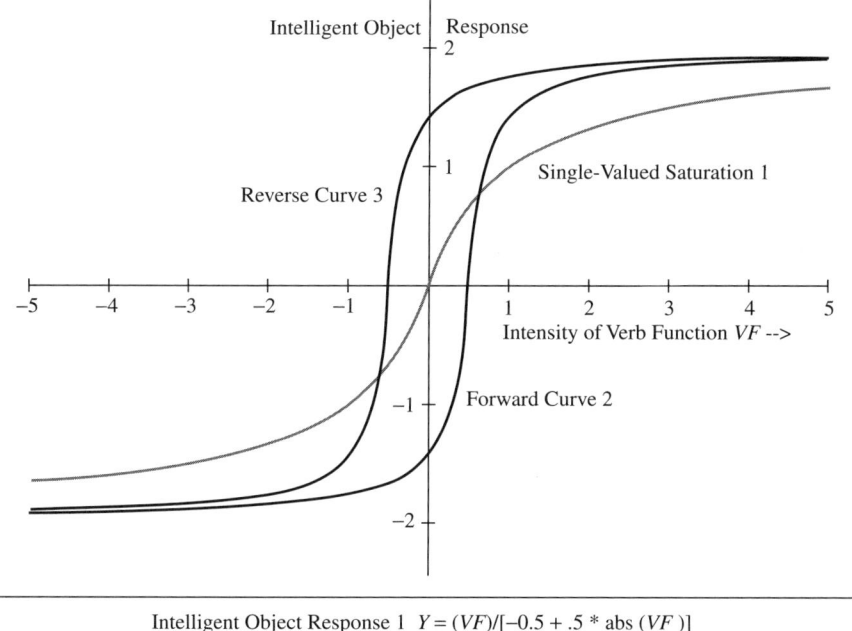

Intelligent Object Response 1 $Y = (VF)/[-0.5 + .5 * \text{abs}(VF)]$
Forward Intelligent Object Response 2 $Y = 5.0 * (VF - 0.5)/[0.5 + 2.5 * \text{abs}(VF - 0.5)]$
Reverse Intelligent Object Response 3 $Y = 5.0 * (VF + 0.5)/[0.5 + 2.5 * \text{abs}(VF + 0.5)]$

Figure 5.A.5 Hysteresis loop generated by combining responses 2 and 5.

For the second type representation, $[Y = X/(a + b|X|)]$, the values of X can range from $-\infty$ to $+\infty$. Any saturation curve shown in Figure 5.A.3 can be displaced by $+t/2$ and then by $-t/2$ to generate a hysteresis loop effect. A typical curve thus derived is illustrated in Figure 5.A.5.

The area in the loop[10] represents the resources consumed by VF to make a (willing or unwilling) intelligent object go through the entire cycle of reversal in its response. The willing/cooperative noun object traces the narrower loops consuming fewer resources (energy) and vice versa.

A.2 OSCILLATORY RESPONSE OF NOUN OBJECTS

Intelligent objects can adjust their response depending on the magnitude and direction of the verb function (VF). Such a response is frequent as a low-level organism responds to the VF (e.g., threaten: fight/flee; hunger: kill/scavenge; acquire: beg/borrow/steal, etc.). The parameters l, m, n, j, and k contribute

---

[10] Such effects also occur in magnetic field problems when flux density B and H are plotted on the Y- and X-axis. The area enclosed in the $B - H$ loop represents the "hysteresis loss" per cycle. The easily magnetized (soft magnetic) materials have lower loss.

to oscillations. The shape and relative magnitudes of a and b (Section A.1.5 and Figure 5.A.5) can contribute to oscillations in the responses of the noun objects (NOs).

There is no particular approach to prevent the instability of the solution in the KPU tracking the response or attributes of NOs. However, underrelaxation of the parameters m, n, j, and k is likely to offer smoother convergence in the response of NOs. Generally, this methodology works satisfactorily in the solution of nonlinear field distribution problems.[11]

A typical oscillatory nature is shown in Figure 5.A.6 for a peaky response of a noun function. Numerous agebraic equations exist that generate oscillatory response functions and are akin to the representation of electrical current in series and parallel R (for resistance), L (for inductance), and C (for capacitance) circuits. In the human/intelligent object environment, such responses are invoked when the VF operates on a resistive, emotional, and/or overly reactive noun object.

Oscillations can be detected riding any of the modes of response presented in Figures 5.A.1–5.A.5. In Figure 5.A.6, exponentially decaying oscillations of a constant frequency are shown atop the response in Figure 5.A.4. In reality, the oscillations do not have to exhibit constant frequency or decay. Both unstablity and divengence can coexist in the chaotic or unruly behavior of noun objects reacting to verb functions.

Figures 5.A.7 and 5.A.8 depict two other oscillatory responses. An exponentially rising response of a noun object function in Figure 5.A.2 has a decaying oscillatory response in Figure 5.A.6.

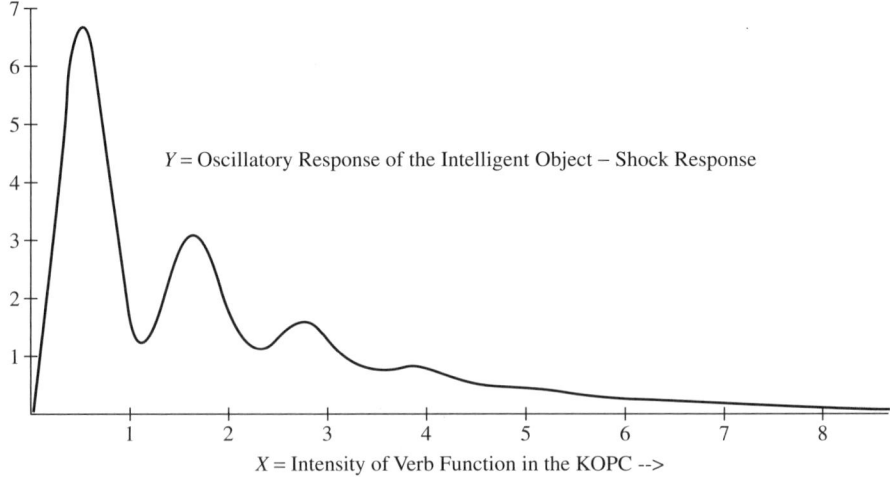

Figure 5.A.6 Oscillatory response (Y-axis) of a noun object with a strong initial response that decays to no response as the verb function intensifies (X-axis).

---

[11] Such as magnetic saturation and the distribution of the vector potential A, or in the determination of the nonlinear varactor capacitance field distribution.

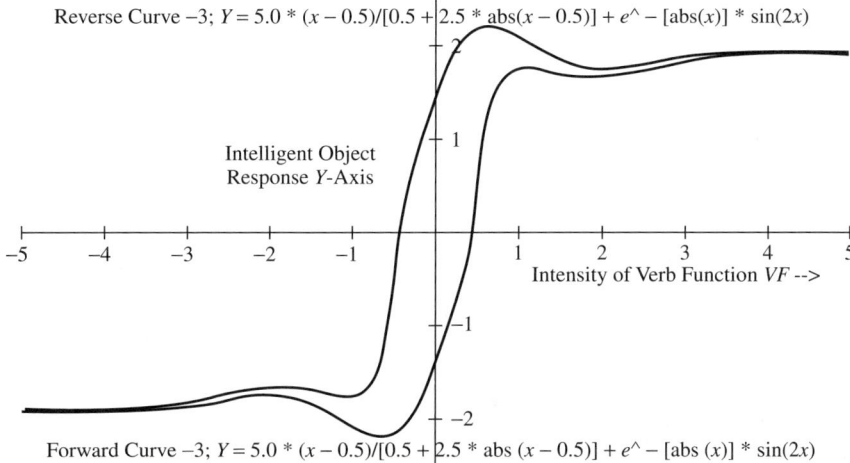

Figure 5.A.7 Oscillatory response (Y-axis) of a noun object trapped in the hysteresis loop by forward and reverse excitation of the verb function.

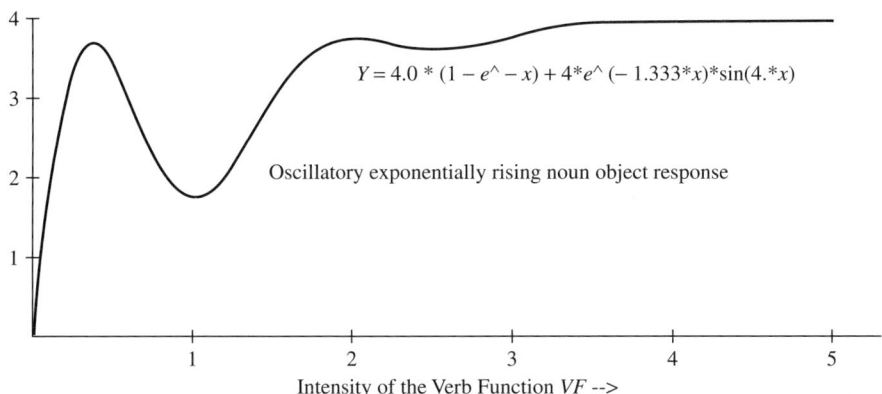

Figure 5.A.8 Oscillatory response (Y-axis) of a noun object with saturation.

A hysteresis loop shown in Figure 5.A.5 can also suffer from oscillatory convergence at the saturation values. Figure 5.A.7 depicts the response of one such noun object. Oscillations in the response generally lead to instability in the solution. Slow and gradual control of the parameters can lead to numerical convergence to the final solution in the KPU.

APPENDIX B

KNOWLEDGE MACHINES FOR HUMAN INTERACTIONS

Human and social interactions are the ultimate manifestations of processes between intelligent objects.[12] These processes are inter-twined and inter-dependent. In perfect bilateral relations, the interactive process is symmetrical and the participation of individuals may be exchanged. However, in reality, any two humans or intelligent objects rarely occupy identical situations and may not be spontaneously replaced, one for the other since experiences and knowledge bases associated with each individual/object offer unique personality signatures, thus making every interactive experience/event highly variable, yet retractable and rational.

The role of knowledge machines (KMs) and humanistic chips is to track the entire scenario precisely and enforce rationality at the microprocess level. Flowcharts, graphs, PERT diagrams, or critical path algorithms make the sequence of steps an interpersonal or a social event. Emotional content and "thrill" are weighed and considered to set the course of events that follow even though they are not measured and quantified.

There are three sections in the rest of this appendix. Section B.1 addresses individual human interaction, and we fall back on the writings of psychologists (Maslow [B.1.1], Fromm [B.1.2], Berne [B.1.3], etc.) and clinical psychiatrists for the basis of this model. Section B.2 addresses labor management negotiations,

---

[12]We do not uplift intelligent objects to the level of humans or downgrade humans to the level of intelligent objects. However, the distance between the two *appears* to be continuous, more easily scaled by processor chips and neural nets rather than by human minds and neurons.

Computational Framework for Knowledge. By Syed V. Ahamed
Copyright © 2009 John Wiley & Sons, Inc.

and we fall back on the writings of Pen, Stevens, Hicks, etc., for the basis of this model. Section B.3 deals with inter-corporate negotiations, and we fall back on the writings of Elas, Zeigler, Roman [B.1.4] etc., for this model. In the later two instances, the interactions assume the role of negotiations and usually result in a legally binding contract that carries substantial economic value. The role of knowledge machines becomes significant when the models presented in this chapter depict the negotiations between nations and institutions.

In the following sections of this appendix, we introduce the role of knowledge machine in human interactions. These machines streamline the essential processes for smooth interactions and allow human beings to enhance the creative component of significant individual, social, corporate, communal, and national interactions. In the industrial environment, we have allowed robots to perform routine production functions. In the accounting profession, we have allowed computers to perform numerical and rule-based functions. In the teaching profession, we have allowed distance-learning methodologies to bring classrooms to remote locations, etc. In the social domain, the role of machines becomes more elaborate because of the immediate interface they provide for human beings. Hence, some of the rudimentary human functions become part of the machine functions. For systems designers, some of the rudimentary machine functions become part of their human functions.

In proposing knowledge machines for social functions, we extensively use artificial intelligence (AI) techniques. Expert systems, pattern recognition, intelligent agents, learning, and robotics are used in combination with rationality, wisdom, and dynamic gaming strategizing derived from human experience. Feedback control, system stabilization, rule enhancement, and dynamic derivation of windows for targeted goal achievement are borrowed form engineering and corporate management and implanted in the traditional AI environment. Such composite software or its hardware embodiment offers new varieties of programmable tools that are humanistic rather than robotic.

B.1 KNOWLEDGE MACHINE FOR HUMAN INTERACTIVE PROCESSES

The dyadic relationship between two individuals, X_1 and X_2, is shown in Figure 5.B.1. The symmetry around the centerline depicts the identical role of two equally positioned parties in the dyadic role. The content of boxes A_1, B_1, and C_1 being distinct from that in A_2, B_2, and C_2 leads to the beginning, middle, and end of the interaction. There are three paths in the diagram—namely, loop 5, 8, 13, 5 (denoting a stable relationship), path 5, 8, 10 (denoting mutual termination of the relationship), loop 5, 12, 15, 5 (for an X_1 negative, X_2 positive relationship). Conversely, three image paths exist for X_2 in the other half of the figure. Largely, the asymmetric relationship is transitory unless the modification, enhancement, and/or adaptation of the reaction of X_1 and/or X_2

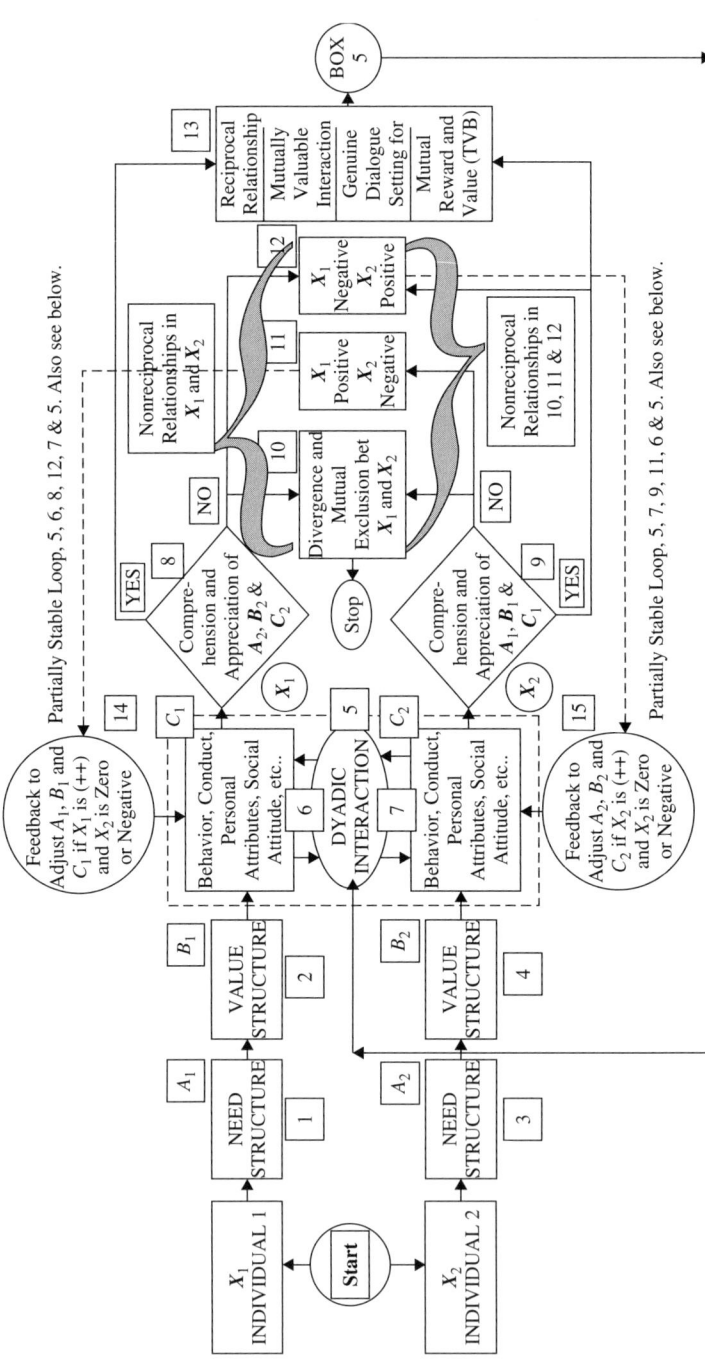

Figure 5.B.1 A systems representation of the dyadic interaction between two individuals X_1 and X_2. The centerline indicates the bilateral nature of the interaction. Stable (Box 13), one-sided (Boxes 11 and 12), and mutually exclusive (box 10) relations are shown after iterative runs through the four dominant loops in the diagram.

220

occur during the traverse of the two asymmetric loops. Positive adaptation leads to lasting and mature relationships and vice versa.

It becomes feasible to convert the movement for X_1 and X_2 in Figure B.1 to numerically controlled processes within the boxes and progress between boxes. The intensity of needs in boxes 1 and 3 moderated by the code of ethics in boxes 2 and 4 leads to boxes 6 and 7 starting the dynamics of interaction in box 5.

The extent of concession (degree of adjustment) of each individual usually bears a monotonic relationship to the intensity of needs in most rational interactions. A quantitative relationship between needs and resources leads to the modality of behavior, conduct, and deployment of personal attributes in boxes 6 and 7. An overlap of appreciation for each other's behavior, sincerity, conduct, and personal attributes occurs in boxes 8 and 9. Economic and utilitarian perspectives lead to boxes 10, 11, 12, or 15. After both individuals weigh and consider the entire process based on laws of marginal utility, the interaction develops a flavor of being stable and sustained or transient and volatile.

It is interesting to note that the equations based on the marginal utility theory dictate the magnitude and direction of force between the boxes of this diagram in Figure 5.B.1. The attractive force of bondage (love) or repulsive force of exclusion (hate) is governed by the ratio of the marginal gain (benefit) derived to the marginal cost (expense) incurred. Some of the equations can become highly nonlinear, causing violent swings in the relationships. Even though Marshall [B.1.4] addressed the notion of marginal utility, clinical psychologists have observed the nonlinear swing of behavior in human beings.

Various personality attributes (concern, responsibility, commitment, and participation in creativity) suggested by Eric Fromm [B.1.2] enter the complexity of how human beings learn to filter out transients and sustain relationships even though the result of instantaneously solved relations may indicate a violent and hateful swing in behavior. Human wisdom prevails, but even wisdom can have a vectorial (i.e., directional and quantitative) representation [B.1.5].

REFERENCES

B.1.1 A. H. Maslow, "A Theory of Human Motivation," *Psychology Review*, 50, 1945: 370–396.

B.1.2 E. Fromm, *Anatomy of Human Destruction*, Holt, Rinehart and Winston, Austin, TX, 1975. See also E. Fromm, *The Art of Loving*, Harper and Rowe, New York, 1954.

B.1.3 E. Berne, *Games People Play: The Basic Handbook of Transactional Analysis*, Ballantine Books, New York, reissued August 27, 1996.

B.1.4 R. Arena and M. Quere (Eds.), *The Economics of Alfred Marshall: Revisiting Marshall's Legacy*, Palgrave Macmillan, New York, 2005.

B.1.5 S. V. Ahamed, "The Architecture of a Wisdom Machine," *International Journal of Smart Engineering Systems Design*, 5, no. (4), 2005: 537–549.

B.2 KNOWLEDGE MACHINE FOR LABOR-MANAGEMENT NEGOTIATION

In this section of the appendix, we present a situation where the "process" of interaction occurs between two social entities: the labor union and the management team. In this quasi-scientific approach, the numerous microsteps follow order and a methodology. Such processes are also encountered in software design and programming that lead to the solution of a problem. The former processes occur between two intelligent parties (objects), and the later processes occur as a result of a series of operation codes on numerical and logical symbols.

Labor management negotiations generally produce a significant document, that is, a "contract" between the labor union and management of a corporation. This rather elaborate and complex process is fist depicted by a systems model for most of the labor-management negotiations in industry. The structural arrangement between various factors that influence labor demands from management is discussed. The model has two feedback loops for the negotiators for either side. Such negotiations may lead to a settlement, an impasse, a strike, or a lock-out. The conditions for the success or failure of the negotiations are delineated. Chamberlain's concept of bargaining power and Hicks's concept of the eventual intersection between union concession curves and employers resistance curves are incorporated in the model. The influence of socioeconomic conditions, the duration of the negotiations, and the way in which this impacts the direction and dynamics of the negotiations are discussed.

B.2.1 Background of Labor-Management Negotiations

The writings of C. H. Stevens [B.2.1], based on the conflict-choice theory of Dollard and Miller [B.2.2] and contributions of Luce and Raiffa [B.2.3], derive from their game theory concepts. Let us shed some light—however dim it may be—on the strategic and communications aspects of the real labor-management bargaining process. In *The Theory of Wages*, Hicks [B.2.4] stated that both labor and management are threatened by a strike or lock-out, and the two are negative goals for both. Both parties approach the terms of the other as the length of the strike (or lock-out) increases. Hicks defines the management wage rate as the monotonically increasing relationship between wages offered by management as a function of the length of the strike. The labor wage rate is defined similarly, as the wages labor is willing to accept as the length of the strike increases. The labor wage rate is perceived as a monotonically decreasing function. From Hicks's perspective, if perfect mutual knowledge of these wage rates exists, there is never any need for a strike or lock-out because of the eventual intersection of the two wage rate relations. In this section of the appendix, we use this concept as the basis of the two symmetric feedback loops developed in the third section. These paths tend to bring both parties toward a successful negotiation in the end.

J. Pen [B.2.5] studied the bargaining mathematical treatment of "ope-limits," and Stevens equated them to utilities. Here, the basis for negotiations is as follows: Each bargainer will strive for a better wage rate W'. If one of the parties is at wage rate $W(<W')$, then the bargaining continues until the expected value of the desired gain balances the fear of total conflict and a loss of W. In this section, we use this concept in the reevaluation of the incremental changes of the net gain every time the inner feedback loop (developed in the Sections B.2.3 and B.2.4) is iterated during the negotiation process in the short run.

B.2.2 Constituents of Model for Labor–Management Negotiations

Maslow [B.2.6, B.2.7] categorized and arranged individual needs into a needs triangle. These concepts can be extended for organizational needs, leading to an organizational structure as depicted in Figure 5.B.2. If organizational goals and needs are correlated to organizational needs, the development and practice of sound technological, economic, and innovative skills in the pursuit of profits and the practice of good social, labor, and community relations force a set of conglomerate objectives of concern to management. In essence, the existence of a needs hierarchy for organizations has been recently accepted to the extent that it is secondary in nature in order for the survival of the organization. This need to survive thus constitutes a dominant underlying goal. If organizational goals and needs are correlated (as they are in individuals), then these goals display the basic principles underlying management literature, and from a needs hierarchy, one may interpret the existence of a local structure [B.2.8] for organizations. The combination of a needs hierarchy and goal structure yields management's expectations [B.2.8] of labor in relation to the compensation that is paid.

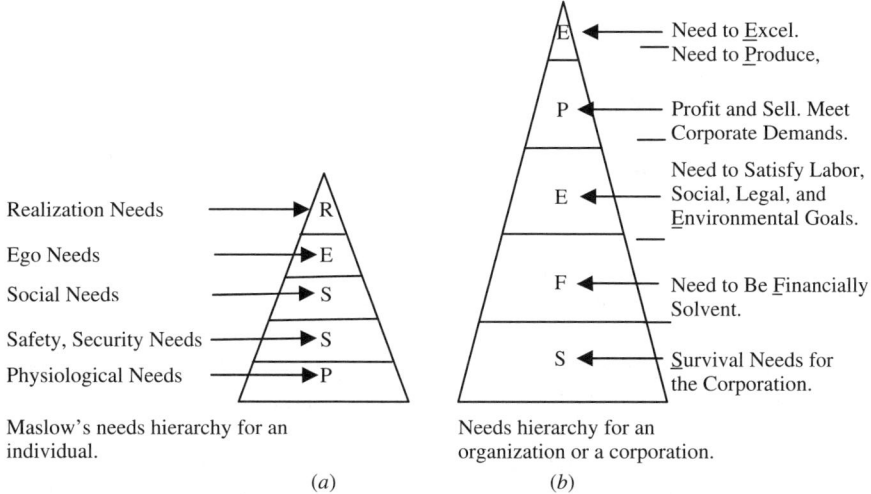

Figure 5.B.2 Comparison between the needs hierarchy of an individual and the needs hierarchy of an organization.

224 HUMANISTIC AND SEMI-HUMAN SYSTEMS

The ratio of the dollar value of those expectations to the compensation yields the demand intensity of management (to be discussed in the next section) negotiated during the bargaining sessions. The needs triangle, value structure, and demand intensity of management do not have to be static, unchanging, and absolute. For the survival of the organization, these parameters change with socioeconomic conditions and the leadership style of company executives.

The institutional writing about labor unions is immense and elaborate, but their discussion of behavioral aspects is limited. P. Henle [B.2.9] reviewed the typical union's bargaining proposals and cast light on its goals. The four major goals documented are (1) sizable income gains, (2) better security against sickness, old age, and unemployment, (3) increased leisure, and (4) safeguards against expanding technology and industrial structure. One may attempt to construct the union's needs triangle and its own goal structure. Being more democratic than bureaucratic, the objective of the union tends to be greatly influenced by the common will of the majority, which blends with the tradition and personality of leaders and transforms itself into the goals of labor. Such goals may be detrimental to the survival of the firm or the industry in general. In one such case, sustained demands for higher pay from coal miners rendered the coal mining industry economically infeasible. On the other hand, labor goals may be moderated by the reality of the situation and react accordingly (e.g., the garment industry in New York City has kept its wage demand low, but maintained the employment of its members and survival of the industry).

If one can accept the existence of a needs triangle for the union and its goal structure—however vaguely defined these may be—then one can interpret the labor intensity of demand as the ratio between the value of compensation (i.e., the extent of needs satisfaction) and value of the services they offer in return.

B.2.3   Demand Intensities of Labor and Management

During bargaining, union and management face each other in a totally monopolistic situation. Union and union alone can provide the labor force in the firm. Management and management alone can provide employment for all the workers. Consider the demands and offers between the two. Management expects and demands that labor participate in the production function, respond to overtime calls, retain good working hours, etc. (the collectivity of which is called the services of labor), for which management is willing to pay certain wages, provide benefits, assure certain security, etc. Labor expects and demands that management pay certain wages, provide benefits, assure security, etc., for which labor is willing to satisfy the labor demand intensity of management (DIM) calculated as the ratio of the dollar value of services expected from labor (as it is computed by management) to the net dollar value of all the compensations offered. The latter are quite accurately evaluated. Similarly, the demand intensity of labor (DIL) can be calculated as the ratio of the dollar value of all compensations offered to the dollar value of the services of labor (as it is computed by union leaders). Hence,

DIM and DIL are defined as follows:

DIM = (Management's perception of the dollar value of services demanded)

÷ (Dollar value of the net compensation offered)

DIL = (Dollar value of the net compensation demanded)

÷ (Union's perception of the dollar value of services offered)

Two perceptual problems and three realistically computable numbers arise in the determination of the DIM and DIL. It can be seen that when these ratios tend toward unity, all negotiations lead to a settlement. Initially, both labor and management tend to bias these ratios in their own favor, and the actual process of negotiation determines the convergence of the ratios and trajectories that they trace in reaching their final value.

B.2.4 Structure of the Model for Lasor–Management Negotiations

Figure 5.B.3 shows the structural arrangement of the various factors that influence the actual bargaining process. The model is symmetrical about the centerline. The common activities and outcomes of the negotiation process are placed along this line. The odd (above the centerline) and even (below the centerline) number boxes represent the goals and activities of management and the union, respectively. An odd-even pair of number boxes for management and the union depicts corresponding functions or goals.

The realistic calculation of DIM and DIL, which is perhaps the most significant step in the negotiation, is depicted in boxes 7 and 8 and made part of an iterative feedback process. Arrows indicate the time sequence of the various activities, and the function of negotiation is represented by a box with a broken line contour. The contents of this box vary considerably, depending on the type and stage in the negotiation process. The nonunique character (like the flow diagram of a knowledge machine program) of the model permits a slight rearrangement of the boxes and their contents for specialized cases such as bargaining under compulsory arbitration.

B.2.5 Negotiation Process and Validation

Management and the union reach the bargaining table with their respective solutions about DIM and DIL. During the first session [B.2.5], management usually receives the union proposals (boxes 10–12). The boxes 8, B, and 1 of the model correspond to this function in light of the preparatory process for the union, in which the demands are spelled out to reflect its needs and goals (boxes 4 and 6). Management also undertakes a similar "preparatory process" [B.2.10, B.2.11], in which the broad objectives of management are spelled out to reflect the union's needs and goals (boxes 3 and 5). In the sessions that follow [B.2.11, B.2.12],

226 HUMANISTIC AND SEMI-HUMAN SYSTEMS

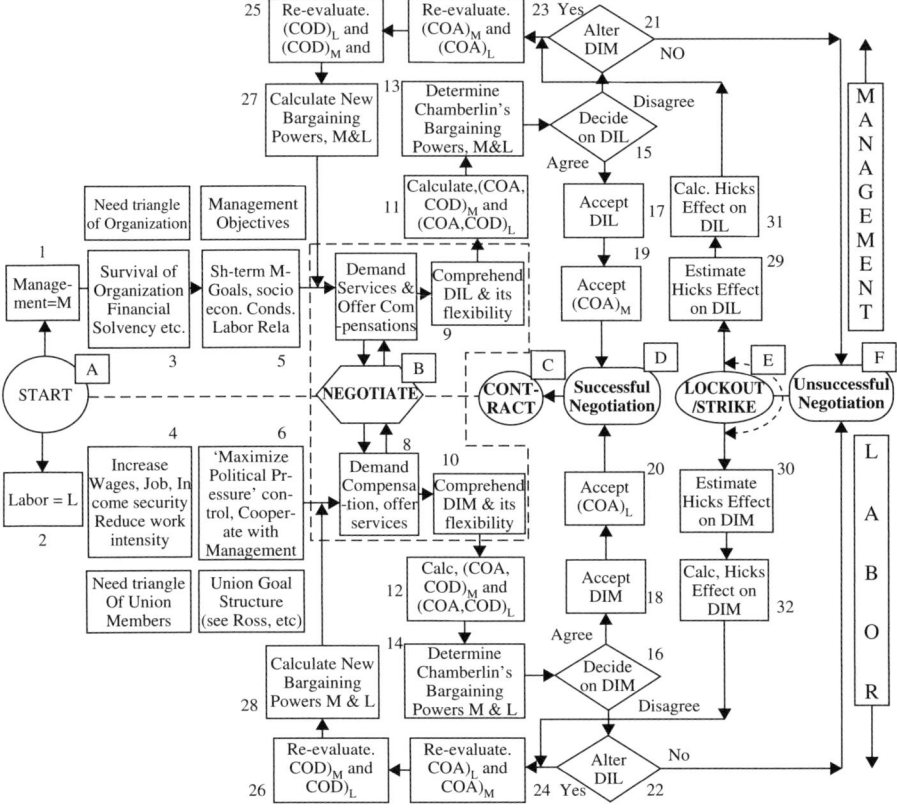

Figure 5.B.3 A systems model for labor–management negotiations.

management classifies and comprehends the demands set forth by the union and interprets the flexibility in these demands (box 9).

The next sessions generally involve the company's counterproposal, and it is in the sessions that follow that actual negotiation (boxes 7–32) takes place. The cost of agreement (COA) and cost of disagreement (COD) are the price tags in dollars computed by each party when an agreement is reached or a disagreement is discussed. This concept is borrowed from Chamberlain's writings [B.2.12]. Ross's effect incorporated in boxes 15 and 16 pertains to the influence of "political pressure" and "orbits of coercive comparison." This effect tends to evaluate the current demands and offers in light of those made by similar unions and managements.

Generally, the union and management traverse their own inner and outer feedback loops (boxes 8, 10, 12, 14, 16, 22, 24, 26, 28, 8 or the loop 8, 12, 14, 16, 22, F, 30, 32, 22, 24, 26, 28, 8 for the union) numerous times before reaching an agreement (box D). Furthermore, the looping may take place for each of the union demands. Only a subjective and numerical processing of this nature for each demand from either party results in a settlement.

B.2.6 Successful Negotiation

It is important to note that labor or management cannot remain trapped indefinitely in the loops, if Hicks's effect is present. Every time either party traverses its own respective loop, it is nudged closer toward agreeing with the other party. In fact, it is expected that the union must traverse the inner loop 8, 10, 12, ..., 28, 8 before it makes a "re-proposal" [B.2.11] in the face of a "counter proposal" [B.2.11] by management, which in turn has to traverse its own inner loop 7, 9, 11, ..., 27, 7 before the counterproposal. These loops (which are programmed in a knowledge machine) yield the flexibilities that are necessary to reach or predict the outcome of the terms of a settlement contract.

B.2.7 Impasse in Labor–Management Negotiations

Next, consider the outer loop for management (7, 9, 11, 13, 15, 21, F, 29, 31, 23, 25, 27, 7) and the outer loop for the union (8, 10, 12, 14, 16, 22, F, 30, 32, 34, 26, 28, 8). If Hicks's effect, that is, the convergent time variations of the proposed DIM and DIL, as shown in Figure 5.B.4, is not present, then the feedback paths become blocked at boxes 29 and 30 and the negotiation process can stop at box F.

However, the path between the boxes F and 29 for management and boxes F and 30 for the union need not be through box E. In these cases, the last contract.

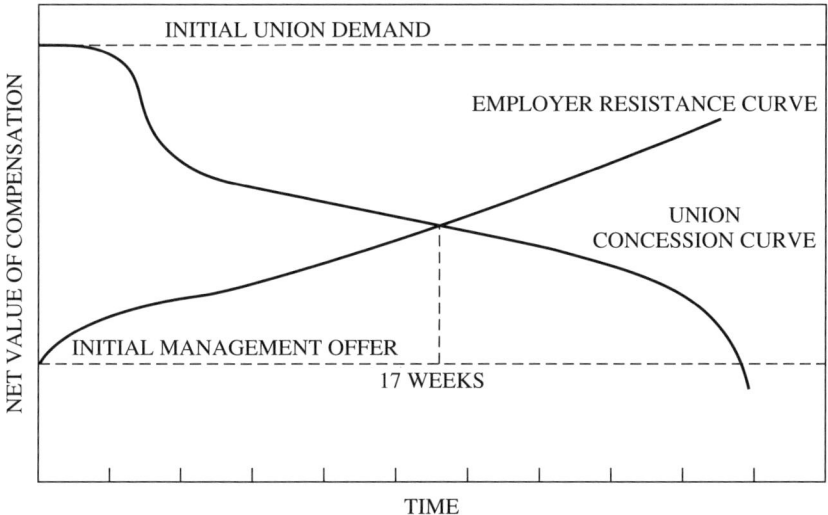

Figure 5.B.4 Hicks's concepts [B.3.4] of the dynamics in bargaining during labor–management negotiations.

81 may be extended or a 90-day cooling-off period imposed. Boxes 29–32 take into account the monotonically time-decreasing nature of the union concession curves and the monotonically time-increasing nature of the employer resistance curves [B.2.12]. These effects become dominant only if a strike or lock-out has occurred, or to some extent if the negotiation has become prolonged and both parties exhibit an inherent aversion toward a real deadlock or an impasse.

When both DIH and DIL become time-invariant and the number of sessions does not alter the perceptual factors in the estimation of DIM and DIL then the COAs and CODs need re-estimation. However, if the CODs can be tolerated by each party, then a no-win condition may persist indefinitely.

Other stable conditions are not feasible in the model and they are not feasible in practice. Any variations of the attitude or dollar values of either party are reflected in the tendencies for time sequencing of the steps in the model. The effects of socioeconomic conditions exert relative influence on the bargaining powers of the union and/or management. Creative aspects of negotiations generally alter the factors or combinations of demands and concessions influencing DIH and DIL, but rarely impact the nature of the negotiating process depicted in this model.

In the cooperative roles between labor and management, it would be desirable to yield to the items demanded by the other, especially when the COA is low. Alternatively, this strategy calls for maximizing the ratio of benefit for the other party to the cost incurred in providing the benefit. Conversely, in the conflict situation, one (rightfully or unjustly) maximizes the ratio of the damage (e.g., wars) inflicted on the other by disagreement to the cost.

In the actual negotiations, it is generally advantageous to be aware of these ratios even if the values are estimated. In some cases, it is necessary to combine two or more demands in these calculations and estimate the combined effect, rather than treating each item independently, and a secure and stronger bargaining position results. In labor–management negotiations, the influence of deadline and the asymmetry of tactical advantages cause considerable swing in positions and relative bargaining pavers. Bluffs generally are squeezed out of the negotiations as the hour of a strike approaches and reality dominates the discussion. Hence, strategies are creatively deployed to make significant gains at the least expense to the other party.

B.2.8 Conclusions for Labor–Management Negotiation

The model presented in this section of the appendix establishes a possible framework for the negotiation process. The two adaptive feedback loops govern the short- and long-term processes of labor–management negotiations. The effect of time, socioeconomic conditions, relative bargaining powers, and the effects of the number and duration of the sessions can be incorporated in the model. This model should be considered as a combination of factors and as an extension of the work of earlier researchers rather than as a replacement for their original contributions.

REFERENCES

B.2.1 C. H. Stevens, *Strategy and Collective Bargaining Negotiation*, McGraw-Hill, Hew York. 1965.

B.2.2 J. Dollard and N. E. Miller, *Personality and Psychotherapy*, McGraw-Hill, New York. 1965.

B.2.3 R. D. Luce and H. Raiffa, *Games and Decisions*, New York, 1957.

B.2.4 J. R. Hicks, *The Theory of Wages*, Macmillan, London, 1932.

B.2.5 J. Pen, "A General Theory of Bargaining", *American Economic Review*, March 1952: 24–42.

B.2.6 A. Maslow, *Motivation and Personality*, Harper and Row, 1964.

B.2.7 T. W. Costello, and S. Zalkind, *Psychology in Administration*, Prentice-Hall, Englewood Cliffs, NJ, 1965.

B.2.8 R. M. Cyert, and T. G. March, *A Behavioral Theory of the Firm*, Prentice-Hall, Englewood Cliffs, NJ, 1965.

B.2.9 P. Henle, "Labor's Future Bargaining Proposals," in *Understanding Collective Bargaining*, E. Marting (Ed.), American Management Association, New York, 1958.

B.2.10 Anonymous author, "Blueprint for Your Bargaining Session," in *Understanding Collective Bargaining*, E. Marting (Ed.), American Management Association, New York, 1958.

B.2.11 Anonymous author, "A Step by Step Guide to the Preparatory Process," in *Understanding Collective Bargaining*, E. Marting (Ed.), American Management Association, New York, 1958.

B.2.12 N. W. Chamberlain., *Collective Bargaining*, McGraw-Hill, New York, 1951.

B.3 KNOWLEDGE MACHINE FOR CORPORATE INTERACTIONS

In this section of the appendix, we discuss the use of knowledge machines and expert system superstructure on a decision support tool in inter-corporate negotiations. The global access provided by the full-fledged knowledge machine provides a basis for national and International validation of inter-corporate negotiations. Such negotiations are common during mergers, trade agreements, and the sale of corporations. This superstructure is built around a model blending a systems approach with a computational framework for the interaction. After chaining the highlights of the negotiation process as sequential and logical events in a systems framework, the monetary value of the assets exchanged and their utilities for the beneficiaries are compared to investigate the conditions for acceptability, rejection, further negotiations, or eventual stalemate.

Expert system rules help to evaluate the actual cost of agreement and cost of disagreement. Segments of the solution space leading to little or no probability of success in the negotiating process are pruned. Expert system software techniques for automatic propagation of probabilities through the inference network are used. Behavioral and perceptual factors are also introduced in the expert system to explain the (nonlinear) irrational attitudes/patterns of behavior. The strength of the systems approach in the representation of the essential chain of events and of the feedback processes in emulating the learning process is fully exploited. Likewise, the strength of mathematical tools to transform the actual composite assets donated from one organization to the utility array of the other by an evaluation matrix is also utilized.

The effects of feedback in the systems approach become an iterative computation of marginal utility in the mathematical framework. The convergence of this iterative process leads to essentially acceptable values in the utility array, and, likewise, instability in the iterative process leads to a divergence in the negotiating process.

B.3.1 Background of Economics-Based Interactions

In a unified approach to the mapping of reality by systems models, the contributions of Elzas and Zeigler [B.3.1, B.3.2] have established the methodology leading to the construction of models for complex human and economic situations. The discussion presented by Elzas [B.3.1, B.3.3] is especially relevant to the particular model presented here. A diversity of corporate, economic, legal, and behavioral forces interacts to produce an inter-corporate decision such as a merger or an acquisition. Such decisions have a critical and long-range influence on the survival of the corporation. Corporate survival is generally influenced by economic stability, product leadership, and both tangible and intangible corporate assets.

Furthermore, the advances in three unrelated sciences have also paved the way for the development of a mathematical and computable basis for human and organizational negotiations. The first one deals with the principles of game theory in economics. The second and third relate to the concepts of systems analysis, expert systems, and a quantitative basis for decision making. The latter fall within the domain of computer science, but with far greater precision and confidence in the decision-making process. The former can be merged to produce a highly legitimate discipline relating to the development of corporate strategy and planning. Each parent science having been endowed with a wealth of ideas supports each other, where one may fail independently. Systems science yields a methodic approach to the pursuit of objectives by a systematic accumulation of the constituents of success in achieving those objectives. Game theory aspects are applicable to decisions that corporate entities may undertake along the pathway to success in achieving their objectives.

Typical examples of games applicable to inter-corporate and social interactions are the zero sum game and the prisoner's dilemma [B.3.4]. Finally, the development of expert systems [B.3.5] facilitates the application of a knowledge base extracted from the experience and wisdom of organizational negotiators. The languages that have emerged from rule-based programming assure that the basic rules and fine points, which finally shape the outcome of a negotiation, are not overlooked and overwritten either by the game theory rules or by an unsophisticated application of the systems considerations. Heuristic rules are incorporated in such languages to deal with exceptional circumstances.

The chief strength of systems science is its capability to separate and seek the conditions leading to a goal on a systematic and day-by-day basis. Generally, in computer science, where systems analysis is extensively used, information is manipulated, sought, used, and processed. In the interaction process, assets, tangibles, and intangibles or organizations or individuals are sought and processed. Systems science can also accommodate most feedback processes effectively. Feedback loops are more or less essential to represent situations where the game participants learn from the interaction to modify their behavioral pattern [B.3.6]. To this extent, systems science can be very effective in adjusting its procedures according to the predictions of game theory. Major feedback loops can be represented in systems notations; the quantity of the feedback signal (stimulus) carried in system loops is determined by the outcome of game theory. In computer systems, the quantity and consequence of this feedback signal are programmed. In organizational and human interaction, both the quantity and consequence of these feedback signals are fuzzy and ill defined. As rationality is assumed to dominate transactions, the behavior pattern becomes more predictable.

The chief strength of game theory lies in its capability to seek and direct behavior and understand the consequence of one player's behavior on the reaction of the other, thus filling the very gap created when one tries to use systems science in its crude form while dealing with human organizational systems.

For the remainder of this section of the appendix, we use inter-corporate negotiations to illustrate the use of expert system decision making for negotiations. Typical of such interactions may be merger and acquisitions, large-order buy/sell negotiations, labor–management negotiations [B.3.7, B.3.8], international development ventures, industry–government interactions, industry–university research, project negotiations, etc.

It is clear that the negotiating process between two organizations or that between two individuals is not radically different. The primary motivation for both may be derived, the basic laws of economic behavior may be anticipated, and the outcome may be that of achieving a "feasible" compromise that "satisfies" a basic "security level" for both parties involved. However, the path- and goal-defining activity for inter-personal negotiation can be highly individualistic, whereas the corresponding activity for inter-corporate negotiation arising from the collective intelligence of a group is likely to be consistent and predictable.

B.3.2 Expert System Concepts

When negotiations take place, a team of experts is often brought in to gather the facts and optimize the decision-making process. Particularly when time is of the essence, as in the Iranian hostage crisis of 1979–81, it would be useful to have much of the expertise incorporated beforehand in a computer expert system. The large amounts of knowledge necessary to reach a successful settlement and the heuristic nature of many of the rules underlying organizational behavior make negotiations an ideal domain for an expert system.

Recently, computer expert system software tools have been developed to encode and deductively combine the heuristic rules an expert in a particular field is assumed to use to solve problems in his or her area of expertise. These computer systems differ from conventional computer programs in that there is a clear separation between the knowledge base of rules and a control module, which decides which rules should be looked at first, and which rules' consequentces should be performed if the antecedent rules of several rules hold simultaneously.

This separation facilitates the addition, deletion, and modification of rules. Typical control mechanisms are forward chaining, in which given pieces of evidence lead to a hypothesis being formulated by the rules, and backward chaining, in which a hypothesis is tested by recursively testing the antecedents of rules whose consequences are the given hypothesis. The user may be queried as to the certainty of some evidence and, in some tools, may respond with a probability value rather than just a true or false value. These probabilities may be propagated through the rule base and then deduce the probabilities of various hypotheses. Some of these tools include explanation facilities for justifying the system's conclusions, and some include sophisticated knowledge acquisition facilities to aid in building the rules. Some include interacting knowledge sources of production rules, each of which is scheduled depending on which data are posted on a common blackboard structure, allowing for the incremental construction of competing hypotheses. We envision a separate primary knowledge source for each side to the negotiations, with a common supervisor. Each of these three would be further refined into specialized knowledge sources for the different types of knowledge. Some allow for inferences involving fuzzy terms such as "high profits" programmed by defining probability distributions for these terms. These features are all relevant to the negotiations scenario.

Guidelines and strategies for the construction of expert system rules are as follows:

1. Organizations/organisms have needs.
2. Society has resources to satisfy organizational/organic needs.
3. Organizational assets are consumed, conserved, or wasted.
4. Organizational/organic motivation and ensuing behavior stem from unsatiated needs, external stimulation (which gives rise to a reaction), and/or internal conditions (which give rise to a motivation).
5. Under complex conditions, the simple maximizing behavior becomes that of "satisficing," the search for an optimal solution becomes that for a "feasible" solution, and the search for certainty in goal achievement becomes that of finding a "security level" for the outcome of the solution.
6. Organizations/organisms learn from interactions with society and from other organizations/organisms and modify their behavior.

Axioms (1) through (4) are self-explanatory, but the effects of (5) and (6) are far-reaching. It is due to such effects and their dynamic combinations that the corporate interactions are neither completely programmed nor completely predicted. Consider the concept of satisficing. This function is human and personality-dependent. It stems from the past experiences of the interacting individuals who represent the corporation and managerial objectives. Such objectives are generally not even documented and lie buried in the collective ambitions of upper management. When two corporations interact, the 'satisficing' effort is sought by two different bargaining parties. Furthermore, the same set of agreements drawn out in successful negotiations by the parties should be satisficing enough for both, and likewise, they should be "feasible" enough for both and finally should provide acceptable "security levels" for both.

Next, consider the concept of one side learning from the behavior of the other side. Numerous factors enter into the learning process. First, the perception of behavior is again a personality-dependent effect. Communications skills of the participants, cultural backgrounds, personal norms of communication and of body language, all play a role in transmitting and perceiving the behavioral patterns of the other side. Second, a feedback process immediately becomes necessary to monitor the present behavior in the context of past learning. Third, the feedback effort has profound effects on the stability of the interacting process itself.

B.3.3 Computational Aspects in Negotiations

B.3.3.1 Needs Vector N Corporate needs like human needs can be classified. To a certain extent, such needs arise from the long-range objectives of top management. The very existence of a stable corporation implies planning and achieving goals of financial solvency, participating in a viable economic activity, and ascertaining the continued existence of the corporation. The basic security needs thus exist and are continuously satisfied. Balanced cash flow requirements, good social and labor relations etc., form a secondary tier of needs necessary for participation in viable economic activity. Finally, the higher needs arise from the long-term goals to grow and expand, earn a reputation, provide a sound emotional and intellectual atmosphere for employees, and satisfy their own higher needs, etc. It is thus feasible to envision a needs structure for corporations like that for individuals. The relative importance of needs becomes important in assigning numbers to the needs vector.

B.3.3.2 Assets Vector A Assets have been classified by the accounting profession. Tangible assets generally satisfy the more basic needs of corporations, even though intangible assets may play a dominant part in the very existence of the corporation. Assets by definition tend to be scarce, and the price of assets depends strongly on the scarcity of such assets in relation to their needs. Corporate assets include productive resources (raw materials, plant, equipment, patents, goodwill, services, etc.), contractual rights to future services and products, and human resources.

B.3.3.3 Consumption Vector C Needs are satisfied by the consumption of assets. These assets may be regenerated over time. The surplus (such as retained earnings of corporations) may be reinvested or accumulated. During a corporate interaction, it thus becomes essential to be aware of the present consumption of the existing assets within one's own corporation producing its basic needs satisfaction.

B.3.3.4 Motivational Vector M The lack of total satisfaction of all the corporate needs produces a sense of motivation to achieve further satisfaction and derive further utility by the consumption of more assets. Such assets may be self-generated or obtained by corporate interactions. The dynamic nature of the needs vector also provides a sense of motivation. Ensuing interactions could be highly streamlined, routine, and stylized, or they could be distinct, unique, and highly customized.

An interaction may develop between two motivated corporations, or between one motivated and one reacting corporation. The former situation leads to a case where one may expect a cooperative game structure during a majority of the interaction process; the latter situation leads to a case where at least one party may hole a cooperative game structure during a majority of the process. This condition itself does not imply that the cooperative or uncooperative attitude of one or the other party will not reverse during the interaction. Frequently, attitudinal and behavioral factors lead to very pertinent changes in the game structure. Legal and social constraints may not only force an interaction from corporations, but under different conditions also specifically prohibit interaction between them.

B.3.3.5 Resource Balance Vector B When assets are retained, they accumulate and grow. Underutilized assets provide a strong incentive for corporate executives to seek some satisfaction of their outstanding needs. If they can be reinvested to generate additional utilization within the corporation or through the standardized and stylized mechanisms of interactions, no major corporate interaction results. However, if greater than normal enhancement or corporate utility is desired, it becomes essential to seek and interact with other corporations (e.g., mergers, acquisitions, etc.).

B.3.3.6 Donation Vector D During interactions, assets are exchanged and one corporation agrees to commit a certain array of assets from the resource balance vector **B**. These donations depend on the cost of the assets and the balance after consumption.

B.3.3.7 Valuation Matrix V The utilization of assets generates utility for the corporation. Assets can be used many ways, and different types of utilities may be derived from the same assets. Hence, by making the valuation (or utilization) function follow a matrix format, one can generate a utility array from the asset vector.

B.3.3.8 Utility Array U This array is obtained by multiplication of the valuation matrix by the asset vector and represents the state of needs satisfaction. If all the needs are completely satisfied, a unit utility array is generated. By default, organizations tend to maximize their utility arrays or generate some particularized weighted sum of all the terms.

B.3.3.9 Marginal Utility Array ΔU This is the change in utility derived by assimilating the donated assets of the other corporation by the recipient organization, and **U** equals the new utility if the donations were made and utilized minus the old utility with no donations.

B.3.3.10 Weighted Marginal Worth ΔW This is a single number derived from the marginal utility array. It indicates the monetary value of the benefits (or losses) derived from the change in utility due to the corporate interaction. Since all the various elements of the marginal utility array (ΔU) do not carry the same importance, it becomes necessary to weigh the relative importance of each element by assigning a weighing factor b_i, so that the net monetary worth of the marginal utility may be derived.

B.3.3.11 Weighted Price of Donations P This is a single number used to determine the cost of the donations vector **D**. Here, the cost for the donor for making such a donation is important. The donation vector **D** has numerous elements and all the elements are not equally expensive for the donor. Hence, it becomes necessary for the donor to attach a weighing monetary value to each element for the donation vector, so that the net monetary value of the entire donation may be aggregated.

B.3.3.12 Perceptual and Behavioral Variables Behavioral variables inherent in the individual negotiators and perceptual variables estimating how particular facts are perceived determine state transition probabilities for the negotiation process, which give rise to a cusp catastrophe model that may be incorporated in the expert system for predicting nonlinear behavioral shifts [B.3.8; Roman and Ahamed, 1985].

B.3.4 Mathematical Representations for the Cost of Agreement, Cost of Disagreement, and Demand Intensity

These representations exist on a symmetrical, bilateral, and reciprocal basis. When the demands of one corporation are listed, the monetary value of the benefit demanded should be divided by the monetary value of the compensations offered to get the demand intensity. (To compute the value of benefits, the expectation of such benefits should be calculated from an estimate of probability if there are uncertainties beyond the immediate negotiation. The monetary value of the compensation offered is well defined, at least during the time of negotiation.)

Where a perfect trade occurs, both demand intensities (DIM1 for Corporation 1 and DIM2 for Corporation 2) are computed simultaneously, and both are I. The negotiation is a completely open and one-step success. When DIM1 as calculated by Corporation I and DIM2 as calculated by Corporation 2 are both greater than 1 (nonzero sum game situation), then the negotiation becomes mutually beneficial win-win situation and completely acceptable to both. The negotiation itself starts to assume its proper dimensions when Corporation 1 maximizes DIM1 and, likewise, when Corporation 2 maximizes DIM2 when there is a zero sum game situation.

The cost of agreement and disagreement arises from a specific set of demands being made at a certain instant of time during the negotiating process. The monetary value of meeting all the demands of the other party are calculated as the cost of agreement. The cost of disagreement would be the total loss incurred in terminating the interaction or agreement to adopt a less desirable contract or interaction.

Step 1:

The negotiating function consists of examining the following events for the two corporations:

Corporation 1
Basic Conditions:
$M^1 = N^1 - C^1$
$B^1 = A^1 - C^1$

Corporation 2
Basic Conditions:
$M^2 = N^2 - C^2$
$B^2 = A^2 - C^2$

Expected Marginal Utility and Worth:
$\Delta U^1 = D^2 \cdot V^1$
$\Delta W^1 = \sum_{i=1}^{i=m} b_i^1 \cdot \Delta U^1$

Expected Marginal Utility and Worth:
$\Delta U^2 = D^1 \cdot V^2$
$\Delta W^2 = \sum_{i=l}^{i=m} b_i^2 \cdot \Delta U^2$

Expected Donations and Price:
$D^1 = D^1(P^1, B^1)$
$P^1 = \sum_{i=1}^{i=n} a_i^1 \cdot D_i^1 \cdot P_i$

Expected Donations and Price:
$D^2 = D^2(P^2, B^2)$
$P^2 = \sum_{i=1}^{i=n} a_i^2 \cdot D_i^2 \cdot P_i$

Step 2:

The result of the negotiating process ends in one of the following five situations:

1. Successful negotiation:

$$\Delta W^1 \gg P^1 \text{ (viewed by Corporation 1)},$$
$$\Delta W^2 \gg P^2 \text{ (viewed by Corporation 2)}$$

2. Agreement, perceived fairness:

$$\Delta W^1 \approx P^1 \text{ (viewed by Corporation 1)},$$
$$\Delta W^2 \approx P^2 \text{ (viewed by Corporation 2)}$$

3. Agreement, genuine fairness:

$$\Delta W^1 \approx P^1 \approx \Delta W^2 \approx P^2$$

4. Exploitation by Corporation 1:

$$\Delta W^1 \gg P^1 \text{ (in actuality)}, \qquad \Delta W^2 \approx P^2 \text{ (in actuality)}$$

Corporation 2 comprehends from its own standpoint that $\Delta W^1 \approx P^1$, or it has no mechanism to reduce or redistribute the terms in D^2 appropriately.

5. Exploitation by Corporation 2:

$$\Delta W^1 \approx P^1 \text{ (in actuality)}, \qquad \Delta W^2 \gg P^2 \text{ (in actuality)}$$

Corporation 1 comprehends from its own standpoint that $W^2 \approx P^2$, or it has no mechanism to reduce D^2 or redistribute the terms in D^1 appropriately.

6. Unsuccessful negotiation:

$$\Delta W^1 \ll P^1 \text{ (estimated by Corporation1)},$$

$$\Delta W^2 \ll P^2 \text{ (estimated by Corporation2)}$$

It can be seen that over a long period of time (implying repeated interactions), situations (3) and (4) are the only stable conditions. Situations (1) and (2) are not stable over a long period since the true perception of ΔW^1 and ΔW^2 in relation to P^1 and P^2 is established, thus leading to situation (3). Constant violation of situation (3) by situations (4) and (5) by one corporation leads to exploitation by that corporation, thus leading to mistrust and withdrawal by the other. This leads to the other stable situation, 6, over a long period of time. The other conditions 1, 2, 4, and 5—even though unstable over a long period of time may find stability over the life span of the contract. The most important factor for the agreement hinges on the approximate equality of the net worth of the increased utility to the price of the donation.

B.3.5 Systems Model of Interaction

Our basic purpose in using systems science is to develop a fairly accurate quantitative basis for the strategy selection, as is done in computer simulations, and to evolve a methodology similar to Program Evaluation and Review Techniques (PERT [B.3.9]) or Critical Path Method (CPM [B.3.10]) for inter-organizational interaction. Figure 5.B.5 represents a structural configuration depicting the sequence of events, their interactions, and the various decision-making processes involved during inter-corporate negotiations. The

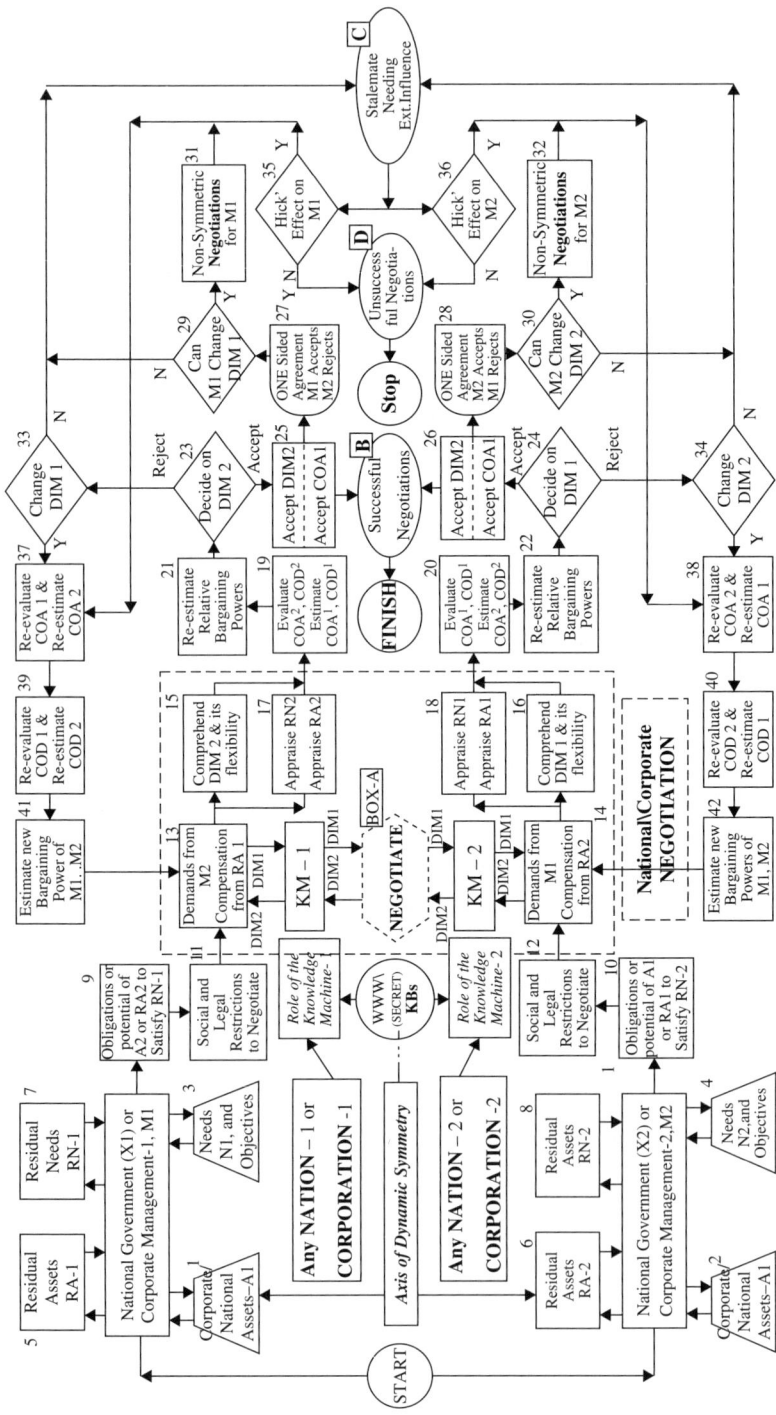

Figure 5.B.5 A structural diagram depicting the sequence of events, their interactions, *and the* various decision-making processes involved during inter-corporate negotiations.

bilateral symmetry is established by the centerline through the boxes A, B, C, and D. The contents of boxes 2, 4, 6, ..., 42 are identical to or the converse of boxes 1, 3, 7, ..., 41, respectively.

B.3.5.1 Sequence of Humanistic Events

When the managements of the two eligible corporations become aware of their needs, they have to initially discover each other. The circumstances or events leading to this mutual discovery are not discussed here since it is outside the realm of the actual negotiation. When it does become apparent that there can be mutual benefit or that a reaction is essential, then the actual interaction starts. The preliminary effort for both managements lies in evaluating their residual needs and residual assets (boxes 5 and 7, 6 and 8). It is here that the actual motivation for interaction is rooted.

Boxes 9 and 11, or 10 and 12 correspond to the investigation prior to the negotiation. Here, there is an evaluation of potential reward by interaction and a determination of the legality and/or social acceptance of the potential negotiation. Boxes 13–17, or 14–18 depict the actual negotiation where there is an exchange of information regarding the reciprocity of assets for mutual needs satisfaction. Also exchanged is a framework of reasoning and fairness in the trade of assets conforming to the social mores and the economic valuation of such assets at that particular time and in that particular place. Boxes 19 and 21, or 20 and 22 establish the costs of agreement if the demand intensities appear reasonable or the costs of disagreement if the demand intensities appear to be hopelessly out of line (i.e., far from their ideal value of unity).

Considerable subjectivity and the valuation of the assets (based on scarcity and potential use) become critically important for reaching the decision in box 23 or 24. Here, all the behavioral traits of the executive decision-making process become active in accepting or rejecting the demand intensity presented from the other side.

If Corporation 2 accepts DIM 1 and Corporation 1 rejects DIM2, we have a situation where a one-sided agreement results and boxes 28, 30, C, etc., are encountered. Conversely, a one-sided agreement for Corporation 1 is equally possible. It is during such a situation that the effects of feedback and learning from the negotiating process become critically important.

B.3.5.2 Sequence of Computational Events

The qualitative tracking of events is reflected in the interpretation of logged human events. However, the quantitative implications of the discussions as proposals, counterproposals, and adjustments to the intensity of demands on both sides are computed by two knowledge machines: KM1 and KM2. The incomplete knowledge of the true assets and donations from the other side imparts a sense of fuzziness on the whole process of inter-corporate negotiation. The firms and rapid response in the computations of the 11 vectors (**N**, **A**, **C**, **M**, **B**, **D**, **V**, **U**, Δ**U**, Δ**W**, and **P**) of Step 1, presented in Sections B.3.5.1–B.3.6.4, offer a well founded basis for either corporation to make its individual moves.

When the two knowledge machines (Figure 5.B.5) are inter-connected, then the entire inter-corporate transaction is a smooth convergence to condition (3) in Step 2. The other possibilities (1, 2, 4–6) are circumvented. When the two knowledge machines are not inter-connected, then the extrapolations and rule-based deductive logic embedded in the two machines are likely to be more accurate than the human comprehension on the spot. Even though the two machines may not entirely replace the human negotiating teams, they provide a quantitatively basis for the proposals, counterproposals, and re-proposals that form the basis of smooth convergence to agreement without hostility. Such convergence may be essential for national and international negotiations.

B.3.5.3 Human Estimation of Feedback Loops

There are two major types of feedback loops in the negotiation process. As the duration of the disagreement between the two corporations starts to increase, the frustration in not achieving the desired goals of both corporations exerts an influence in relaxing their respective demands. This tendency was first detected by Hicks in labor–management negotiations (see Section B.2; Roman [B.3.8]). Feedback loops depend on the duration for the parties to traverse them. The effects of the first type of loop are generally felt during the interaction. It provides a minute-to-minute, hour-to-hour, or day-to-day correction for demands and expectations. The continuity of the negotiation is rarely lost. The second type of loop provides for correction of a long-drawn stalemate situation where a major overhaul of demands and expectations becomes essential. In the former type of correction procedure, small and minor concessions are made. In the latter type of correction procedure, the effect of long-term needs still unsatisfied provides a major restructuring of the demands made on the other side.

B.3.5.4 Computed Stability of Feedback Loops

The conditions for the stability of feedback loops are documented in control theory [B.3.11]. In the human environment, such conditions are not as well documented even though some guidelines exist (see E. Berne [B.3.12]). False perceptions and deceptions only cause instability and either oscillatory or smooth divergence. Honesty, truth, and trust are likely to reinforce stable and sustained feedback loops. As a guideline, the priming of the two knowledge machines needs to based on truth, virtue, and beauty (TVB), as discussed in [B.3.13].

Either one of the knowledge machines primed with deception, arrogance, or hate (DAH), as discussed in [B.3.13], can add to oscillations of all 11 vectors (N, A, C, M, B, D, V, U, ΔU, ΔW, and P). Such oscillations lead to convergence, divergence, or a stalemate between the two corporations. When both machines are primed with DAH, the threats and bluffs (even blows and wars) between the two parties only become intensified.

It is possible to envision and computationally deduce a mode of corporate interaction ranging from a smooth agreement to corporate war. However, when both machines are deployed to yield a smooth agreement, considerable human effort, time, and energy are spared and the entire process may prove to be far more economical. The TVB- and knowledge-based machines with good deductive and predictive algorithms generally avoid a stalemate before it becomes a problem.

B.3.5.5 Stalemate and Its Consequences When a stalemate condition forces a Hicks's effect on the demands from the corporation, then there is a correction for the DIM1 and/or DIM2, thus edging toward a gradual convergence to box B, a successful end to the negotiations. On the other hand, the absence of a Hicks's effect leads each corporation holding dogmatically to its own demands and gradually diverging to box D, thus terminating the negotiation.

The effect of time without a conclusion of the interaction also motivates corporations to look for many solutions, one of which is the search for an alternate, more flexible corporation to interact with. The termination of the interaction with one may be only the start of an interaction with another. The reaction to a stalemate situation depends on the personality and style of upper management. The outcome in such cases further depends on evaluating the bargaining powers of individual corporations. Federal agencies and social organizations (with the aid of courts or legal mechanisms) can hold an enormous amount of bargaining power. In such cases, the cost of disagreement becomes not only high but increases with time.[13]

B.3.6 Pruning and Predicting the Outcome

B.3.6.1 Smooth Convergence A smooth convergence of negotiation essentially entails no feedback loops in the block diagram (Figure 5.B.5), where there is an explicit change of DIM1 or DIM2. This essentially eliminates the loops 13, (15, 17), 19, 21, 23, 33, 37, 39, 41, 13 and its image loop 14, (16, 18), 20, 22, 24, 34, 35, 40, 42, 14, or the loops passing through boxes 29 or 30. The Hicks's effects discussed in Section B.3.5.2 and depicted in boxes 35 or 36 are also eliminated. The expert system rules will guide users toward this type of a convergence process in expediting the negotiation. Pruning becomes imperative in the search for an optimal and expedient solution.

B.3.6.2 Smooth Divergence The prerequisites for smoothly diverging negotiations are weak motivations and/or lack of compensating assets that are offered. From a systems perspective, this condition exists if the demand intensities of both corporations are so high that the negotiation ends in box D via the paths 13, (15, 17), 19, 21, 23, 33, box C, 35, box D and its image loop 14, (16, 18), 20, 22, 24, 34, box C, 36, box D. The demand intensities DIM1 or DIM2 never change in boxes 29 and 30 or at boxes 33 and 34.

From a mathematical perspective, the smooth divergence occurs if $\Delta W^1 \gg P^1$ as seen by 2 and also $\Delta W^2 \gg P^2$ as seen by 1; thus, there is no incentive to alter P^1 by 2 or conversely P^2 by 1. The expert system will evaluate these quantities and guide both sides accordingly.

---

[13]When corporations and/or unions are penalized (fined) on a daily basis for noncompliance on a code, court order, etc.

B.3.6.3 Oscillatory Convergence This type of outcome of negotiation is by far the most commonly encountered agreement between parties. Here, demands are explicitly spelled out in proposals. Initially, they are quite high but do taper down. Lowering of demands from their artificially high initial values is a common feature of such an oscillatory process. Oscillations could be weak or strong, overdamped or persistent. Weak oscillations occur when one party fails to maintain long and meaningful resistance to the demands of the other party, and the second party easily convinces the first to assume its position. From a systems viewpoint, the demand intensity of the first corporation DIM1 is high, but the second corporation convinces the first that the cost of agreement on its part would be much too high; the first corporation thus reduces its own DIM. From a mathematical perspective, this situation calls for a relatively high value of ΔW^1 as seen by the corporation at a disproportionately low value of P^1, thus making π^1 excessive; the first party therefore either reduces ΔW^i by reducing D^1 or increases P^1 to increase the π^2 profit of Corporation 2.

On the other hand, strong oscillation occurs when there is a severe lasting conflict over demands of major importance to either corporation. Very severe and strong oscillations can lead to divergence, thus exposing each corporation to the risk of wasted efforts and unsatisfied needs. Persistent weak oscillations are sometimes tolerated because the parties cannot agree on every microscopic detail of every issue. At this level, the compromise achieved by equalizing the weighted averages of the costs involved may be logical. Issues dealing with valuable assets (assets being too close to nonnegotiable assets) lead to persistent oscillations. Issues of major importance involving highly valuable (bordering on being nonnegotiable) assets are a major source of oscillatory divergence leading to unsuccessful negotiations. Here, both dimensions of negotiations (behavioral and economic) become entirely important to alter the course of agreement.

In an oscillatory convergence of negotiations, both parties eventually arrive at box B (Figure 5.8.5). The feedback loops in Figure 5.B.5 through boxes (13, 15/17, 19, 21, 23, 33, 37, 39, 41, 13), or (13, 15/17, 19, 21, 23, 25, 27, 29, 31, 37, 39, 41, 13) and its corresponding image loops (14, 16/18, 20, 22, 24, 34, 38, 40, 42, 14), or (14, 16/18, 20, 22, 24, 26, 28, 30, 32, 38, 40, 42, 14) should be traversed at least once for an oscillation to be present. Furthermore, the more times these loops are traversed, the more persistent these oscillations are going to be. Feedback loops via box C may also be traversed during the feedback process. The knowledge machine and expert system keep track of long-drawn oscillatory negotiation and suggest cooling-off periods further used to reevaluate their positions. The knowledge machine-based expert system would eliminate those sources of oscillation due to the behavioral dimension of negotiation if they were present.

From an economic view point, let π^1 and π^2 be the total profit levels for both corporations. Their sum is viewed as being essentially constant; however, their relative values take an oscillatory path before becoming acceptable. In a zero sum game, such oscillations arise due to the incomplete knowledge of probabilities and costs. These probabilities and cost levels are altered due to the negotiation process and due to mutual learning effects. A further cause of these oscillations is incomplete analysis of the organization about the details of expected and donated asset levels. Costs and prices having been derived by nonlinear relations to the scarcity of assets can be another major, reasonable, and persistent source of oscillations, not only in evaluating the cost of agreement and cost of disagreement for the Company itself, but also for the other party.

B.3.6.4 Oscillatory Divergence From a systems perspective, oscillatory divergence usually implies one party reaching box B and then the other party increasing its demand intensity (DIM), due to a reversal of position on the part of the first party. During divergent oscillations, the expert system would indicate to users modified demand intensity and suggest concessions to make the cost of agreement attractive to control the divergence in the negotiating process.

B.3.7 Conclusions for Corporate Negotiations

Inter-corporate negotiations are based on the behavioral foundations of organizations. The cornerstones in such behavioral foundations are economic, accounting, social, and transactional laws. First, the economic basis of donations and expectations, proposals and counterproposals, demands and concessions stems from the marginal price theory of assets. When conditions of bilateral monopoly are encountered between informed parties, and the zero sum constraint on profits exists in its entirety, the fair distribution of profits forms an equilibrium condition. When these conditions are not ideal, the maximization of profit (defined as the difference between the expected net worth of donations received and the price paid for the compensating donations) is a popular form of economic activity in negotiations. Cooperative or noncooperative game structures are chosen by individual parties depending on relative knowledge levels.

Second, accounting principles become dominant in evaluating the allocations and/or availability of corporate assets. Such assets being actively exchanged and bartered during negotiations are the cards that lie in the hands of corporate players.

Third, social laws set the outer limits on the negotiation per se. The legal framework defines which assets may or may not be negotiated, the types of prices, the use of assets by either corporation, etc. This aspect of negotiation becomes particularly important in international negotiations.

Finally, the transactional aspect of a negotiation is perhaps where the greatest amounts of effort are expended. While making demands and offering compensation, economic justification is necessary. The basis of the transaction rarely stops here. Each party maximizes expected profit after partially establishing a knowledge base for itself and for the other side. Full exploitation of the positions is expected, and each transaction (proposal and counterproposal) is thus a manifestation of the corporation in the other three positions: economic, accounting, and social.

All these aspects can be incorporated in an expert system for negotiations. Such a system may include a selection of the following components:

1. The systems and mathematical components we have modeled.
2. One or several interacting knowledge sources implementing the items listed in 1, together with a control module for logical inference.
3. Other expert system features available in current software tools including automatic propagation of probabilities, fuzzy reasoning, explanation facilities, and knowledge acquisition facilities. Least commitment control strategies adaptive learning, constraint satisfaction, dependency-directed backtracking, criticality levels, etc., are also applicable.
4. Behavioral and perceptual factors incorporated perhaps in a cusp catastrophe model built into the system. The resulting system can calculate costs, probabilities, and bargaining powers; monitor oscillations; suggest cooling-off periods and concessions; and learn and predict.

With many major breakthroughs evolving in the technologies, humans are still at the frontier of the complex process they command during their association with others. The paths and mazes through the labyrinth of association and interaction have been explored by philosophers with a candlelight. We believe the techniques discussal here will provide a more practical scientific light.

REFERENCES

B.3.1 M. S. Elzas, "System Paradigms as Reality Mapping," in *Simulation and Model Based Methodologies: An Integrative View*, NATO, ASI Series F, Computer and Systems Science, Vol. 10, ISBN 978-3540128847 Springer-Verlag Berlin Heidelberg, 1984.

B.3.2 B. P. Zeigler, *Theory of Modeling and Simulation*, John Wiley & Sons, New York, 2000.

B.3.3 M. S. Elzas, "Concepts For Model-Based Policy Construction," in *Simulation and Model Based Methodologies: An Integrative View,* NATO, ASI Series F, Computer and Systems Science, Vol. 10, ISBN 978-3540128847 Springer-Verlag Berlin Heidelberg, 1984.

B.3.4 O. Stuart, "Effects of Programmed Strategies on Cooperation in the Prisoner's Dilemma and Other Mixed-Motive Games," *Journal of Conflict Resolution*, 15, 1971: 225–259. See also M. Pilisuk, S. Kiritz, and S. Clampitt," "Undoing deadlocks of Distrust: Hip Berkeley Students and the ROTC," *Journal of Conflict Resolution*, 15, no. 3, 1971: 81–95.

B.3.5 F. Hayes-Roth, D. Waterman, and D. Lenat, *Building Expert Systems*, Addison-Wesley, Reading, MA, 1985; E. Rich, *Artificial Intelligence*, McGraw-Hill, New York, 1985.

B.3.6 S. V. Ahamed, "A Systems Model for Labor-Management Negotiations," in *Proceedings of the 1984 Summer Computer Simulation Conference*, 1984, pp. 943–947.

B.3.7 S. V. Ahamed and E. G. Roman, "Extensions in the Application of Expert System Concepts to Labor-Management Negotiations," in *Proceedings of 1985 Summer Computer Simulation Conference*, 1985, pp. 702–703.

B.3.8 E. G. Roman and S. V. Ahamed, "An Expert System for Labor-Management Negotiations," in *Proceedings of the 1984 Summer Computer Simulation Conference*, 1984, pp. 1226–1229

B.3.9 F. S. Hillier and G. J. Liebermann, *Introduction to Operations Research*, McGraw-Hill, New York, 2002.

B.3.10 A. Kaufmann *Critical Path Method*, Routledge, New York, 1969.

B.3.11 H. Ozbay, *Introduction to Feedback Control Theory*, CRC Press, Boca Raton, FL, 1999.

B.3.12 E. Berne, *Games People Play*, Ballantine Books, New York, reissued, August 27, 1996.

B.3.13 S. V. Ahamed, *Intelligent Internet Knowledge Networks*, Wiley-Interscience, Hoboken, NJ, 2006.

CHAPTER 6

INFORMATION AND KNOWLEDGE FILTERS

CHAPTER SUMMARY

Traditional signal filter theory is extended into the knowledge domain. The separation between poles and zeros is recast into roadblocks and highways for the flow of undesirable and desirable information, respectively. Information and knowledge are characterized by the objects that they holds. In a sense, the quality of knowledge is reflected in the objects that are held. Once identified, such objects in the input information-blocks offer clues about the desirability or undesirability of the bodies of information or segments of knowledge. The sensitivity of the information and knowledge filters now becomes an adjustable parameter by specifying the objects, object groups, their relationships, their attributes and their relationships.

The filtering of information and knowledge starts to become a distinct discipline in its own right. On the one hand, the standard procedures of parsing and, lexical, syntactic, and semantic analysis in the compilation of higher-level computer programs start to offer clues for the de-segmentation of knowledge. On the other, the typical steps of statistical, syntactic, and contextual analysis in computer vision for disassembling complex scenes start to offer new methodologies. In unison, the two procedures converge in initially disassembling elements of knowledge. After the due processing in the knowledge domain, reassembling and recasting the seminal objects, their attributes, and their relationships into more desirable compositions becomes a science of creating new knowledge from

---

Computational Framework for Knowledge. By Syed V. Ahamed
Copyright © 2009 John Wiley & Sons, Inc.

old knowledge. Purifying the contents of embedded knowledge suddenly appears as a distinct scientific feasibility.

To some extent, musicians, painters, and artists practice their unique talents as they create new masterpieces. The various designs of signal filters, such as low-pass, high-pass, bandpass, etc., offer new insights into the design of information filters. In a truly generic sense, knowledge equalizers and knowledge recovery systems can be designed much like transmission line equalizers and timing recovery circuits in receivers. This chapter offers insights into feedback and feed-forward to stabilize and amplify certain desirable contents and common principles in the sciences, arts, and philosophy.

6.1 JUNK INFORMATION AND HYPE KNOWLEDGE

Filtering out impertinent and insignificant information is essential for intellectual sanctity. The rejection of irrelevant and retention of appropriate information have become instinctual over the evolution of species. This strategy is most appropriate in the modern knowledge society. In lower species, survival depends on quick selection (for prey, water, safety, etc.) and the immediate response that follows. In higher species, the selection is an intellectual procedure and sophisticated game for maximizing the goals and returns of investments. In operations research (OR), the maximization of the objective function (OF, a composite representation of numerous goals being simultaneously pursued) involves programming and computational techniques. Tools, and strategies exist in both the disciplines of Management Science and Operations Research.

In a broad sense, selection is mostly instinctual and partly learned. Inherited or acquired systematically, selection entails filtering out distracting objects from pertinent objects to derive an economic benefit from such activity. Predators practice selection as they hunt, farmers practice it as they pant, humans practice it as they find a mate, etc. In most instances, selection enhances the probability of success in achieving the final goal since the basis of selection is to be optimal in reaching the initial goal(s). Initially, the basis of selection may be subjective, but it soon becomes logical, rational, and then quantitative. When the art of selection is pursued on a mathematical basis, filtering becomes a science and the logical steps in the selection process become goal-oriented and programmable.

Filtering and selection are inter-twined. Filtering out irrelevant information facilitates the final selection of the most relevant and most significant information. After filtering, attention and resources are focused on the essential and away from trivia. Filtering has a technical connotation and selecting has a humanistic context, even though machines can be highly selective in the objects that they process and the attributes of objects that will be processed. For example, some types of noise in communications systems (such as echoes and reflections) are systematically filtered out from the received signal. A human being routinely rejects unfamiliar faces in a crowd while looking for a familiar face and then compares the faces that are initially selected before final identification or rejection.

Pattern recognition is a segment of the science in artificial intelligence (AI). Recognizing patterns within objects and their attributes is currently automated to the extent that machines can search and find a needle in a haystack quite mechanically and much more efficiently than most human beings. Syntactic, semantic, and statistical PR techniques start to become precise and quantitative, and the machine generally presents a confidence level in the final object recognized or identified.

Routine searching usually entails finding an operative (noun) object (such as a fish in a lake), and an action search involves a (verb) function (such as swimming) associated with the (noun) object (such as a swimming fish in contrast to a dead fish). Noun objects have a set of attributes (or adjectives such as their color, black) that qualify the (noun) objects. Verb functions also have attributes (or adverbs such as rapidly); thus, a human being may be looking for "a rapidly swimming black fish." Attributes facilitate the search process and reinforce the final selection from an array of similar objects (all colors of fish) or verbs (stationary, dead, spawning, etc.) with similar attributes (adjectives such as dark gray, almost black, patchy black, etc., and/or adverbs such as chasing, escaping, drowning, etc.). For this reason, we initially discuss searching, filtering, pattern recognition, and selection in a general sense and then concentrate on the special features of information and knowledge[1] $(I \ll \gg K)$ filtering.

Whereas selection involves the separation of desirable objects from a group of objects, filtering involves removing the undesirable objects noise and echoes from a group of objects (such as the received signals). Generally, the filtered objects (signals) retain the properties of the original objects (transmitted signals) to offer the utmost economic value or most relevant information. The pursuit of the most desirable objects, object sets, or signals can be quite intense as the original objects or signals become weak in comparison with the contaminating objects or noise. The contamination and corruption of objects and signals in most instances, are a natural and inevitable phenomenon. Thus, the rules of selection (or filtering) are customized to the character of the original objects or signals and the nature and character of the corruption and contamination of such signals. When the nature of the object or signal degradation is deterministic, filtering can also mean restoration of the object or signal. In the context of the communications sciences, such restoration is known as equalization because the nature and extent of degradation are known from the laws of physics or from statistical observations of the signal degradation.

In configuring filters for the information and knowledge $(I \ll \gg K)$ domain, the hardware and software are sensitive to the unique characteristics of the operative noun objects and the verb functions embedded within a sample of $(I \ll \gg K)$. Removing the contamination and restoring the degradation of the original information and a body of knowledge become the functions of the $(I \ll \gg K)$ filters.

The challenge is formidable at first, but AI systems can address the question of corruption of information from one timeframe to the next, or from one situation

---

[1]Information and knowledge under process and undergoing change are introduced and symbolized as $(I \ll \gg K)$ in [1]. We use the same symbol to form a conceptual link between information and knowledge, and in the knowledge domain to process and modify the transitions.

to another almost as efficiently as human beings can analyze the incremental change in structure of information, body of knowledge, collectivity of concepts, or axioms of wisdom. The requirements of such hardware devices, systems, and machines sometimes overlap with the type of analogue and digital filters used in communications systems. We also fall back on the human aspects embedded in the very human activity of recognizing a face in the crowd or taking appropriate steps to avoid danger in daily life.

In this chapter, we review some of the filter design principles from communications systems and then suggest the basis for treating the frequency domain as a two-dimensional space (of operative nouns and verb functions) in the information domain. Some methodologies (such as low-pass, high-pass, bandpass, band-reject, etc.) are applicable and some techniques (such as Fourier, Laplace, Z-transforms, etc.) are not immediately applicable. A knowledge machine is necessary to process samples of *(I «» K)* in the knowledge domain.

6.2 DESIGN OF CONVENTIONAL SIGNAL FILTERS

Filtering techniques in engineering applications are well studied and highly refined. The design of filters for processing signals has become an effortless, inexpensive, and routine activity. Numerous filter designs and computer-aided design (CAD) packages serve the industrial applications. If active filtering concepts are enhanced to the status of signal processing, then many more fields of human endeavor, such as acoustics, sonar, seismology, biomedical engineering, EKG and EEG tracking systems, even radar and weather tracking systems, stand to gain from the contributions to signal process engineering.

In earlier days, filter designers deployed lumped LC (inductance and capacitance) components to permit or block certain bands of frequencies. Even though they still exist and can constitute sophisticated filters, these devices rank low in comparison with the modern breed of active filters. The newer filters generally deploy integrated circuit (IC) components for very accurate control of the transfer functions and feedback for stability. The main features that the new filters offer include last-minute optimization of characteristics, very low cost, and optimality of design. Optimization stems from being able to use a minimum numbers of active elements (IC amplifiers and laser-trimmed RC components), as opposed to the bulky inductive components of earlier designs.

High-precision components are sometimes essential for critical applications. However, sensitivity analysis of components narrows down the number of trimmed resistors and capacitors to one or two elements, thus reducing the overall cost of very prolific devices to a few pennies in some instances. A large number of very practical devices (from hearing aids to night vision cameras) benefit from such advances in digital signal processing.

Digital signals represent sequences of symbols or numbers. To a limited extent, traditional numerical techniques such as interpolation, integration, and differentiation can process digital signals. However, programmable chips and their digital codes bring adaptability to a much higher level. The processing of such signals or clusters in a digital format starts to deviate from conventional selection, filtering, or shaping of analogue signals. The science of digital signal processing deals with tools and techniques [(Z-transforms, fast Fourier transforms (FFTs), and field programmable amplifier chips)] generally not used in earlier era (using Laplace transforms and frequency domain techniques) or mid-era (using pole/zero filter adjustments in the 's'-domain) designs. Over the generations, the design of filters has become a fully evolved mathematical science.

Filtering can also be used to clean up (i.e., selectively enhance) information-bearing signals and remove contaminating noise and signals. Such applications arise in communications systems where channel distortion, fading, coupled noise, and cross-talk can finally drown information signals below the level of detection. Signal restoration generally demands processing in at least two dimensions for restoration *(1)* of the signal amplitude and *(2)* of the relative phase of the signals. These two are rarely precise, thus degrading the restored signal from the original signal. In image processing for graphics and video applications, the restoration can become multidimensional to restore hues, colors, intensity brightness, and their synchronicities. Satellite photo enhancements and space probe images also demand much more sophisticated digital signal processing than what earlier filter techniques had to offer. Digital signal-processing algorithms are generally highly sophisticated.

Hybrid (analogue and digital elements in) filters are used in some instances. The critical question about filtering is not how the most optimal transfer function can be realized as a device or circuit component, but the choice and suitability of technology to realize the device most economically. One of the contenders is (was) the charge coupled device technology that is (was) used in some specialized but a limited number of applications.

Information-bearing signals are equalized and purified to be able to yield their properties after having suffered as much as 80 dB (10,000-fold) loss in the magnitude of optical signals. Such remarkable gain in optical devices (with erbium doped fiber) is feasible with the highest quality of components in optical networks. Signals from distant-galaxies or from the oceans-below, can also be restored by customized IC components.

When the extent and nature of degradation are known, equalizers and filters can undo the degradation and restore the signal. Signal recovery is rarely perfect and a certain amount of residual (tell-tale) effects linger on. However, the confidence level in the recovery of the signals is generally very high, with an estimated average data error or bit error rate approaching one bit for every 10^{10} bits that are processed.

Tunable and self-regulating filters sometimes deploy scan, verify, and learning techniques borrowed from artificial intelligence (AI). When the chips are programmable, the code that drives such chips controls the response from such systems. This extension of control capability adds more dimensions in the discipline of digital signal processing. The number and nature of applications start to multiply and the expansion becomes geometric.

6.2.1 Types of Traditional Signal Filters

Wave filters [1] are generally frequency-selective networks that transmit (sinusoidal) waves within frequencies of certain bands and reject (sinusoidal) waves within frequencies in complementary bands. Such frequency bands may be continuous over the entire band of interest or located in narrower bands where the information in most crucial. The simplest of such wave filters are low-pass, high-pass, and bandpass. The low-pass filter permits signals in a band below the cutoff frequency of the low-pass filter to be transmitted, and conversely, the high-pass filter transmits above its cutoff frequency. The bandpass filter transmits a signal within its low- and high-frequency cutoff points. Since these precise locations of the cutoff points and the flatness of the amplification or attenuation can be adjusted, frequency-emphasizing and frequency-rejection filters can also be built.

The image parameter filter generally consists of a cascade of two port sections. The impedances are matched at the junctions. If the matching of impedance also occurs at the terminals, the resulting image attenuation will approach zero in the pass-band. However, due to the resistive nature of the cascaded sections, the attenuation will be small but finite. In filter design, one of the approaches is to minimize the insertion loss in the pass-band and maximize the insertion loss in the band-reject region. Since it becomes expensive to have uniformly low value for insertion loss in the pass-band and a uniformly high value in the band-reject region, a certain amount of ripple in the insertion loss is tolerated in both regions.

6.2.2 Active Filters

Active filters circumvent the need for LC components. These components cannot be miniaturized beyond a certain limit to be included in IC chips. They are replaced with grouped resistors, capacitors, and variable-gain devices in active IC devices. Such IC elements perform as well or better than the LC-based filter devices. The active devices can be cascaded. This type of design calls for a cascade of isolated second-order sections (or biquads). Generally, operational amplifiers are also deployed in individual building blocks.

Active filters may also be realized from a LC filter structure. The inductor is replaced by a gyrator-capacitor combination or by general-purpose impedance converters and frequency-dependent negative resistors. These designs use operational amplifiers in their final realization. Coupled filters are feasible. First- and second-order sections are coupled to offer the desired characteristics obtainable with simulated LC filter structures. The final design considerations call for minimum power requirement, simple tuning and production techniques, and moderate tolerance requirements.

6.2.3 Digital Filters

Digital signal processing and digital filtering share considerable overlap. The continuous time domain signals are represented by closely sampled discrete time domain symbols. These symbols or numbers are defined at discrete times, thus, the independent variable that is mathematically processed by the digital signal processors (DSPs) takes only discrete values. For speech and video applications, the representations may be continuous or discrete with very close equivalency. Continuous time and continuous amplitude signals are simply the analogue signal handled by earlier types of filters.

Signal processing is essential in almost all areas of science and technology to acquire the most exacting measure of information embedded in signals. Such signals abound in nature and knowledge. Most applications are also the beneficiaries of signal processing where data are sensed and acquired. All the advantages offered by programming and controlling of digital systems become available in most applications. Filtering, processing, and enhancements of digital signals can be quite elaborate.

Digital signal processing thus deals with transforming the discrete samples of numbers or symbols into the retrieve the most precise (although not entirely precise) information and knowledge embedded in the signals surrounding the objects bearing information and knowledge.

6.3 SIGNAL WAVES AND KNOWLEDGE FLOW

Signals have patterns and embedded wave shapes. Signals convey raw information by time domain variations of the embedded patterns and wave shapes. Fourier transforms provide a basis to slip into the frequency (s-) domain from the time (t-) domain. Inverse Fourier transforms provide the reverse path back into the time domain. The effects of filter components are thus conveniently investigated in the frequency domain. To remain in the time domain entirely, we exploit the z-domain transformations. The signal properties can thus be computed accurately as they propagate through any physical media.

Data and symbols are most easily manipulated and processed by machines. Like signals, data and symbols also carry raw information, but they can be stored and manipulated and processed by machines. Digital signal processing is also feasible. However, raw information held in data and symbols can be processed to yield a context or basis for derived information just as information can be processed to yield knowledge.

The prime recipients of data and symbols are machines, and the prime recipients of information and knowledge are human beings. Processing is the forte of machines and interpretation is the forte of humans. To bridge the gap, we interject information filters and knowledge machines that alter the flavor of information and knowledge just as wave-shape filters and digital signal processors will alter signals and symbols. The purpose of both is to enhance the generic information content, reject the noise and corruption, and make the signal-to-noise ratio as high as possible and error rate as low as possible. The basis for exploring the similarities and exposing the dissimilarities is shown in Figure 6.1 for wave shape filtering and in Figure 6.2 for information and knowledge filtering.

THE *(I ≪≫ K)* FILTERS   253

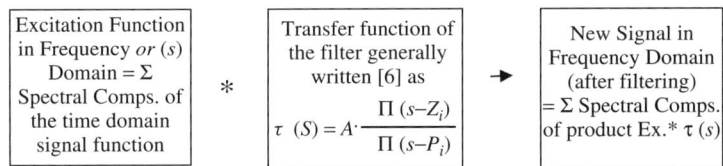

Figure 6.1 Typical filtering of signals in the frequency domain by any linear lumped parameter finite filter (bandpass, band-reject, low-pass, or high-pass) can be represented by the poles (P_i) and zeros (Z_i) embedded in the filter. $\tau(s)$ represents the transfer function and Π the product symbol. The value of i ranges between 1 to m for the m-poles and zeros in the filter. A is a numerical constant determined by the parameter values in the filter.

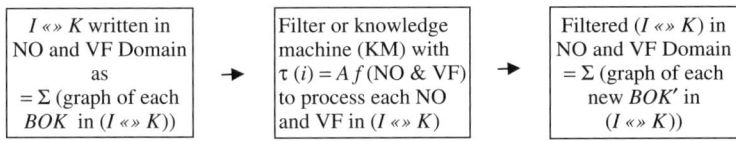

Figure 6.2 A proposed filtering methodology for filtering noun objects and verb functions in a sample *(I«»K)* in the information domain. Frequency is mapped onto a plane of noun objects and verb functions, and the information filter performs a group operation on these noun objects and verb functions, thus altering the flavor of *(I«»K)*. The program in the filter or knowledge machine selects the appropriate noun objects and/or verb functions, their attributes, and interrelations thereof, and modifies the information contained in *(I«»K)*. $\tau(i)$ is an information-based transfer function that modifies (enhances/degrades) the noun objects and verb functions and their interrelations embedded in *(I«»K)*. The symbol body of knowledge *(BOK)* represents each embedded body of knowledge in *(I«»K)*: BOK' is processed body of knowledge.

6.4 THE *(I ≪≫ K)* FILTERS

The tracking of signals and symbols is quite precise and accurate at this initial stage of filtering with s- and z-transformations. However, as can be seen from Figure 6.2, the processing of knowledge and information by filters and KMs starts to grow fuzzy for two reasons: *(1)* The laws for documentation in the *(I «» K)* are not as precise as the laws for building data structure and symbols, and *(2)* the mathematics for handling *NOs* and *VFs* are not as precise as the laws for handling frequency and time domain operations for signal wave shapes.

If the inadequacy of precision in handling the *(I «» K)* domain is tentatively overlooked, then the basis for the functionality of the information filters can be derived from the platform for wave shape filters. It becomes necessary to extract the *NOs* and *VFs* for any particular sample of *(I «» K)* before designing the filter. In Sections 6.5 and 6.6, we propose one such methodology to determine the operative noun objects and their associated verb functions.

A filter information or knowledge machine thus acts as a selective barrier (programmable template) to undesirable nouns (such as deception, falsehood, etc.) or verbs (such as falsifying, cheating, or unethical conduct) for band-rejection type of filters. Conversely, the machine may enhance positive objects and verbs for bandpass type of filters, etc. The constant A is a programmable number to control the extent of amplification or attenuation of the preselected *NOs* and/or *VFs*. The positive use of such filters is to enhance the social value of knowledge (such as the notion of beauty in classical poems, the rhythm of a classic symphony, or the art in classical paintings) or to impede its negative impact (such as the killing power of a new bomb, advertisement of pornography, or claim to fame through cruelty). The negative use of such filters is to falsify the representation of reality (as in news media reporting) or to exaggerate claims beyond reality (as in advertising). The flavor of information is tainted by the programmed control of the filter.

Filtering of information becomes considerably more complex than filtering of wave signals based on their instantaneous frequency. The design techniques for later types of filters have been optimized. They provide a limited *conceptual* framework for developing information filters. The commonality and the design approaches may be merged in a new dimension of design.

6.4.1 Symbols as Objects and Functions

Particularly important are digital filters due to their capacity to treat symbols individually and collectively. Even such filters fall short in their capacity to perform *(I « » K)*-based functions incisively. On the one hand, if symbols from the digital filters' perspective are redefined to represent noun objects *(NOs)* and verb functions *(VFs)*, then some of the digital filter design methodology becomes applicable to "information filtering." On the other, if authenticated information and knowledge (as opposed to gossip and hearsay) are defined to be the cumulative results of objects undergoing scientific, social, commercial, business, etc., processes, then *(I « » K)* processing at any given instant of time may be considered as "digital filtering." Even though some experts may object to the stretching of definitions to this extent to find commonality, we use the overlap to import some of the traditional filter design configurations into *(I « » K)* filters.

The basic configuration of an information and knowledge filter is shown in Figure 6.3. Samples of input are taken from specified knowledge banks or from local or Web bases. Any individual sample $X = (I « » K)_i$ is matched against a programmable filter template T that has three major components: the criterion C for filtering (e.g., 90% match for noun objects, 60% match) for verb functions, etc.), a template (e.g., three noun objects are truth, virtue, and beauty, and three verb functions are adhere, practice, and express) and the tolerance limit (e.g., that the sample has noun objects in an acceptance range of above 66% for noun objects and an acceptance range of above 33% for verb functions) for the sample matching the filter[2] template.

---

[2] A sample X_1 that has a sentence such as "truth adhered, virtue practiced, and beauty expressed are good" will pass the filter and becomes, Y_1 with a acceptability of 100%, whereas a sample sentence

THE $(I \ll \gg K)$ FILTERS 255

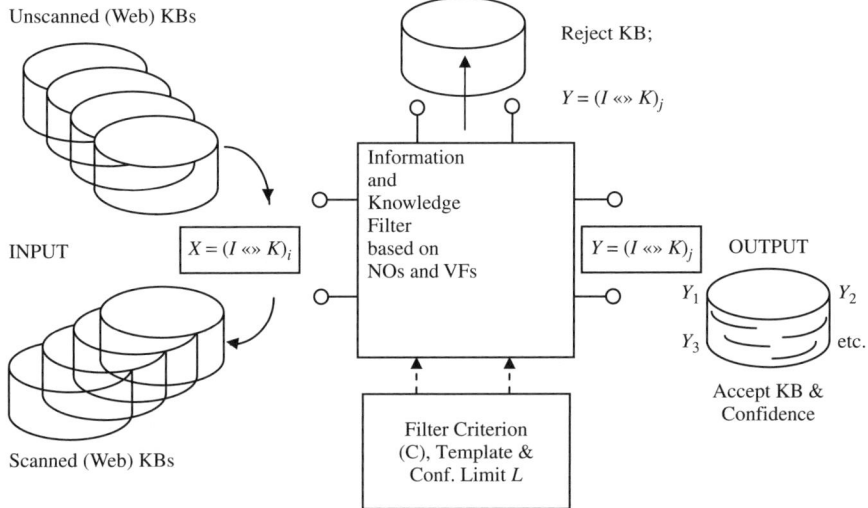

Figure 6.3 Basic configuration of an information and knowledge $(I \ll \gg K)$ filter. The output $Y = f(C, X)$ is a function of input X and filter criterion C. The filter criterion could be band-reject, high-pass, low-pass, band-pass, etc., features that are typical wave-shape filters. In this case, the filters transform noun objects, verb functions, attributes, etc., in each body or module of knowledge, the inputs $X = (I \ll \gg K)_i$, and select and/or process $(I \ll \gg K)$'s accordingly to satisfy the criterion and confidence limits.

The criterion for the template can be based on the noun objects, verb functions, and their interrelationships. A stringent criterion on the objects (and their adjectives) and verbs (and their adverbs) will block (like an extremely narrow-band bandpass filter) most of the samples and vice versa.

The output of the filter $Y = (I \ll \gg K)_j$ offers the selected samples and a map for the match and mismatch for each sample that meets the selection criterion and the confidence level in the selection procedure. The contents of knowledge and wisdom bases result from authenticated and final filtering from global digital filters. "Information digital filters" thus become knowledge machines; "knowledge digital filters" thus become concept machines; and "concept digital filters" thus become wisdom machines. Whereas signal and digital filters are cast in silicon and pentium, information, knowledge, concept, and wisdom machines are complemented by human beings, their imagination and creativity. Human beings become an integral part of such machines. Whereas older computer systems minimize the role of operator control on the computed results, the proposed new machines emphasize and integrate the creativity of participating humans. Human

X_2 "honesty practiced, virtue claimed, and beauty unraveled are not so good" will barely pass the filter (66% for two matches for noun objects and 33% match for one match for verb functions) and become Y_2 but only by satisfying the minimum requirement. In the later case, there only two out of three nouns match (i.e., (virtue and beauty)/(truth, virtue and beauty) and only one out of three verb match (i.e., (practiced)/(adhered, practiced and expressed).

thought is still remains the pinnacle of the new information digital filter that can be a full-fledged computer system in its own right. Machines perform the programmable functions and human perform the creative and artistic (inspirational and imaginative but scientifically valid tracking and input) functions.

6.4.2 Role of Humans

The programmable components of such hybrid (human–computer) systems implement the methodologies that are feasible in digital filter design theory from a scientific and mathematical perspective. The norms of human etiquette from Internet wisdom bases serve to provide a basis for machine-assisted decision support systems. For a unique and innovative solution, the thought process of participating humans becomes inter-woven in the search for the highest grade of inventions that serve humanity. It is not the intent to burden the human being with machine functions, but it is the goal to drive the machine to pursue all or most of the programmable human functions.

For instance, if a human being pursues truth in any Internet (noun) object (such as a cure for cancer) by a new drug treatment (a verb function), then the knowledge machine ("digital filter") will remove most errors and doubts about such a claim by scanning, searching, and filtering out inaccuracies in the original statement (i.e., a cancer cure by a new drug treatment). The filtering is facilitated by a Web search for any counterclaims or repudiations. Furthermore, the new machine offers documented proofs to authenticate or repudiate such a claim. The confidence level of such complex object functions in all the knowledge banks around the world will be computed. Similar drugs would also be scanned and searched, and the closeness of the original claim to other similar drugs will also be documented, adding a new dimension to the task that was initiated. The search for truth in any statement analyzed by the filter the validity of any object/verb will be substantiated or repudiated quickly, efficiently, and accurately by mathematical and statistical processes by the object and number processors within the filter.

Strings of microcommands to perform complex humanistic problem tasks will initiate a precise response with established mathematical processes to solve the original problem. Simple interrogation of a complex question would result in an expert solution as far as Web banks contain authentic information or knowledge. This methodology differs from expert system (ES) solutions because of the newer filter techniques borrowed from digital filters, whereby operative noun objects and critical verb functions are treated as "objects" in their own right.

As an example, if a musician is trying to compose (verb function) a new symphony (noun object), or an artist is trying to paint (verb function) a new scene (noun object), etc., then the knowledge machine would scan all the prior works of the individual, search, write down an objective function of the contribution, and then go on to optimize such an objective function by adjusting the previously used noun objects (musical instruments, players, etc., or paints, colors, strokes, etc.), their attributes, the verb functions of each noun object, and the mode of each verb function, and individually and collectively write a new symphony or paint a new canvas. The machine does not infringe on human freedom but adds a new dimension to human creativity. Intellectual space opens up like outer space. It is not that the human being is going to act as a machine, but that the machine is going to act humanistic. The human thus has a greater chance of being a super-human by developing a new dimension of creativity that neither the human nor the machine can independently create. Much as the potential of a human being as an explorer is enhanced by a spacecraft rather than an aircraft, the creativity of the human as an inventor is enhanced by a knowledge machine rather than a computer.

6.4.3 Precise Solutions and Confident Results

Precision is generally associated with numerical solutions. In computer systems, numbers are processed by numerical processor units (NPUs). For higher precision in numerical solutions, double (or more) precision processors are deployed. Thus, the accuracy is limited by the representation and processing of numbers in the NPU. In most cases, algorithmic accuracy is taken for granted.

In the processing of knowledge, precision is not limited by the NPU, but by the algorithms and methodology for processing the (noun) objects and (verb) functions. The subjectivity of individuals is replaced by how such similar objects have been treated by the scientific communities and cultural environments around the globe. In a sense, the new machines have a built-in mechanism to deal with subjectivity, individual bias, or prejudice by looking up the Web treatment of such (noun) objects and (verb) functions. A world-wide standard thus evolves for objects and functions and their relationships.

For example, if gay marriages in Far Eastern cultures are to be examined, then the knowledge machine investigates all aspects of gay marriage in all cultures to establish a norm and then focuses on Far Eastern cultures to identify special connections, influences, and ramifications. All aspects are investigated, irrelevant aspects are filtered out based on any given criteria, and only significant aspects are (pattern) recognized, identified, and reported. Filtering, however, need not be the final phase and a certain amount of pre-filtering of knowledge from Web knowledge bases can save considerable execution time for the knowledge machine. Such approaches are plentiful in signal processing and recovery. Other applications in herbal medicine, psychology, and music compositions, etc., are also feasible.

In dealing with the processing of $(I \ll\gg K)$, the algorithms for processing the objects and functions are not standardized. Mathematical procedures such as multiply, divide, exponentiation, series expansions, etc., for any $(I \ll\gg K)$ object do not have a precise psychological and social parallel. However, it is possible to train a knowledge machine to fragment $(I \ll\gg K)$ into more primitive objects and functions that constituted the original $(I \ll\gg K)$. It is possible to foresee that the human mind will benefit from such augmented capability of the machine.

For example, if a cure for leukemia is being pursued, the machine will be able to consolidate the signs, symptoms, nature, causes, etc., of leukemia from all authenticated knowledge bases and the confidence level in the conclusions gathered by the knowledge machine. As an additional example, if NPU design is being pursued, the machine (if so instructed) will be able to decompose the design to a gate level and indicate the sensitivities of the RC components as the NPU performs its numeric functions.

The knowledge machine readily provides statistically backed hindsight from a global and numerical perspective from the Web knowledge bases, Further, it provides insight based on the insight of humans (such as Bardeen and Brattain in developing the transistor in 1948), a streak of creativity from humans (such as evinced by Einstein in developing the concept of relativity in 1905), a glimpse of inspiration from humans (such as by Arnold at Dover Beach in 1851), or even a ray of enlightenment from humans (such as expressed by Buddha in expounding on nirvana in 500 BC). The iterative cycle between a human and a machine can run in reverse if the human being drives the machine in an extrapolative mode and assumes a meditative posture. Forward and backward iterations can only strengthen the ultimate search for scientific truth (as in the formulations of Bardeen, Brattain, Einstein, etc.), human virtue (as in the teachings of Buddha and Gandhi, etc.), and classic beauty (as in the words of Arnold and Tagore, etc.). Integrated behavior of machines simply becomes reversible cascading of information and knowledge filters in knowledge machines.

In dealing with *(I «» K)* processes, the subjectivity of human handling of *(I «» K)* is replaced by an objective measure of how the noun objects and verb functions [embedded in the body of knowledge *(I «» K)*] have been handled in the past and documented in the knowledge bases. In a sense, the machine quantitatively suggests to the human being that his or her use be consistent with the documented utilization and context of noun objects and verb functions under normal usage. However, the human being has the option to use noun objects and verb functions in a more creative and constructive way than ever documented before. Even though the knowledge machine may not be directly useful to a human being in developing an invention, it can analyze any invention to any degree of detail as to what was known and what has been invented (even if it was a false invention or a stolen idea).

6.5 THE DESIGN OF *(I ≪ ≫ K)* FILTERS

The basis of the design of a signal filter is its sensitivity to frequency in the frequency domain (in a traditional sense) and response to the unit function in the time domain. The two are intricately inter-woven, and the Laplace and Fourier transforms permit the conversion in these two inter-dependent parameters.

In the *(I «» K)* domain, such a precise mathematical inter-dependence does not occur but the structure of language has well-defined rules. Three seminal rules are as follows:

1. All information domains have one common basis of representation and that is the language (without regard to what language) which bears the sample *BOK* or *(I «» K)*.

2. As the next stage of redefining the *(I «» K)* space, the language that carries *(I «» K)* has the structure and numerous-variable format of sentences.[3]
3. All sentences have nouns, verbs, and embedded tense within the active verb.

6.5.1 Knowledge Filter Concepts

When frequency is reconstituted as a three-dimensional space rather than the conventional one-dimensional axis, then the concepts of poles and zeros of traditional filter theory become applicable to the design of knowledge domain filters. The placement of poles and zeros becomes crucial to the conventional designs of filters.

6.5.1.1 Specific Filter Profiles In the realm of knowledge, the signal represented (as an algebraic function of the Fourier components of signal levels and frequencies) becomes the entropy E of information entering the knowledge filters. The poles and zeros will be dictated by the attenuations/enhancements of the three characteristics of TVB or DAH being represented in the digital snapshot of the instantaneous entropy of the knowledge signal. The premise of this extension of conventional filter theory into the information and knowledge domain is represented in Figure 6.4.

It appears naïve to attempt to build a discrete element *(I «» K)* filter based on Figure 6.4, unless the transfer functions τ are generated by humans. Such functions are typically accomplished by media reporters who can accentuate or negate certain features of an event: that actuality (the t-component), benevolence (the v-component), and/or elegance (the b-component) are totally obscured to the observer of transformed *(I «» K)'*. In the political arena, this practice is flagrant to the extent that any event can be manipulated and transformed by human beings to suit the agenda of politicians. Such practices are indicative of human filters

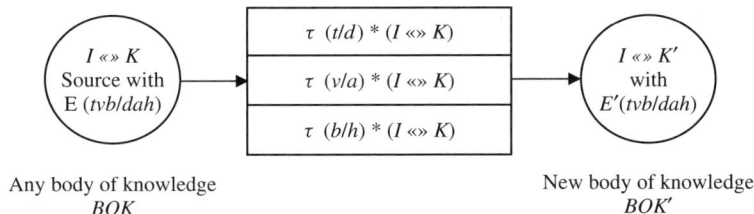

Figure 6.4 An overly simplified diagram of an information filter. *E(tvb/dah)* represents a map or graph of entropy for *(I«»K)*. This topology alters the balance between t/d (i.e., truth/deception), v/a (i.e., virtue/arrogance), and/or b/h (elegance or/hate) in any *BOK* or *(I«»K)* as it passes through the filter.

---

[3]Headings in documents may not have sentence structures but usually designate operative noun objects and/or verb functions for the text that follows. Scanning the headings usually offers a definite set of noun objects and verb functions.

(and reporters) with poles at t, v, and b. Machines can imitate such traits by enhancing or amplifying the knowledge operation codes (*kopcs*) for d, a, and h and/or by blocking the t, v, and b *kopcs* to be executed

6.5.1.2 Generic Information Filter Profiles

It can be seen that the three-dimensional (*tvb/dah*) space which replaces the frequency axis can be any hyper-dimensional space with any required set of parameters. For example, if the body of knowledge (*BOK*) input in Figure 6.4 is a digital snapshot of the entropy of a university, then the parameter for estimation can be its location, rating, student population, faculty, housing, etc. The transfer functions can indeed accentuate the strengths of the university and negate its student violence, excessive tuition, poor library, lack of computer facilities, etc. In most cases, these parameter cannot be held stationary but need continuous updating from Web bases. In the political arena, the news media around an unworthy President boosts the insignificant successes and willfully hides the flagrant failures. The output of a knowledge machine whose norms are based on statistical achievements of previous Presidents, will be weigh the t/d, v/a, and b/h ratios accurately and block any falsified *BOK*. Programmability and Internet access for the parameters are built into any design of generic-level knowledge filters.

Any initial *BOK* has noun objects and verb functions. It also has a message to convey. The graph of embedded nodes (noun objects) and the links (verb functions) connecting the nodes, their attributes, etc., carry the entropy E of *BOK*. This graph emphasizes the message to be conveyed.

The positive characteristics necessary to effectively and optimally carry the message are isolated and the negative characteristics that negate the flow of the message are also isolated during the first step. The positive characteristics are accentuated by the zeros (i.e., the information impedance is zero or the filter amplifies the positive). The negative characteristics are attenuated by the poles (i.e., the information impedance is infinitely high or the filter (effectively blocks the negative) in the message. The transfer function of the information filter thus contains the embedded zeros and poles that accentuate the positive characteristics and attenuate the negative characteristics of the message respectively. Numerous configurations and combinations of traditional filters exist, and it appears feasible to conceive of configurations for the corresponding information filters. However, micro knowledge computers and knowledge processor units may replace the lumped components or active filter elements in traditional signal filters. Such filters have embedded processors rather than customized IC chips in active filters.

At a conceptual level, we present the profile of a knowledge-level filter in Figure 6.5. In this diagram, we have chosen only to accentuate the TVB characteristics. Conversely, it is equally possible to build information filters to accentuate the DAH characteristics, or to select any number (rather than three) and any selection of characteristics, such as truth and rhyme (see Appendix A).

In comparison with the classical filter design, the information filter in Figure 6.5 corresponds to a multiple pole filter at numerous frequencies for each

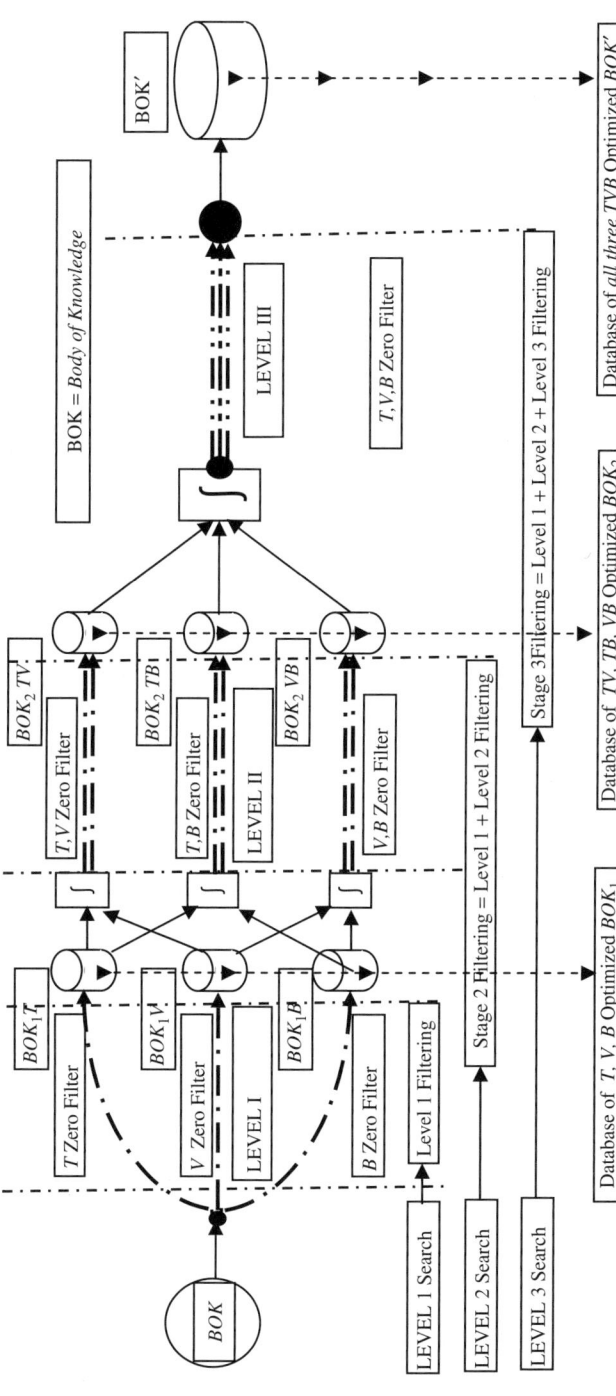

Figure 6.5 Schematic of an information and knowledge ($I \leftrightarrow K$) filter to enhance the entropy of any body of knowledge by maximizing three positive traits (T, V, B) with three levels of filtering. At the first level, the noun objects and verb functions are aligned and adjusted to maximize T, V, or B in BOK_1. At the second level, the previously optimized noun objects and verb functions are realigned and readjusted to maximize TV, VB, or TB together in BOK_2. Finally, all three TVB's are relaxed, realigned, and readjusted to maximize the total entropy of noun objects and verb functions in BOK'. Any number and combinations of traits may be chosen to maximize the entropy of BOK to derive BOK'. Also, when the filters are replaced by a knowledge machine, then the single knowledge machine can be used sequentially instead of the seven filters in a pipeline configuration as shown.

of the noun objects and verb function embedded in the *BOK,* thus enhancing the TVB characteristics that are associated with each of the *NOs* and *VFs*.

In the current case shown in Figure 6.5, the maximization of (positive) entropy is assumed to occur as the *NOs, VFs,* their relationships and attributes are relaxed. The maximization of TVB occurs as the *NOs* and *VFs* are rearranged by altering the information content around each NO and VF, their attributes, and cross-linkages. These relationships are not algebraic but depend on social and cultural rules. For this reason, an iterative relaxation of successive solutions appears more feasible than an algebraic solution. In addition, the relationships are estimates rather than being numerically precise, thus leading to an approximate solution rather than an exact one. Numerous solutions are possible based on the estimates and preferences of the users. The processes of information and knowledge filtering thus become a matter of personal choice. However fuzzy the exact solution for any individual, the basic laws of economics and expectations are applicable in the information/knowledge processing in any cultural or social setting.

6.5.1.3 Building Blocks of Knowledge Filters There are two aspects to consider: (1) the cascading of numerous filter sections to build a multiple-zero, multiple-pole filter, and (2) the output function of the filter derived as the product of excitation and the overall transfer function of the filter. We consider each of the two aspects as follows.

1. In traditional filter designs, the overall transfer function of the filter is the complex product of the transfer function of each of the cascaded sections at each frequency. The generation of the overall transfer function is a precise and mathematical process. Discrete component knowledge filters are impractical, if not impossible. In the knowledge and information domain, the function of any particular filter section can be replaced by set of programmable knowledge-processing systems (KPSs) or KPUs. When the overall transfer function of the filter is necessary, it becomes feasible to "convolute" the transfer functions of each KPS or KPU. The cascading of the knowledge operation in each KPS or KPU (constituting the overall knowledge process) is not a precise complex multiplication, but an optimization process for each noun object embedded in the input *BOK.*

2. In traditional filter designs, the output of the filter is the product complex excitation function and the transfer function of the entire filter at each frequency. The generation of the overall output is a precise and mathematical process.

The precise complex domain products of the excitation (generally evaluated by the fast Fourier transform routines) with transfer functions that are necessary to evaluate the output functions become inapplicable in the knowledge domain.[4] Excitation functions are *VFs* in the knowledge filter, and the input *NOs* and *VFs* are the operands. In the information and knowledge domain, a series of filter *VFs* operate on (excite) a series of *NOs* and *VFs* (from the input *BOK*) to generate a new complex object output *BOK'* with its own newly generated *NOs* and *VFs*. If the KPU processes are conceived as a series of convolutions between the series of *VFs* (an array) on a series of input *VFs* and *NOs* (a row), then the knowledge program can be viewed as a series of convolutions and the output is a pseudo-matrix of *VFs* and *NOs*.

The effects of (1) and (2) can also be combined into one optimization procedure in the knowledge domain. If an objective function for the output of the knowledge filter is assigned, then each of the filter *VFs* on each of the input *NOs* and *VFs* can be iteratively relaxed until the objective function is maximized. Since the knowledge domain can be nonlinear, the relaxation of *VFs* should be carried out in each direction individually and in combination. This aspect is illustrated in Figure 6.5 and utilized in Appendices A and B at the end of this chapter.

The precision of transfer functions and definitive relations between frequency and the numeric component values computed from signal filter design theory become impractical in the knowledge domain, but offer a methodology for the organization of numerous KPUs to constitute the sections of an information filter. Estimations and probabilities enter the knowledge filter design, since the entropy of any *BOK* at the left side of Figures 6.4 and 6.5 is not amenable to precise measurements, and the entire entropy depends on numerous input noun objects *(NOs)*, their attributes, lateral and bilateral relations between objects, and their attributes embedded in the *BOK* in addition to the numerous verb functions *VFs*.

When the numerous filter sections (that accomplish the overall transfer functions) are replaced by KPSs or KPUs, it becomes feasible to transform the entire entropy profile of the input *BOK* to the desired entropy profile of the output *BOK'*.

6.5.1.4 Configuration of a Generic Knowledge Filter
The schematic shown in Figure 6.5 can be transformed into the device layout depicted in Figure 6.6. The transfer function generally has numerous poles associated with each noun object, each verb function, and each relationship.

In a sense, humans are better equipped for this global optimizations than machines. However, machines can be more exhaustive in the relaxations of the *NOs, VFs,* their relationships and attributes. In addition, the machine can (locally and globally) search for similar relaxations and combinations of objects, verbs, and relationships. Though not genuinely creative, the humanoid machines can be tireless in all the activities that make humans creative. In the case of a wisdom machine (Chapter 2), the switching of functions from machines to humans occurs at this level.

---

[4]The exact procedure and validation of these convolutions call for additional investigation. In the interim, to emulate these such as the one complex [as in $(\cos(\omega t) + j \cdot \sin(\omega t))$] domain multiplications [2] in traditional filter theory, we propose a convolution suggested in Figure 3.2. By deploying this strategy, the transform of the entropy of the input *BOK* to its desired profile of the output *BOK'* may be achieved.

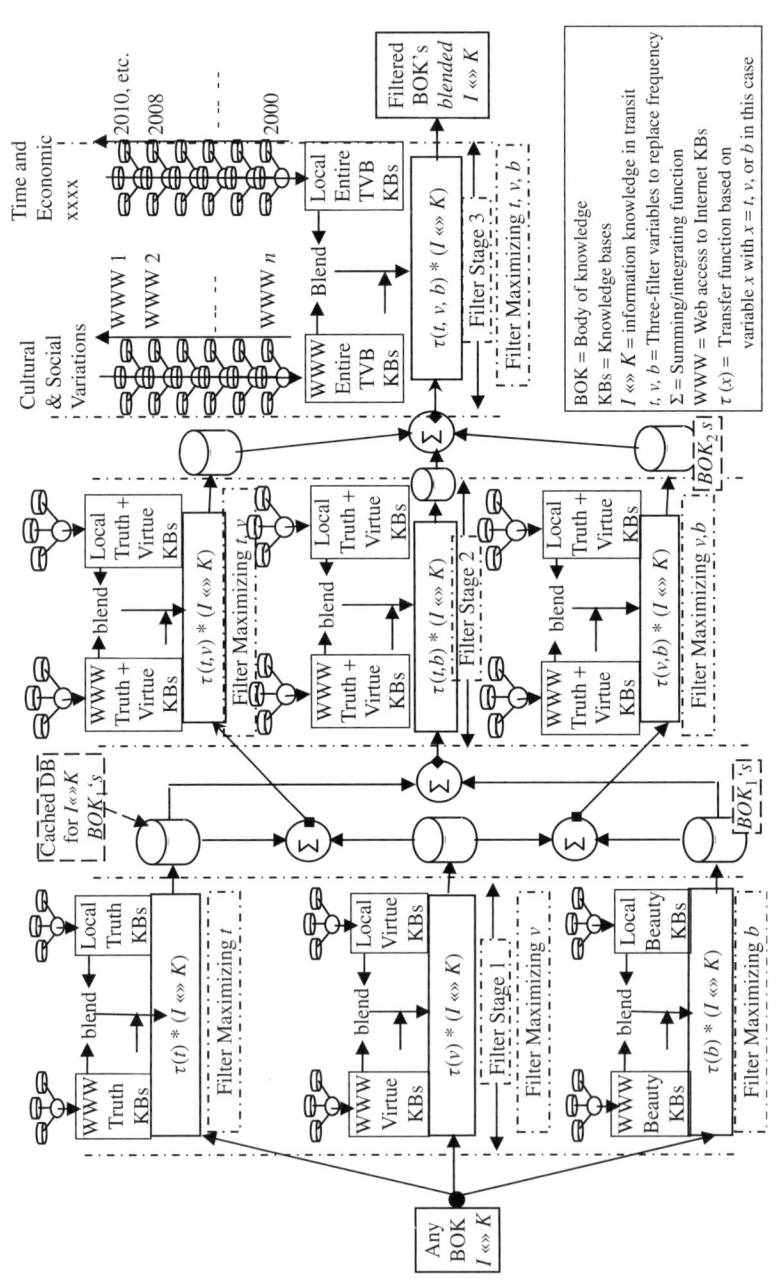

Figure 6.6 A three-stage filter for $(I \Leftrightarrow K)$ to accentuate the TVB, that is, poles are at all t, v, b's for all noun objects and verb functions.

Numerous clusters of outputs appear possible in each of the seven filters $\tau(t)$, $\tau(v)$, $\tau(b)$, $\tau(t,b)$, $\tau(b,v)$, $\tau(v,b)$, and $\tau(tvb)$] of the generic knowledge filter in Figure 6.6, since the maximization process is fuzzy within each of the filters. An order of priority can be assigned in the forwarding of $(I \ll \gg K)$ to the next filter, or all the possible clusters from the previous filter(s) can be examined. In addition, all the filtering processes of all the prior filters need to be reexamined in view of the current filtering process.

For example, at third-level filtering, the optimality of the *NOs, VFs,* their relationships and attributes at both the first and second level need examination in view of the maximization of all the TVB characteristics at the third level. Tedious as the processes are, tireless are the machines that accomplish these functions.

Feedback and feed-forward techniques may become necessary to assure convergence of the solution at the final output of the information filters. it is possible that a multiplicity of equally acceptable solutions may result from the information and knowledge filters (see Appendices A and B).

6.5.2 Noun Objects and Verb Functions

This structure of $(I \ll \gg K)$ is instrumental in making the filtering process amenable to developing a basis for performing some of the (information and knowledge) functions in the object-oriented programming (OOP) environment. If the noun objects are treated as objects and the verb functions are treated as one or more instructions for performing an object-level function, the sample $(I \ll \gg K)$ becomes the input to a higher-level OOP application. The complex application-level functions are performed by a series of macro- and microcommands performed on the noun objects, verb functions, and their attributes. The current computers and OOP compilers (with an additional class of $(I \ll \gg K)$ domain-level routines) will provide a software solution to the $(I \ll \gg K)$ type of filter problems.

This structure for processing $(I \ll \gg K)$ is instrumental in making the filtering process comparable to basic-level assembly instruction to the CPU of any computer. This basic code has two components, namely:

1. Opcode [which is loaded in the instruction register (IR) of any CPU]
2. Operands (which are brought into the *A, B,* etc., registers of the CPU)

If the hardware for processing $(I \ll \gg K)$ is now equipped with a knowledge instruction register (KIR) see Figure 1.2) and knowledge operand registers (KOPRs), see Figure 1.2), then the knowledge processing unit (KPU) can function in a knowledge machine (KM) or computer in the same fashion as a CPU would function in a conventional computer. The verb functions at their microinstruction level would flow into the KIR and the noun objects or their constituent would flow into the KOPRs. The architecture of the KM can thus be derived from the architecture of a typical computer.

6.5.3 Scanning for Operative Nouns and Verbs

If the operative (core) nouns and (active) verbs are isolated (as symbols are isolated in traditional computer programs), then the structural relation between:

1. Each noun object and other noun objects
2. Each verb function and other verb functions
3. Attributes of each noun object and attributes of all noun objects
4. Attributes of each verb function and attributes of all verb functions

is established for each sample of $(I \ll\gg K)$. Such structural relationships are represented as one or more graphs for nouns, verbs, attributes, etc. A series of such graphs of objects, verbs, and their attributes is a unique signature of *BOK* sample. A procedure of this sort is used in identifying and comparing genome sequences and chains. It is also used in image processing, string searches, and magnetic resonance imaging (MRI) techniques to separate malignant and benign tissues in cancerous areas.

6.6 CONFIGURATION OF $(I \ll\gg K)$ SYSTEMS

6.6.1 Application Constraints

The activities of humans and social organizations are based on their needs however obscure they may be.[5] Hence, it becomes necessary to provide a basis for $(I \ll\gg K)$ machines to satisfy human beings and social organizations that use machines to satisfy their own immediate (individual, social, corporate, etc.) needs. In fact, these machines should operate within the constraints of such diversified needs well documented by psychologists and social scientists.

6.6.2 Design Concepts for $(I \ll\gg K)$ Systems

Information $(I \ll\gg K)$ is built around objects embedded within any body of knowledge. Such objects have structure and attributes that become important in the selection criteria for filters. Selection based on objects alone may not be sufficient for effective filtering. Hence, we import the concept of scanning and timing information from digital filters for most applications such as speech, video, EKG, EEG, MRI, etc. In the $(I \ll\gg K)$ domain, the concept of scanning frequency or timing information becomes irrelevant because the $(I \ll\gg K)$ scans are accumulated in the knowledge bases at any given time, even though they may be updated from time to time.[6]

---

[5]The implication is that all human activity is goal-directed, and the goals are to satisfy one or more needs within human beings. An extended needs hierarchy [5.4] encompasses the needs beyond those proposed by Maslow and extends into the more abstract nature of mind that searches and unifies information and knowledge under a global umbrella of human wisdom and ethics.

[6]The results of filtering can become time-dependent if the updating of a knowledge base is continuous or too frequent (as in stock market updates).

CONFIGURATION OF $(I \ll \gg K)$ SYSTEMS

A few basic concepts become essential to design and build the $(I \ll \gg K)$ filters. These concepts are derived from truisms in human behavior, and we organize the status of information and knowledge in the human mind. The symbol $\alpha \rightleftharpoons \beta$ indicates that both α and β interact until a stable condition is reached. Accordingly, these truisms may be written and symbolized as follows:

$$\text{Human beings and social entities} \rightleftharpoons \text{Needs}$$

1. Human beings and social organizations have needs:

$$\text{Knowledge} \rightleftharpoons \text{Objects and the processes they undergo}$$

2. Knowledge is built around "objects" that play a role in resolving needs:

$$\{Humans/machines\ perform\ VFs\ on\ NOs\} \rightleftharpoons$$
$$\{\text{Secondary objects for human satisfaction}\}$$

3. *VFs* processed on *NOs* result in secondary objects that also satisfy needs:

$$\text{Secondary objects} \rightleftharpoons \text{Process VF on NO} \qquad (a)$$
$$\text{Human satisfaction} \rightleftharpoons \text{Secondary objects} \qquad (b)$$

4. Human beings and/or machines can process *NOs* and perform *VFs* and generate secondary objects, such as robots that can produce "objects" for human consumption or a new $(I \ll \gg K)$ "object" in the knowledge domain.
5. *VFs* and appropriate *NOs* can be cascaded to perform large complex human and social function like micro- and macrosteps in software systems. *VFs* and *NOs* can be assembled like the assembly-level programs for VLSI chips and computers.
6. Steps 1 through 4 form a quasi-stable dynamic loop that varies distinctly from individual to individual, corporation to corporation, society to society, nation to nation, culture to culture.

These first four truisms can be graphically represented in Figure 6.7 as a state and transition diagram. The first three truisms are represented by the inner triangle 123.

The three vertices 1, 2, 3 are supported by three sets of Web-bases containing categories of human and social needs, and the primary and secondary objects to satisfy such needs. The global nature of human and social responses to needs is included by addressing the Web bases that focus on objects which are associated with the needs of individuals and social entities.

Once immediate needs are satisfied, the system modifies the two vertices (objects) 2 and 3 and the six links (processes) *VF1* and *VF1'*, *VF2* and *VF2'*, *VF3* and *VF3'* of this transition diagram to traverse from needs to satisfying them

268 INFORMATION AND KNOWLEDGE FILTERS

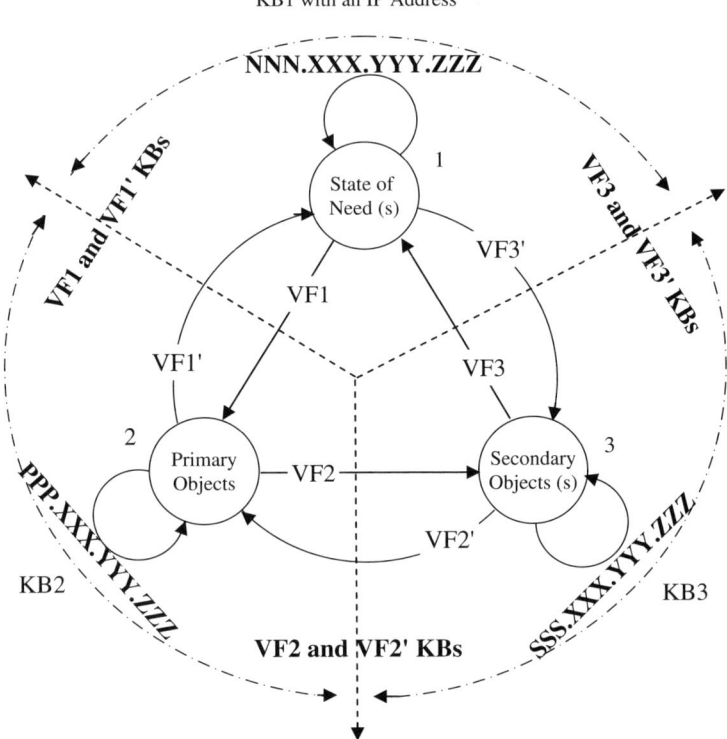

Figure 6.7 A state and transition diagram for humans or social entities in a state of satisfying one or more needs. N signifies need(s), P primary objects associated with need(s), and S secondary objects to gratify the need(s). Node 1 = state of need(s); node 2 = primary objects for the need(s); node (3) = secondary object(s) that are deployed to satisfy the need(s); $VF1$ = process to find and select; $VF2$ = deploy and adjust; $VF3$ = fully or partially extinguish the need(s); $VF1'$ = learn and adapt; $VF2'$ = negotiate, adjust, and adapt; $VF3'$ = learn and adapt.

quickly and optimally and with minimal effort, thus maximizing the *marginal utility* of all resources. A condition of dynamic equilibrium is reached and modified only by the changing needs and contents of the three knowledge bases: KB_1, KB_2, and KB_3.

While Figure 6.7 depicts a state diagram, it is possible to map the three states of nodes 1, 2, and 3 into the memory of a computer. It is also feasible to depict the processes that alter these states (in the memory) by the programs executed in the knowledge processor units or by a knowledge machine (KM). The realization of a physical system that acts very much like a human being or social entity is shown in Figure 6.8.

In order to make the system generic and universal, the *NOs* and *VFs* are used in Figure 6.7 rather than primary/secondary objects, and *VF1* and *VF1'*, *VF2* and *VF2'*, *VF3* and *VF3'* in Figure 6.8. Needs are drawn from needs structures, and

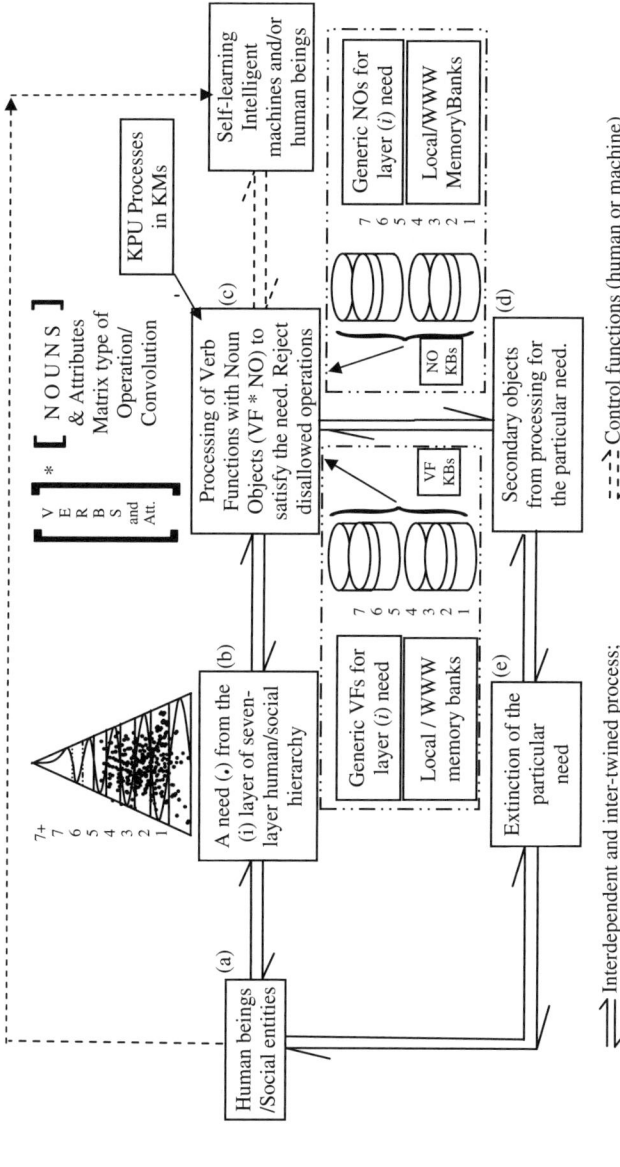

Figure 6.8 A repetitive quasi-stable loop (a) through (e) in solving the simple and complex needs of human beings or social entities. The role of intelligent machines is modified and enhanced by the level of expertise and technology of the individual or social organizations. Operative noun objects and verb functions become crucial in needs-resolution.

the *NOs* and *VFs* from knowledge bases around the world. Such a generalization leads to a more universal KM capable of addressing a wide spectrum of needs. Local and global objects are also blended to offer latitude in selecting the most appropriate objects for any need even though it may be obscure.

The attributes (adjectives) of *NOs* are also incorporated in the derivation (enhancement, filtering, clarification, deduction, problem solving, etc.) of new knowledge by processing the old knowledge in conjunction with Web knowledge bases. In the same vein, the attributes (adverbs) of *VFs* are also incorporated to alter the flavor of processes that bind together objects to form a new body of knowledge. Local and global *VFs* are also blended to offer latitude in selecting the most appropriate objects for any obscure or derived need.

The truisms 1 through 4 presented in this section are represented as a flow diagram in Figure 6.8. This figure also contains the quasi-stable dynamic loop (a) through (e) to bring the overall process to a logical end. From a longer perspective, individuals and social organizations process information and knowledge to derive the most gratification for present and future needs.

The entire knowledge process becomes an exercise in the marginal utility theory based on the highest gratification for any incremental resources expended. In a mathematical sense, the process is an attempt to optimize the performance of this loop. Computers (robots) and software systems (management information systems, SAP, Oracle PeopleSoft, etc.) assist human beings in satisfying their needs in the most optimal way based on the immediate environment of the problem.

6.6.3 Application for Filtering and Comparing

In applying the entire set of truisms 1 through 6 in Section 6.5.2 to filters, the process stops at step 4. The *NOs* and *VFs* are first determined from the template for the filter.

A similar set of *NOs* and *VFs* are determined for a comparison sample. In order to select or reject a sample, an inclusive comparison must be made between the template *NOs* and *VFs* and the sample *NOs* and *VFs* to find a fit:

- If the sample has the same (or dictionary-based) *NOs* as the template
- If the sample has the same (or dictionary-based) *VFs* as the template
- If the same structural relations exist between all the *NOs* within the sample and template
- If the same structural relations exist between all the *VFs* within the sample and template
- If the same attributes exist for every NO within the sample and template
- If the same attributes exist for every VF within the sample and template
- If the same structural relations exist between all the attributes of *NOs* within the sample and template
- If the same structural relations exist between all the attributes of *VFs* within the sample and template

It can be seen that the process may become infinitely long and time-consuming if the attributes of attributes are included in the comparison process. However, every step enhances the fit and increases the confidence level in the selection/rejection procedure. At a predefined level of confidence, the filter moves on to the next sample—either to select or reject the sample on the basis of the template.

6.6.4 Low-Pass and High-Pass Information Filters

The process of filtering in the $(I \ll \gg K)$ domain has three additional variations. If objects in the sample are less than objects in the template (e.g., metals whose atomic weight is less than that of silver, ailments that are less severe than cancer, etc.), then the choice of a low-pass filter can be pursued by selecting samples that satisfy this (less than) requirement. Similarly, if objects in the sample are greater than objects in the template (e.g., metals whose atomic weight is less than that of silver, ailments that are more severe than cancer, etc.), then the choice of a high-pass filter can be pursued by selecting samples that satisfy this (greater than) requirement.

A bandpass filter selects objects in a range of pre-selected attributes. It is also evident that the selection may be based on verb function rather than noun objects. All three versions of information (low-pass, high-pass, and bandpass filters) may be programmed into a knowledge system that can be forced to serve as an information filter.

A new dimension in the processing of information and knowledge can be introduced in programming the knowledge machines by imposing restrictions on noun objects and verb functions. Whereas such filtering criteria are easily deployed by human beings, knowledge machines can also perform similar functions through a programmable profile of the filter template.

However, in the $(I \ll \gg K)$ domain, noun objects and verb functions have a special bondage like samples and timing have in digital filters. Functions and actions that the noun function experiences or undergoes can be strongly coupled. Whereas mathematical rules apply to the transformation of domains (frequency to time and time to frequency) in signal processing, syntactic rules of the language apply to the bond of nouns objects and verb functions, and semantic rules of knowledge programs apply to the bond of every noun object with other noun objects and to the bondage of every verb function with other verb functions, etc. This assertion is based on the observation that almost all $(I \ll \gg K)$ have operative noun objects, and objects undergo change. Internal changes or externally administered changes on the object functions now become the verb functions. Syntactic and semantic rules apply in identifying the unique signature of any $(I \ll \gg K)$ sample.

If the filtering is based on both the objects and the change they undergo in the body of $(I \ll \gg K)$, then the scanning, searching, recognition, and finally selection of the exact fit for the sample $(I \ll \gg K)$ with the filter template can be made exact and precise. To some extent, it becomes necessary to tie the concept of time in digital filter theory with the concept of change in digital knowledge theory. Static noun objects become dead objects and may generally have a historic value that is not critically important to the dynamic nature of $(I \ll \gg K)$ which are frequently scanned, searched, recognized, and selected.

6.6.5 Design Steps for ($I \ll \gg K$) Filters

Much like signal filters, information and knowledge $(I \ll \gg K)$ filters can have numerous configurations. To contain the problem, we suggest a conceptual bridge that treats frequency in the signal space as two sets of entities: noun objects and verb functions in the $(I \ll \gg K)$ space. The methodology can thus be focused around objects and the verbs embedded in the virtual $(I \ll \gg K)$ space. Bandpass information filters will remove samples of $(I \ll \gg K)$ that do not have noun objects and verb functions of interest in the design of such filters. Low-pass filters will pass samples that contain noun objects that are lower in rank (such as crude oil in comparison to refined petroleum products, or gemstones in comparison to diamonds, etc.). Such low-pass information filters may also be instructed to pass verbs that are lower in rank (such as driving being rated under flying, cleaning of items under cleansing of items, gems under diamonds, etc.). A similar philosophy of filtering applies to high-pass filters.

6.6.6 Design Methodology

To evolve a design methodology, we propose the following seven-step process and an eighth iterative procedure

1. Define an objective function (OF) for selection/rejection based on the noun objects *(NOs)* and filter verb functions *(VFs)* that the filter will handle.
2. Define the boundary for scanning global noun objects *(GNOs)* and global verb functions *(GVFs)* from where the samples will be drawn.
3. Initiate a local or Web search within boundary (2) based on the *NOs* and *VFs*.
4. Perform an AI-based pattern recognition (PR) procedure based on the *NOs, VFs,* and their attributes. Generate a confidence level for the accuracy of the subjective type of steps in (3) and the precision of numerical operations in PR (syntactic, semantic, and statistical) for each of the samples $(I \ll \gg K)$ drawn from *GNOs* and *GVFs*.
5. Filter out the possible contamination, corruption, and degradation of information-bearing *NOs* and *VFs* from the final list of samples $(I \ll \gg K)$ existing in knowledge banks within the boundary conditions for the search.
6. Grade the samples of $(I \ll \gg K)$ drawn from a local or Web search, select or reject the particular $(I \ll \gg K)$, and then place the sample in the appropriate bin for more advanced filtering steps.
7. Perform ah AI-based learning procedure for the methodology to recognize patterns of objects and their inter-relationships based on the criteria for selection/rejection.

8. Reiterate steps (1) through (7) based on learning until no improvement in the confidence level results.

A filter configuration is shown in Figure 6.3 to select samples of *(I «» K)* that will be processed by the filter and compared against the filter-specific *NOs* and *VFs*. The OF or the template is constructed from the relationships between the *NOs, VFs,* their relationships, and the attributes and each of their inter-relationships. In a sense, the degree and extent of filtering can be made so precise that no sampled *(I «» K)* is a perfect match or so general that every sample will be a good enough match. The OF will be an initial gateway to the selection of samples.

At the PR stage of the filter, the OF becomes a precise mathematical template for the measurement and selection of all samples based on a given confidence level. The matching process becomes a numerical procedure based on the degree of match, thus leading to a confidence level in the entire filtering procedure. The established techniques for pattern recognition (statistical, syntactic, string-based, and semantic) become feasible in the context of measuring the sample against the template.

Learning and refinement of the techniques for building the OFs, and drawing samples from the local or Web knowledge bases, become desirable to improve the performance and accuracy of the *(I «» K)* filters.

An alternative approach to building a *(I «» K)* filter is shown in Figure 6.6. Samples for selecting and rejecting *(I «» K)* items from local and Web knowledge banks are based on criteria from a problem posed by the user or as a minor step in the functionality of an overall knowledge machine. Noun objects, verb functions, their relationships and attributes are all used in the PR, filtering, and select/reject functions of the filter.

6.7 SELECTION OF NOUN OBJECTS AND VERB FUNCTIONS IN SAMPLES

When searching and PR concepts are applied to the identification of texts randomly selected from books or from any scripts, the task of searching for commonality starts to become necessary. In such a case, the OF would be to write down the extent of overlap between the text items. If the contents are arranged according to a hierarchical pattern, then the main heading, subheadings, etc., would function as anchor *NOs* and/or *VFs*. The methodology for the systematic breakdown of information blocks *(I «» K)* is shown in Figure 6.9.

Performing a PR analysis (Section 6.4.3) would yield a quantitative measure of the overlap at the main heading level. Performing a similar PR analysis with a lower level of subheadings will further affirm/deny the initial finding or invoke further search via filtering of the texts until the required level of confidence is reached. At lower heading levels (starting from the block level down to sentences and an individual phrase), the confidence level in the selection/rejection procedure

274 INFORMATION AND KNOWLEDGE FILTERS

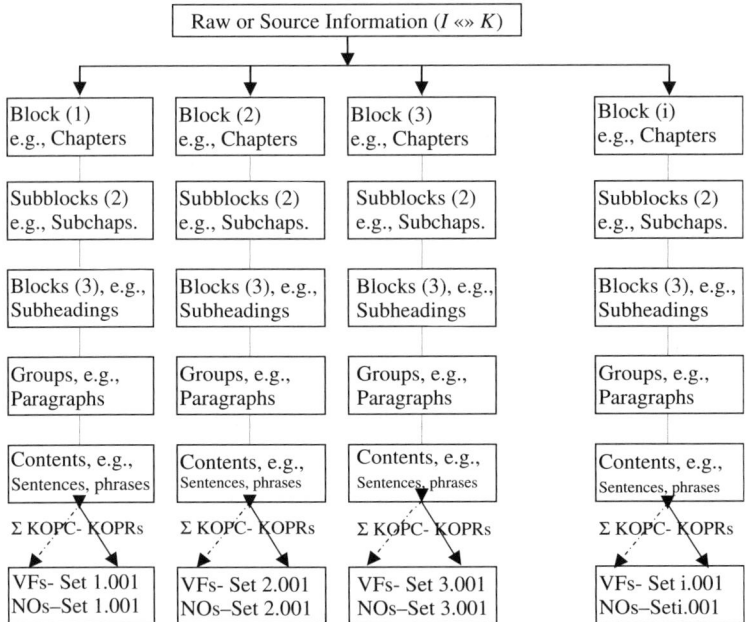

Figure 6.9 Segmentation of information into appropriate verb functions and noun objects from a sample (I «» K).

can only increase. Only if the two texts are identical can a confidence level of unity in the comparison be achieved.[7]

A binary comparison of files is now feasible when the files are treated as binary strings and compared at bit levels. Word comparison is feasible when files are compared in a verbatim mode. In this section, we have proposed a methodology that is more flexible than the bit-level or word-level comparison. Instead, the operative *NOs* (or their synonyms) and critical *VFs* (or their synonyms) and their structural relations are matched to numerically evaluate the conceptual overlap between various segments of *(I «» K)*.

Typical deployment of the selection methodology for three positive *NOs* (truth, virtue, and beauty) and three negative *NOs* (deception, arrogance, and hate) are presented in the two Appendices C and D. Examples for filtering samples drawn from typical text material and writing are also shown. The knowledge machine will handle longer and a greater number of samples drawn from Internet traffic.

---

[7]The publishing industry has assigned world-wide ISBNs for books to be compared in a digital format, and within the classification, a search comparison of the title of books generally suffices. In the Internet era, segmentation of information is rampant and special PR algorithms become necessary. Even in such cases and in most instances, the PR process for comparing *(I«»K)* samples is limited to one or two levels. In this section, we have developed a core methodology to compare many segments of information and knowledge to any depth of search.

6.8 SYSTEMS FOR $(I \ll \gg K)$ FILTERS

The automated system for filtering consists of the basic hardware—i.e., the KPU, object and program memories, input/output (I/O) subsystems and bus structure with appropriate switching capability—and the essential software—i.e., the compilers to process the input queries, build (derive) objective functions (OFs), process the input samples from local knowledge bases, or the Internet, etc.. An knowledge based operating system would also be necessary to ensure the associated I/O and job-monitoring functions.

A viable architecture for a system to accomplish filtering functions is shown in Figure 6.10. The input at box 1 to the systems [such as the operative *NOs, VFs,* type of filter (low-pass, high-pass, band-pass, etc.), Web knowledge bases to be scanned, noun–verb relations, hierarchies of nouns and verbs, etc.] lead to an OF in box 2. The reason for this OF is to establish a benchmark for the extent of fit between the user specifications and embedded information or knowledge in any sample of Internet information derived from Web or local knowledge bases. As the OF becomes elaborate, the selection of sample(s) gets more stringent. The lowest order of the OF is to find word/phrase matches or mismatches. This mode of operation is typical of Internet search engines.

When the entire graph of a knowledge tree is matched and a quantitative measure of the number of branches, twigs, or even leaves that match/mismatch is tracked, then the extent of match/mismatch can be a machine-driven process.

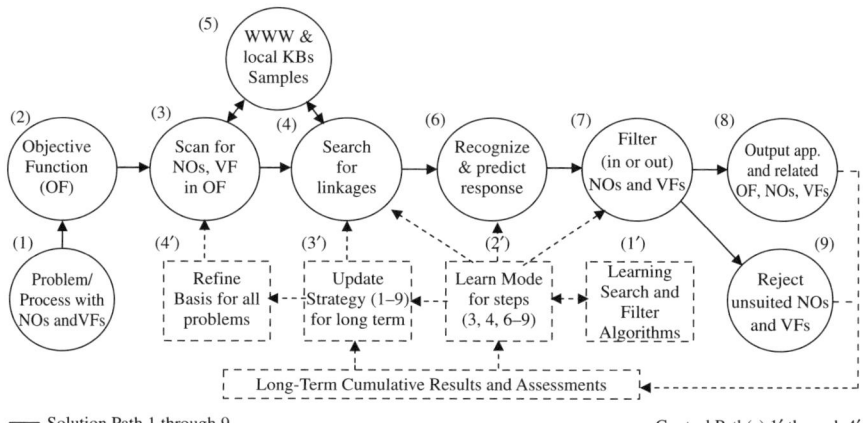

Figure 6.10 Basis for building a $(I \ll \gg K)$ filter for selecting and rejecting items from local and Web knowledge banks based on criteria from a problem posed by the user or as a step in the functionality of a larger knowledge machine. Noun objects, verb functions, their relationships and attributes are all used in the pattern recognition, filtering, and select/reject functions of the filter. *NOs* = filter noun objects; *VFs* = filter verb functions; KB = knowledge bases; OF = objective function derived from the problem input such as low-pass, bandpass, high-pass, bands-reject, etc.

The filter for *NOs* and filter for *VFs* and their inter-relationships are extracted (box 3) to make a genome pattern of the filter.

The samples are derived from Web or local knowledge banks or libraries (box 5) that undergo scrutiny for sample noun objects (SNOs) and sample verb functions (SVFs). The genome type of recognizing and matching of the four key elements (*NOs, VFs,* SNOs, and SVFs) constitutes the basis for the initial acceptance/rejection of the samples. The extent of matching can be made arbitrarily deep to become as selective as the user may decree. For example, the four elements are matched first, then the linkage between each of the first two elements is matched next, then the next two elements, then the four elements in partial unison and total unison, etc.

As a next step, if the filtering does not produce a satisfactory match, then the synonyms of the four key elements may be matched, each individually and group-wise as shown (box 8). The process can be pushed down one more level by examining the synonyms of synonyms, etc.

The filter will continue to process when the sample objects are sparse and the filter is too stringent to examine even a trace or hint of match in any language or from one language to another. The extent of filtering can thus be forced to suit the knowledge and content of information without regard to language. The pattern-matching process now becomes "concept matching" based on the concepts embedded in the *NOs* and *VFs,* and their inter-relationships, rather than the words themselves.

The control of the filter is programmed or made artificially intelligent by incorporating the learning mode based on the results of the procedures in boxes 3, 4, 6–9. Over a period of iterative trials, the filter will learn to pick the right moves in selecting samples from large knowledge bases like a computerized chess machine. Knowledge filters tend to be more elaborate and complex compared to wave filters, and the results can only be fuzzy rather than perfect because of the ambiguity of the words in *NOs* and *VFs.*

The filter will also yield different results depending on the deployment of the keywords and the dictionaries used to derive the synonyms. It is interesting to note that the same samples analyzed by the same filter template would start to produce different results based on dictionaries from different decades. The extent of such a difference becomes a measure of cultural change in the use of words in any language over the decades. A numerical measure of the cultural change over the decades becomes feasible. For example, if truth, virtue, and beauty are three noun objects (see Appendix C) used to analyze samples writings (say, from Einstein from Aristotle), then 1980s dictionaries would offer more commonality between the two writings than a 2002 dictionary, indicating more rapid changes that have taken place in recent times.

An alternate arrangement to accomplish the information filtering is shown in Figure 6.11. In this configuration, a problem is offered to the systems, and the system finds the *NOs, VFs,* and associated interrogation(s). The answer is synthesized based on the bond between the three [i.e., *NOs, VFs,* and interrogation(s)]. Complex queries are degenerated into a sequence(s) of simpler queries that are

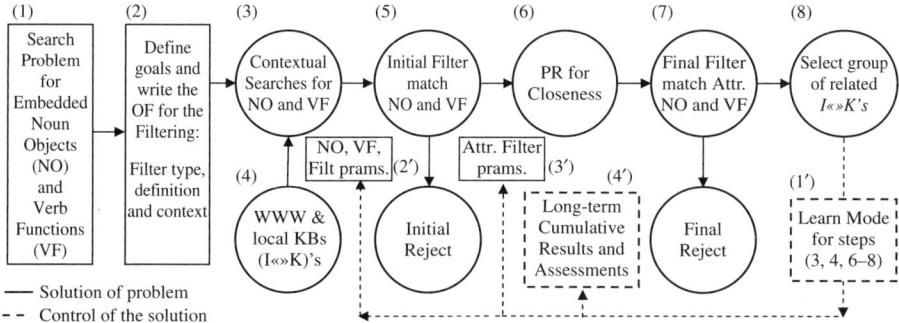

Figure 6.11 An alternative basis for building a *(I «» K)* filter to respond to queries or search problems. The system searches local knowledge banks and/or the Web based on criteria from a problem posed by the user. The whole procedure may be a step in the functionality of a larger knowledge machine. Noun objects, verb functions, their relationships and attributes are all used in the pattern recognition, filtering, and select/reject functions of the filter. OF= objective function.

resolved initially, and then the responses are integrated to establish a linage of subsets of solutions to complex queries. In box 3 the context is based on the query type. For instance, if the query is "when," only time-sensitive information about the *NOs* and *VFs* is gathered for final filtering or for the response. If the query is "how," information based on logical linkages (between *NOs*, *VFs*, and *NOs* and *VFs*) is gathered. A query "why" will initiate a reason-based information-gathering strategy and a query "what" will initiate backward pointers to *NOs* and *VFs*. The responses are collected for initial rejection, final rejection, and then the final selection to respond to the problem with the most significant response and its numerical score of significance. Other responses may be collected in a tentative database for further analysis.

6.9 TWO-PORT KNOWLEDGE NETWORK

Two-port network theory has been an invaluable approach to simplifying network calculations in circuit and communication networks. In a simplistic sense, two-port theory correlates the input and output parameters (such as currents, voltages, transmitted and reflected power, impedance mismatch effects, etc.). In this section, we extend the notion of this classic theory to the domain of knowledge networks.

In the knowledge domain, verb functions, noun objects, their convolutions, the object attributes, (adjectives, if any) verb modifiers (adverbs, if any), and the overall entropy of any body of knowledge take up the role of input and output parameters. The flow of entropy, much like the flow of energy, can be smooth, bumpy, reflective, and/or resistive through different two-port knowledge networks. Perhaps the role of the characteristic impedance [defined as the square root of $(A*B/C*D$, from traditional circuit analysis) of the electrical network] is hardest to analogize in the knowledge domain.

6.9.1 Two-Port Electrical Network Theory

A two-port network in an electrical engineering environment is shown in its simplest form in Figure 6.12. Two directions are shown, and the condition for the transfer of maximum power occurs if the source, sink, and characteristic impedance are the same within the bandwidth of the signal from the source. Generally, this is not the case, and this gives rise to reflections of the signal, thus restricting the flow of power.

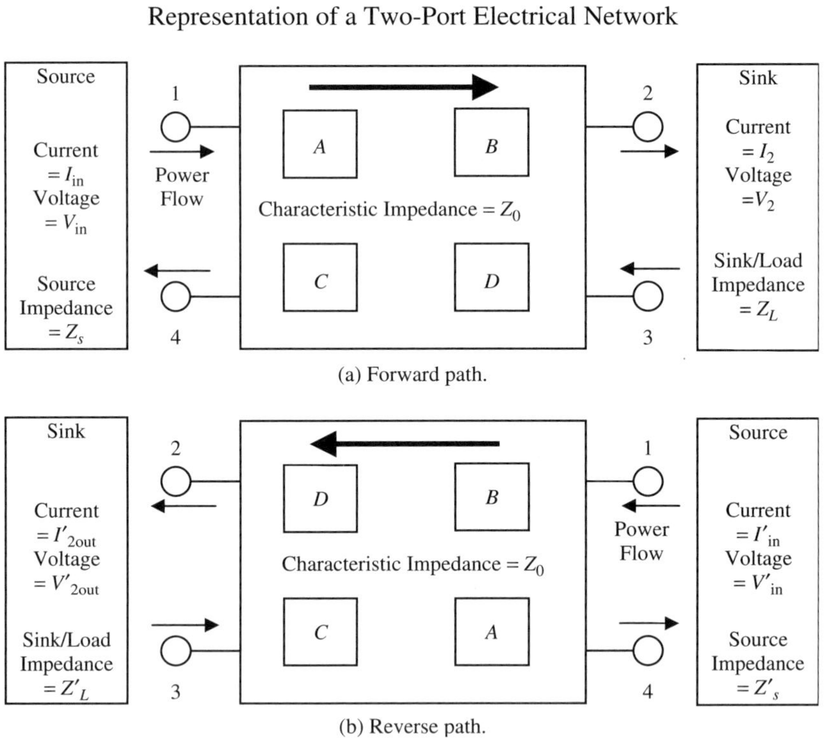

Representation of a Two-Port Electrical Network

(a) Forward path.

(b) Reverse path.

Figure 6.12 Mathematical representation of a two-port electrical network to transfer power from any source with a source characteristic impedance of Z_s to any sink/load with its own impedance of Z_l. The characteristic impedance of the media is Z_0. Typically, all three impedances are frequency-dependent, and the condition for maximum power transfer is that all three impedances are the same at each Fourier component of the signal V_s. This condition makes the reflections zero at the input and output ends of the network. The external losses will be losses in the media that are totally characterized by the A, B, C, D] matrix. The $ABCD$ matrix is generally complex and facilitates the computation of input impedance between the two ports. See [8].

For the forward path, the matrix equations that relate the input parameters (voltage and current) to the output parameters can be deduced [8] as

$$\begin{bmatrix} V_1 \\ I_1 \end{bmatrix} = \begin{bmatrix} A & B \\ C & D \end{bmatrix} \begin{bmatrix} V_2 \\ I_2 \end{bmatrix}$$

Written alternatively,

$$V_1 = AV_2 + BI_2$$
$$I_1 = CV_2 + DI_2$$
$$V_1 = AI_2 Z_L + BI_2$$

since

$$V_2 = I_2 Z_L$$
$$= (I_2(AZ_L + B)$$

$$\frac{V_2}{V_1} = \text{Transfer function } (\tau) \text{ of the two-port network} = \frac{Z_L}{AZ_L + B}.$$

Similar equations can be written for the reverse path. It is apparent that in knowledge networks, information can flow in either direction (e.g., in dialogs and in query-response modes) as in the duplex mode of communication.

6.9.2 Configuration of a Two-Port Knowledge Network

In order to facilitate the methodology for relating the input and output parameters in the knowledge domain, a two-port network is shown in Figure 6.13. The condition for the flow of information through the network is that there is potential knowledge difference between the two sides of the network. Any given body of knowledge (*BOK*) at the input is based on four basic parameters:

1. A series or sequence of verb functions ($\sum VFs$) embedded in *BOK*
2. A series or sequence of convolutions ($\prod^*$) embedded in *BOK*
3. A series or sequence of noun objects ($\sum NOs$) also embedded in *BOK*
4. Tense of each of the verb functions in item 1

To simplify building the two-port network, we consider only the first three parameters at the input side. In the knowledge machine, the transfer of information is optimized as it adaptively processes a 3×3 matrix with nine coefficients, *A* through *I* (see Figure 6.13), to suit the source knowledge characteristics and *BOK*. In the same vein, it also re-optimizes the nine coefficients to suit the output knowledge to suit the recipient characteristics and *BOK'*. This is a process of negotiation, adaptation, learning, and creativity, much like the human

280 INFORMATION AND KNOWLEDGE FILTERS

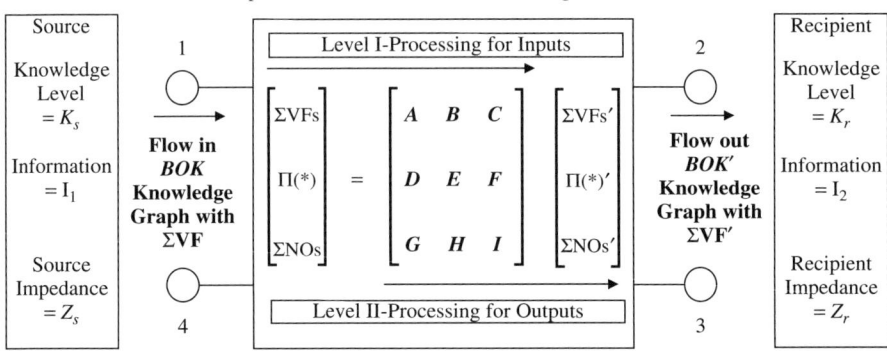

Figure 6.13 Representation of a two-port knowledge network. Nine characteristic coefficients ($A, B, C; D, E, F; G, H, I;$) are optimized to maximize the flow of information from left to right. The optimization criterion (such as maximize TVB) is that the entire graph of *BOK* be accurately mapped to the graph of *BOK'*. The following equations explain the steps involved in building the two-port knowledge network.

and corporate interaction processes discussed in Appendix B of Chapter 5. The quasi-mathematical input/output relations of the two-port knowledge network are represented as

$$\left\{ \begin{bmatrix} \sum VFs \\ \prod^* \\ \sum NOs \end{bmatrix} = \begin{bmatrix} A & B & C \\ D & E & F \\ G & H & I \end{bmatrix} \begin{bmatrix} \sum VFs' \\ \prod^{*'} \\ \sum NOs' \end{bmatrix} \right\}$$

For the two-port network, the basic theme for any element of knowledge (ΔK), or the information contained within this element, is that it is a

VF(verb function) *(convoluted) with NO(noun object)

Stated alternatively,

$\Delta K = \text{VF}^* \Delta \text{NO}$, or $\Delta K = \Delta \text{VF}^* \text{NO}$, or $\Delta K = \Delta VF^* \Delta NO$,

or $\quad \Delta K = \Delta \text{VF}^* \text{NO} + \text{VF}^* \Delta \text{NO}$

The three parameters *VFs*, *s*, and *NOs* are appropriately handled by the two-port network. In this network, the nine coefficients *A* through *I* are input- and output-sensitive. If the inputs, processing, and output requirements are matched appropriately then the flow of information from input to output can be maximized.

The next step in building the network is depicted in Figure 6.14. The source and recipient characteristics are shown and a set of adaptive processors included in the network. The adaptation becomes a series of multiple adaptations. At least three stages can be readily identified:

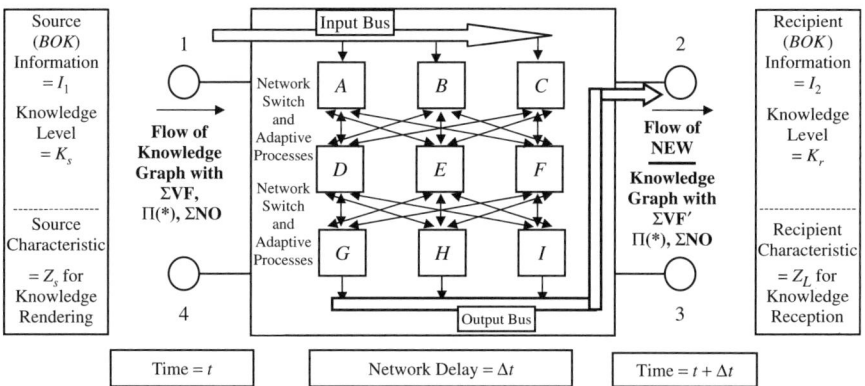

Figure 6.14 Configuration of an architectural and a mathematical representation of a two-port knowledge network to process and convert knowledge from any given source with source characteristics to a recipient with its own characteristics. The [ABC, DEF, GHI] matrix within the network intelligently adapts itself for the maximum transfer of knowledge from the source to the recipient. The parameters can be made dynamic and time-sensitive. Intelligent knowledge processing makes it possible to import the artificial intelligence principles (adaptation, learning, expert systems, pattern recognition and vision, intelligent agents for brokering the rate and contents of knowledge flow, etc.) into the knowledge-processing domain.

First, coefficients A through I are adapted by individual relaxation, then by group relaxation to receive the graph of *BOK* (see Figures 2.11a and b) into the knowledge caches (see Figure 6.15) of the network.

Second, the coefficients A, B, C; D, E, F; and G, H, I are adapted by group relaxation, then by group relaxation of the entire set (A through I) to dispatch the graph of *BOK'* to the recipient.

Third, the finetuning of all nine coefficients to place the "zeros/poles" of the network so all knowledge-centric *VFs,* ∗s, and *NOs* may be effectively "transferred/blocked" from the input to the output.

The process is repeated for every increment of knowledge flow until the two port knowledge network performs as well as an optimized electrical network.

In the social domain, networks do not exist in isolation. Unlike individual voltages and currents in isolated electrical circuits, the parameters *VFs,* ∗s, and *NOs* in social settings are generic and universal. For this reason, the local parameters (*VFs,* ∗s, and *NOs*) from the *BOK* and global parameters for the same or similar parameters (*VFs,* ∗s, and *NOs*) from Web knowledge bases need to be coordinated throughout the design of the two port knowledge network.

The coordination of local and global parameters and three knowledge caches are shown in Figure 6.15, where two filters are cascaded in stages I and II. As a first step, the initial *BOK* is scanned into the three-parameter input knowledge caches. Local optimization for the coefficients A, B, and C is initiated in conjunction with similar parameters (*VFs,* ∗s, and *NOs*). The seven layers

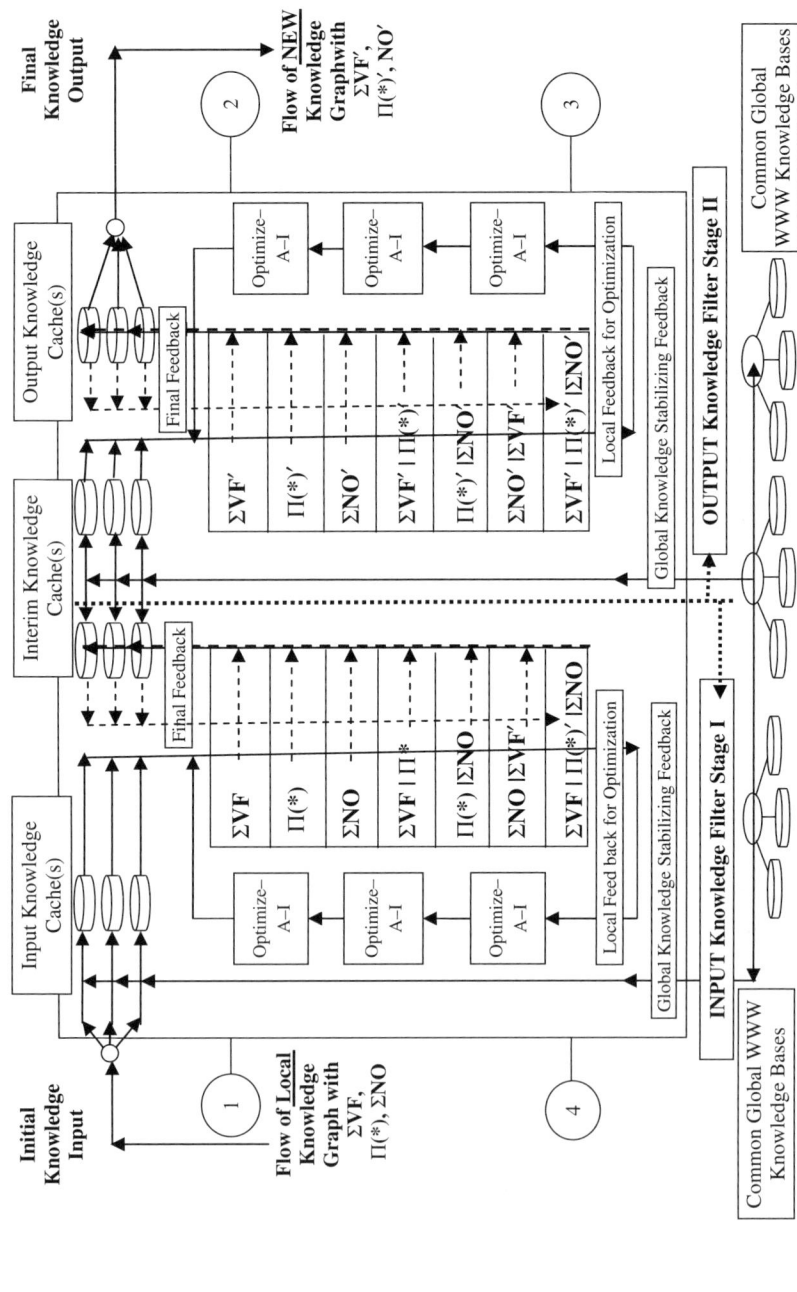

Figure 6.15 Two-stage filter for a two-port knowledge network. Stages I and II are matched for maximum flow of information based on any predefined criterion.

(see Figure 6.15) in each stage depict the independent optimizations, given $\sum VFs$, $\prod^* s$, $\sum NOs$, and their combinations, that are embedded in *BOK* and *BOK'*. The output from the stage I filter is temporarily stored in the interim knowledge caches, and the entire process is repeated and optimized by the stage II filter to match the characteristics of the recipient for *BOK'*.

6.9.3 Mathematical Implications of the Two Port Network

Conventional mathematical treatment of the input/output relations is inapplicable to the three parameters *VFs'*, *s'*, and *NOs'* at the input or the three parameters *VFs'*, *s'*, and *NOs'* at the output. The nine variables *A* through *I* are neither real nor complex entities. However, these variables assume the role of linkers and transformers in the knowledge domain, since they link th three input parameters to the corresponding three output parameters. In addition, they conceptually transform the roles and relationships between input and output parameters. To proceed further with the functionality of the nine variables *A* through *I*, we start the mathematical operations in a traditional sense and modify them to suit the knowledge domain by starting with the conventional two port network relations:

$$\begin{bmatrix} \Sigma VFs \\ \Pi(*) \\ \Sigma NOs \end{bmatrix} = \begin{bmatrix} A & B & C \\ D & E & F \\ G & H & I \end{bmatrix} \begin{bmatrix} \Sigma VFs' \\ \Pi(*)' \\ \Sigma NOs' \end{bmatrix}$$

This matrix equation is first written in a simpler form by stripping the $\sum$ and $\prod$ symbols and then rewriting them as shown below and as depicted in Figure 6.16:

$$VFs = A \cdot VFs' + B \cdot (*)' + C \cdot NOs'$$
$$(*) = D \cdot VFs' + E \cdot (*)' + F \cdot NOs'$$
$$NOs = G \cdot VFs' + H \cdot (*)' + I \cdot NOs'$$

6.9.4 Change of Operators (+, ×, /, and =)

The conceptual similarity between traditional mathematics and knowledge operations ends at this stage. The three traditional operators—plus or +, multiply or ×, divide or /, and equal to or = —need to be reexamined in the context of knowledge processing.

The traditional plus or + operator conveys numerical addition. The numbers increase or decrease by the same incremental unit. In the knowledge environment, the influence of numerous parameters is combined (rather than added). The influence can increase or decrease but not necessarily by the same unit as it does in a numeric sense. To convey this concept, we propose that the + operator be replaced by ╬ (double vertical and horizontal with a character code of 256C).

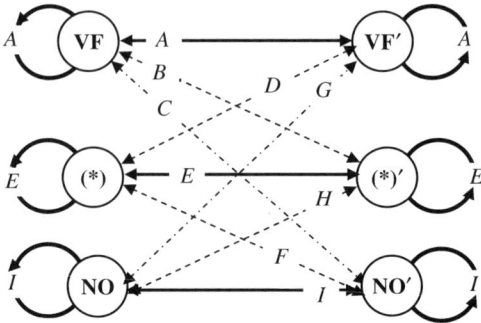

Figure 6.16 State diagram to depict the variations of the nine variables A through I. These variables (*ABC*, *DEF*, and *GHI*) are computed from the six estimated parameters VF, ∗, and NO at the input and VF', ∗', and NO' at the output of the two port knowledge network. The dominant relations exist via the variables *A*, *E*, and *I*. Six secondary relationships exist via the cross-occur influence between the input/output parameters. The arrows go from left to right if the input parameters are held constant, and vice versa.

In a similar sense, the multiply or × operator enhances or reduces the values in a linear scale, whereas in the knowledge domain, the values may adhere to a pre-set scale depending on the nature and character of the VF and/or NO. For example, hearing may increase or decrease in decibels or on an exponential scale. To convey this concept, we propose that the × operator be replaced by ⁞⁞ (two broken bars, each with a character code of 00A6).

The divide by or / operator is replaced by two fraction slashes or // (or two fraction slashes, each with a character code of 2044) and conveys the concept that the two entities are related by one or more variables *A* through *I*. For example, the dimensionality of the variable *C* forces it to relate *VFs* to *NOs* because (*C* ⁞⁞ NO') becomes a part of VF.

In the same vein, the equal to notation or = becomes imprecise in the knowledge domain, and it conveys the net effect of the right-hand side of the equation being approximated as parameters (VF, ∗, or NO) with those on the left-hand side. To convey this concept, we propose that the = operator be replaced by ‖ (double vertical bars with a character code of 2551).

In the knowledge domain, the input/output approximations may be rewritten as

$$\text{VFs} \parallel A \parallel \text{VFs}' \ \#\!\!\!+ \ B \parallel *' \ \#\!\!\!+ \ C \parallel \text{NOs}'$$

$$* \parallel D \parallel \text{VFs}' \ \#\!\!\!+ \ E \parallel *' \ \#\!\!\!+ \ F \parallel \text{NOs}'$$

$$\text{NOs} \parallel G \parallel \text{VFs}' \ \#\!\!\!+ \ H \parallel *' \ \#\!\!\!+ \ I \parallel \text{NOs}'$$

The knowledge equations above can be reinterpreted to indicate the qualitative relationships between VF, ∗, and NO. They signify the underlying relations between the inputs, the outputs, and the nature of the nine variables *A* through *I* that relate the inputs and outputs. For example, if the bond between VF, ∗, and NO (at the input) is to be explored, then the output parameters (VF' ∗', and NO') play a significant role. Stated alternatively, knowledge input to the two port knowledge network has to be tailored to conform to the nature and character of the knowledge output to be delivered. This truism is written as

$$VF //NO \parallel \{A :: VFs' \pm B :: *' \pm C :: NOs'\}$$
$$//\{G :: VFs' \pm H :: *' \pm I :: NOs'\}$$

and restated as

"The expression of knowledge depends on the nature of the recipient."

This inference is universal[8] across all intelligent communications systems, which adapt the expression of knowledge but do not necessarily alter the contents and concepts embedded in the numerous bodies of knowledge. The evaluation, estimation, optimization, and fine tuning of the nine variables *A* through *I* are an essential function of intelligent knowledge-communications systems.

Other truisms such as "convolutions (i.e., ∗s) necessarily finetune *VFs* with *NOs* at both the input and output sides," "the nature of convolution of *VFs* and *NOs* depends on both the characteristics of the source and the receptor," etc., can also derived by writing down the relation between *VFs* with convolutions (∗s), or *NOs* with *VFs* and their respective convolutions (∗s).

---

[8]In all instances, communication depends on the nature and characteristics of the source and the recipient. When an elephant rumbles, when a bee stings, when a lion roars, when a preacher preaches, etc., knowledge or embedded information is suitably tailored to its receptors. When the two port network is not intelligent (e.g., the atmosphere or a twisted wire-pair, etc.), it becomes transmission media. When the two port network in intelligent (e.g., satellite stations or intelligent digital subscriber line modems, etc.), the adaptive elements within the network optimize their properties to suit the input and deliver the output according to a specific output model. In human communications, the role of the (intelligent) negotiator (Appendix B in Chapter 5) is to establish an economic basis for communication that is viable and acceptable to both the source and receptor of knowledge, goods, commodities, or even intangible assets. When the source and receptor have numerous sections, each having a two port network representation, then the flow of information, knowledge, concepts, wisdom, or even ethics through that knowledge trail (see Chapters 2 and 3) is facilitated by cascading the transfer functions of each of the elements. The two port theory thus becomes as applicable in knowledge theory as it is in communications theory. In knowledge theory, knowledge is based on *VFs*, *'s, and *NOs*, whereas in communications theory, information is based on bits generally represented as electrical or optical signals.

286 INFORMATION AND KNOWLEDGE FILTERS

6.9.5 Intelligent Two-Port Knowledge Networks

When intelligence is involved in transforming bodies of knowledge, a good estimate for the variables A through I is desirable. The precise computation may be as feasible as initially determining the value of π. An accurate estimation can evolve over a period of time. For example, consider a *BOK,* "electrons move in vacuum tubes" at the input of the two-port network. This statement can be reinterpreted as *BOK',* "the nature of a gaseous medium in the vacuum tube can be deduced by the current in a vacuum tube" at the output of the two-port network. In this case the network facilitates the communication between one party (an engineer) and the other (a physicist) by knowing the most significant *VFs* and *NOs* of the two parties. Other assertions such as the effects of impurities in silica glass, or for the truth of *BOKs* from Buddha or Gandhi, can also be deduced by appropriate selection of the variable A-I in the two port knowledge networks (see Appendix A). In fact, when an intelligent knowledge two-port network acts as an intermediary between any two bodies of knowledge *BOK* and *BOK'* shown in Figure 6.17, then the role of A through I can be depicted by the forward arrows in the figure.

6.9.6 Needs, Knowledge, and Innovations

Needs, knowledge, and innovations form a spiral of social progress. These three elements pull away from the origin of thought but give it shape and consistency, much like gravity, mass, and electromagnetism in galactic space, creating an ever-changing universe. Over time, a complex balance is achieved and an oscillatory equilibrium in the shape of a pyramid with four panels, as shown in Figure 6.18.

Needs, knowledge, and innovations had maintained a dynamic equilibrium and a modest rate of expansion for the prior two centuries. The computer society, an information explosion, and knowledge highways have stilted the equilibrium in favor of information, which is a precursor to knowledge. The bridge between information and genuine knowledge has become fragile to extent that junk information and hype knowledge dominate the networks. This glut of information and dearth of knowledge have prompted the "information opportunism" that permeates the society and kills creativity in the population. It appears that a new perspective on the entire knowledge trail (Figures 2.5, 3.13, and 4.9) will restore the dynamic balance for the pyramid in Figure 6.18.

6.10 KNOWLEDGE GATES

Gating of binary data is essential in almost all CPU functions. The basic gates are AND, OR, and NOT. Other gates such as NAND, NOR, EXCLUSIVE OR, and EXCLUSIVE NOR, and multiple-input single-output gates are derived from these three basic gates. Their functionality is shown in Figure 6.19.

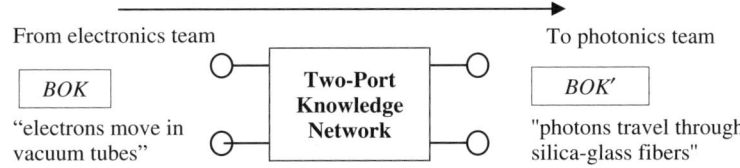

(a) Conversion of a *BOK* to *BOK'* by a two port knowledge network.

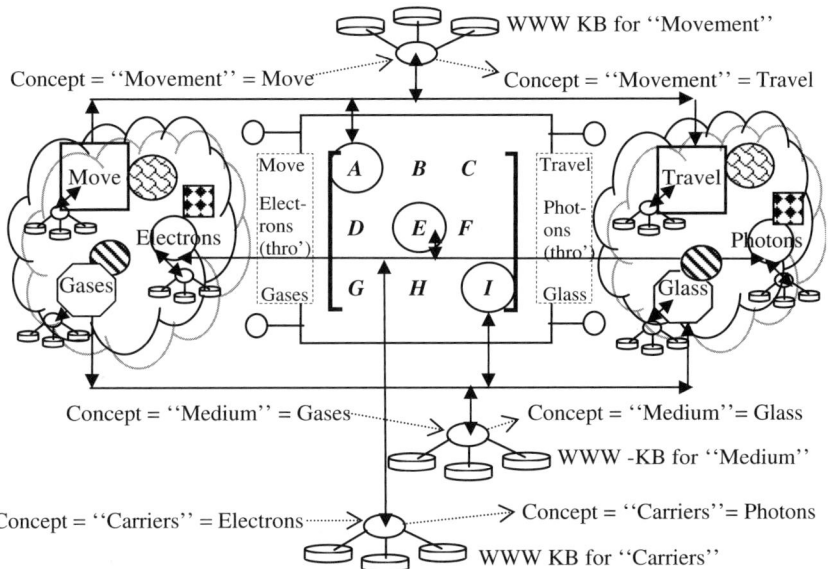

(b) The conversion of a *BOK* = "electrons move in vacuum tubes" to *BOK'* = "photons travel through silica-fibers." Three secondary concepts (VF = movement, NO_1 = carriers, and NO_2 = medium) are involved.

Figure 6.17 *A*, *E*, and *I* assume the primary role in this conversion of *BOK* to *BOK'*. For more complex conversions, the knowledge machine offers output such as "since the type of gas in the vacuum tube influences the flow of electrons, the impurities in the glass can influence the flow photons."

The three basic gates AND, OR, and NOT need to be reframed in the knowledge domain for the KPU functions and for the logic of the three parameters. Unlike the two-level input–output for binary logic gates, knowledge logic gate have a variety of inputs (like VF, ∗, NO, verb function adjectives, and noun object attributes) and different levels of outputs, but they are of the same variety as the input. The truth tables become complicated but not formidable. For illustration, we take two like parameters: one from the local knowledge *BOK* and the other from a Web knowledge base of a similar nature, for example, a noun object such as "rose" from *BOK* and "hibiscus" from the Web knowledge base. The AND gate will offer the features (attributes) that are common to both. In order

288 INFORMATION AND KNOWLEDGE FILTERS

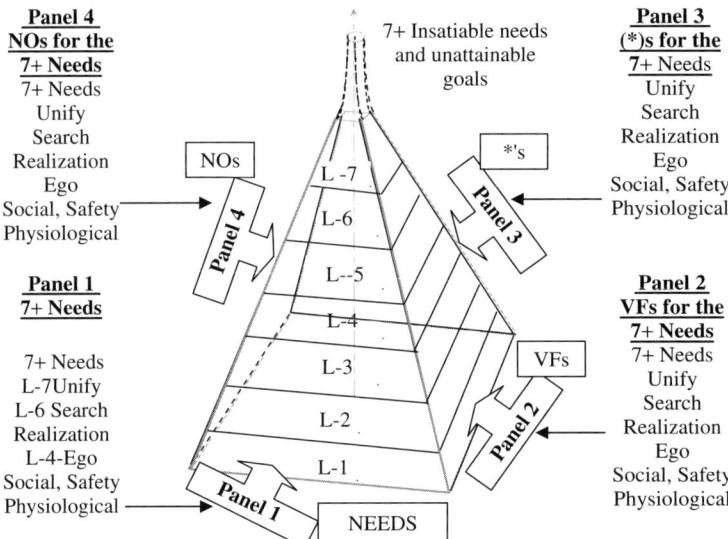

Figure 6.18 The influence of human, corporate, and national needs on the growth of knowledge for the social entity. The space inside the pyramid is the room for innovations by admixing the *VFs*, *s, and *NOs* specific to the needs of the social entity.

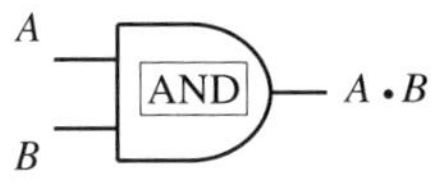

	A B A•B	Threshold
A —[AND]— A•B B	0 0 0 0 1 0 1 0 0 1 1 1	for All Signals $0.0 - < 0.5 \to 0$ $> 0.5 \to 1$
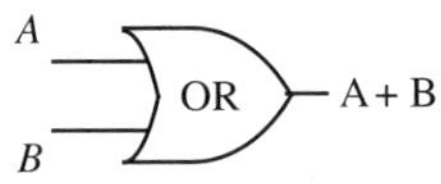 A —[OR]— A+B B	A B A+B 0 0 0 0 1 1 1 0 1 1 1 1	Threshold for All Signals $0.0 - < 0.5 \to 0$ $> 0.5 \to 1$
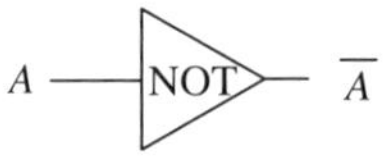 A —[NOT]— $\overline{A}$	A $\overline{A}$ 0 1 1 0	Threshold for All Signals $0.0 - < 0.5 \to 0$ $> 0.5 \to 1$

Figure 6.19 Truth table for the three traditional basic binary gates.

to be numeric, if the output is an integer number 32 bits long, then the ratio 12:2 indicates that 12 attributes are common to the 2 flowers. Identical logic applies to the operators (convolutions) ∗ or noun objects.

Accordingly, we present three tables in Sections 6.10.1, 6.10.2, 6.10.3 for each of the three basic gates (AND, OR, and NOT) to deal with the three parameters (*VFs*, ∗s, and *NOs*) in the knowledge domain.

6.10.1 Knowledge-Based AND Gate

The exactitude of binary truth tables is lost in the knowledge-processing domain. Instead of a binary choice, a string of threshold-based decisions starts to surface. The threshold level can be internally computed or a user choice. The gate output has three possible options: accept, consider, or reject the gated output. The decision tables for a knowledge-based AND (KAND) gate to offers choices rather than a binary output or two choices 0 or 1. The table of selection and decision alternatives (depending on the threshold values) is shown in Figure 6.20. The function of this gate is to select the lower of the inputs A and B and find the range of thresholds as the criterion for the decision to reject, consider, or accept the output. It also marks the input conditions that yielded the output from the gate. Rather than being a numerical comparison, it is based on a combined qualitative and quantitative estimate of the two signal strengths.[9]

The KAND gate is especially useful in choosing between two sets of parameters whereby each is intricately tied to a second parameter. The thresholds are the limits of rejection, consideration, or acceptance of the output. The gate also tracks the input conditions that yielded the choice. For example, if a bride to be is considering a groom with a minimum educational threshold of a doctoral degree

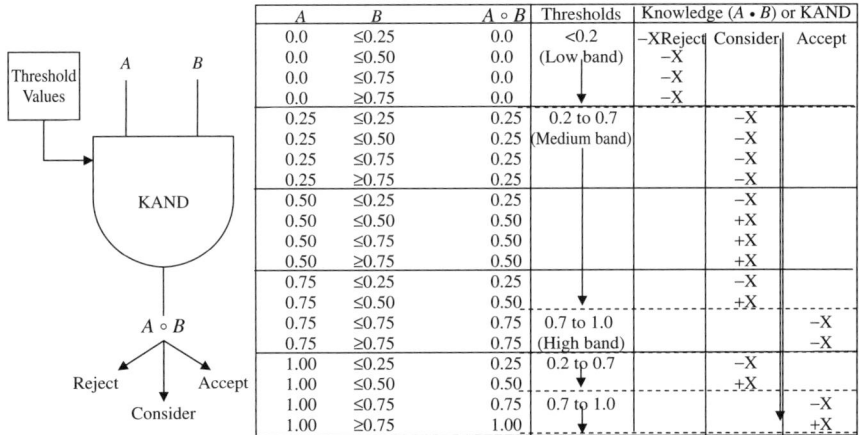

Figure 6.20 In the knowledge-based AND gate (KAND), the lower of the two inputs is chosen as the output. It is not a dot or · function since objects and parameters rather than binary levels are involved. This function is useful for choosing alternatives that minimize the maximum (or the minimax business strategy) expected threat/exposure. Both numerical and qualitative aspects of parameters are involved. A weighted average of the estimate is represented as A and B. The knowledge-based AND is useful for the comparison of local knowledge objects and parameters, such as VF, ∗, or NO, with other parameters.

---

[9]Such gates can be built as combinations of operational amplifiers and binary gates. This breed of knowledge gates calls for the a unique admixture of operational amplifier and binary gates technologies.

and a "good" family background (with a threshold of 0.5 compared to her own family rating of 1.0), then all suitors with at least Ph.D./M.D./D.Phil./D.Sc., etc., degrees and a good+ family background will score ≥ 0.5 and not be rejected by the KAND gate. The KAND gate offers these possible groom choices as the output. If these are the only two threshold criteria, then the process stops here and the suitor with the highest degree and best family sets will be selected. However, reality can be more complicated than this, as discussed in the following sections.

One of the useful deployments of this gate is for minimizing the maximum exposure to danger or damage. The exposure to minimum or maximum danger becomes an instinctual basis for economic behavior. In this sense then, the bride to be will select one of the grooms who poses minimal threats to her own social/financial security, holds a doctoral degree and being from a family with a score of ≥ 0.5. The KAND gate offers this choice as the output.

6.10.2 Knowledge-Based OR Gate

A similar table for selection and decision alternatives is shown in Figure 6.21. The output of this gate also has three possible options: to accept, consider, or reject the gated output. The function of this gate is to select the higher of the two inputs A and B and also mark the input conditions that yielded the output from the gate. The knowledge-based OR (KOR) gate is especially useful in choosing between two sets of parameters when a choice is imminent. The thresholds are

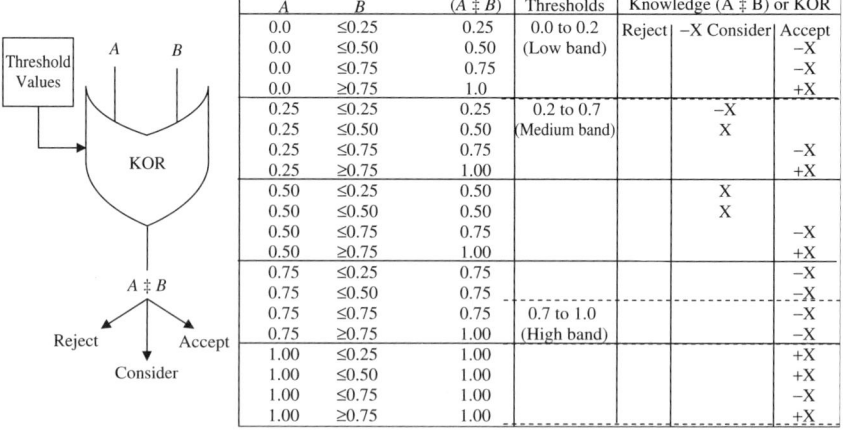

A	B	(A ‡ B)	Thresholds	Reject	−X Consider	Accept
0.0	≤0.25	0.25	0.0 to 0.2	Reject	−X Consider	Accept
0.0	≤0.50	0.50	(Low band)			−X
0.0	≤0.75	0.75				−X
0.0	≥0.75	1.0				+X
0.25	≤0.25	0.25	0.2 to 0.7		−X	
0.25	≤0.50	0.50	(Medium band)		X	
0.25	≤0.75	0.75				−X
0.25	≥0.75	1.00				+X
0.50	≤0.25	0.50			X	
0.50	≤0.50	0.50			X	
0.50	≤0.75	0.75				−X
0.50	≥0.75	1.00				+X
0.75	≤0.25	0.75				−X
0.75	≤0.50	0.75				−X
0.75	≤0.75	0.75	0.7 to 1.0			−X
0.75	≥0.75	1.00	(High band)			−X
1.00	≤0.25	1.00				+X
1.00	≤0.50	1.00				+X
1.00	≤0.75	1.00				−X
1.00	≥0.75	1.00				+X

Figure 6.21 In the knowledge-based OR gate, the higher of the two inputs is chosen as the output. It is not a plus or + function since objects and parameters rather than binary levels are involved. This function is useful for maximizing the minimum (or the maximin business strategy) expected gain/benefit. Both numerical and qualitative aspects of objects and parameters are involved. A weighted average of the estimate is represented as A and B. The knowledge-based OR is useful for the comparison of local knowledge objects and parameters to select the more appropriate object or parameter, such as VF, ∗, or NO.

the limits of rejection, consideration, or acceptance of one choice over the other. The gate also tracks the input conditions that yielded the choice.

For example, if a bride to be is considering two suitors with a maximum expected life-long earnings potential of 0.75 (the threshold for financial security compared to her own family fortune of X.0 M), then all suitors with expected lifetime earnings of 0.75X M will not be rejected by the KOR gate. If there were no criteria for education and family (as in Section 6.10.1), then the process stops here and the suitor with the highest life-long earnings potential will be selected. However, reality can be more complicated than this, as discussed in the following sections.

One of the useful deployments of this gate (see footnote in Section 6.10.1) is for maximizing the maximum expected benefit. In most species, the choice for maximizing the minimum expected benefit becomes an instinctual economic behavior. In this a sense, then the bride to be will select one of the grooms who maximizes her potential benefit by offering an expected financial security of at least 0.75X M. The KOR gate offers this choice as the output.

6.10.3 Knowledge-Based NOT Gate

The representation of the knowledge-based NOT (KNOT) gate is shown in Figure 6.22. In binary gating, the choice is simplified since there are only two states for each binary bit. In the KNOT gate, when the object or parameter A is identified to fall within a certain range, then the NOT functions identify all the other ranges for the parameter. NOT (KNOT) is useful for eliminating certain knowledge objects and parameters A, such as particular VF, ∗, or NO. The same reasoning applies to the output of NOT (KOR) or the KNOR gate.

In this sense, then the bride to be who wants to marry a groom who is not diabetic will use the KNOT gate to remove all possible grooms who have every

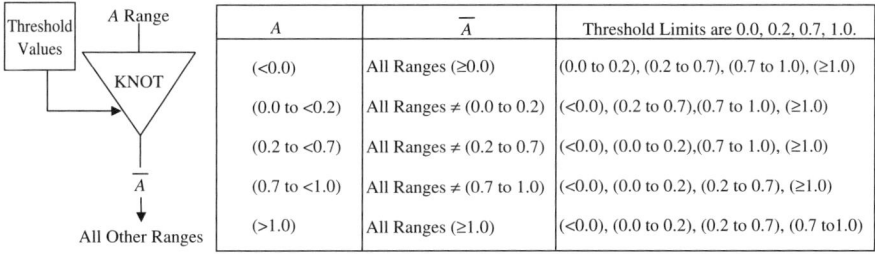

Figure 6.22 In the knowledge-based NOT gate, when the object or parameter A is identified to fall within a certain range, then the NOT functions identify all the other ranges for the parameter. NOT (KNOT) is useful for eliminating certain knowledge objects and parameters A, such as particular VF, ∗, or NO. For instance, the NOT(KAND) gate would offer all other choices that the KAND gate would offer, that is, if the output of the KAND gate is +ACCEPT, the NOT(KAND) would be (−,0, +) REJECT, (−,0, +) CONSIDER, and (−,0) ACCEPT. The same reasoning applies to the output of the NOT(KOR) gate.

possible strain of diabetes and select all the grooms who may have everything but diabetes. However, if the threshold is chosen to be ill-health rather than diabetes, then the KNOT gate will select only healthy grooms and the knowledge machine will provide the conditions for being healthy based on local and/or Web knowledge bases.

6.10.4 Practical Use of Knowledge Gates

6.10.4.1 The Bridal Dilemma In continuing with the bridal choice example from Sections 6.10.1 and 6.10.2, the best choice from all the suitors will consist of a two-step procedure: (1) selecting from the general population of all suitors those who offer a minimum of a master's degree and are from a good+ family background with a score of ≥ 0.5, and (2) selecting from the pool (knowledge base) of eligible suitors, a groom who maximizes the expected financial security greater than 0.75X M. These two steps can be accomplished by two knowledge gates. Three flow diagrams for this approach to solving the bridal choice problem are shown in Figure 6.23.

In reality, a large number of gates including the normal AND, OR, and NOT gates and traditional arithmetic processor units, and ALU processors may be necessary to solve knowledge problems. In the knowledge domain, the arithmetic domain becomes a subdomain. Knowledge gates facilitate the functionality of human preferences and decision-making processes, such as range selections for

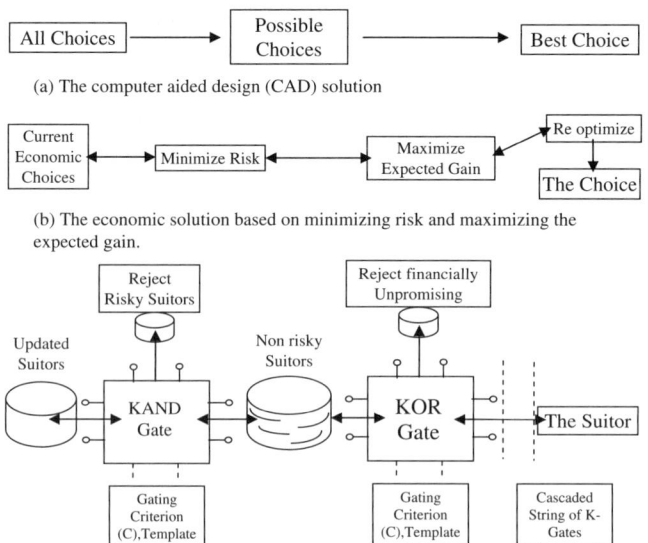

Figure 6.23 A string of cascaded gates for accomplishing knowledge functions. Complex knowledge functions need a flow diagram of gating KAND, KOR, and KNOT operations. The gating criteria depend on user preferences and the problem at hand.

input and output parameters, minimizing maximum risk, maximizing minimum expected gain, etc.

6.10.4.2 Corporate Executive Choice

As another example, consider two applicants X and Y for an executive position. Applicant X has an M.B.A. from a less renowned institution, and applicant Y has an M.S. from a better institution. If only one needs to be selected, then the applicant noun object (NO with M.B.A.) needs to be KORed with the second applicant noun object (NO with M.S.).

If the norms call for a master's degree from an established institution, then there are two thresholds: one for a master's degree (0.45, say) and a second one for the institution (also 0.45). Next, if the selection team establishes that an M.B.A. has a rating of 0.75, whereas an M.S. has a rating of 0.60 in comparison with the current executive educational rating (C) of 1.00, and if the better institution is rated as at 0.80, whereas the lesser institution is rated at 0.55 with respect to the current executive experience C of 1.00 in the corporation, the selection criteria can be written as follows:

Applicant X: M.B.A. rating $= 0.75$;   threshold $= 0.45$;   institutional rating $= 0.55$

Applicant Y: M.S. rating $= 0.60$;   threshold $= 0.45$;   institutional institutional rating $= 0.80$

Current executive C rating $= 1.00$;   used as a standard;   institutional institutional rating $= 1.00$

Three steps to resolve the choice are shown in Figure 6.24. 4-KAND, 4-SUB, 2-MPY, and 1-KOR instructions are necessary in a knowledge program for the executive choice problem. It can be seen that a program in traditional languages will duplicate the functionality of this k-program. However, the program control and attributes of the input parameter can be most easily modified in the k-program with simplistic k-gates. For example, if the applicant X has an M.B.A. and work experience similar to that of C, then the degree rating for A would be enhanced from, say, 0.75 to 0.85 and the final choice could be different. Conversely, if the institution for Y suffered a recent scandal involving substance abuse, then the instruction rating for Y would go down from, say, 0.80 to 0.70. The outcome of the k-program could be different, etc. Complex knowledge programs can be resolved and reiterated in the knowledge machines quite easily. As can be seen, the KPU in the knowledge machine can closely emulate human behavioral decision-making processes.

6.10.4.3 General Knowledge Programs

It can be seen that the knowledge gates permit the admixture of NO, ∗, and VF parameters freely (as a human mind would accomplish), and the gating processes would become very close to human

294 INFORMATION AND KNOWLEDGE FILTERS

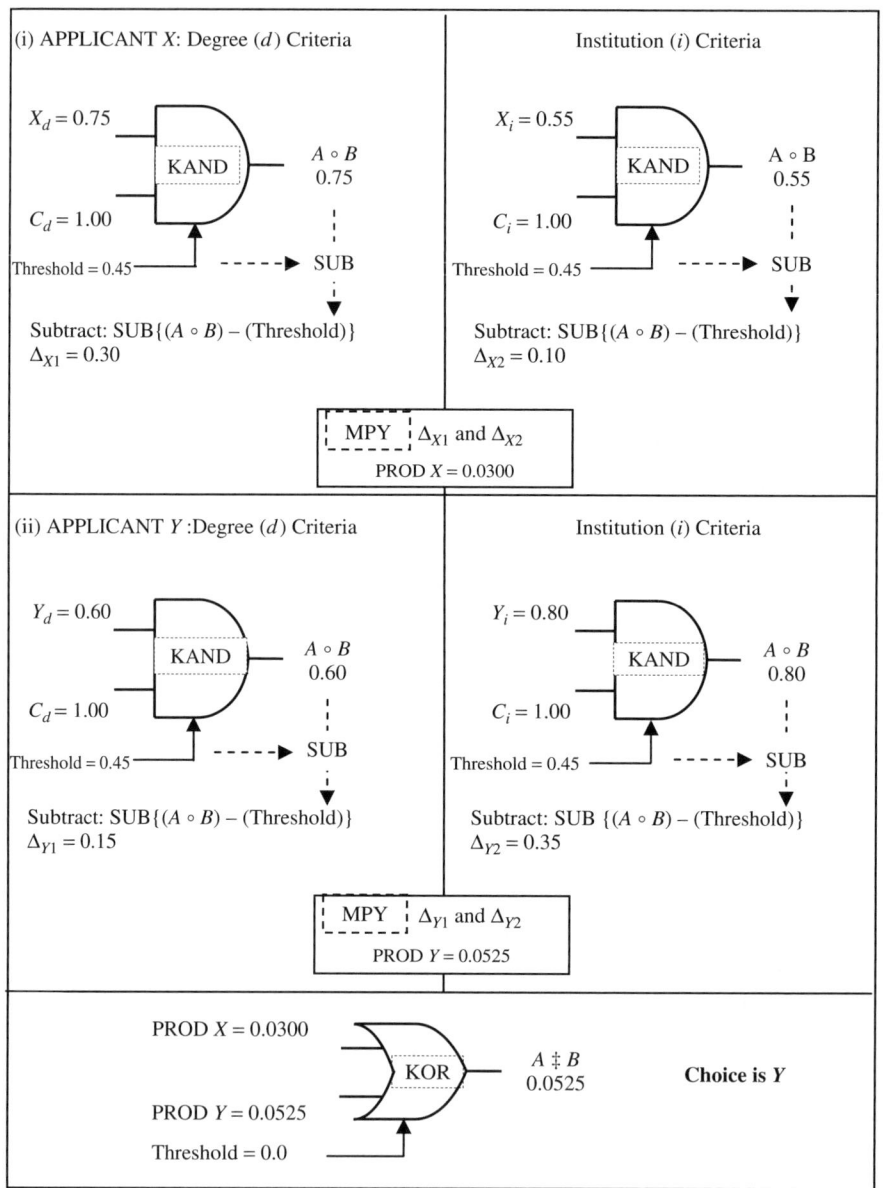

Figure 6.24 The structure of a 4-KAND, 4-SUB, 2-MPY, 1-KOR knowledge program for the executive choice problem.

decision-making functions. When a "consider" output appears from the gate, then other parameters, attributes of noun object, adjectives of verb functions, steps in the ∗ or convolution processes become necessary. Such reinforcing computations can be done internally within the k-program, obtained from more extensive Web knowledge base searches, or external data may be furnished by the user.

Programming in the knowledge languages (domain) can be more powerful and challenging than programming in the object-oriented languages (domain). It can be foreseen that knowledge-level programming will be more human-and society-oriented than object-oriented programming. Once established as any of the normal programming languages, knowledge-level programming can address intricate human problems as effectively as Fortran programs can address scientific and engineering problems.

The functionalities of the three types of basic knowledge gates are shown in Figure 6.25. For the binary gate, "true" output of the OR gate occurs when either or both inputs are true as compared to the output of the AND gate that occurs when both the inputs are true. True output from the OR gate is three times more probable than true output from the AND gate. These outputs from the OR gate are generally higher than the corresponding outputs for the AND gate. For the NAND or NOR gates, the thresholds remain the same but the decision or accept/reject is reversed. For the simple NOT gate, the decision for the parameters (VF, ∗, and NO) will be in favor of all parameters except the original one. The KPU processes call for greater computational capacity compared to those for a conventional CPU.

6.11 CONTAMINATION OF $(I \ll \gg K)$

The sources of contamination of data in most data-handling systems are well documented. Data contamination can arise during any process or at any point, from their source to their destination. Any ill-functioning gadget (from a contact to a chip), any illogical step in a code (from a binary instruction to a subroutine), any link (from a mere network connection to a global optical link), etc., can cause contamination from a single bit error in the total content of data and data structures. The nature and remedies for data contamination have been studied well by systems designers. Network security and data protection are disciplines in their own right.

Knowledge is contaminated by human beings (such as media editors, advertisers, reporters, agencies, etc.) and knowledge handlers, including contaminated knowledge machines. The systematic corruption of knowledge is a willful act (like planting viruses and worms in networks) and it can become insidious such that the recipient is unable to detect and correct any errors and corruption (like Trojan horses). Detection and correction of error in the knowledge domain are unlike detection and correction (by error-correcting codes) in the data domain. Human beings innately sense knowledge that is illogical, incoherent, and ill

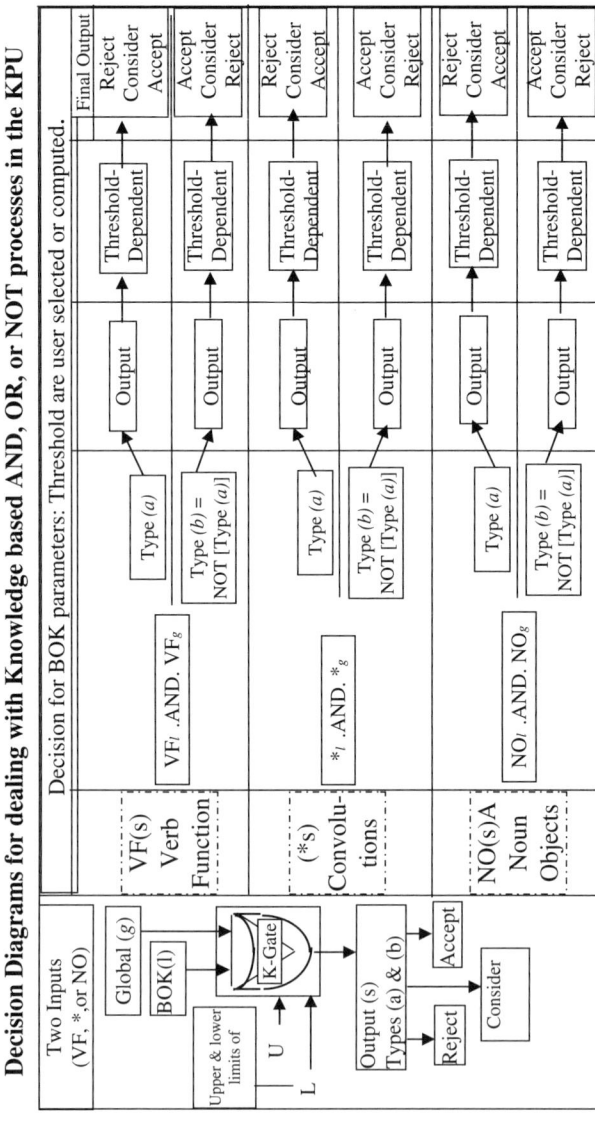

Figure 6.25 The generic functionality of the three basic knowledge (KAND, KOR, and KNOT) gates for the three *BOK* parameters (*VFs*, *∗s*, and *NOs*).

conceived. In essence, a subconscious content search in invoked, concept verification is pursued, and a numerical estimation is attempted in real time before the element of knowledge is accepted, marginalized (subject to further consideration), or declined.

In the network domain, the detection and correction of block data, encoders, and decoders depend on parity bits. The content of the data block (of bits) and parity block (of bits) is algebraically related such that the decoder can verify the correctness of the data block at the receiver end.

If knowledge detection and correction are to be attempted, then the "parity" block of knowledge can be and should be deduced from the "knowledge" block. In reality, such parity blocks are not appended to the knowledge block. However, repetitions and reaffirmations are commonly practiced in spoken and written languages, leading to a nonalgebraic but almost human way of verifying the previously stated knowledge blocks.

This latter practice permits the *detection* of inconsistencies (errors) in the knowledge domain. Correction becomes the next goal. Algebraic laws for the generation of parity bits in the knowledge domain are not immediately feasible; however, an information filter is able to extract a concept based on the *NOs* and *VFs* from a sample of $(I \ll K)$. If the concept is a precursor or restatement of a partial or entire axiom of wisdom generated and stored in the Web wisdom bases for the known knowledge of the world, then the correctness of the "data" block in the knowledge domain can be tentatively affirmed.

Contrarily, if the extent and nature of the contradiction between the knowledge block of the sample $(I \ll K)$ and known axioms of wisdom are evaluated (by the information filters discussed in Section 6.4 and 6.5), then error correction in the knowledge domain may be attempted. Fortunately, knowledge and wisdom banks (written, documented, or preserved) exist around the world, and knowledge validation and correction (if not incisive) are still feasible.

6.12 DECONTAMINATION OF KNOWLEDGE

In the hierarchy of knowledge, data reside at the lower level. From a theoretical standpoint, scientific methodologies (algebraic coding) exist to correct error ridden data. To enhance the signal to noise ratio techniques such as repeatering, enhancing signal power, amplification, filtering, exist.. It is somewhat misleading to refer to the interpretation of signal-to-noise ratios, computation of the probable bit-error rates, error correction, and coding theory as information theory. It is the data that are protected and not the information. Information is derived from data, and thus, the correction and safeguarding of data is are not decontamination of information.

Corrupted binary data are generally rectified in networks and communications systems by appropriate codes and the noise-suppression techniques to maintain a high signal-to-noise ratio, generally in excess of about 40 dB. The nature of contaminants (such as cross-talk, pickup, echoes, etc.) is well studied and are suitably blocked. Network security measures ensure reliable firewalls between data and sources of contamination.

In the information and knowledge domains, the scientific methodologies to rid noise (gossip, hearsay, or casual remarks) from a given body of knowledge have yet to be documented, even though repetitious and identical phrases may be detected. When a content check reveals a statement or phrase that is totally contradictory to the contents of established knowledge bases (such as "enjoy a cigarette," without reference to the risk of smoking), then the statements or phrases are tagged as false or deceptive. A series of such tests to validate the concepts in a body of knowledge will verify the extent of truth (universality), social value, and elegance. Systematic exclusion of deceptive information from magazines, papers, and media becomes a major step in the decontamination of knowledge. It becomes feasible to program the filters to decontaminate knowledge from scientific journals, media magazines, literature, etc., based on scientific truth, social value, artistic content, etc., by using the information filters presented.

A predictive filter configuration for information and knowledge is shown in Figure 6.26. The filter compensates distortion and delay in society (or any public media) that is likely to degenerate information by altering its composition. In this particular case, the attributes corrected are truth, virtue, and beauty (T, V, B, or any attributes of information). The characteristics predicted[10] are distortions in the (T, V, B) content of information or the NOs and VFs that imply the TVB characteristics. Corrections for attenuations and the delay of critical information are fed back at the source (see Figure 6.26). When the parameters of the three timing correction coefficients $\xi'(T)$, $\xi'(V)$, $\xi'(B)$] and amplifier coefficients $\mu^{-1}(T)$, $\mu^{-1}(V)$, $\mu^{-1}(B)$] are appropriately tuned to the society and culture where the information is propagated, the reconstructed information at time t' at t', T', V', B' will be the same as or have a composition more attractive than the original information at time $= t$ at t, T, V, B. The filter thus processes and corrects any information degradation during its transmission and correction through the media. Due to the highly variable and dynamic nature of the parameters in society, validation of the coefficient of the filter may be necessary based on the prior behavior of society. The human touch of gauging society and its traits will force the stability of the feedback process built into this type of information filter.

---

[10]In the conventional filtering of signals, this methodology is used in most transmission systems. For example, in DSL systems, the transmitted signal is pre-emphasized to undo the delayed echoes caused by line discontinuities and bridged taps [5]. The distortion was designed to undo the effects of a subscriber loop with the statistical configuration of discontinuous loops with bridged taps. In a sense, the received signal is close to being perfect for a large number of loops. The most distorted signals are received for the loops that deviate the most from the averaged loops and these are relatively few. In the information domain, the distortion that is perceived by media networks which practice the art of deception, arrogance, and hate, require an estimation. At the source, such an averaged distortion can be compensated by restating the T, V, and B in the message as many times as the media stations are most likely to distort, etc. Such practice is ever so common during elections and in corrupt societies.

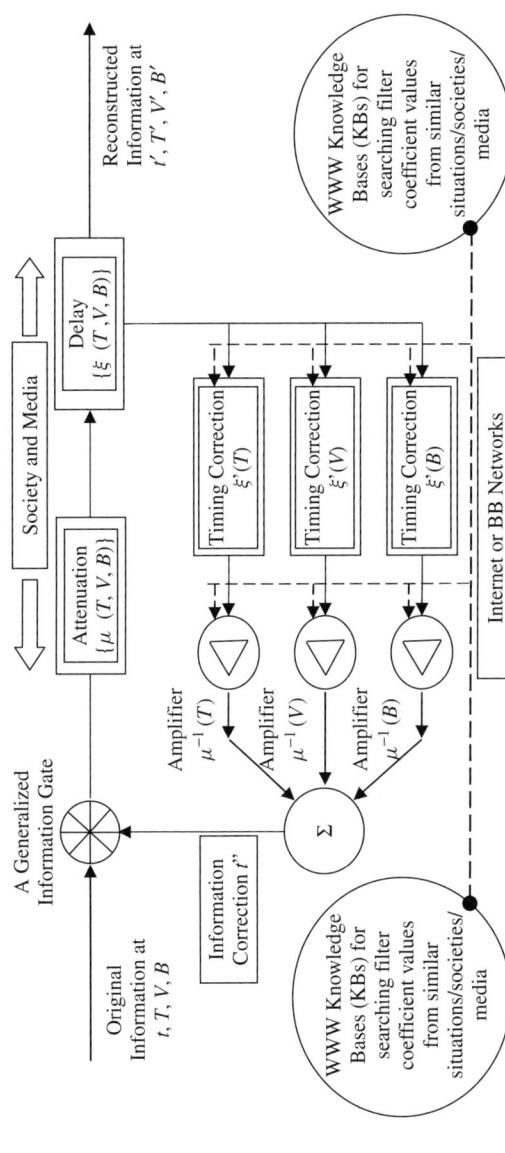

Figure 6.26 Configuration of an information filter to compensate for a distortion-prone society that degenerates information by altering the compositions of information of truth, virtue and beauty (T, V, B, or any attributes of information). When the parameters of the three timing correction coefficients $\xi'(T)$, $\xi'(V)$, $\xi'(B)$] and amplifier coefficients $\mu^{-1}(T)$, $\mu^{-1}(V)$, $\mu^{-1}(B)$] are appropriately tuned to the society and culture where the information is propagated, the reconstructed information at t', T',V',B' will be the same as or have a composition more attractive than the original information at t, T, V, B. The filter thus processes and corrects most degradation. Due to the highly variable and dynamic nature of the parameters in any society, Internet validation for the coefficients of the filter may be necessary.

It is also possible to decontaminate graphics, images, and scenes using such a filtering methodology based on image files in any accepted and acceptable formats (jpeg, tiff, eps, etc.). Dictionaries of graphical objects, their synonyms, and scenes are not available at present. However, it is possible to program machines to generate image bases by specifying graphical objects (such as war scenes, the setting sun, beaches, or even rain drops, etc.) from the painting of artists, images of photographers, etc. The process can be totally mechanized.

Decontamination of soundtracks and audio files in movies is a necessary form of art. Filtering of audio signals (to focus on selected sounds such as whale songs, bird calls, voice authentication, etc.) has been well documented. Computer-generated music can be as fascinating as composed music.

The semantic and syntactic rules of music composition offer new dimensions for human creativity to compose new rules of composition rather than merely cutting and pasting segments of music. In the same vein, it remains to be seen if computer-generated knowledge can compete with classic textbooks or at least decontaminate tainted books.

Knowledge filtering by machines offers new dimensions for human thought. When the science of classical and digital filtering is pushed in this direction, it is likely to generate new knowledge of greater social worth than the original knowledge. The tools for the enhancement of knowledge can be mechanized by the machines supplementing content and linkage to knowledge from local and Web-based Knowledge bases. The art of blending knowledge can become a science of artificial intelligence as pattern recognition and implanting intelligent agents in a multiple-domain (i.e., DDS or LoC classification system) knowledge environment.

6.13 CONCLUSIONS

The methodology for filtering information presented in this chapter permits machines to condense large bodies of information or knowledge to precise statements that reconstruct knowledge based on the filter template or problem criterion. The steps are based on recognizing the operative nouns (objects), associated verbs (processes), and their inter-relationships. When the user specifies the template or criterion, then user specifications become dominant. When the machine is offered the freedom to verify or solve a problem, then the local knowledge bases or authenticated Knowledge bases around the Web knowledge bases are scanned to determine the most robust relationships between noun objects and verb functions in the context of the problem at hand.

Knowledge filters have the potential to purge false and erroneous statements and claims from information. Mathematical cleansing becomes a mechanized procedure of knowledge machines (KMs). When the criteria for the truth of every statement become too stringent, the KMs yield no results because absolute certainty is illusive. When the confidence limit for statements is set at zero, the KMs will fail to filter. Any information will just passes through them. When the directionality of discipline is defined by its Dewey Decimal System (DDS) or Library of Congress (LoC) classification and a confidence limit is reasonable, then the machine can output a scientific document from raw data or observations. It is possible to alter the flavor of information by knowledge machines. It is possible to purify knowledge by specifying that every statement be strenuously checked (i.e., have a higher confidence level) for truth, universality, and accuracy. In the extreme case, KMs can be forced to offer only the classical equations in physics and science as the final output of all information and knowledge. By relaxing the constraints on KMs, the user can generate masterpieces from manuscripts or symphonies from sounds.

The range of a search can be made narrow and constrictive or broad and global. The criteria for selecting or proposing solutions can also be slanted toward universal truth, social benefit, and/or enduring beauty. Conversely, it is possible to generate propaganda to entirely deceive and mislead society and nations. The destruction of culture from corrupted knowledge is as feasible as the enhancement of values by rational thinking written down as notions in concepts and axioms in wisdom. The enhancement or reversal of culture can be made artificial by these new machines that can prime the psyche of the masses.

In a totally abusive mode of functionality for the filter or the machine, the user can select information or propose solutions slanted toward deception, arrogance, and hate to drown the good senses of the populous. The KM works equally diligently in propagating deception instead of truth, arrogance instead of beauty, and hate instead of social benevolence. The user and machine in conjunction become an access of evil like some politicians who abuse social systems to bring wars and suffering to humanity.

The filter methodology suggested here is an initial step toward "concept recognition" and its validation with concepts that are documented in Web knowledge bases. To this extent, the filter can extract seminal ideas and concepts in any body of knowledge and then verify its validity. The key to the methodology is the isolation of operative nouns and verbs and generation of a graph showing the inter-relationships. Such a graph becomes a signature of the document. The signatures of other samples can thus be compared to mathematically determine the extent of match/mismatch, thus offering a numerical measure of a match based on nouns, verbs, their relationships; synonyms of nouns, verbs, their relationships; synonyms of synonyms of nouns, verbs, their relationships, etc., after removing circular synonyms to prevent endless loops in the operation of the machine(s). Even this process can be mechanized. The machine is capable of substantiating every step on a numerical basis. There is no subjectivity in filtering, solutions, or conclusions returned from such machines.

Linkages to human needs and motivation provide the behavioral context as the machine attempts to resolve individual or social problems. The major thesis in attempting such solutions is that almost all human/organizational behavior is goal-directed, and such goals stem from human/organizational needs and these are well documented in the literature. Learning and adaptability based on artificial intelligence are implicit functions of these intelligent machines.

REFERENCES

1. G. Moschytz and M. Hofbauer, *Adaptive Filter: Eine Einführung in die Theorie mit Aufgaben und MATLAB Simulationen*, 1st ed., CD-ROM, Springer, 2000. Also see G. Moschytz, *MOS Switched-Capacitor Filters: Analysis and Design,* IEEE Press, Piscataway, NJ, 1984.
2. S. G. Krantz *Function Theory of Several Complex Variables,* 2nd ed., American Mathematical Society, AMS, Providence, RI, 2001.
3. A. H. Maslow, "A Theory of Human Motivation," *Psychology Review*, 50, 1943: 370–396. Also see A. H. Maslow, *Motivation and Personality,* Harper & Row, New York, 1970, and A. Maslow, *Farther Reaches of Human Nature,* Viking Press, New York, 1971.
4. S. V. Ahamed, *Intelligent Internet Knowledge Networks*, Wiley-Interscience, Hoboken, NJ, 2006.
5. S. V. Ahamed, "Simulation and Design Studies of Digital Subscriber Lines," *Bell System Technical Journal*, 61 (6), July–August 1982: 1003–1077.
6. William Reville, "Life and Works of John Tyndall," University College, Cork, Ireland, http://understandingscience.ucc.ie/pages/sci_johntyndall.htm.
7. S. V. Ahamed and V. B. Lawrence, *The Art of Scientific Innovation: Cases of Classical Creativity*. Pearson Prentice-Hall, Upper Saddle River, NJ, 2004.
8. S. V. Ahamed and V. B. Lawrence, *Design and Engineering of Intelligent Communication Systems,* Kluwer Academic Publishers, Boston, 1997.

APPENDIX 6.A

A TWO-STAGE TR FILTER

As a practical example of the information filter that is able to enhance, reconstruct, and amplify information depending on the user's interaction, we illustrate the classical words of Henry Higgins: "The rain in Spain falls mainly on the plain." This particular body of knowledge (BOK) has only two aspects: truth (T), in the statement and rhyme (R). The context of the filter may be represented as shown in Figure 6.A.1, based on Figures 6.5 and 6.6. The filtering process can be seen as a three-step process involving T-Zero filtering, R-Zero filtering, and TR-Zero filtering.

6.A.1 T-Zero Filtering and Output BOK_1-T

This filtering enhances truth in every possible way, ranging from an average rainfall of about 600 mm over large areas of Spain, recent droughts (in 2005), water rationing, dying fish in river beds (in 2005), the influence of the Atlantic Ocean on Spain's climate, the effect of Hurricane Vince (in 2005) that produced less than 2 in. of rain, the killer rains of Spain (in October 2007), to the fact that the rain in Spain is not enough as desertification continues. A large collection of factual, truthful, and serious information is collected in the output of the T-Zero filter, and it offers zero (close to zero) impedance for all factual information from knowledge bases as they pertain to the *NOs* (rain, Spain, and plain) in *BOK*: accurate data on related objects like mist, drizzle, sprinkle, downpour, torrent, storm, etc. It also collects information on a VF (falls) and related *VFs* like mists, drizzles, sprinkles, pours, torrents, storms, etc., classifying nouns distinctly from verbs, with their differences separated.

Computational Framework for Knowledge. By Syed V. Ahamed
Copyright © 2009 John Wiley & Sons, Inc.

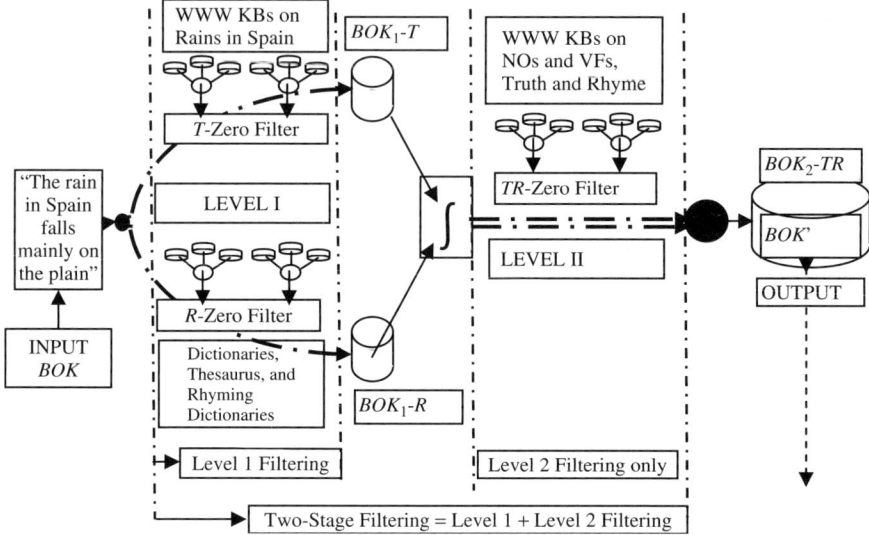

Figure 6.A.1 Configuration of an information filter to enhance T (truth) and R (rhyming patterns) in the original words of Professor Henry Higgins.

The output file $BOK_1 - T$ will contain the truth regarding all noun words, their synonyms and thesaurus words, and verb functions, their synonyms and thesaurus words. More than passively collecting facts, the file also collects statements that may appear truer or more accurate than the original BOK—for instance, the connection between the rain in southern Spain, the North Atlantic Oscillation (NAO), and El Niño (ENSO) effects; the winter and spring correlation that varies among the three; and NAO and ENSO that had an effect on the precipitation in this area during 1997, etc.

Many scores of outputs are possible even with simple seed words (*NOs* and *VFs*) like "rain," "plain," "Spain." The objective function for the filters that favor given directionalities for optimization establishes the choice of *NOs* and *VFs* to further investigate in collecting the outputs of level I filters. The process continues beyond level I filters, thus targeting the zone of optimization for the outputs of the filters.

6.A.2 *R*-Zero Filtering and Output BOK_1-*R*

The main *NOs* in *BOK* are rain, Spain, and plain. The rhyming dictionary offers other secondary[11] *NOs* (such as Cain, gain, main, Maine, pain, train, grain, vein,

---

[11] Other words such as bane, cane, Dane, lane, mane, pane, plane, sane, vane, wane, rein, and vein are not included to limit the possible choices for the machine. The contents of the BOK_1-R database may indeed have additional access to these words and generate greater diversity from the filter. Multisyllabic words such as terrain, constrain, featherbrain, quatrain, refrain, and restrain also add greater diversity to the possible contents of the truth database BOK_1-T.

wain, Wain, etc., but without classifying fain, lain, vain, since these are verbs or adjectives) to the *R*-Zero filter. It also searches for secondary verb functions (such as calls and palls but without classifying balls, galls, halls, malls, and walls, since these are noun objects) to match the main VF (falls).

With an enhanced group of *NOs,* the machine can make sensible phrases such as "the rain in Spain can be a pain," "the rain of Spain may not occur in Maine," "the pain is never in vain," "there is no plain in Maine to have the rain of Spain," "the terrain of Maine cannot sustain the rain on the plain," etc. It is possible to see that information can be constructed (fabricated) with any desired level of confidence based on the functioning of mechanistic pattern-recognizing (PR) and pattern-generating filters. Such functions are frequent in AI (PR and robotic) machines.

6.A.3 *TR*-Zero Filtering and Output *BOK$_2$-TR*

In this filtering mode, the machine optimizes the available *NOs* and *VFs* (each independently and both together) to satisfy the *T* and *R* objective functions that output the truth with the highest degree of confidence and yet retain a rhyming pattern (in this case) at the output, such as:

"Global warming is likely to change the rain in Spain" (confidence level of 80%)

"Continued El Niño effects will damage the vegetation of the plain in Spain" (confidence level of 50%)

"Greenhouse gases will adversely affect the rains that fall on the plain in Spain (confidence level of 85%)

"It is plain that Maine is not Spain to have the rain falling on the plain of Spain (confidence level of 100%), etc.

These are more true than rhyming. The logical extensions of truisms at the filter *BOK$_1$-T* are simply generated by enhancing the emphasis on the *T* (truth) database rather than the *R* (rhyme) database.

The converse scenario is equally viable at the *TR*-Zero filtering level. Much as a great variety of true statements can be generated, a large variety of rhyming statements can also be generated. To limit the styles of the rhymes, it is possible to attach a post-processor at the output of *BOK'* that emulates the style of Shakespeare, Omar Khayyám (a poet), Alfred Marshall (an ancconomist), or any unknown poet. The styles are borrowed from the KBs (or even books) of the writers. We provide some examples of the output at *BOK$_2$-TR* as follows:

Case 1: Shakespearean-Style Post-Processor at the Filter Output:

Caesar: Grow Spanish blossoms in Rome.

Brutus: The rain in the plain of Spain does not fall in Rome; Yet we can still create a misty syndrome; That the Spanish blossoms may still bloom in Rome; For we have enough slaves who built the hippodrome; So my lord, we can grow Spanish blossoms in Rome.

Case 2a. Omar Khayyám:

O my love! Should you and I bitterly complain; To know the secrets for the rain in a distant plain; That lies far, far yonder of Maine; Should we not grow the blossoms of Spain in Maine.

Case 2b: Omar Khayyám:

The rain falls in a distant plain in Spain; And having fallen brings the blossoms in the main; The thirsty soul still longs for evermore; Alas! Rain from heaven does not fall in Maine.

Case 3: Alfred Marshall:

What! The rain in Spain in the plain of Maine? The terrain cannot sustain, your work will be in vain; The utility of such a garden will fetch disdain; Even if there are all the resources from the wealth of Maine.

Case 4: Unknown Poet:

The blossoms of Spain cannot bloom in Maine; The beauty of the plane is lost in the terrain; Alas! The truth is simple and plain; The rain in Spain falls mainly in the plain.

Silly and comical as they may sound, the output examples (with the words of Henry Higgins at the input) still offer a purpose for information filters. If the words of the scientist John Tyndall[12] (also spelled Tyndell) about "light transmission optics" (1854), or the words of Alexander Graham Bell about the "photophone" (1880), are introduced as the input at *BOK* in Figure 6.A.1, then the entire evolution of fiber-optic technology or of the television industry can

---

[12]From [6]: "In 1869 Tyndall discovered the scattering of light by dust and large molecules, now known as the Tyndall effect. He noticed that a beam of light, visible as it passed through ordinary laboratory air, disappeared when it entered a flask of pure filtered water. He now passed a light beam through filtered air and got the same result—no beam. He deduced that light bounces off little particles and into our eyes, allowing us to see the beam. He found that different-sized particles scattered light in different ways. Visible light contains 7 different coloured lights, ranging from red to violet, but is colourless when the 7 lights are jumbled together. Tyndall suggested that the sky is blue because molecules in the atmosphere preferentially scatter the sun's blue rays. Tyndall later showed that "optically pure" (i.e., extremely filtered) air contains no micro-organisms. He compared what happened when he left fish or meat sit in such pure air or in ordinary air. The preparations in ordinary air gradually went putrid and maggoty, but the preparations left in pure air did not deteriorate. These studies extended Pasteur's demonstrations that life can only arise from pre-existing life and ruled out any spontaneous generation of life.

be traced through the knowledge caches in the intermediate BOK_i's of an information filter that has taken more than a century to offer the innovations of the twentieth and twenty-first centuries. But this is hindsight.

It becomes logical to pursue scientific S truth (rather than virtual truth T), economics E (rather than virtue V), and technology T (rather than beauty B) in building an information filter for practical and engineering applications. The ideology of *SET* goals rather than TVB goals are discussed further in Appendix 6.B.

However, to be futuristic and predictive, the knowledge domain filters need knowledge bases that deal with future human needs (that drive society) and futuristic machines (that serve society). In perspective, basic human needs are not likely to change drastically over the next few years. Higher needs do evolve differently for every individual and higher society. The newest innovation of new machines point to the greater computing power and capacity of current computer chips. One area of the intersection of humans and machines occurs when a machine directly satisfies individual and social needs. In envisioning the synergy between the two major forces within the society, it appears that knowledge and wisdom machines which directly address and effectively solve individual and social problems are likely to bring about the next wave of socioeconomic change.

APPENDIX 6.B

A THREE-STAGE SET FILTER

Science, economics, and technology (SET) are chosen for building this filter (see Chapter 2 for the many flavors of wisdom). When the findings (see footnote in Appendix 6.A) of John Tyndall 1820–1893) are introduced at the input of a filter shown in Figure 6.B.1, then numerous logical extensions start to appear at BOK_1-T. The configuration presented in Figure 6.B.1 offers a methodology for combining science, economics, and technology in a three-stage information filter. The first stage of filtering tries to maximize the three attributes—scientific truth S, the economics E of communications systems, and technology T—independently by examining the knowledge bases (or libraries) in local and global bases. In the same vein, the second level of filtering enhances the two out of three attributes in three independent filters. Finally, the third-stage filter maximizes all three attributes simultaneously.

These filters are knowledge-processing systems (KPSs) in their own right. They arrange and rearrange the objects, their relationships, and their attributes such that a local or global maximum occurs. When the problem is complex, a series of local maximum peaks may be located. Such peaks are stored in the cache databases at the output of each filer or the KPS. These cache databases offer an initial set of variables for the next stage of filtering. In the optimization process at the second level, the KPS may override the findings of the first stage of filters. In a sense, the final filters have the option of discarding the findings of earlier filters. The earlier filters only provide an initial point for the next level of optimization by the KPS.

Computational Framework for Knowledge. By Syed V. Ahamed
Copyright © 2009 John Wiley & Sons, Inc.

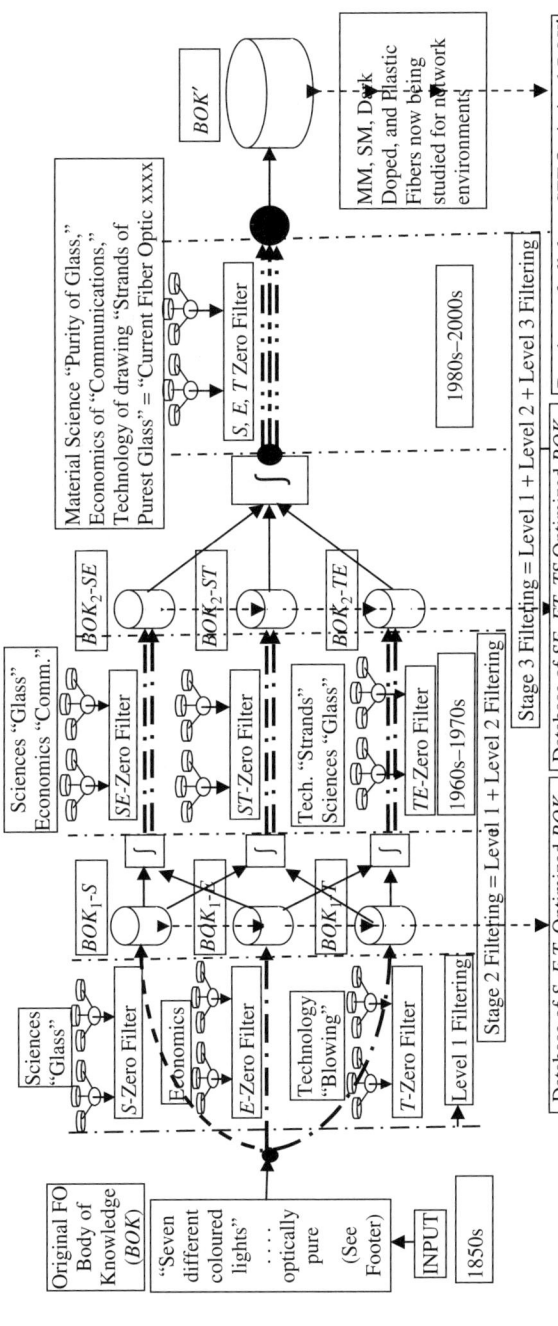

Figure 6.B.1 Configuration of a three-stage information filter to enhance scientific truth S, the economics E of communications systems, and technology T to be able to draw optically pure glass strands (see Dow Corning patents for purifying glass and pulling fiber materials with a dual core). Level I filtering offers the words of Tyndall new meaning and new possibilities to search Internet knowledge bases. Level II and III filtering stages offer combined strategies in glass purification, economics of communications, and technology for the highest-quality multicore fibers. For the input to this filter see the footnote in Appendix 6.A.

6.B.1 S-Zero Filtering and Output BOK_1-S

This filter enhances and permits the scientific aspects of Tyndall's original writings to pass through to cache $BOK_1 - S$, and it also verifies the validity of every similar object and verb function and indicates if the original statements of Tyndall could not be made truer, qualitatively more accurate, and quantitatively more precise. For instance, the "seven different coloured lights" would be transformed and rewritten as "light of numerous wavelengths."[13]

Furthermore, the "optically pure" would be transformed to "glass with extreme purity" to reduce the attenuation and dispersion characteristics of pure glass fiber (see Dow Corning patents of the 1960s [7]). The optical purity[14] also get interpreted on a wider scale as to the source of optical purity or laser sources and then onto the single-mode fiber (see Bell Labs patents of the 1970s [7]). The list goes on and on. Whereas humans may tire out, the information filters and KPSs will not stop deriving new knowledge from old information.

The search in the knowledge bases is expansive until all the leading pointers reach a dead end. When the search of primary *NOs* (and their attributes) and *VFs* (and their variations) does not offer any venues for optimal solutions, the synonyms and thesaurus words need to be included. The machine has an ability to reprocess the entire procedure with new variable NO and VF sets. The methodology is to trap the machine in a mode that offers some solution which is better than no solution. The chances of the machine offering a nonsensical solution are minimized by a well-written knowledge program.

Level I filtering offers the initial finding of Tyndall [6] new meaning and new venues for searching local or Internet knowledge bases. Some of the inputs of level II filters are derived from the technologies to draw optically pure glass strands (see Dow Corning patents for purifying glass and pulling fiber materials with a dual core). Level II and III filtering stages offer combined strategies in glass purification, economics of communication, and technology for the highest-quality multicore fibers.

6.B.2 Generality of Information Filtering

Information and knowledge filters built with knowledge-processing systems can assume many configurations, depending on the initial number of noun objects and (possibly) the verb functions. In a sense, wave filters can also be built in many configurations. The commonality between the two filters (wave and information) is the enhancements/attenuation of the characteristics of the information content that passes through both filters. These "poles and zeros" can be tailored and adjusted based on application requirements. To duplicate the capacity of

---

[13]This leads directly to the concept of wave division multiplexing (WDM), a mode of communicating numerous independent channels over one physical fiber.

[14]This leads directly to the concept of laser sources and single-mode fibers. Terms such as "optical purity" are so potent and encompassing that they apply to the light traversing in the glass medium and to the composition of the glass itself.

information filters, they should be implemented as programmable (and adaptive) digital filters. In hindsight, the capacity of a KPS in an information filter can hardly be duplicated in a digital filter, unless the digital circuitry in the later filter approaches the complexity of a modern digital computer.

It is appropriate to use one adaptive KPS rather than numerous systems. When only one KPS is used, then the system is deployed sequentially to perform level I, II, and III functions. The output after each sequence is held in the corresponding cache databases until it is reused in its own sequential turn. This configuration corresponds to the simplistic architectures of the older single-processor computers.

APPENDIX 6.C

A PRACTICAL INFORMATION FILTER

In this appendix, we illustrate that information filters are generic by examining the statements of Plato and Einstein. They can provide a qualitative and quantitative evaluation of the inputs with respect to the well-defined characteristics of information or knowledge. In this case, truth, virtue, and beauty in the input are highlighted. This methodology provides a basis for evaluation and comparison.

6.C.1 Application to Three Positive Noun Objects

The suggested approach is generic and offers a methodology for evaluating the concepts embedded in almost any segment of information and knowledge. To illustrate the techniques, we select three positive noun objects: truth, virtue, and beauty (TVB) listed in this section and three negative noun objects: deception, arrogance, and hate (DAH) listed in Section 6.C.4 of this appendix. Section 6.C.4 is brief by choice. The operative nouns are selected to force a knowledge machine (KM) to function as a quantitative indicator of the positive concepts TVB or DAH embedded in any $(I \ll \gg K)$. It is evident that these operative nouns could be any number of concept-bearing words (e.g., nonviolence, humility, justice, freedom, democracy, etc., or even war, violence, weapons, greed, selfishness, etc.) and their structural relationships in a socially acceptable format. During the execution phase, the KM would not be fooled by the deceptive users through arbitrary overuse of positive or negative concept-bearing words. Nonsensical statements would carry no weight since the KM constantly verifies the use of such words in knowledge bases around the world. The content, quality, and sophistication of

---

Computational Framework for Knowledge. By Syed V. Ahamed
Copyright © 2009 John Wiley & Sons, Inc.

the use of language in $(I \ll \gg K)$ will all contribute to the numerical evaluation of any $(I \ll \gg K)$.

The tagging of each of the three positive noun objects,[15] TVB, is treated as follows:

1. *Truth* has this set of synonyms: (fact, reality, certainty, genuineness, precision, legitimacy, veracity, and honesty). Each of the synonyms has its own synonyms: (information and detail); (actuality, authenticity, truth, certainty, and veracity); (confidence, conviction, faith, belief, and assurance); (authenticity, reality, substance, legitimacy, and validity); (accuracy, exactitude, exactness, case, correctness, and meticulousness); (legality, authority, and authenticity); (reality, actuality, genuineness, sincerity, and truth); (candor, integrity, dedication, loyalty, devotion, and uprightness), respectively. Removing the words that contribute to circular definitions and words that have a meaning very close to that of neighboring words, we get a set of nine words (noun objects or concepts) for truth as follows:

    (Certainty, fact, genuineness, honesty, legitimacy, precision, reality, veracity)

2. *Virtue* has this set of synonyms: (asset, quality, caliber, merit, value, and worth). Each of the synonyms has its own synonyms: (benefit, advantage, feature, quality, skill, and talent); (excellence, superiority, exceptionality, selection, and high quality); (righteousness, integrity, honesty, morality, and uprightness); (quality, ability, completeness, and level); (value, advantage, good point, worth, and plus); (price, cost, charge, rate, assessment); (value, merit, appeal, significance, attractiveness, importance). Removing the words that contribute to circular definitions and words that have a meaning very close to that of neighboring words, we get a set of seven words (noun objects or concepts) for virtue as follows:

    (Asset, caliber, goodness, merit, quality, value, worth)

3. *Beauty* has a set of distinct synonyms: (loveliness, attractiveness, goodness, prettiness, exquisiteness, gorgeous, splendor, and magnificence). Without actually writing the synonyms of the synonyms, we present a set of seven words (noun objects or concepts) for virtue, after removing the words that contribute to circular definitions and words that have a meaning very close to that of neighboring words, as follows:

    (Loveliness, splendor, attractiveness, good looks, prettiness, exquisiteness, gorgeous, magnificence)

---

[15] Based on the 1980 edition of the *American Heritage Dictionary*. Earlier and later dictionaries offer slightly different concepts associated with TVB. This incremental change over a period of time in concepts associated with these three words, truth, virtue, and beauty, from one generation to another offers a glimpse into the slow but finite shift (drift or movement) of values within society.

6.C.2 Filter Template Based on TVB

The filter F template has three noun objects (NOs) as follows:

FNO01 Truth
FNO02 Virtue
FNO03 Beauty

We chose the comparison of $(I \ll \gg K)$ between fragments of writings by Plato (sample 1) and by Einstein (sample 2).

Sample 1 *Plato*: Virtue is the very health, beauty, and strength of the soul, while vice makes the soul sick, ugly, and weak. ... what is beautiful, then, must lead to virtue and what is ugly to vice. ... what is beneficial must be beautiful; only the harmful is ugly.

Quoted from Plato, *The Republic,* translated by Richard W. Sterling and William C. Scott, New York, W.W. Norton, 1985.

Sample 1: Identifying NOs

NO01 Virtue
VF01 is the
NO02 soul.

Attribute NO03, Atr01: health, beauty, strength

NO04	Vice
VF02 makes	
	sick, ugly, weak.
NO05	What is beautiful
VF02 leads to	
NO06	virtue.
NO07	What is ugly
VF03 leads to NO8 vice.	
NO09	What is beneficial
VF04 must be	
NO10	beautiful.
NO11	Only the harmful
VF05 is	
NO12	ugly.

Matching of *NOs*, Filter *NOs,* and Sample 1 *NOs*

Truth   Match 0.0

Synonyms: certainty, fact, genuineness, honesty, legitimacy

Synonyms of synonyms: certainty (an established fact, quality if being certain); fact (true and accurate, real and demonstrable existence); genuineness (sincere, frank); honesty (honesty); legitimacy

    Virtue   Match 1.0 (FNO02, SNO01, SNO06); virtue and virtue

Synonyms: asset, caliber, goodness, merit, value, worth

    Beauty   Match 1.0 (FNO03, SNO05, SNO06); beauty and beautiful

Synonyms: loveliness; splendor; attractiveness; good looks; prettiness; exquisiteness; gorgeous; magnificence

Extent of 2.0 out of a maximum of 3
Further analysis for truth since there was no match
No match for FNOs and synonyms of truth; score = 0.0
No match for FNOs and synonyms of synonyms of truth; score = 0.0
Total extent of match is $2.0 + 0.0*0.5 + 0.0*0.25 = 2.0$

Sample 2 *Einstein* The ideals which have lighted my way, and time after time have given me new courage to face life cheerfully have been kindness, beauty, and truth. Quoted from http://www.finestquotes.com/select_quote-author-Albert%20Einstein-page-0.htm.

Sample 2: Selection of NOs

(SNO01)   Ideals
(SNO02)   my way,
(SNO03)   new courage
VF01 have lighted
VF02 have given
VF03 have been
(SNO04)   kindness
(SNO05)   beauty
(SNO06)   truth.

Matching of *NOs* between Filter *NOs* and Sample 2 *NOs*

Truth   Match 1.0 (FNO01 and SNO06) truth and truth

Synonyms: certainty; fact; genuineness; honesty; legitimacy

    Virtue   Match 0.0

Synonyms: asset; caliber; goodness; merit; value; worth. No match

Synonyms of synonyms: asset (valuable possession); caliber (degree of worth, distinction); goodness (virtuousness, kindness, benevolence, good part of something); merit (value, excellence, superior quality, character deserving approval/ disapproval, intrinsic right/wrong of any matter)

Beauty   Match 1.0 (FNO03 and NO05), beauty and beauty

Synonyms: loveliness 0.0; splendor 0.0; attractiveness 0.0; good looks 0.0; prettiness 0.0; exquisiteness 0.0; gorgeous 0.0; magnificence 0.0

> Extent of 2.0 out of a maximum of 3
> Further analysis for virtue since there was no match
> Synonyms match for virtue; score = 0.0
> Synonyms of synonyms match for virtue; score = 1.0 (kindness and kindness)
> Total extent of match is 2.0 + 0.0*0.5 + 1.0*0.25 = 2.25

6.C.3 New Filter Template Based on TVB and Soul

The filter template has four *NOs* as follows:

>   FNO01   Truth

Synonyms: certainty; fact; genuineness; honesty; legitimacy

>   FNO02   Virtue

Synonyms: asset; caliber; goodness; merit; value; worth

>   FNO03   Beauty

Synonyms: loveliness 0.0; splendor 0.0; attractiveness 0.0; good looks 0.0; prettiness 0.0; exquisiteness 0.0; gorgeous 0.0; magnificence 0.0

>   FNO04   Soul

Synonyms (from a 2006 dictionary): spirit; essence; heart; core; character; psyche; person
   Synonyms (from an *American Heritage Dictionary* of the 1980s): vital principle; immaterial entity; spirit; person; vital part; emotional nature

Sample 1: Plato Analysis Comparing the *NOs* of sample 1 from Plato, with those of sample 2 from Einstein, we see that sample 1 has a match of 3.0 (virtue, beauty, and soul) and sample 2 has a match of 2 (truth and beauty). The word "soul" does not appear among the synonyms of any SNOs for sample 1 or SNOs for sample 2. However, in the refinement process if the synonyms for truth, virtue, beauty, and soul are considered in the selection procedure for matching, then the numerical process becomes the following: Sample 1: Mismatch (truth and each NO of sample 1)

Match between (truth and synonyms of soul (according to a 2006 dictionary): No match

Match between synonyms for truth and synonyms for soul (according to a 1985 dictionary): No match

Match between synonyms of synonyms for truth (according to a 2006 dictionary): (information and detail); (actuality, authenticity, truth, certainty, veracity); (confidence, conviction, faith, belief, assurance); (authenticity, reality, substance, legitimacy, validity); (accuracy, exactitude, exactness, case, correctness, meticulousness); (legality, authority, authenticity); (reality, actuality, genuineness, sincerity); (candor, integrity, dedication, loyalty, devotion, uprightness)

Match between synonyms for soul (spirit; essence; heart; core; character; psyche; person): No match

Match between synonyms of synonyms for truth and synonyms for soul (according to a 2006 dictionary): for truth, (information and detail); (actuality, authenticity, truth, certainty, veracity); (confidence, conviction, faith, belief, assurance); (authenticity, reality, substance, legitimacy, validity); (accuracy, exactitude, exactness, case, correctness, meticulousness); (legality, authority, authenticity); (reality, actuality, genuineness, sincerity); (candor, integrity, dedication, loyalty, devotion, uprightness); for soul, (vital principle, immaterial entity, spirit, person, vital part, emotional nature): No match

Match between synonyms of synonyms for truth and synonyms of synonyms for soul (according to a 2006 dictionary): No match

Match between synonyms of synonyms for truth and synonyms of synonyms for soul (according to a 1985 dictionary): synonyms of principle are (basic truth, rule, law); (standard conduct); (basic source). Match 1.0 (truth with truth); synonyms of entity are (fact of existence), synonyms of spirit (loyalty, courage, significance)

Total extent of match is $3.00 + 0.0*0.5 + 0.0*0.25 + 1.0*0.125 = 3.125$

Sample 2: Einstein Analysis

Matches FNOs and SNOs. Truth with truth and beauty with beauty. Score = 2.0
Mismatches: virtue and soul
Synonyms match:

Virtue = caliber; goodness; merit; value; worth

Soul = vital principle; immaterial entity; spirit; person; vital part; emotional nature: No match

Synonyms of synonyms match:

Virtue (FNO02): : asset (valuable possession); caliber (degree of worth, distinction); goodness (virtuousness, kindness, benevolence, good part of something); merit (value, excellence, superior quality, character deserving approval/disapproval, intrinsic right/wrong of any matter)

Ideals:: Model of perfection, excellence, or beauty; ultimate objective; worthy principle: Match = 1.0 excellence and excellence.

Courage:: quality of mind; resolution; bravery

Soul (FNO04):: vital principle (basic truth, rule, law); immaterial entity (basic source); spirit (loyalty, courage, significance); person; vital part; emotional nature; principle (standard conduct):

Match = 1.0 courage and courage

Total extent of match is $2.0 + 1.0*0.5 + 1.0*0.5 = 3.0$

In essence, although Plato's statement (statement 1) has an extent of match = 3.125, Einstein's statement (statement 2) has almost the same information and knowledge *(I «» K)* content at an extent of match = 3.0. Whereas Plato approached the current filter template via his major concepts of virtue, beauty, and soul and incrementally (0.125) via the concept of truth, Einstein approached the current template via his major concepts of truth and beauty and twice fractionally (0.5) via excellence (from virtue) and courage (from soul).

This methodology of analyzing *(I «» K)* offers a clue to the common concepts embedded in different bodies of knowledge. It appears that if commonality exists in two scientific papers, documents, books, etc., then this methodology will eventually lead to common major, fractional, and incremental concepts that will unify the two knowledge objects.

APPENDIX 6.D

APPLICATION TO THREE NEGATIVE NOUN OBJECTS

This methodology for analysis is applicable based on the three negative noun objects: deception, arrogance, and hate (DAH). When the procedure in Appendix 6.C is applied to these noun objects, the sets of eight, nine, and six words (noun objects or concepts) for D, A, and H can be derived as follows:

Deception: dishonesty; trickery; ruse; sham; fraud; con; cheating; pretext

Arrogance: conceit; haughtiness; egotism; superiority; pride; overconfidence; superciliousness; self-importance; condescension

Hate: abhorrence; detestation; odium; revulsion; disgust; extreme dislike

In this section, we have not presented a detailed analysis of matching the filter template FNOs with SNOs. However, we submit that a knowledge machine or knowledge filter can perform the analysis and indicate the underlying concepts of commonality (or exclusion for that matter) between any given sets of *(I « » K)*. Statements by things and criminals can be used as inputs for the DAH template. An object-processing unit of a KPU [4] (also discussed further in Chapter 8) will perform such functions.

Computational Framework for Knowledge. By Syed V. Ahamed
Copyright © 2009 John Wiley & Sons, Inc.

CHAPTER 7

PROCESS AND CHANGE OF ENTROPY

CHAPTER SUMMARY

The processing of knowledge implies a change in its entropy. If the incremental change of entropy is tagged as the effect of any knowledge process in a knowledge-processing unit (KPU), of a knowledge machine (KM), then the chain of cause–effect can be established for the execution of any knowledge program on any number of bodies of knowledge. Knowledge objects can be passive, intelligent, and/or reactive. Active objects can respond in a linear or quasi-linear, or nonlinear fashion. They can also react to operators *(kopcs)* aimed at changing the entropy of such intelligent objects. Saturation and hysteresis can also be detected.

The variation in the response of the objects to the operators is classified in this chapter. The direction of response of both objects and operators can vary from positive, to neutral, to negative. The nature of response can also vary from linear to nonlinear. Numerous scenarios of interaction are classified and quantified by constants of linearity, saturation, reciprocity, etc. The machine attempts to stay as close to the reality of the situation as dictated by the input parameters. When the input data are insufficient, the machine examines the history of the object "behavior." The machine also attempts to keep track of large changes in entropy, since the change of entropy demands an expenditure of energy.

Computational Framework for Knowledge. By Syed V. Ahamed
Copyright © 2009 John Wiley & Sons, Inc.

The numerous types of knowledge operation codes for knowledge objects are identified. These elementary instructions for objects correspond to assembly-level instructions for numbers and logical entities in the central processing units (CPUs) of any ordinary computers. Numerous types of single- (knowledge) instruction single-object (SISO), single-instruction multiple-object (SIMO), multiple-instruction single-object (MISO), and multiple-instruction multiple-object (MIMO) instructions are introduced in this chapter. Knowledge-level instructions are inherently more complex than CPU instructions because any object can also have a number of attributes and the dependencies associated with it. Knowledge instructions can (and should) execute attribute operations (such as find commonality, find relations, find similar attributes, etc.).

7.1 ACTIONS AND ENTROPY

Knowledge domain functions bring about an incremental change of entropy in the body of knowledge or knowledge-centric objects, however small or large it may be. Major knowledge domain programs (treated as collections of executable binary operation codes or *kopcs* in knowledge processor units) cause large changes in entropy, and small microscopic changes occur due to the execution of any such *kopcode* or *kopc*. A null *kopc* causes zero change in the knowledge operands. In a symbolic representation of the concept, it is possible to state

Execution of any real $kopc \rightarrow$ Change of entropy($\Delta E =$ finite)

Null $kopc$ or pseudo-$kopc \rightarrow$ Zero change of entropy($\Delta E = 0$)

Furthermore, from traditional assembly-level programming in computer science, only selected *opcs* can be executed on the operands classified as integer numbers, floating point numbers, exponential numbers, logical entities, input/output (I/O) parameters, etc. In the knowledge domain, a more sophisticated strategy is necessary. The knowledge operands are objects, (simple objects, objects with attributes, objects with attributes of attributes, etc.). For this reason, classification of the objects and respective *kopcodes* (*kopcs*) also needs a finer degree of grouping and classification. In order to break down the layering of *kopcs*, a classification of object types and attribute spaces also becomes necessary.

7.2 KNOWLEDGE-CENTRIC OBJECTS

Knowledge-centric objects (KCOs) are objects around which knowledge is focused. They represent the main or derived operands in a knowledge processor unit (KPU). As a tree, their representation becomes simple, as shown in Figure 7.1

322 PROCESS AND CHANGE OF ENTROPY

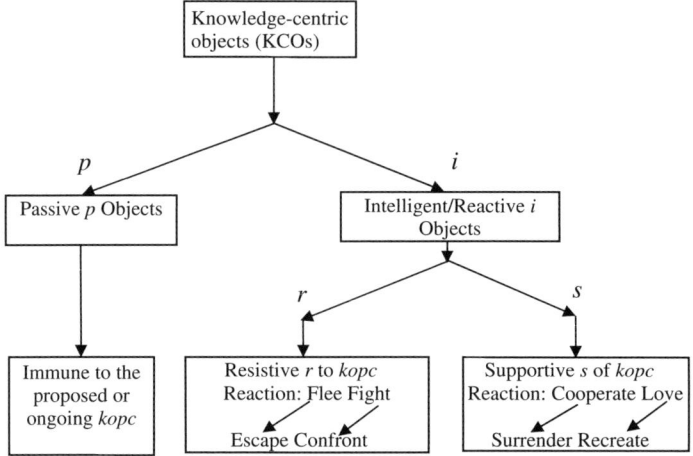

Figure 7.1 Overview of objects from reality that become operands in the knowledge processor unit environments of knowledge machines.

Two major segments of KCOs result at the second level: passive KCOs and reactive KCOs. Passive KCOs are immune to the *kopcs* performed. Reactive KCOs respond to any knowledge *(kopc)* process in either an assistive or a resistive mode.

7.2.1 Passive Knowledge-Centric Objects

Passive and nonresponsive KCOs are most abundant in computing environments. For example, in the computational environment, numbers retain their identity in a numbering system as the CPU performs arithmetic *opcs* on numeric operands. Likewise, the logical and graphical objects retain their identity as the arithmetic logic units (ALUs) and display processor units (DPUs) execute logical operations and graphical display functions. In a sense, traditional computing is the passive processing of immune entities. GPUs on the other hand can alter the entropy of passive graphical objects or images from one frame to the next.

7.2.2 Reactive or Intelligent Knowledge-Centric Objects

Reactive KCOs like intelligent operands respond to the *kopc* process (on them) in the KPU and can learn from prior operations and become resistive or assistive to the proposed change of the *kopc*. The extent and magnitude of the response of the resistive objects can be variable, ranging from fleeing to fighting. The fight can be violent enough to bring chaos to the KPU.

For example, when a predator (noun object) is to catch (verb function) a prey (intelligent object), the flee instinct (a segment of a knowledge process) is invoked. When a lion challenges another in a dominance fight, the injuries on the objects can be fatal, violent, and brutal. to one or both. In cases where an antivirus program is attempting to clean up an infected program, the virus may disable the operating system from regaining control of the computer, disable the protective algorithms that prevent deadlocks causing a massive crash, or simply issue a "format" command to the hardware for the secondary memory systems, thus erasing all protective data. In this case the viral programs maximize the minimum (see Sect 6.10.2) damage and/or maximize the maximum damage.

Reactive KCOs like intelligent operands respond to the process in the KPU and can learn and adapt from prior operations. Such operands can become assistive and supportive of the proposed change to be brought about by the execution of the *kopc*. The extent and magnitude of the response can be variable, ranging from surrender to love. The extent of cooperation can vary significantly.

For example, when a leader (noun object) is to reform (verb function) a society (intelligent object), the cooperative instincts between the two (a segment of a knowledge process) may be invoked. When a leader (e.g., Lincoln) proposes to abolish slavery in the United States, the reaction of the numerous States was varied ranging from confrontation to cooperation. In fact, both resistive (the Southern states) and assistive modes (the Northern states) were invoked. The extent of adaptations of the leaders (noun objects) in administering and executing the appropriate guidelines (*kopcs*) in an appropriate sequence (knowledge programs) to the nation (an intelligent operand) could have possibly taken society to a satisfactory outcome, i.e., the abolition of slavery (successful execution of a long and tedious knowledge program). Other examples of direct resistance occur in society. For example, during the Vietnam War, Johnson's policies of aggression were vehemently resisted in the form of antiwar protests, campus riots, and draft dodging.

The KPU processes will generate new knowledge from the old KCOs or change in the entropy of the KCOs under process. In fact, it can be seen that almost any object is a candidate that can undergo any knowledge process. If there is any thing that is known (knowledge-centric) to exist, then there is (or has to be) a knowledge trail (in time), leading to the present status of the KCO. Similarly, if there is any law of extrapolation, the status of the KCO and its entropy can be predicted in time. Any other objects influencing the KCO also alter the entropy of the object. If a KCO is dynamic, then the *kopc(s)* on the KCO operand will process the KCO such that the change in entropy is a reflection of the reality that the KCO would have undergone in the natural sequence of events changing the entropy of the KCO.

In dealing with KCOs, appropriate priming of operands from knowledge bases based on the Dewey Decimal System (DDS) or Library of Congress (LOC) classification will prevent mismatches between *kopcs* and *koperands*. The processing of intelligent objects needs special consideration since mismatch of the executable codes on a possible operand can produce unpredictable (silly or garbage) new knowledge. Much like a "garbage in-garbage out" scenario, we could encounter a "non serse in-non serse out" or "silliness in-more silliness out" scenario in the knowledge domains, unless the knowledge system is super intelligent.

Two countermeasures are feasible: The knowledge compiler can block the generation of an executable binary knowledge code for inappropriate operands or an intelligent KPU hardware can perform a "sanity" check before any execution based on a lookup table. The latter approach may prove to be more effective, since the machine can import and export many millions of *kopcs* and *koperands* under network (Internet) control.

Figure 7.2 depicts the net changes of entropy ∂E for any KCO over a duration ∂t. The change as it occurs in nature or the environment of the KCO is depicted by the outside major loop *acda'*. The inside major loop *ab12ea'* depicts the change in the knowledge machine (KM) environment for the same KCO. As the KCO experiences the change in its entropy due to natural events, the processes within the KPU also alter its entropy. The tracking of natural/environmental events that cause the change in entropy is maintained by the knowledge program encoded as a series of machine-executable *kopc* instructions. The reality check at *RC* in the machine retains the accuracy and authenticity between the real world of events and the execution of *kopcs* in the KM. It simply becomes an intelligent KM.

Information that is gathered around any specific *KCO (a)* in Figure 7.2 undergoes changes in the entropy, depending on the complexity of events in the environment, and is reflected by the complexity of the knowledge software and programs in the KM. The newly generated status of *KCO (a')* is thus computable (by the exact change in its entropy). Knowledge processing thus retains both flavors of information processing and numerical computing in any KM environment. During the execution of the knowledgeware (KW) in a KM environment, it is highly desirable, if not imperative, that the newly machine-generated knowledge objects be vanishingly close to real objects in nature. The reality checks in box *RC* in local loops *c-KM-1* and *d-KM-2* ascertain that the major loops are stable over a period. The operating system of the knowledge machine KM prevents the execution of unnatural or illegal *kopcs* on the machine. Thus, the machine does not slip away into the unchartered territory of the knowledge domain by the constant monitoring by the reality check (in *RC*). Such precautions are generally taken in monitoring most quasi stable systems.

However, in forcing the KM to exercise a reasonable amount of "artificial creativity" in addition to artificial intelligence, the human counterpart of the KM may program an additional degree of freedom in the machine to search for KCOs that are not entirely real in the pursuit of final KCOs that can be (possibly) synthesized. For example, if the KM is searching for a cure for cancer, the user may permit the machine to investigate the molecular structure of drugs that do not exist with the hope that they may be synthesized at a later stage of research by the same machine.

As an additional example, if the machine is investigating elementary particles that can (may) travel faster than light, then the user may instruct the machine to overlook the laws of physics or an axiom of wisdom (such as equation, $E = mc^2$), or include the forces of negative gravity with a prespecified confidence level. The machine now largely performs one of the functions of the mind. The tightening or relaxing of constraints for processing knowledge is an option of the user, much

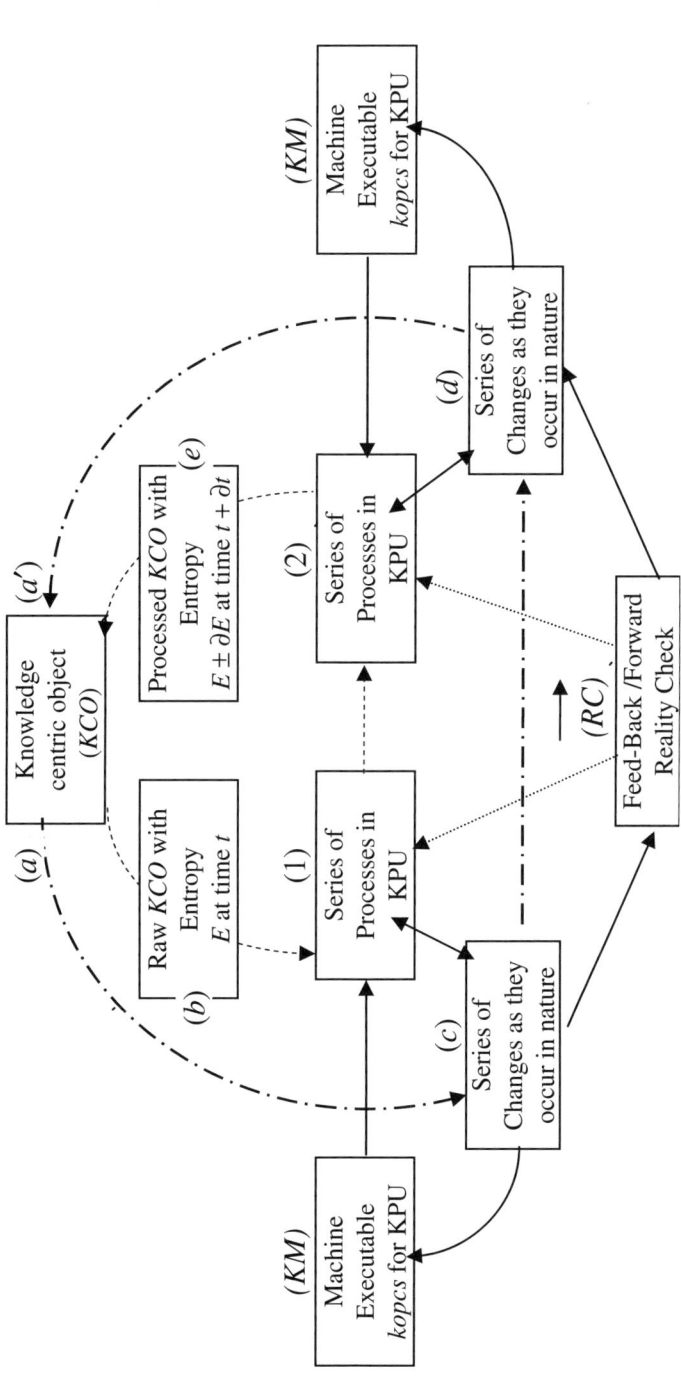

Figure 7.2 Entropy gathered around any *KCO* (a) is altered in nature over time. This change of entropy is tracked in the knowledge processor unit by a series of *kopcs* that the machine executes. Two major loops *acda'* (in nature) and *ab12ea'* (in the machine) are shown. When changes in nature are accurately reflected by the *kopcs* (in *KM*), then the change of entropy computed by the machine is the same as what has occurred in nature. To maintain the accuracy and authenticity of the natural events with *kopcs* in the knowledge processor unit, a reality check is invoked at the box *RC* at numerous test points in the program. In addition, the energy for the executing the operation on real objects in the physical world is also tracked in one of the peripheral CPUs. The rate of expenditure of energy (or resources) leads to the power requirement or the net expenditure of resources over the life time of any individual, (social or political) project.

326 PROCESS AND CHANGE OF ENTROPY

as a coach can instruct a team to invoke a predictive response to the opposing team's offensive and defensive strategies before they are even implemented.

7.3 CLASSIFICATION OF KNOWLEDGE-CENTRIC OBJECTS

Objects without attributes are the simplest KCOs to process, even though the real world is too complex for such simplification. However, in the context of any given problem, objects may be artificially deprived of their attributes. Such gross simplifications provide a niche in most rudimentary knowledge-processing environments.

For example in an aura of generality, it may be stated that "liquor shall not be offered to children." When interpreted by a human being, this minutia of information/knowledge implies the most inclusive definitions of the three keywords ("liquor" as the noun object, "offer" as the verb, and "children" as the adverbial enhancement of "offer"). The noun object "liquor" implies all types of liquors and intoxicating drinks that have their our attributes, "offered" implies a sale, gift, present (even one forcibly ingested), and "children" implies infants, toddlers, very young and older children. In such a broad and general statement, the objects although complex are forcibly simplified.

In the real world, almost all KCOs have a single attribute (SA) or multiple attributes (MA), and in rare cases the KCOs do not have attributes (NA). There are three major types of KCOs, as depicted in Figure 7.3. The KCOs range from

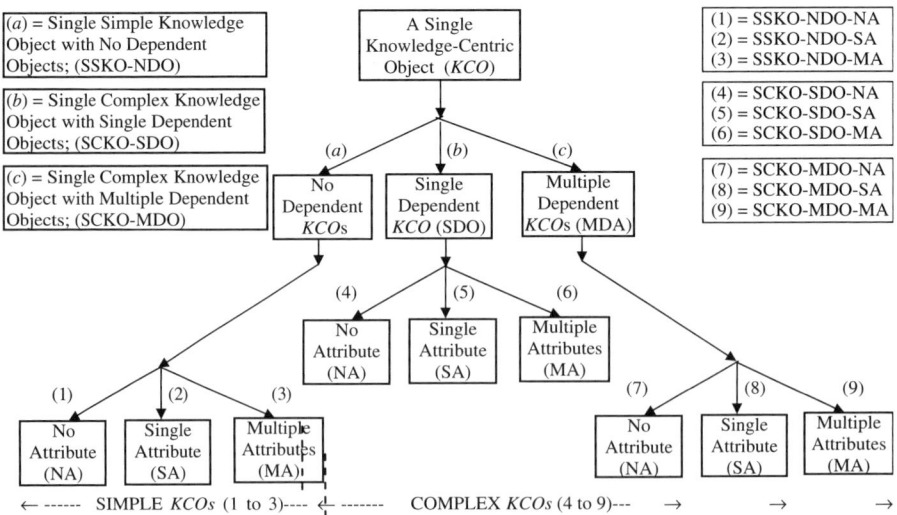

Figure 7.3 Classification of knowledge-centric objects. The tree structure is two levels deep, and there are three branches [(no dependent objects, single dependent object, and multiple dependent objects) and (no attributes, single attribute, and multiple attributes)], leading to nine types of objects and types of *kopcodes*.

the simplest [see (*a*)] to the multiply complex [see (*c*)]. The requirement for complexity is that the KCO have one or more other KCOs that depend on the primary one, and the change in entropy of one will result in a change in entropy of the other.

For example, the rainforest (KCO) affects land erosion [adjective of land (another KCO, land)], and this in turn affects the long-term landmass (another KCO) to support the rainforest (primary KCO). In a sense, any knowledge processing of one KCO will alter its entropy. It will also affect the entropy of other dependent KCOs. The execution of certain *kopcs* [such as ALTER (the rainforest)] will influence the entropy of that particular rainforest.

In reality, there are only complex KCOs shown in branch (*b*). When *b*-type KCOs lose their complexity, they degenerate into branch (*a*) or *a*-type KCOs, and when multiple *b*-type KCOs are grouped together, they cluster into branch (*c*) or *c*-type KCOs. Such classifications, even though they may appear simplistic, bear an influence on the design of the KPUs that have to handle the three types of objects (in the hardware).

In addition, the design of the KPU to handle the three KCOs at the first attempt can be considered as either of the following: a degenerate *b*-type KPU to handle an *a*-type KCOs, or a recursively deployed *b*-type KPU to handle *c*-type KCOs. This concept is discussed in detail in Sections 7.9 and 7.10. In many cases, superposing numerous processes of type 2, type 5, or type 8 KCOs (Figure 7.2) will not be the same as KCOs of type 3, type 6, or type 9.

For example, if the *kopc* ALTER, from the previous example, is changed to REDUCE, then the reduction of one KCO (rainforest) will influence the other two KCOs (land erosion and landmass), and the resulting reductions of all three KCOs will be greater than the independent execution of the *kopc* command REDUCE on each of the three KCOs independently.

7.3.1 Single Simple Knowledge Object (SSKO)

Objects may be independent of all other with any independent identity, or objects may have strong dependencies. Entirely independent objects are hypothesized and become almost mythical like zero, infinity, virtue, beauty, ratio of circumference to diameter of a circle, etc. Even though they may not be tracked by a machine to perfection, the machine can determine them to its limit of precision. A level of confidence is associated to the extent that the computed value is the limit imposed on the machine in the design of the numerical units (ALU and NU), or the extent of reason in the logic unit (LU) of the sequential deductions of the embedded processes. Thus, we start the classification of the objects since the KPU for the new machine should be designed to handle (almost) any type of objects. As will be seen in the following section, recursive processing can reduce the many designs of the KPU handling complex objects. In the simplest case with one *kopc* operating on one single simple(st) object, the knowledge process is represented as follows:

CPU → Operation   Single Operand
         Code *opc*   *Operand*

KPU → | Single *kopc* | Single-Simple Knowledge Object   No Attributes |

Written as
   kopc SSKO

Example Instruction to KPU
   Define Science

7.3.2 Single Simple Knowledge Object, Single Attribute (SSKO-SA)

In generalizing most real objects, single or multiple attribute(s) may become essential. The execution of KPU instruction and control signals from the knowledge instruction register will depend on the type of operand objects. This philosophy of design is prevalent in traditional CPU design methodology since the same operation code (e.g., add, mpy, div, etc.) depends on the type of operands (e.g., integers, floating point, exponential double precision, etc.).

For example, in a command to a KM, it may be stated that "hard liquors shall not be offered to any rowdy individual whose blood alcohol level exceeds 3 percent." When interpreted by a human being, this information/knowledge implies very specific and highly directed instructions. In such a specific statement, the identification of the noun objects (rowdy individual and 3% blood alcohol level) and verb (offered, under force, coercion, sale, etc.) requires selective filtering to meet the constraints in the statement.

If the knowledge operands are to have attributes, then the type of instructions and associated control signal in the KPU hardware start to differ from those discussed earlier. This permits designers to track such a conceptual difference in methodology. The knowledge process associated with SSKO-SA is represented as follows:

KPU → | Single *kopc* | Single-Simple Knowledge Object | Single *Attribute* |

Written as
   kopc SSKO-SA

Example Instruction to KPU
   Define Science   Nuclear

7.3.3 Single Simple Knowledge Object, Multiple Attributes (1−*n*)

This type of instruction is based on the instruction presented in Section 7.3.2. When the single simple object has numerous attributes $(1 - n)$, then instruction can be decomposed into a series of n-instructions of the type shown there. However, there is a substantial amount of risk since objects and attributes may be and usually are inter-dependent.

KPU → | Single *kopc* | Single Simple Knowledge Object | SA-1 | SA-2,… | SA-*n* |

Written as
    kopc SSKO-MA(*1– n*)

Example Instruction to KPU
    Define Science   Nuclear (of) French, British, U.S., etc., origins

In the knowledge domain, the independence of attributes cannot be assumed. In contrast to the computations of plain old computer systems, this inter-dependence can (and usually has) ramifications. For example, if the rate of change of velocity is being computed for a given mass m, then the equation of motion[1] applies and mass is held constant and is independent of the matter that contains it.

However, in a real context and knowledge domain, if the HDL cholesterol count is being derived, then the levels of LDL and triglycerides cannot be held as constants. The three types of measurement and analysis of LDL, HDL, and triglycerides need to be made on the same sample of blood drawn at the same time and generally by the same test machine. Synergy and timing become a major concern for *kopc* functions on knowledge objects. As an additional example, if the financial health of an organization is being evaluated, other factors (such as sales, production, and impending taxes) cannot be ignored, or held as stationary. The time dependence of related objects and their attributes become highly influential in almost all knowledge-processing environments. In some cases, it becomes imperative to consider the effect of a single *kopc* on a group of clustered KCOs rather than the effect on each independently and then linearly added together. Iterative processing becomes essential.

7.3.4 Single Complex Knowledge Object, Single Dependent Object

Real and practical objects rarely exist in isolation and most real-life objects have ties, however strong or weak they may be with other objects. It is only in relation to other objects that any one object retains its identity with attributes of its own. Primary objects may be linked to one or more inter-dependent objects. For example, a CEO (primary object) exists typically because of an organization (single secondary object) and vice versa, etc.

KPU → | Single *kopc* | Single Complex Knowledge Object |
                                               | Single Dependent Knowledge Object |

Written as
    kopc   SCKO
              SDKO

Example Instruction to KPU
    Define   Nuclear Science
               Radioactive Materials

---

[1] Such as force = mass × acceleration or $F = m \cdot a$.

7.3.5 Single Complex Knowledge Object, Single Dependent Object, Single Attribute

In a network of related objects (like the members of an organizational team or pieces of an aircraft, etc.), relationships should be assigned a numerical weight (ranging from 1 to 0.0) to denote the extent of inter-dependency or the bond between objects. Such a numerical value is essential in the KPU because the machine alters the nature and flavor of objects and their attributes in the process of executing a *kopc*. The execution of any *kopc* on a primary object has its ramification on the secondary or dependent objects. Such effects are executed by the knowledge machine, (KM) and the execution of the *kopc* is legitimized. In a sense, the parent–child link of any object to its antecedent and descendent objects is also analyzed by the knowledge-processing software systems or through the *kopc*'s decomposition into microcommands in any standard control memory configuration.

For example, readjusting the wing span of an aircraft influences the aerodynamic stability (secondary object 1) and wind drag (secondary object 2) due to a change in size, the additional fuel (indirect secondary object due to 2) needed to fly any predetermined distance (fixed object), etc. As is traditionally done, such effects may be computed by computer-aided design (CAD) software for aircraft with human feedback.

On the other hand, the problem may also be handled by a KM. The programming of such details to force the KM to automatically trace the antecedent and descendent objects to any level, the attribute of primary and secondary objects, the attributes of objects, the attributes of attributes of primary and secondary objects to any given depth (as may be uncovered in knowledge bases around the Web, on any newly written laws of aerodynamics), is the responsibility of software designers.

KPU → | Single *kopc* | Single Complex Knowledge Object | Single *Attribute* |
| Single Dependent Knowledge Object | Single *Attribute* |

Written as
   kopc SCKO SA
           SDKO SA

Example Instruction to KPU
    Define   Nuclear Science   Nuclear Power Plants
             Radioactive Materials Controlled Fission

7.3.6 Single Complex Knowledge Object, Single Dependent Object, Multiple Attributes

A university would exist because of students, faculty, and staff (multiple inter-dependencies), etc. An independency may become essential to the existence of the primary object. For instance, matter exists because we can measure it, numbers exist because we can perceive them, a spouse exists because of the existence of the othermate, society exists because of the members that make it up, etc. Complementary relations abound, but in a general sense any primary object may be related to any number of secondary objects, like relationships within a family group.

KPU → | Single *kopc* | Single Complex Knowledge Object | SA-1 | SA-2,... | SA-*n* |

Written as
    kopc   SCKO-MA(1− *n*)

           SCDO-MA(1− *n*′)

Example Instruction to KPU
    Define   Nuclear Science (of) French, British, Indian, etc., origins
                  Uranium Source   French, British, Indian, etc., origins

7.3.7 Single Complex Knowledge Object, Multiple Dependent Objects ℓ

When numerous *m* complex objects are encountered, each with $\ell(i, j)$ inter-dependent objects, then the representation becomes two-dimensional. Parameter $\ell(i, j)$ represents the *j*-dependent objects associated with each of the *i* (from *1* to *m*) objects. The representation is as follows:

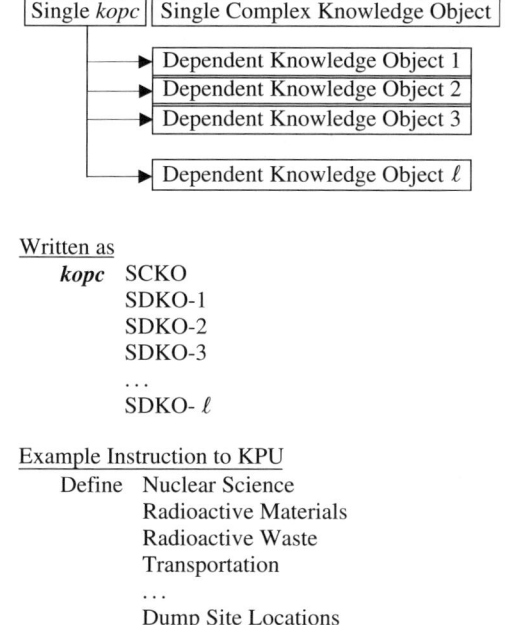

Written as
    kopc   SCKO
           SDKO-1
           SDKO-2
           SDKO-3
           ...
           SDKO- ℓ

Example Instruction to KPU
         Define   Nuclear Science
                     Radioactive Materials
                     Radioactive Waste
                     Transportation
                     ...
                     Dump Site Locations

7.3.8 Single Complex Knowledge Object, Multiple Dependent Objects ℓ, One Attribute

In most cases, complex objects [i.e., interdependency(ies)] and attributes are both encountered. In order to deal with such objects, the KPU in a KM environment needs adequate addressable object registers or cache memory space to execute a *kopc* and then trace its effect on the primary object, secondary objects, and also single or multiple attributes. Such elaborate inter-dependencies are rarely encountered in traditional computer environments. However, some software programs automatically update (Symphony and some spreadsheet programs) the dependent variables when certain changes in parameters influence the derived variables. In [1], such a strategy was suggested to stabilize systems when there are unexpected swings in the data sensed throughout the environment.

To deal with such situations, *three* considerations are necessary for the design of the KPU. *First*, the addressing capability of the dependent objects and their respective attributes needs consideration; *second*, the change in the entropy value of the primary and dependent objects needs to quantified; and *third*, the confidence level for the execution of the *kopc* and produced the computed change in entropy.

The execution of the single *kopc* on the single knowledge object with ℓ-dependent objects with a single attribute each is represented as follows:

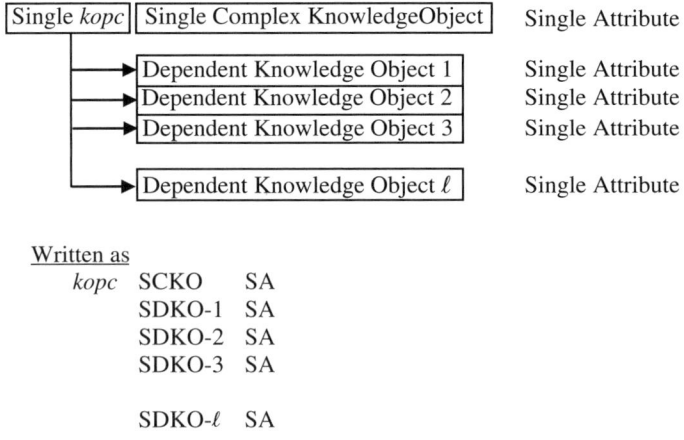

Written as
   kopc   SCKO   SA
            SDKO-1 SA
            SDKO-2 SA
            SDKO-3 SA

            SDKO-ℓ   SA

Example Instruction to KPU
   Define   Nuclear Science   Nuclear Plants
               Radioactive Materials   Controlled Fission
               Radioactive Waste   Collection
               Transportation...   Hazards Materials
               Dump Site Locations... Permission to Dump

7.3.9 Single Complex Knowledge Object, Multiple Dependent Objects ℓ, Multiple Attributes n

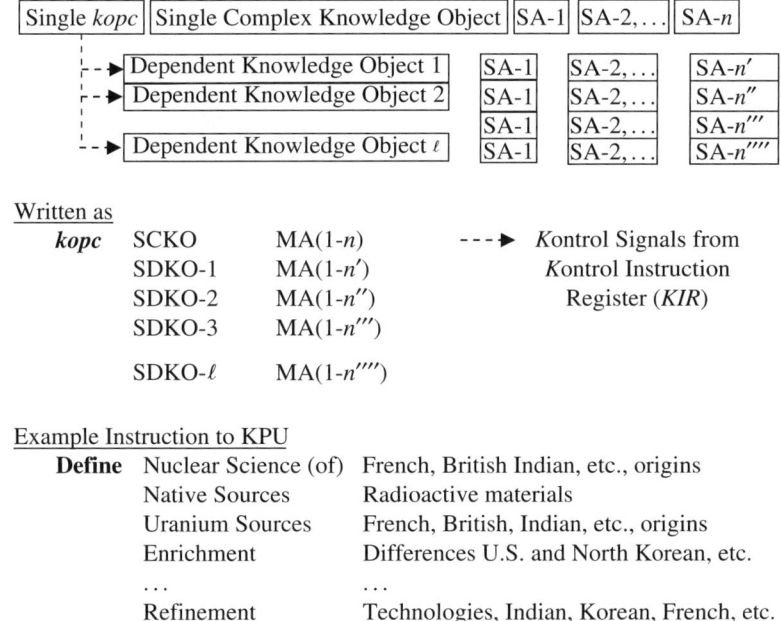

Written as
    kopc   SCKO   MA(1-n)   ---▶ Kontrol Signals from
            SDKO-1 MA(1-n')   Kontrol Instruction
            SDKO-2 MA(1-n'')   Register (*KIR*)
            SDKO-3 MA(1-n''')
            SDKO-ℓ MA(1-n'''')

Example Instruction to KPU
    Define Nuclear Science (of) French, British Indian, etc., origins
                 Native Sources   Radioactive materials
                 Uranium Sources   French, British, Indian, etc., origins
                 Enrichment   Differences U.S. and North Korean, etc.
                 ...   ...
                 Refinement   Technologies, Indian, Korean, French, etc.

7.4 CLUSTERS OF COMPLEX KNOWLEDGE-CENTRIC OBJECTS

It is logical to extend the object type to groups or clusters of single objects, thus making such clusters multiple complex objects. The user can now command the KM to execute a *kopc* on a cluster of complex objects. For example, if a branch of an organization needs to be relocated, then each of the m employees of numerous ranks is affected. Each member is a KCO in his or her own right and has ℓ multiple complex dependent objects (such as spouse, children, house, relatives, etc.) with n numerous attributes (such as job of spouse, schools for the children, their cars, their own friends and relatives, etc.). Thus, when the KM is instructed to study the impact of the relocation and report on it (*kopc*), then the effects of *kopc* on the primary objects, derived objects, the attributes of each object, and its ripple effects on all the inter-related objects directly and via the attributes will be investigated and documented.

In general, the execution of k complex *kopcs* on $m \times \ell \times n$ objects/attributes will become multiple-complex but not chaotic. The need for keeping the addresses of all the objects and attributes becomes demanding but manageable. Such complex addressing tasks have been accomplished in the billing systems [2] of most automated public telephone systems. Dedicated computer billing systems track the "call data records" (CDRs) of all customers of all telephone services, all the

334 PROCESS AND CHANGE OF ENTROPY

billable services that have their attributes, and promotion codes and special rates for independent and "bundled services." The clustering can be based on numerous criteria and for this reason the types of clusters can also have at least nine classifications as delineated in Section 7.4. We present only two such examples in Sections 7.4.1 and 7.4.2.

7.4.1 Multiple Complex Knowledge Objects *m*, Multiple Dependent Objects ℓ, Single Attribute

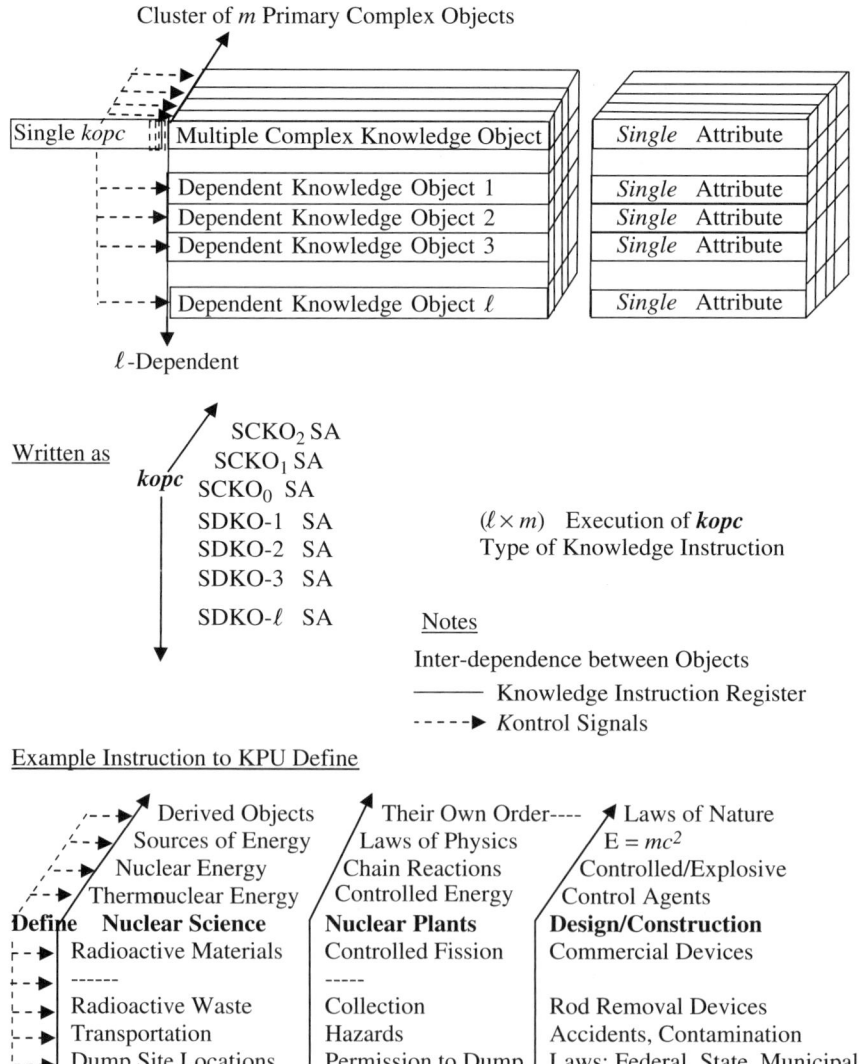

7.4.2 Multiple Complex Knowledge Objects *m*, Multiple Dependent Objects ℓ, Multiple Attributes *n*

This particular type of instruction is one of the most generic knowledge instructions that can be represented in the two-dimensional pages of a book. Consider the most general *kopc* that influences m primary objects, each with ℓ_i secondary objects associated with the ith primary object. Each of the primary and secondary objects has $n_{i,j}$ attributes. The representation is still feasible by having n single three-dimensional matrices on the single attribute matrices presented for the *kopc07* type of knowledge instruction.

When the representational scheme is extrapolated for this *kopc08* instruction, then there will be n third-level attribute matrices on the left-hand side of the diagram. When these matrices are drawn to a much smaller scale, the diagram has the following representation:

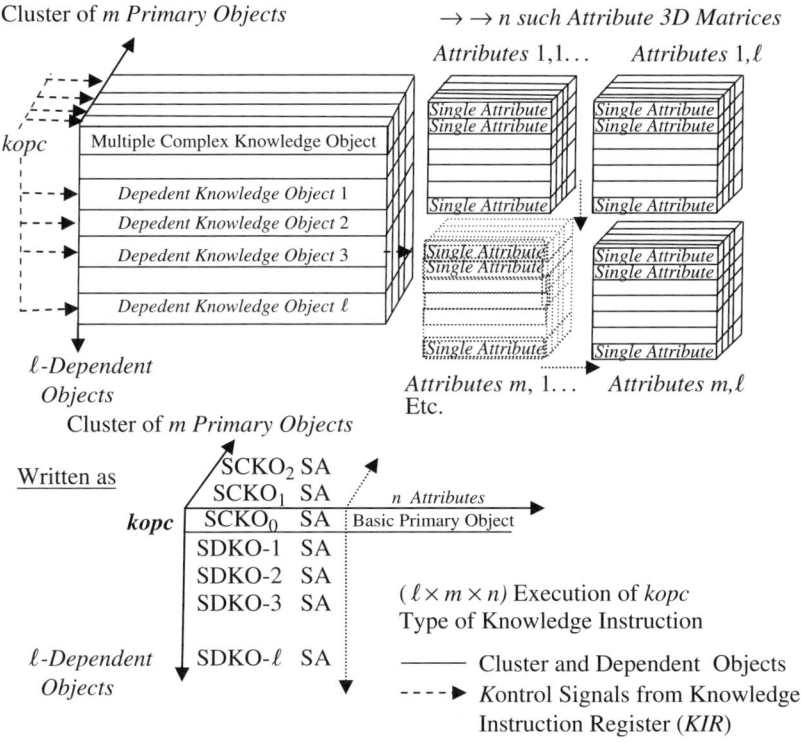

The representation of the instruction to the KPU in the two-dimensional plane of the paper is not feasible, but it can be seen that the cache memory of the KPU contains the representation, provided the address and linkage are properly established. Such addressing capability is one of the functions of the compiler of the knowledge programs for the knowledge machines. The responsibility for maintaining appropriate linkages can be delegated to the (assembly-level) programmers of the knowledge machines or handled by the machine itself through a self-optimizing compiler.

7.5 SINGLE-PROCESS *KOPCODES* FOR GENERIC KNOWLEDGE-CENTRIC OBJECTS

Single-process *kopcodes* force the KPU to perform a single *kopc* instruction (Sections 7.5–7.7) and one generic to the SP architectures presented in [3]. In particular, *kopc01* of generic type 01 serves a class of knowledge domain (KD) instructions of one primary object without any attributes. Then in a hierarchical sense, class 2 may be reserved for KD instructions for objects with one layer of m attributes, and class 3 may be reserved for objects with m attributes that have n attributes each.

7.5.1 Single-Process *kopcodes* for Simple Objects

The three *kopcodes* (*kopc01–03*) are specific to the type of objects represented as (1), (2), and (3) in Figure 7.3 and expanded upon in Sections 7.4.1.1–7.4.1.3.

7.5.1.1 SP-SSKO-NA, Class 1 kopc01 Instruction
A single-process on single simple knowledge object with No Attributes (SP-SSKO-NA), or machine-executable real *kopc* on single simple object operands with no attributes, this is a single *kopc* on a single object. Single-processor (SP) instructions, such as where, why, how, when, find, define, find known attributes, dependent objects, or explain, operated on one generic object (single object), such as health, hospitals, markets, universities, publishers, authors, etc., will be executed in the KM and the appropriate information (graphic, textual, logical, numeric, etc.) will result.

7.5.1.2 SP-SSKO-SA, Class 2 kopc02 Instruction
A single-process single knowledge object with a single attribute (SP-SSKO-SA), or machine-executable real *kopc* on single object operands with one attribute, this is a single *kopc* on single simple object. Single-processor instructions such as where, why, how, when, find, define, find known attributes, dependent objects, or explain, operated on a single object, such as where in tropical regions, when in hospitals, bull markets for stocks and bonds, why in universities for research, publishers of books in the sciences, authors of textbooks in digital formats, etc., will be executed in the KM and the appropriate information (graphic, textual, logical, numeric, etc.) will result only for objects with a specific attribute (SA). It is possible to view the attributes of objects as adjectives for noun objects and the different variations of *kopc* as the adverbs of the verb functions the KPU will perform on the noun verbs.

7.5.1.3 SP-SSKO-MA, Class-3 kopc03 Instruction In the developed human languages, numerous adjectives and adverbs may be strung out in long sentences. However, very long compound sentences may become incomprehensible due to the limited memory and attention span of human beings. In the KM, such limitations do not exist as long as the addresses of noun objects and their attributes are properly linked. Similarly, *kopc* (verbs) and their modifiers (adverbs) may also be handled to any depth.

In the simple case where multiple attributes of objects exist and enter the knowledge-processing domain, the instruction format of the *kopc* needs modification by stringing such attributes in the command structure of the instruction. Similarly, the assembler of the knowledge level instructions has to assign appropriate locations for each of the attributes. To ascertain that all the functions are accurately performed, the assembler performs two basic functions: the address and location in the cache or main memory(ies) of the KM, and the actual attribute.

For example, if the instruction to the KM is to find all scientists who are over forty years, live in the NE corridor of the United States, and have a Ph.D. degree, then the four filters (scientists, 40+, NE of U.S., Ph.D.) of all generic objects will be automatically performed by the machine. For this reason, the decoding of the class-3 *kopc* is representative of simple filtering functions (see Chapter 6) performed by the KM.

7.5.2 Single-Process *kopc* for Single Complex Object

Complex objects have inter-dependencies on other antecedent or descendant objects. For example, an island may exist only if there is water (or other media) surrounding it, a town may exist typically, only if there are inhabitants, a mind may exist only if a body exists to host it, etc. Hence, the KM will evaluate the effects of executing a *kopc* on all dependent objects. It can be seen that a complex object without inter-dependencies will degenerate into a simple object. By the same token, a complex object with ℓ dependencies can be treated as ℓ simple objects and the simple object instructions of Section 7.5.1.1 may be repeated ℓ times, unless there are inter-dependent interactions between attributes.

However, some inter-dependencies will have a circular effect on each of the ℓ objects. Iterative processing (see Section 7.3.3) will have to be invoked until a convergence of all ℓ objects is reached due to the processing of *kopc* on any one of the ℓ objects.

The following three *kopcodes* (types *kopc 04–06*) are specific to the type of objects represented as (4), (5), and (6) in Figure 7.3 and expanded upon in Sections 7.5.2.1–7.5.2.3.

7.5.2.1 SP-SCKO-NDO-NA, Class-4 kopc04 Instruction A Single-process single complex knowledge object with no dependent objects and with no attribute (SP-SCKO-NDO-NA) or machine-executable real *kopc* on single complex object operands with no attributes, this is a single *kopc* on a complex object (that has one dependent object associated with the complex single object).

338 PROCESS AND CHANGE OF ENTROPY

Single-processor instructions such as where, why, how, when, find, define, find known attributes, dependent objects, or explain, operated on a single object, such as health in tropical regions, hospitals for malignant cancer, French markets for stocks and bonds, British universities for research, world-wide publishers of books, U.S. authors of textbooks, etc., will be executed in the KM and the appropriate information (graphic, textual, logical, numeric, etc.) will result only for both primary and dependent objects. It can be seen that the adjectives which modify nouns can make objects complex or objects that have close links to other objects, such as faculty and university(ies), rainfall and thunderstorms, can also make objects that display a strong inter-dependence and relationship.

7.5.2.2 SP-SCKO-SDO-SA, Class-5 kopc05 Instruction
A single-process single complex knowledge object with a single dependent object and with no attribute (SP-SCKO-SDO-SA), or machine-executable real *kopc* on single complex object operands with one attribute, this is a single *kopc* on a complex object (that has one dependent object associated with the complex single object). Single-process instructions such as where, why, how, when, find, define, find known attributes, dependent objects, or explain, operated on the single complex object with one attribute, such as the health of diabetics in tropical regions, hospitals in Europe for malignant cancer, French American markets for stocks and bonds, directory of British universities for research, world-wide Italian publishers of books, U.S.-born textbook authors of Chinese heritage, etc., will be executed in the KM and the appropriate information (graphic, textual, logical, numeric, etc.) will result only for both primary and dependent objects. It can be seen that the adjectives which modify nouns can make object processing complex or objects that have close links to other objects, such as faculty and university(ies), rainfall and thunderstorms, can also make objects that display a strong inter-dependence and relationship. However, a specific attribute for any object offers more programmer control for enhancing the filtering or processing in the KM environment.

7.5.2.3 SP-SCKO-SDO-MA, Class-6 kopc06 Instruction
A single-process single complex knowledge object with a single dependent object and with multiple attributes (SP-SCKO-SDO-MA), or machine-executable real *kopc* on single complex object operands with multiple attributes, this is a single *kopc* on a complex object (that has one dependent object associated with the complex single object and each of the objects may have multiple attributes). Single-process instructions such as where, why, how, when, find, define, find known attributes, dependent objects, or explain, operated on the single complex object with numerous attributes, such as the health of diabetics and asthma patients in tropical regions, hospitals in Europe for malignant lung cancer in advanced stages, French American market traders in the Middle East for stocks and bonds, etc., will be executed in the KM and the attributes' specific information (graphic, textual, logical, numeric, etc.) will result only for both primary and dependent objects. It can be seen that the adjectives which modify

nouns can make object multiply complex for objects that have close links to other objects, such as faculty and university(ies), rainfall and thunderstorms, can also make objects that display a strong inter-dependence and relationship. However, multiplicity of attributes offers more flexibility for enhancing the filtering or processing in the KM environment.

This type of a knowledge operation is equivalent of macro utility in normal computing environment. At such a detailed level, the knowledge compiler software can be used to execute such knowledge operation codes.

7.5.3 Single-Process *kopc* for Multiple Complex Objects

When complex objects have a string of complex objects associated with them, then processing becomes further complicated. These steps can be completed by the knowledge machine compiler software or a set of recursive routines that the user may develop to suit the application. In addition, some inter-dependencies will have a circular effect on each of the complex objects. Iterative processing (see Section 7.5.2) will have to be invoked until a convergence of all these objects is reached due to processing of this type of *kopc* on any of the complex objects.

The following three *kopcodes* (types *kopc 7–9*) are specific to the type of objects represented as (7), (8), and (9) in Figure 7.3 and expanded upon in Sections 7.4.3.1–7.5.3.3.

7.5.3.1 SP-SCKO-MDO-NA, Class-7 kopc07 Instruction A single-process single complex knowledge object with multiple dependent objects and with no attributes (SP-SCKO-MDO-MA), or machine-executable real *kopc* on single complex object operands with multiple multiply dependent complex objects without any attributes, this type of *kopcode* (*kopc07*) is specific to the type of objects represented in branch (7) of Figure 7.3.

7.5.3.2 SP-SCKO-MDO-SA, Class-8 kopc08 Instruction A single-process single complex knowledge object with multiple dependent objects with a single attribute (SP-SCKO-MDO-SA), or machine-executable real *kopc* on single complex object operands with multiple multiply dependent complex objects, each with a single attribute, this type of *kopcode* (*kopc08*) is specific to the type of objects represented in branch (8) of Figure 7.3.

7.5.3.3 SP-SCKO-MDO-MA, Class-9 kopc09 Instruction A single-process single complex knowledge object with multiple dependent objects and with multiple attributes (SP-SCKO-MDO-MA), or machine-executable real *kopc* on single complex object operands with multiple multiply dependent complex objects, each with multiple attributes, this type of *kopcode* (*kopc09*) is specific to the type of objects represented in branch (9) of Figure 7.3.

As can be seen, the complexity of instructions increases from the simplest objects like numbers, logical entities, symbols, etc., to those for complex objects with multiple dependencies with multiple attributes. These instruction sets are next examined in Section 7.6. Both single-object and multiple-object processing architectures appear feasible. All the advantages of multiple processing in traditional computing environments that host multiple CPUs become evident in multiple-object-processing knowledge machines (KMs) that host multiple KPUs.

7.6 MULTIPLE-PROCESS INSTRUCTIONS

Nine further generic instruction (*kopc10–18*) possibilities corresponding to instructions in Sections 7.5.1.1–7.5.3.3 also exist, and they are multiprocess instruction sets that perform a multiple *kopc* on single simple objects through multiple complex objects. These instructions are not depicted in Figure 7.3.

Initially, multiple instructions can be viewed as a loop of the instructions presented in Section 7.5.1.1–7.5.3.3 for each object. However, innate nonlinearities and inter-dependencies between objects, their relationships, and their attributes can render such a process inaccurate and perhaps insufficient. We present a subset of three of these final sets for *multiple* processes for *multiple* complex objects in Sections 7.6.1–7.6.3 corresponding to class-16 *kopc16* to class-18 *kopc18*. The other instructions, *kopc10–15*, are simpler versions of the following.

7.6.1 MP-MCKO-MDO-NA, Class-16 *kopc16* Instruction

A multiple-process multiple complex knowledge object with multiple dependent objects and with no attributes (MP-MCKO-MDO-NA), or machine-executable real *kopc* on single complex object operands with multiple multiply dependent complex objects, without any attributes.

7.6.2 MP-MCKO-MDO-SA, Class-17 *kopc17* Instruction

A multiple-process multiple complex knowledge object with multiple dependent objects and with a single attribute (MP-MCKO-MDO-SA), or machine-executable real *kopc* on single complex object operands with multiple multiply dependent complex objects, each with a single attribute.

7.6.3 MP-MCKO-MDO-MA, Class-18 *kopc18* Instruction

A multiple-process multiple complex knowledge object with multiple dependent objects and with multiple attributes (MP-MCKO-MDO-MA), or machine-executable real *kopc* on single complex object operands with multiple multiply dependent complex objects, each with multiple attributes.

When hardware complexity needs to be curtailed, then iterative software loops cycle through the series of instructions as given in Section 7.5.1.1–7.5.3.3. Each will suffice until all the objects, their inter-dependencies, their attributes, and their relationships iteratively converge to their stable values or limits.

On the other hand, a set of multiple KPU architectures can be deployed. When each KPU is capable of handling the instructions given in Sections 7.5.1.1–7.5.3.3, then multiple objects are processed simultaneously, but the parameters in one KPU can influence the process in the rest of the KPUs. In a sense, the entire collections of KPUs will process an entire group of objects simultaneously and produce a composite output from the KM that holds the parallel-processed KPUs.

7.7 PASSIVE *KOPCS*

Passive *kopcs* do not materially alter the entropy of knowledge-centric objects (KCOs); instead, they scan and examine these objects and extract a snapshot, status, or condition, obtain a pulse, etc., without directly or indirectly altering, modifying, enhancing, rearranging, etc., the KCOs, the dependent objects, and/or attributes. In contrast, active *kopcs* alter the nature of KCOs. Active *kopcs* are deployed to derive new knowledge from old knowledge (subject to the constraints of reality) in order to examine the newly created KCOs and their behavior in the application.

Most of human communication encompasses passive functions. In the structure of information and knowledge [1] depicted in Figure 7.4, the human activity (gossip, hearsay, comments, judgment, etc.) dealing with information at the lower levels is equivalent to the passive *kopc* functions of the KM.

The passive verb commands to the KPU yield observational results as the outputs from a set of input KCOs. In the real world, a passive command is a snapshot

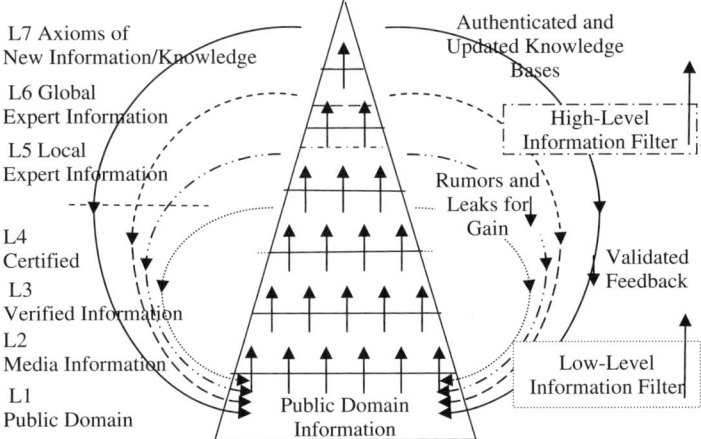

Figure 7.4 Levels of Information with upward flow reinforcing the scientific process to enhance truth, universality, social value, and elegance in information. Vertical movement involves active *kopcs* to shift the entropy, and horizontal movement involves passive *kopcs* in the knowledge machine.

of the environment (e.g., a photograph without altering its subjects/objects). In a strict sense, the entropy of information in the set of KCOs either individually or as a cluster of dependent or inter-related primary objects is altered by the execution of passive *kopcs*, but the change occurs within the level of accuracy of the KM to register such a change.

When a passive *kopc* (e.g., filming a movie) is performed on a fixed KOC (e.g., sunset) in a passive environment (spectators only), the resulting change of entropy may be too minute to be perceived, unless the level of precision of measurement is artificially raised to record the effect of the passive *kopc*. In some applications, such precision may not be desirable or warranted. For example, sensing systems [4] do not alter the entropy of objects. However, the robotic feedback to stabilize the objects that *may* follow will alter the entropy and attributes of the objects and their relationships.

In the movie example given, filming may need simply to convey the descriptive or passive message (the attributes of a sunset) without invoking a response from objects. The filming will not affect the nature of entropy (the colors, setting, or other attributes) of the sunset. In contrast, if the *kopc* calls for filming the sunset with numerous filters and exposures to capture its attributes, then the different entropies will be flimed and need to be tagged.

In the KM environment, total passivity of any *kopc* is rare unless the *kopc* is a pseudo-*kopc* (i.e., an operation to prepare the hardware of the KM for the *kopcs* that will follow), is very observational, or gathers information without feedback. When users approach the KM merely to "read" some value from the set of input KCOs, however insignificant, it is usually the first attempt to study the environment. Such searches are routine in Internet surfing. During the activity that follows, the user may (and usually does) alter the attributes of the primary object (e.g., sunset from different angles, in different landscapes, involving seascapes, etc.), and secondary objects (filter colors during filming, post-picture processing, frame superpositioning, etc.).

In running the application programs (APs) on KMs, passive *kopcs* are generally well inter-spersed with active *kopcs* to achieve a desired set of results from a set of input KCOs. For example, the management team of a corporation (the primary KCO) may instruct a report-generating management information system (MIS [5]), (the passive *kopc*) to study the current status of the corporate environment by using a KM and then utilize those findings to improve, enhance, boost, etc. (active *kopcs* to the KM) the productivity, cost reduction, sales improvement, etc. (attributes) of the corporation.

7.7.1 Report Generation *kopcs* on Passive Knowledge-Centric Objects

Report generation (RG) is an established mode of passive MIS function. Probing, examining, and defining do not intentionally alter the status of passive KCOs. Instead, they scan and detect the status quo of KCOs, their attributes, their settings, and their environments and correspond to the gossip, chitchat, admiration, criticism, blame, condemnation, etc., expressed in human transactions. Such human activity invokes a similar response from "other" passive humans, and the entire activity results in the low-level (L1 and L2) functions shown in Figure 7.4. It is therefore implied that RG searches carried out on passive KCOs do not result in changes of entropy for the objects that are searched.

It is possible to invoke such a response from a KM, and it corresponds to a Google search on the Internet. The exercise simply gives back word-based search results. To determine the relevancy to the broader overall picture or the context of numerous prior (interpolative) or impending (extrapolative) searches needs additional human information processing. Limited constraints are possible by specifying the type or dates for searching the databases.

7.7.2 Report Generation *kopcs* on Dynamic Knowledge-Centric Objects

Dynamic objects can respond to socioeconomic changes and time. When such changes occur, then passive *kopcs* perform a dynamic tracking of the changes in the KCO and tag the environmental conditions. The RG function for a meteor, a thief, an unstable human, etc., needs a log of every passive *kopc* that sensed the status and corresponding conditions. It is also implied that the execution time of a normal *kopc* is short in relationship to the execution time of a similar *kopc* for the dynamic KCO.

Intelligent KCOs (like human beings and animals) may not only alter the results of a *kopc*, but also foil the execution of the next process in the *kopc* sequence of functions. Such situations also arise in war when an intelligent spy plane automatically changes its course as a missile approaches or when a commander will predicatively switch strategy depending on a possible attack from the opponent.

7.8 ACTIVE *KOPCS*

Active *kopcs* perform functions and change the entropy within KCO(s). The nature and/or attributes, inter-dependencies, and relationships between primary and/or secondary objects may be altered in the resulting KCO(s). Such active *kopc* functions are akin to those within intelligent, adaptive, and self-regulating MISs that respond to an unhealthy corporate environment. As another example, the reflex actions of the human body can provoke a flee response (active *kopc*) and avert (environment attribute for the operand "body") danger. A computer antivirus program can chase (invoke active *kopcs*) an infected program to eradicate it (operand) on the Internet, etc.

In most instances, active and passive *kopcs* are used in coordinated KPU command statements. In effect, the well-programmed set that generates an efficient knowledge program consists of both active and passive *kopcs*. In traditional computing, optimizing compilers (e.g., Fortran V compilers) follow the approach of generating more effective codes for executing frequently used application and utility programs.

7.8.1 Active *kopcs* for Passive Knowledge-Centric Objects

These instructions are akin to routine corrective measures in a production environment. After scanning the inventory of raw materials, an inventory control program [6] may reorder low inventory items or postpone next delivery overstocked items. As an additional example, if a conflict in scheduling is observed, the resource allocation programs in an automated MIS [5] environment may readjust the assignment of personnel and/or space. Operating systems in routine computer environments automatically perform such monitoring functions.

When minor readjustments in the environment are necessary, feedback and correction mechanisms are invoked by executing active *kopcs* on passive KCOs. The net effect of these instructions redistributes the entropy of information between and around the KCOs to regain the stability of the knowledge environment. In an all-human environment, the redistribution of power and responsibility within the management of a corporation is routinely practiced to enhance productivity or achieve insidious corporate goals.

In a knowledge environment, these instructions perform important functions quite routinely. For example, the *kopc* command to the KM can be downloaded as an object (bearing a predefined digital identifier). Examples of such *kopc* objects are autoconvert DOC format to PDF format, autoconvert TIFF format to JPG format or CorelDraw (CDR) format, or even just translate an html from one language to another. It is seen that these *kopc* instructions alter the attributes (and thus the overall entropy) of operands (files).

In a research environment, these instructions can have powerful overtones. For example, if the machine is instructed to change the molecular structure for a solvent to dissolve diamonds, the machine may (and perhaps will) come up with a null answer after long and hard searches of all the knowledge bases around the world, from prehistoric times to now, and extrapolate the incomplete finding of all efforts to "find solvents for diamonds." It will indeed not leave any word or verse unturned in the knowledge banks on the Internet.

The programming skill of knowledge programmers will materially influence the change in entropy of the resulting objects of the KPU and thus the outcome of the KM. Such instances are dominant in traditional programming environments where skillful programmers deploy the full potential of hardware that executes routine programs.

If a passive object (e.g., a movie) is reviewed (a passive *kopc*), then the feedback effects via the public (active KCO) may alter the attributes (such as popularity, sales, profitability, etc.) of the movie. Ripple effects via intelligent objects that follow can have dominant impacts. As another example, the use of a certain commercial product by a dignitary may alter its sale, even though the product remains unaffected. However, commercial advertising of the luminary's use of the product can have a secondary effect on its potential sales. This feedback occurs through the public, which should be considered an intelligent object in the environment.

Intelligent objects that occur in the knowledge operand set (e.g., the moviemakers, directors, and/or staff) can alter the entropy of dependent objects, or the environment of the operands. The feedback that alters other passive *kopcs* thus has a compound effect in altering the overall entropy of complex objects. Such feedback can and sometimes does destabilize the primary KCO (movie): It may lose popularity, experience reduced sales, or even be barred.

In perspective, looking back at the deployment of traditional computers in industrial environments, the acceptance of KMs is likely to be resisted in certain social environments, much as medical information systems and informatics were rejected by the medical community in the 1980s and 1990s. But, it is also possible that simpler KMs are likely to serve individuals and gain credence quickly, as personal computers demonstrated during the 1970s and 1980s.

7.8.2 Active *kopcs* for Reactive Knowledge-Centric Objects

These types of commands perform active processing on and from the KCOs and alter the structure of the derived or resulting objects. In traditional computing systems, the newly computed value replaces the old value and the older arithmetic and logical entities are discarded. Much like executable operational codes (*opcs*) that will alter the values of variables in traditional computing environments, active *kopcs* will alter the attributes of derived objects.

For example, if the *kopc* command in a KM calls for altering the attribute A_i of an object O_j, then the multidimensional matrix for the object O_j will carry the new attribute. As an additional example, if the *kopc* alters the dependency relation between the object O_j to other objects ($O_{j'-j''}$), then the structure of relations is altered to reflect the execution of the *kopc* in the KPU. When the KCOs react to the execution of the active *kopc*, iterative processing becomes essential.

7.8.3 Active *kopcs* for Intelligent Knowledge-Centric Objects

If the KCO set ($O_{j-j}, \ldots$) is intelligent, then the execution of one active *kopc* on one object will shift the attributes of one or more objects in the set. The balance between objects, attributes, and attributes of attributes is likely to be altered. For example, if one member of a closely knit family suffers a loss, then the level of empathy among other members is likely to be affected, causing a secondary effect to readjust the loss effect of the primary object, altering the balance of relations between members of the family. Regaining the stability is the common goal of all intelligent KCOs.

In such instances, an iterative execution of the *kopc* (on the object) is necessary until convergent values of the new parametric relations are reached for the entire object set. For the intelligent object set, a change of status in one active *kopc* can cause a ripple effect for other objects. The execution of such instructions is common in specialized software packages. For example, the effect of fluctuations in currency conversion rates alters the future revenue stream or debts of a corporation. The balance sheets are affected, and this reflects a new debt/asset ratio and hence the willingness of banks to lend monies to the corporation. A new stable scenario evolves for the primary object (the corporation). Sometimes, spreadsheet programs (XL, Symphony, etc.) have built-in routines to track and update all the parameters to reflect such changes. In word-processing applications, adding or deleting material from chapters brings about changes in a book's page numbers, and the table of contents will then need to be updated. In such cases, active *kopcs* (such as, updating a table of contents) are executed on two inter-related intelligent objects (a book and its table of contents).

In the knowledge domain, the computation and tracking of such ripple effects can become complex since some of the object attributes and relationships are not numeric and precise. In a sense, the calculations will be based on estimates [7] of numbers rather than their numerical values. In econometrics and weather prediction, such estimates are routinely used in forecasting. The semi-numerical tools and algorithms [8] to deal with such fuzzy object matrices require formal computational methodologies.

7.9 EXECUTION OF MULTIPLE-PROCESS INSTRUCTIONS

A multiple-process instruction commands the KPU, typically execute numerous (related) single processes on any one cluster, or (related) cluster, of simple or complex KCOs. It is logical that the instructions presented in Sections 7.2–7.4 be executed in the same KPU environment either sequentially or in parallel on single or multiple KCOs. Based on this detail of designing the code (sequential or parallel, for a simple KCO or complex KCOs, with or without attributes), a large variety of KPU hardware architectures will result. In a sense, to build a viable KM, the complexity of the initial designs needs to be moderated. Such designs can offer a modest set of KPU instructions, like the IAS [9] machine offered to rudimentary computers of the 1950s and 1960s. However, the concept of stored program control (SPC) was firmly established in the 1948–1949 period. In the knowledge domain, the evolution of the KM is likely to closely parallel the evolution of the computer systems from 1950s to 1990s.

In the proposed *kopc* instruction code, instruction length and the resulting assembly-level commands can become quite complex. In order to import the SPC concept in the knowledge domain, the memories to store programs and objects can are the same by an appropriate of set internal addressing capabilities in the KPU environment. For example, if the n attributes of ℓ-dependent KCOs and m-related primary objects are allocated to the "cells" of a three-dimensional matrix, then a variable word-length memory will accommodate the KPU instructions and KCOs. The traditional 3-D memory design in CPU environment now becomes a 3-D array of 3-D (objects, dependent objects, and attributes) memories in the knowledge environment. An intelligent memory management system can simplify the need for complex memory hardware.

7.10 ITERATIVE AND REFLEXIVE PROCESSING

When dependent objects are also affected by the execution of the *kopc*, which will in turn affect the primary object under process, iterative processing becomes necessary. Such processing is different from recursive programming since iterative programming sends the process to the primary object, which affects the secondary object, which may also be affected by the processing of other secondary object(s), which may indeed be all objects undergoing the *kopc*. Multiply recursive loops within each iteration may become essential.

Process instability is likely under such conditions. To circumvent the problem, iterative processing with under-relaxation of the objects, dependent objects and attributes, is proposed. The effect of *kopc* on objects is first evaluated; a superposed solution effect of *kopc* is then determined. Next, the ripple effect of executing the *kopc* on the new objects is determined until the net change in net entropy (from one iteration to the next iteration) of the entire complex object set converges to zero. Such procedures a common in numerical relaxation techniques.

An additional level of complexity further exists. In conventional computing systems, the process time is dramatically faster than the time of natural change of the symbols under process. In the knowledge environment, the objects (especially if they are intelligent) can undergo changes as the process is taking place. For example, if a nuclear plant is likely to melt down due to elevated reactor temperatures during operation, then the selection of a computer-controlled stabilizing mechanism may be too slow to avert the disaster (e.g., the Chernobyl radioactive fallout). Other examples abound. In the case of incident in Bhopal, India, or Hurricane Katrina in the United States were disasters human control were too sluggish to control the result or the evacuation. In such cases, the reflexive reaction of the KPU and/or predictive processing in the KPU is recommended, in which the process of control is derived from a look-up table rather than risk gross dysfunction in the environment.

7.11 MACROINSTRUCTIONS FOR KNOWLEDGE PROCESSOR UNITS

Most multiple instructions in traditional computing facilities are assembled as a series of single, simpler, or preprogrammed instructions or instruction sets. For the KPU environments, dealing with simple objects involves the SP instructions discussed in Sections 7.5.1.1–7.5.2.6. Macroinstructions are derived from these. In order to deal with complex objects, complex objects instructions need to be formulated (whenever possible) from the single process instruction set, and then macroinstruction's are reassembled from these new instruction sets.

When the application is for super-complex objects (such as human organs, Ebola virus, cancer, etc.), then specific instruction sets need to be created. Such specialized macroinstructions have been devised for determining the boundaries of malignant cancer by using magnetic resonance imaging (MRI) techniques. Many intelligent MISs (such as Oracle PeopleSoft or SAP) have built-in algorithms that make routine decisions (e.g., reordering inventory items) as reflexive functions. Hospital environments also respond to emergencies in a preprogrammed fashion.

Generally, more complex multiple instruction sets can also be disassembled to single instruction sets (Sections 7.5 and 7.6). The capacity of the KPU to handle the *kopc* usually sets the limit for the breakdown of machine-executable operation codes. However, complexity does not have to stop at the highest levels, that is, only one *kopc* on a set of m primary objects, each ℓ_i-dependent objects with $n_{i,j}$ attributes (adjectives and/or adverbs), but it can follow a set of nested instructions of the *kopc09* type. The KM is thus forced to perform additional executions (for the primary instruction) to suit the level of complexity. However, any KPU design that can execute the *kopc09* type of instruction may be reactivated for the other lower-level (*kopc01–08*) instructions. The knowledge program continues to be executed.

7.12 KNOWLEDGE PROCESSOR UNIT ARCHITECTURES FOR *KOPC01–12* INSTRUCTIONS

KPU architectures were initially disclosed in [10] and elaborated upon in [1, 11], they are presented in Section 8.4.5. The KPU in [9] is capable of handling some of the preliminary *kopc01–03* types of *kopc*s. To implement the KM in an object-oriented hardware, addressing all the dependent objects and attributes of all the objects requires meticulous care. When the object set becomes large and their attributes also grow to be a larger set, then some algorithms to address management become essential, especially for the *kopc04–08* types of *kopc*s.

Knowledge processing is based on the processing of knowledge-centric objects or KCOs. Hence, the KPU architecture is derived from one or more object proce (OPUs). Conceptually, the architectures of the object processors can be arranged as single-process single-object (SPSO), single-process multiple-object (SPMO), multiple-process single-object (MPSO), multiple-process multiple-object (MPMO). and pipeline object processors. These architectures are presented in detail in [1] and briefly summarized here for the sake of completeness.

7.12.1 Single-Process Single-Object Processors and Machines

SPSO processors are akin to SISD processors for traditional CPUs. When the object is immune to processing, the OPU hardware is simplified to traditional CPU hardware with an instruction register (IR) to hold the operation code, data register (DR), memory address register (MAR), program counter (PC), and a set of A, B, and C registers. The CPU functions according to the fetch, decode, execute (FDE) cycle in the simplest case.

KNOWLEDGE PROCESSOR UNIT ARCHITECTURES FOR *KOPC01–12* INSTRUCTIONS

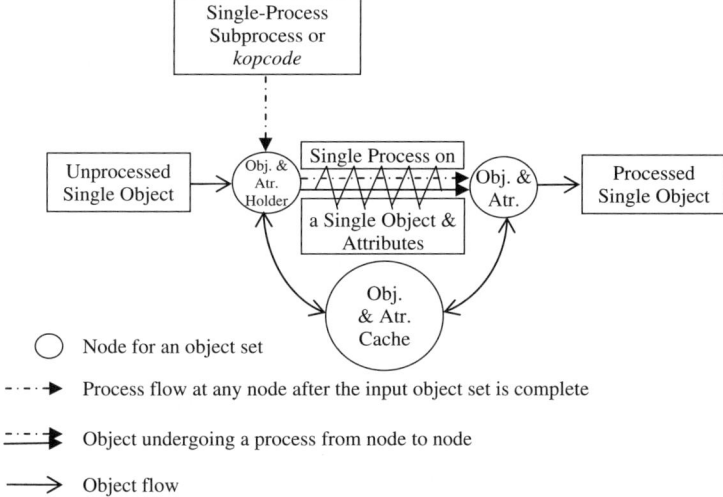

Figure 7.5 Simplified representation of a single-process single-object processor that operates on a single object and its attributes. When placed in Figure 7.6, the system works on a simple von Neumann object computer.

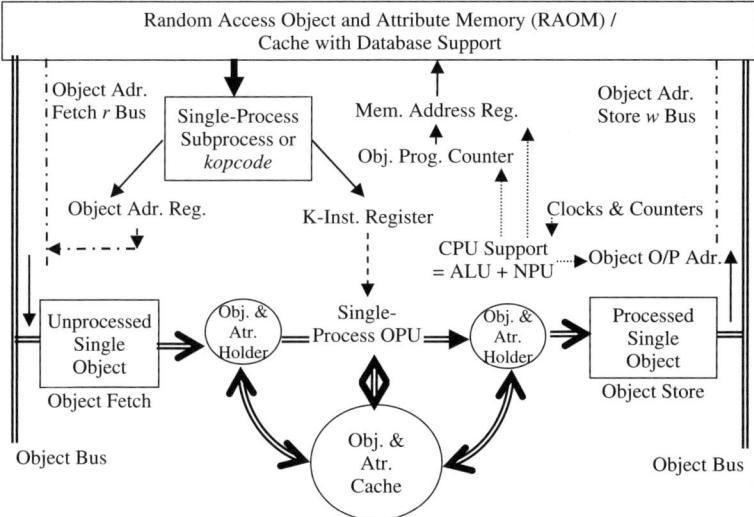

Figure 7.6 Simplified representation of a single-process single-object object machine. The configuration of a single-processor object-processing unit is presented in Chapter 8.

The SPSO architecture (Figure 7.5) becomes more elaborate to accommodate the entire entropy of the object that is under process. The entropy can have a series of other dependent objects, their relationships with the main object, its attributes, and dependent object attributes. The simplest handware configuration of a KM based on SPSO processor is shown in Figure 7.6.

The format of the objects and attributes can be alphanumeric (descriptive) rather than symbolic. Primary and secondary objects and their attributes can be local and predefined in the simplest cases, or they can be Internet-based and fetched from Web knowledge banks. The execution of a single *kopc* can influence the entropy of the single object via the secondary objects and attributes. In essence, numerous caches are necessary to track the full effect of a single *kopc* process on the entire entropy of the single object in the SPSO processor.

It can be seen that the machine configurations for other types of architectures, that is, SPMO, MPSO, MPMO, and pipeline object processors, can be derived by variations similar to those for the SPSO systems. In the following sections, we present the change for an MPMO type of object processor and object machine.

7.12.2 Multiple-Process Multiple-Object Processors and Machines

The configuration of an MPMO object processor is shown in Figure 7.7. This dual-object bus structure holds numerous single-object processors between the input object bus and the output object bus.

The program bus feeds segments of object programs to the program caches, and the systems operate as numerous independent object processors under the control of an independent operating system.

An enhanced architecture to implement the MPMO processor is shown in Figure 7.8. Both local and global programs can be executed by permitting universal access to program, knowledge, and technology bases. The role of the operating system becomes crucial in coordinating the numerous 1 through n programs to be accomplished on 1 through l objects, and each program can have any number of associated tasks.

From a design perspective, the optimal KPU design would depend on the sophistication of the technology being deployed in the actual KPU hardware. Such an approach has been effectively used in traditional CPU designs ranging from multiple instruction single data (MISD) and multiple instruction multiple data (MIMD) [12] to numerous microprogrammable architectures [12]. A very large number of multiple-process instructions are possible.

The conventional indirect addressing (available in most assembly language instruction sets) via the base sector does not offer enough flexibility. Double, triple-level nested addressing via active memory locations is likely to offer the latitude to access secondary objects, their attributes, the attributes of attributes, etc. It must be appreciated that nested address capability in the KM can provide quick access to the objects and attributes in the cache memory of the KPUs. Furthermore, the operation of knowledge programs can be prevented from turning chaotic by robust addressing algorithms. In fact, such addressing capability offers the basis for the development of knowledge-processing theory as distinct from complexity theory [1].

For example, if the KPU command turns specific, such as "Hard liquors may be offered to intoxicated Native Americans in order to deprive them of the capacity to be judicious about the sale of their lands to British Army officials who have

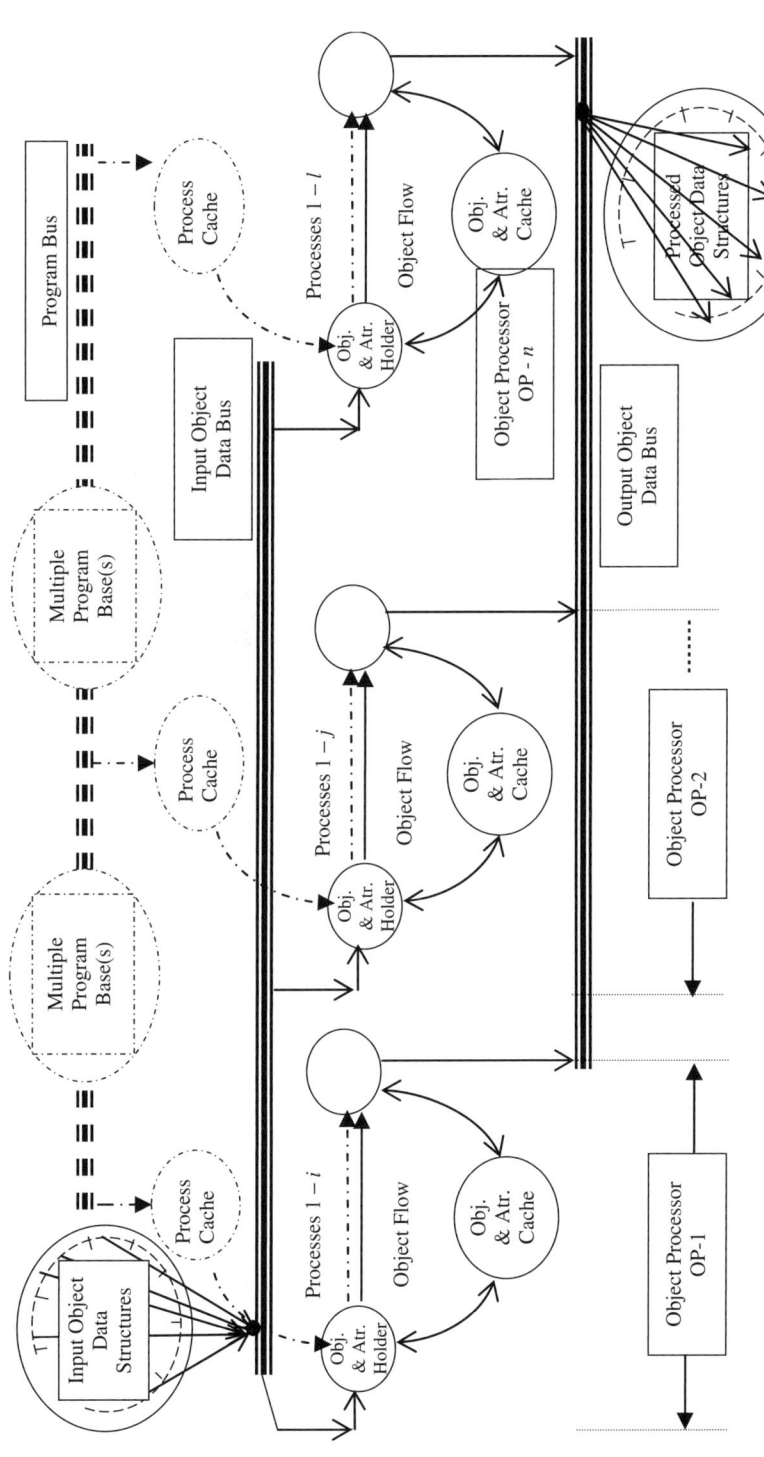

Figure 7.7 Multiple-process multiple-object (MPMD) architecture. Multiple programs are executed multiple objects in multiple processors at the same time. This architecture is akin to traditional multiprocessor computer systems with the exception that multiple attribute processing can also take place with each of the multiple object processors. Such application of the multiple-process multiple-object hardware exists if a search firm seeks multiple executives from multiple databases of applicants based on numerous attributes or requirements.

351

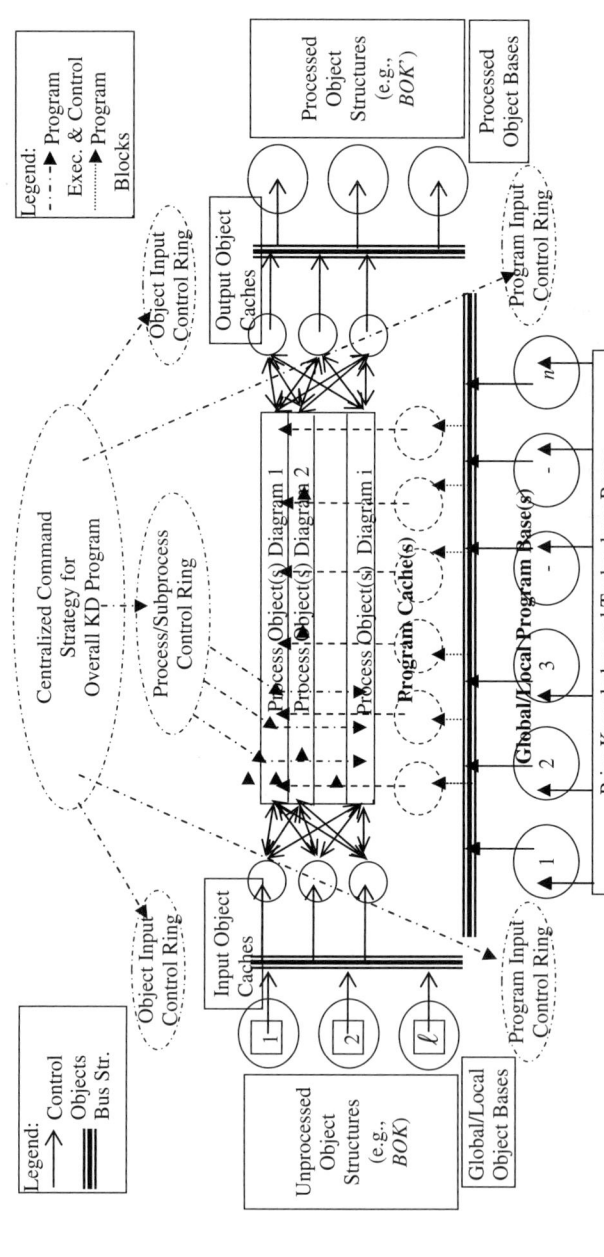

Figure 7.8 Process hierarchy of multiple objects in a multiple-process multiple-object configuration from left to right. Local and Internet access is also shown in this machine configuration. Caches and processes are shared, and the operating system allocates the sources for complex knowledge programs to pass through the machine. The configuration of the machine shown will be similar to that of a multiple-instruction multiple-series object machine.

enjoyed the hospitality of the Native Americans." When this specific information/knowledge is interpreted by a human being, it implies very specific and highly directed instructions. The identification of the noun objects (intoxicated Native Americans) and verbs (offered, forced, coerced, sold, etc.) with adverbial modifications (by British Army officials, for the purpose of depriving "them of the capacity to be judicious about the sale of their lands") needs at least three additional selective filters to meet the constraints in the statement. The attributes of attributes are already implied in the terms "intoxicated" and "capacity to be judicious about the sale of their lands."

In a general sense, the machine should be capable of tracking the attributes of attributes to any depth (from 0 to i) and of exploring the n_{i+1} attributes of n_i attributes of any object (multiple primary objects, dependent secondary objects, attributes of objects, attributes of attributes of all objects, etc.). However it would be derivable to build the simplest KMs first in order to gain enough experience indesigning knowledgeware and knowledge based operating systems.

7.13 CONCLUSIONS

In order to build a suitable set of operation codes to process objects, objects are classified and grouped in this chapter. Classification is based on dependencies that follow and grouping is based on attributes. Objects are classified into three broad categories: (1) simple and independent (i.e., no dependent objects are influenced by this object), (2) complex with one dependent object, or (3) complex with multiple dependent objects at the first tier. Objects are further classified at the second tier into three groups: (i) objects without attributes, (ii) objects with a single attribute, and (iii) objects with multiple attributes. A set of nine instructions becomes evident. In the simplest case, the object processor becomes a simple von Neumann CPU, where numbers and logical entities can be processed by the instruction set of the von Neumann CPU.

The simplest single objects (with no dependent objects and no attributes) have the simplest operations, and conversely, the most complex single objects (with multiple dependencies and multiple attributes) have cumbersome operations. Such tedious operations can be performed in hardware, software, or firmware (with control memory-based object processors). In this chapter, a set of object operation codes for hardware that can handle single complex objects with multiple dependencies and multiple attributes are presented.

Operations on multiple complex objects (e.g., a human group) with multiple attributes still need particular attention. Such examples are common in everyday life (e.g., teach a group of talented students). It appears logical to write libraries of knowledge macros (e.g., develop customized classes) to perform such elaborate operations (present, repeat, illustrate, image, project, convince, etc.) on groups (talented students) or clusters of complex objects. When the inter-dependencies (retention and imagery, presentation and learning, repetition and convincing, etc.) are linked to one another, iterative processing (also discussed in this chapter) will become necessary to perform the operation and reach an accurate result.

In this chapter, knowledge functions are introduced as ordered, sequenced and structured object operations. When the structure of knowledge macros is firm and established, then the pre-assigned knowledge functions can be executed on simple or complex objects. Direct hardware command or the execution of inter-dependent software routines will accomplish such knowledge functions on inanimate or animate object operands.

REFERENCES

1. S. V. Ahamed, *Intelligent Internet Knowledge Networks*, Wiley-Interscience, Hoboken, NJ, 2007.
2. J. P. Hunter, and M. Thiebaud, *Telecommunications Billing Systems*, McGraw-Hill, New York, 2002.
3. S. V. Ahamed, "Architectural Considerations for Brave New Machines," in *Proceedings of Artificial Neural Network Engineering Conference*, Vol. 15, 2005, pp. 71–80. See also *Journal of Business and Information Technology (JBIT)*, Academy of Business and Information Technology, Indiana, PA., Vol. 6, fall 200, pp. 1–28 2006.
4. S. V. Ahamed, et al., "Sensing and Monitoring the Environment Using Intelligent Networks," U.S. Patent Publication US2003/00441013, U.S. Cl. 706/59, February 27, 2003.
5. *SAP Solutions and eLearning*, SAP (UK) Ltd., Feltham, Middlesex, England, 2002. See also R. Rowley, *PeopleSoft Volume 8, A Look Under the Hood of a Major Software Upgrade*, White Paper, PeopleSoft Knowledge Base, 2005; R. Davis, "Application of Meta Level Language to the Construction, Maintenance, and Use of Large Knowledge Bases," in *Knowledge-Based Systems in Artificial Intelligence*, McGraw-Hill, New York, 1982.
6. K. C. Laudon, *Management Information Systems*, Prentice-Hall, Englewood Cliffs, NJ, 2003. Or, consult any textbook on operations research.
7. W. Feller, *An Introduction to Probability Theory and Applications*, Vol. 2, John Wiley & Sons, Hoboken, NJ, 1971.
8. D. E. Knuth, *The Art of Computer Programming*, Vol. 1–3, Addison-Wesley, Reading, MA, 1998.
9. A. W. Burks, H. H. Goldstine, and J. von Neumann, *Ordinance Department Report*, U.S. Army, Washington, DC, 1946. See also G. Estrin, "the Electronic Computer at the Institute of Advanced Studies," in *Mathematical Tables and Other Aids to Computation*, Vol. 7, IAS, Princeton, NJ, 1953, pp. 108–114.
10. S. V. Ahamed, "Architecture for a Computer System Used for Processing Knowledge," U.S. Patent No. 5,465,316, November 7, 1995. Also see "Knowledge Machine Methods and Apparatus," European Patent No. 146248, U.S./05.11.93, Denmark, France, Great Britain, Italy, December 29, 1994.
11. S. V. Ahamed and V. B. Lawrence, *Intelligent Broadband Multimedia Networks*, Kluwer Academic Publishers, Boston, 1997.
12. H. S. Stone, et al., *Introduction to Computer Architecture*, Computer Science Series, Science Research Associates, New York, 1980. See also J. P. Hayes, *Computer Architecture and Organization*, 2nd ed., McGraw-Hill, New York, 1988.

CHAPTER 8

KNOWLEDGE SYSTEM ARCHITECTURES

CHAPTER SUMMARY

Knowledge-centric objects and their representations have been introduced in Chapter 5. The need for specialized registers, object bus structures, cache and memory modules, and data (object) base managers is introduced and VLSI implementations are discussed in this chapter. Under detailed scrutiny, a knowledge processor unit (KPU) starts to resemble a small computer, and a knowledge machine starts to resemble a small localized VLSI network with object banks, and attribute memories. The typical input system needs full-fledged input access to world knowledge banks, and the output system also requires access to high-capacity data base management systems. These concepts are introduced and then enhanced in this chapter to modify the designs to process medical objects and educational material with complete processing capability. The fundamental differences between computers and knowledge, medical, educational, or simply object-oriented machines are explored in this chapter.

In this chapter, we also introduce the functional requirements of low-level logical units, integer processors, signal processors, digital signal processors, communications processors, etc. The hardware aspects are emphasized such that the *hardware* of more advanced object processors may be derived. The central processor unit (CPU) cycle is transplanted by a knowledge processor unit (KPU) cycle.

Computational Framework for Knowledge. By Syed V. Ahamed
Copyright © 2009 John Wiley & Sons, Inc.

The schema of the simplest register-register (RR) instructions and the corresponding CPU cycles are carried through to "knowledge-based" object register to object register (OROR) instructions, etc. The more complex memory-memory (MM) type of instructions for the CPU environmently by need considerably more sophisticated hardware in the KPU environment. Some of the most complex instruction sets (e.g., for array processing, matrix operations, etc.) require more elaborate sets of coordinated inventions to be implemented in the knowledge domain.

8.1 FROM THE CENTRAL PROCESSOR UNIT TO KNOWLEDGE PROCESSOR UNIT

Adaptation is an essential art of survival. Learning and adaptation are also the essential components of progress. Modern knowledge-based societies demand the highest levels of adaptation for a better life. The passive attitude toward learning and adapting to the Industrial Revolution of the late eighteenth and early nineteenth centuries appears inadequate for dealing with the dynamic shifts that the knowledge society has resulted in over the last few decades. The technological and engineering classwork and textbooks of the mid twentieth century appear insufficient to deal with multiprocessor Pentium machines and fiber-optic highways next generation embedded in the knowledge machines. More than tools and techniques, the adaptation of a new philosophy and attitude may be essential.

To bridge the gap that the knowledge revolution has brought about, we propose a conceptual ladder. This ladder becomes a trail as it is stretched horizontally with knowledge explosion in every direction. This ladder when slanted against human intellect becomes a seven-node graph, with binary data at the first level and human values and ethics at the seventh. In hindsight, movement along the trail and a climb up the ladder became both evident in the six decades following 1945 (von Neumann's machine) and 1948 (transistor). As for foresight, the vastness of the trail ahead and the blue skies above both become even more challenging.

Precise predictions of future innovations are of no avail due to innovations ahead and technological uncertainties. However, conceptual pathways are clear. In order to let progress begin, we traverse the conceptual domain rather than untie inventions and dismantle past technological roadblocks. As was suggested earlier, if the binary bit conveys the status of reality (however fragmented it may be) and if human values and ethics become the goal, then the trail ahead and the climb above have several intellectual stops, however fuzzy they may be.

A conceptual breakthrough is a rarity. The womb of concept is the human mind, and the pristine goal of social betterment brings the highest rewards. However, the human mind can be an equally effective host for the deceptive nuclei of evil. The temptation to embrace negative wisdom can retard as easily as the determination of positive wisdom can propel progress along the knowledge trail. The genesis of negative wisdom is much like the birth of a Trojan horse that can efficiently ride high-speed computer networks. Such a trespass shatters the serene path to longer-lasting axioms of positive wisdom. Since wisdom is authentic and universal, the effects of negative wisdom can only cause disruption rather than alter the course of positive wisdom. Events and incidents since the days of Buddha and Confucius document the movement along a trail of human betterment. In most cases, positive wisdom survives despite deception in society. In some cases, the corruption of society and deception of the masses can last longer than a lifetime, leading to an era of oppression, arrogance, and cruelty.

It is our contention that computers and wisdom machines can automatically sense signs and trends more keenly than humans and thus warn social agents and operators of impending disasters. Such precautions are much like the national security systems that alert a nation to an impending act of terrorism, or like an airline security check that can detect explosives before they are carried onboard. The detection and termination of disruptive forces are better than the cure of their ill-effects, much like early diagnostics and elimination of malignant tissues in a human body is better than the cure of a cancer.

Being able to learn from history, we look back and find that the pursuit of wisdom has been alive and well since the days of Socrates and Aristotle. Inspiration and insight dominated the direction of famous Greek literature that had (and still has some) political and social significance in chosen directions. Human values remained in vogue and held in esteem by intellectuals and political leaders. During the intermediate period, scientists and philosophers occupied that role as human beings of values, integrity, and morals.

In recent decades and with the advent of the knowledge society, the sensing and monitoring of the knowledge and information domain need the processing power of microchips and digital capacity of optical networks. In addition, a new breed of politicians and a new creed of business ethics and morality have both evolved and grown stable like a mature form of social malignancy. Being able to finance and attract scientific talent, ethics and values have eroded to give way to prosperity, and the accumulation of wealth as a symbol of prestige rather than the integrity of humans. Karl Marx, for one, launched bitter complaints about this trend in the materialistic dimension. His efforts to uphold human dignity (that comes from human values) brought about the recognition of labor unions and workers' rights in the corporate world.

8.2 THE PHILOSOPHIC DIMENSION

The total independence of human thought offers intellectual freedom and an opportunity to leap into new domains of knowledge. This mode of behavior offers the thrill of creating breakthroughs, inventing, and innovating as the highest level of mental activity. The joy of being born free also brings social isolation. Sometimes, it brings individualistic arrogance that alienates others from offering joy and comfort at the social, psychological, and personal level. The paradox of being free while still remaining connected, enjoying individual freedom with social dependence, and gift of imagination despite the constraints of reality—all point toward the doors of wisdom, order, and unification (see Chapter 3).

The pinnacles of truth, beauty, and virtue (TVB) lie alongside the trail of constructive knowledge. Ironically, the chaotic cliffs of fall are dug deep in deception, arrogance, and hate (DAH) and lie alongside the trail of destructive knowledge. If an upward triple helix of TVB occurred, then the illusive downward triple helix of DAH lurks only a slip, trip, and fall away. The two spirals come alarmingly close to each other. Unnoticed, human slips (or deliberately implanted bugs and viruses in the modern information age) can undo decades of spiraling climb to attain the goals of an educated society. Pharaohs, Romans, and Incas each have left their own legacies stating as much.'

A knowledge machine (KM) with its knowledge bases (KBs) primed with the discipline and order based on a triad of TVB navigates itself higher, based on the steps of rationality, logic, and mathematics. Conversely, a KM with its KBs primed with DAH finds irrationality, lack of logic, and inequalities everywhere, only to trip and collapse because of the inadequacy of truth, logic, and scientific methodology. Fortunately, rationality, truth, and computation have a firmer and larger union space compared to the space of union between deception, irrationality, and temptation.

Human perception by itself becomes frail and uncertain in an unchartered knowledge domain. Knowledge processing and artificial intelligence in a KM can still offer a good grip on the complexity that surrounds objects and their relationships. Such key objects become the building nodes, and the processes they undergo become the links in the graph of knowledge. Such objects and their inter-relating links constitute knowledge represented as a well-balanced graph. Knowledge suddenly starts to have structure, form, and beauty.

Deep in a human mind, the webs of neurons encompass every new body of knowledge. Deep in the memory of the KM, there are graphs of inter-related objects. Their processes and manipulations give rise to new structures of knowledge. The capacities of the mind and machine are willfully synergized to make a quantum leap into the knowledge domain, much as mathematicians made when symbols and numbers evolved.

Enlightenment has dawned since the days of early human contemplation. In the past, the paths to knowledge, wisdom, and virtue led to an area of convergence of purpose. Such a purpose has a philosophic goal of betterment of humanity and preservation of values. The questions of human knowledge puzzled Aristotle as much as the questions of human bondage puzzled Albert Schweitzer or the questions of enlightenment puzzled Buddha. There is ample documentation from the days of scriptures of the positive politics of nations that set forth the methodologies to realize such sound social and rational notions.

The constraints visible in any snapshot of a time period are the mindset of the population and limitations of the technological feasibility of that era. From the days of Aristotle, knowledge, practical considerations, and productive aspects were plentiful but the technology was primitive. Knowledge,[1] again, was regarded

---

[1] Taken from Aristotle's writings. There are other significant contributions of philosophers from many ages and many cultures. The technical discussions are unaffected by the philosophy of these writers.

as theoretical (as reflected in philosophy, physics, and mathematics), practical (ethics and politics), and productive (arts and rhetoric). Abstractions[2] (ethical, physical, and logical) were abundant (Aristotle's prior analytics and posterior analytics for deductive reasoning, and then his analysis of "expressions" as representing substance, quantity, quality, relation, place, time, position, state, action, and affection), but the global cohesion of concepts was curtailed from a modern perspective.

In the knowledge society, freedom of thought has surpassed the outer limits of conceptual achievements. Numerous breakthroughs in both the technologies and methodologies embedded in computation, communication, command languages, and organizational integration leave no excuses to lie asleep during the information revolution. The convergence of numerous paths (especially those in philosophy and pure sciences), during medieval times has reappeared as the convergence of numerous modern disciplines (especially the humanities, economics, object-oriented programming, knowledge processing, broadband networking, high-speed switching, etc.) in the past few decades.

8.3 ITERATIVE CYCLIC SOCIAL PROCESSES

Society is far more ancient than machines. Social networks existed long before data networks. Instinctual creativity runs deeper in the mind than the recent rash of inventions in the information age. In a sense, the pursuit of novelty without finding its link to the human and social side would appear as shallow as the Internet is in evolution. In this section, we explore the needs that motivate human beings, which are more primitive than the binary sequences that activate computer systems. Two aspects, that is, human affairs and human transactions (see Figure 8.1), become dominant in a social setting.

Social needs (e.g., communication, negotiation, compromise, adaptation, etc.) are more primitive than human transactions represented as data structures and constitute information. Accordingly, in Section 8.3.1, the awareness of and knowledge on human affairs and human transactions are discussed; Sections 8.3.2 to 8.3.5 address these topics in the *social* domain. Their importance in the *scientific* domain in reintroduced in Section 8.4.

8.3.1 Awareness and Knowledge

The monumental advances in computer, communications, and information technologies offer unique opportunities for knowledge and wisdom machines to address social and global problems. Modern machines can process knowledge as easily as computers can process numbers. Fiber-optic networks offer (almost) unlimited bandwidth. Context-related information processing is as easy as a

---

[2]Only to be consistent, we fall back on Aristotle's writings. There are other significant and equally valid contributions from philosophers from many cultures and many civilizations.

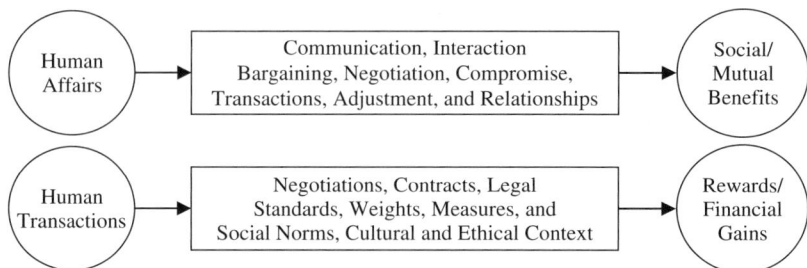

Figure 8.1 Human affairs arising from human needs the result in transactions that lead to the satisfaction of a need. The process repeats itself for each outstanding need.

UNIX-based search. With such strides in technology, information glut almost drowns an ill-prepared mind. The Internet challenges the freedom of the human soul to remain free and soar above ignorance. Without such machines, the human mind, destined to reign free and leap over the mountains of apathy, becomes trapped in trivia. It becomes a paradox that the information society which offers freedom also corrupts the senses from being socially responsible and astute. It robs the mind of global harmony and wisdom of peace. The bias inherent in the communications media that disseminates information become insidious within it.

In the industrial societies that drive the modern world, profit is the dominant motive. That motive is well camouflaged and almost invisible to lazy eyes. Social viruses such as deception in advertising, misrepresentation of facts, and willful distortion of truth are taken casually in most societies. The practice of swaying public opinion based on false advertisement and politically motivated promises challenges the foundations of honesty and integrity for the entire society. In an urgent sense, the tolerance of such practices expresses a willingness to slide down into greater deception. The defense against such a downward spiral lies in education and the reinforcement of ethical conduct in business, education, the government, and the community.

On the one hand, the truth, virtue, and beauty embedded in information and knowledge enable the mind to grow and migrate toward peace, social justice, and wisdom. Yet, on the other, the deception, arrogance, and hate that lie hidden in information and media clips cause the mind to spiral deep down into the prehistoric caves of self-preservation. Such attitudes and accompanying behavior are common in individuals, corporations, cultures, and nations. The reactions to perceived threats become real in well-armed nations that initiate war amidst a mist of fear without a drop of threat.

The role of the Internet in bringing information and knowledge to every member of an educated society has increased rapidly over the last decade. During the past few years, the four directions of human achievement have been the Internet, machine intelligence, knowledge, processing and global networks. These dimensions appear unrelated at first glance, but it soon are becomes clear that they are indeed inter-woven: They blend into each other in a synergistic and harmonious way without boundaries. The four-dimensional space and our way of thinking are unified freely, encompassing the Internet, intelligence, knowledge, and networks.

At the pinnacle of the sciences resides natural intelligence ready to tackle any problem of which the mind can conceive. At the intermediate level, the role of artificial intelligence (AI) becomes dominant in carrying out slightly higher-level tasks, blending machine intelligence with previously programmed human intelligence. At the lowest rung resides the encoded machine intelligence that is cast in silicon and Pentium and that carries out mundane tasks synchronized to the cycles of a cesium clock.

At the next pinnacle of human interaction resides the Internet, ready to interface with any Website in its validated TCP/IP format. Intelligence and the Internet are fused into next-generation internets that still lurk as of 2009. At the intermediate level, all digital networks are well synchronized and properly interfaced to accept information in an orderly and hierarchical format and to take on the optical paths or the ether space. At the lowest rung of networks reside analogue networks, cast in copper, space, and any conceivable media in an unsynchronized and imprecise fashion just to get across information.

At the third pinnacle of human endeavor reside values, ethics, and social benevolence. Here, the process of unification starts to have significance. the Internet, Intelligence, knowledge, and networking commingle to produce the composite fabric of human bonding and long-lasting human values. At the lowest rung of knowledge are binary bits, somewhat like the raw cells in a body. Bits soon blend to become bytes and data blocks, data become information, information turns into knowledge, knowledge gives rise to concepts, concepts lead to wisdom, and wisdom offers values and ethics. The migration along the knowledge trail (Figure 2.5a) is evident in almost all civilized societies.

At the final stage of human involvement, computer applications exist to serve human beings; but such machines have their limitations. If the traditional seven-layer Open System Interconnect (OSI) model is downsized to three layers and then topped with four additional layers (global concern, human ethics, national environment, and economic/financial basis), then the new seven-layer network model would encompass the networks as they exist now and include human values in their makeup. The convergence of the four dimensions (the Internet, intelligence, knowledge, and networks) starts to become evident at the top. Currently, at the application layer (AL), the OSI model falls short of satisfying any human needs. At the lowest rung is the physical layer (PL) with six higher layers to new constitute the OSI model.

In order to bring about synergy at the highest levels of human thought and to move ahead in the social dimension, the current technology, enhanced awareness of the environment, and the search for knowledge and wisdom become essential. A judicious choice toward integrity and ethics becomes equally important. Human beings retain mastery over machines and machines are highly stylized but mindless slaves. Humans choose the trails for the future, and machines simply scour the path.

8.3.2 Communications and Interactions

Communication enables societies to function smoothly and coherently. Amidst the innumerable modes of communication, the multimedia digital mode stands out as the most prominent in recent history. It is robust, secure, and media-independent. Human beings have preferred the simultaneous duplex mode since the beginning of civilizations. Interactivity is thus inherent.

In most cases, communication originates from need and human needs can be highly variable. The flexibility built into the structure of a language makes it versatile, able to communicate effectively with anyone, at any time, in any place, in response to any need. Most languages have rules of syntax and some variability based on semantics. The basic module of communication is the sentence that has three absolute requirements: one or more nouns (pronouns included), one or more (explicit or implicit) verbs, and tense (which brings in the time factor). Sentence types and structure can offer enormous flexibility in communication.

Communication has numerous variations, ranging from human–human to computer–computer communication. Almost all forms of effective communication involve interactions with a series of activations and responses. Human–computer processing almost always implies input and output processors. Graphical user interfaces provide easy access to novice users to deploy complex application programs.

The use of signal processors in providing an interfaces, processing, and feedback control signals has been in vogue since the 1960s. Small systems can have the flexibility of providing digital control and field programmability through the use of field-programmable arrays (FPAs). Such FPAs may serve a large number of applications, ranging from security systems to equalizers. Typically, application requirements and cost become the constraints in designing the capability, nature of interaction and its extent between any parties involved. The structure and format needed for communication to deal with most applications and media have evolved sufficiently to provide easy and error-free access to visual, auditory, and multimedia types of applications.

In dealing with human and machine interactions, humans drive the process and the machine responds according to the programmed modes of response. Intelligent machines would act in an adaptive fashion and yet not be creative. To this extent, the machine will only respond but not behave. Software designers still lack the skills to program based on behavioral intelligence rather than artificial intelligence.

In dealing with human interactions, the dynamic response can be highly variable and usually becomes complex. The nature of the transactions and attitude of the two parties cause uncertainties in the direction and result of the process. We addressed human–human communications in Chapter 5 and imported the concept of intelligent content processing in Appendices A and B there.

8.3.3 Bargaining and Negotiation

Bargaining and negotiation involve uncertainty and predict the behavior of the human parties involved. When behavioral patterns are not programmable, the maximization of expected utility under uncertainty becomes an economic solution. Generally, economic solutions do not benefit from the influence of human intuition, learning, and adaptability based on human expression and emotion. For this reason, most of the bargaining and negotiations are done by human beings who have the capacity to sense, feel, and compute.

With total rationality under conditions of exact knowledge about the costs and benefits is involved, bargaining and negotiations are unnecessary. When the costs and benefits are not known, estimation plays a role. When extraneous factors are involved, the probability of future events plays a role. When the parties involved are intelligent and responsive (i.e., learning, remembering, and predictive with individual profiles), then the behavioral mode in invoked. A very rough estimation with a sense of doubt results in a guessing game rather than economic analysis. In Appendix B and Section B.3 of Chapter 5, some of the factors dealing with machine-aided interactions are presented as a flow diagram (see Figure 5.B.5) rather than a mathematical derivation.

In dealing with goal-directed interactions [such as labor–management negotiations (Section 5.B.2 of Chapter 5) and other cooperative interactions], two steps are generally used. The *first* step consists of offering concessions with changes in the demands made, and the *second* step consists of accommodating (e.g., substituting, postponing, foregoing, etc.) the demands in consideration of the concessions made by the other party. Bargaining and negotiation also invoke creativity from the parties. In Section 5.B.3, an economic basis for corporate/national negotiations is presented.

8.3.4 Compromise During Transactions

Adaptability and compromise generally precede agreements and transactions. A framework for computing the cost of agreement or cost of disagreement in relation to the demand intensities of the two or more parties involved is presented in Sections B.2 and B.3 in Appendix B of Chapter 5. These considerations become important as transactions come to a close and result in an agreement, partial agreement, postponement, or disagreement. In labor–management negotiations (see Figure 5.B.3), these three situations are classified as "contract, cooling off, or lockout/strike." An economic/mathematical approach for navigating through the behavioral aspects of negotiations is presented in Sections B.2.5–B.2.7 and again B.3.3 at the end of Chapter 5.

8.3.5 Adjustment and Relationships

Developing an entirely mathematical basis for the adjustment and prediction of relationships becomes a monumental task. It can vary significantly from one set of entities to the next and becomes situational. Computers can function in a very narrow sense and supply an economic, a financial, or a social framework. The human side of adjustment can be partially reinforced by knowledge machines (see of Appendix B of Chapter 5) that deal with rational and intelligent objects which are of the same importance to human beings as numbers and ratios are to accountants.

The emergence of the knowledge age has altered the dimensions of human intellect and behavior. If human beings are labeled by their knowledge, a quotient is developed that bears the norm of an informed individual who knows every relevant piece of knowledge which a totally networked society may provide. A totally informed individual will be a knowledge genius even though the traditional IQ test may label that individual as less than an intellectual genius. In a sense, the machine equivalent of such a human would be a computer with complete memory capacity, memory access in an entirely networked machine. The behavioral intelligence (BI) of humans should be considered an extended form of intelligence when behavior is inferred based on patterns of behavior accessed from high-speed networks and processed by knowledge machines.

8.4 THE SCIENTIFIC DIMENSION

The sciences have inculcated a sense of truth in society. Just as numbers do not lie, the laws of arithmetic and algebra do not flounder. However, processors can have errors and programs can have bugs. When human intuition is clad in concepts and the senses wear a robe of sanity, the scientific dimension extends everywhere in the domain of knowledge. The synergy between mind and machine opens up new vistas in the computability of (or the entropy of) knowledge. Individual estimates of entropy in any body of knowledge (BOK) can vary drastically based on the individual perception and prior knowledge that offer a common array of noun objects (NOs) and verb function (VFs), but the knowledge banks that feed knowledge machines (KMs) are established from authoritative sources and agencies. For this reason, the computation of entropy in a BOK by a KM is likely to be accurate and unbiased.

8.4.1 The Search for Precision and Infallibility

Processors in computers have allowed human beings to be scientific, logical, and precise. Scientific rigor is enforced by programming sequences, steps, and methodology. Inference capability is embedded in program steps. Overall validity and inter-relationships between the steps in the program, and precision are achieved by making the numerical processes accurate by enhancing the binary bits to represent numbers. Overall precision can be pursued to a certain extent in the computational environment. In the overall process of finding a solution, machines are fallible at any of the long chain of processes before the solution is computed and then optimized.

In the computational domain, absolute precision and total infallibility are both ideal pursuits. The limit of precision is generally limited by the number of bits in each word and the words could be concatenated. Double and extended precision machines have been built or their compilers are generally available. A stage of economic compromise is reached until the effort to gain additional accuracy does not justify the reward of harnessing the added accuracy. For a vast number of scientific applications, the traditional 32 or 64 bits machines will suffice. The 128- or 256-bit machines are hardly needed even for specialized applications.

Traditional processors execute machine language instructions. Fallibility generally derives from within the firmware or the executable code that drives the hardware. Hardware connectivity can also lead to erroneous results. The hardware logic simulation programs usually verify and validate the numeric code and logic embedded in VLSI chips. Over the last few decades, programmers have systematically isolated and effectively debugged the instruction sets, privileged-instructions, firmware, code, macros, and libraries that are the inherent attire for the basic hardware of any machine. Such layers of code have been cleansed of any bugs underlying the code used in most technical and scientific applications. When machines and codes are developed for special applications (such as space travel, weather tracking, human factors engineering, etc.), then thoroughness in ascertaining hardware functionality and software testing become essential.

8.4.2 Machine Accuracy and Human Adaptability

In the context of this chapter, when a machine and human being search for solutions to complex problems, absolute infallibility is an illusive goal, with a qualified pursuit of the most accurate solution a realistic objective. However, machines process scientific relations, logical inferences and numerical computations at high speeds and humans validate the results wisely most of the time. Together, machine accuracy and human adaptability reach a working balance that offers quick and satisfactory results. In this chapter, a wide variety of balanced solutions are investigated in medical applications and object environments, and knowledge domains are sought.

The processing environments for central processor units (CPUs), digital signal processor units (DSPs), medical processor units (MPUs), object processor units (OPUs), and knowledge processor units (KPUs) are discussed in Sections 8.5.1–8.5.5. All these environments have definite limits of scientific rigor, logical accuracy, and numerical precision. These limitations become evident as the processors are studied in detail.

8.5 VARIETIES OF PROCESSORS AND MACHINES

Computers and IT platforms have propelled social progress immensely over the last few decades. The expansion has been almost explosive in the data, information, and knowledge domains. The progress beyond knowledge towards concepts, wisdom, and ethics has been slow at best. The absence of financial incentives in reaching the final three stages has made the human effort scarcely concept-, wisdom-, and ethics-oriented. In Figure 8.2, a global overview of the migratory path toward the lofty goals of values and ethics is depicted. In the current social setting, the mental constriction appears to occur at the knowledge node because knowledge machines are not plentiful as computers that process data and information.

In Figure 8.2, the developmental hierarchy of machines is depicted form left to right. Knowledge, concept, and wisdom machines are in the evolutionary stages, even though the Internet and Web knowledge bases are well established. There are a profusion of processors in the market. Processor glut is real since it is possible to build a specific processor for every function and every subfunction. Designs and architectures abound as well. Ranging from a simple logic circuit to complex array processors, their cost, complexity, and speed can vary by many orders of magnitude.

Largely, it is not economical to design a customized processor for every small function, but instead to deploy mass-produced general-purpose processor chips. With some code implanted in these general-purpose chips, most of the application and industrial requirement can be met.

The current technology can be used to integrate the hardware/software/firmware for object machines (OMs), KMs, concept machines (CoMs), and wisdom machines (WMs). These new machines are discussed in Sections 8.5.7 to 8.5.10. The most unpredictable link in the chain is the role and attitudes of the next generation of human beings and their predisposition to truth, virtue, and beauty (TVB) in a positive sense. Conversely, the human predisposition to deception, arrogance and hate (DAH) in a negative sense is also equally unpredictable.

In the rest of this section, we present the most common processors that may be modified to serve as knowledge processor units. Generic architectures (Section 8.5.1) already exist for central processor units, communications processors, input/output processors, graphics processor units, array processors, digital processor units, etc.

8.5.1 Central Processor Units and Generic Computers

Central processor units (CPUs) have been the hardware for numeric, arithmetic, and logic functions since the von Neumann machine [1] in 1945. The architecture of the (Institute of Advanced Study (IAs) machine [2] was the blueprint for many more machines that were built in the decade following the IAS machine's favorable reception by computer manufacturers. The earlier CPU architectures have been described in detail [3].

The organization of the primitive von Neumann machine with three basic elements—a CPU with a data-processing unit (DPU) and program control unit (PCU), main memory (M), and an array of input/output (I/O) devices—is shown in Figure 8.3. This is perhaps the earliest and simplest design for a digital computer.

A second-generation machine with additional I/O devices and a more elaborate CPU is shown in Figure 8.4. Index registers, input-output processors IOPs,

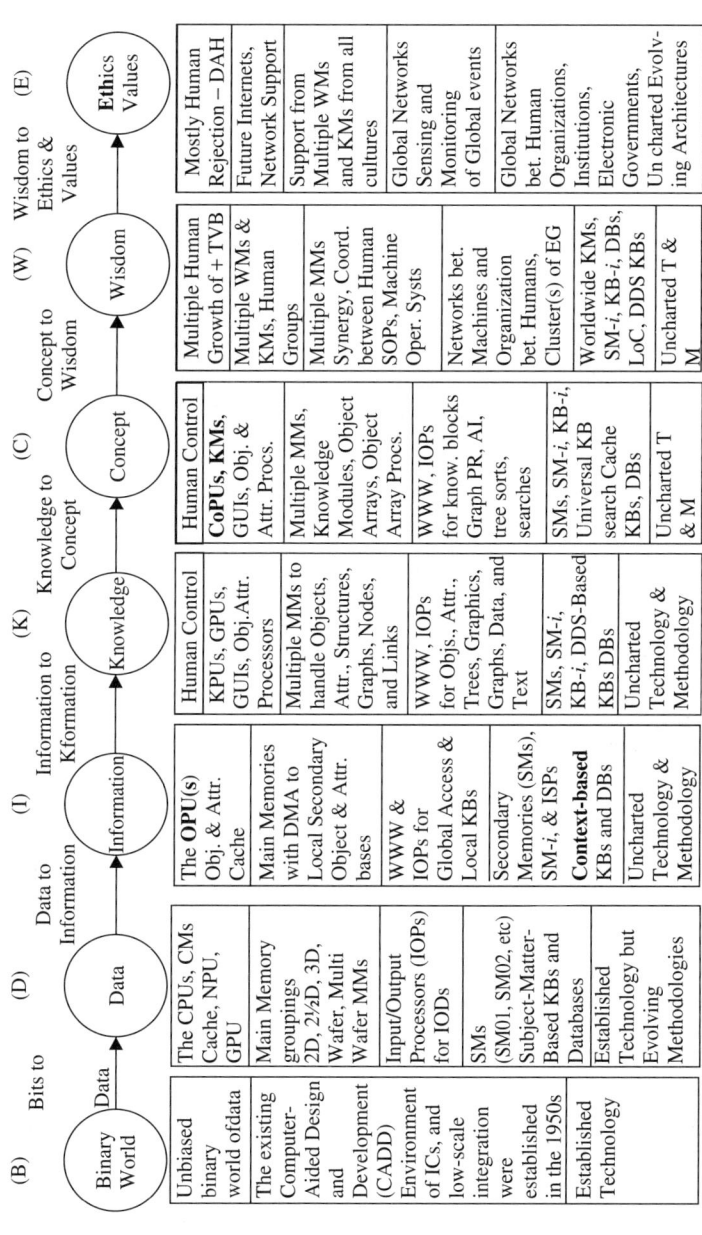

Figure 8.2 Evolutionary path from binary data to human ethics and values. The algorithms, methodologies, devices, and architectures on the right-hand side of the figure remain uncharted.

367

368 KNOWLEDGE SYSTEM ARCHITECTURES

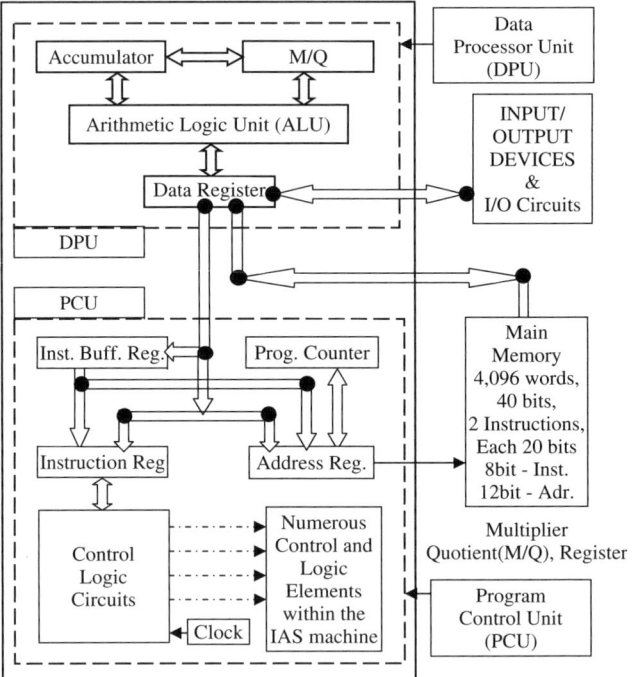

Figure 8.3 A simplified architecture of the first-generation IAS machine representing its basic components. The features of stored program control can be implemented in this configuration of the central processor unit (with a program control unit and data-processing unit), main memory, and input/output circuits and devices.

and memory control units MCUs (see the figure) were some of the added features of the disk, and drum storage devices increased storage and access capacity immensely. Together with the increased processing power of the CPU, the second-generation machine facilitated these types of machines through the late 1950s and early 1960s.

Larger machines were also introduced after the 7090 and 7094 machines. The 360 (1964) and 370 series (1970) and other series were introduced as late as 1985. Compatibility with these machines had almost become a requirement for most computer manufacturers around the world. The typical configuration of a third-generation machine is shown in Figure 8.5.

The switching function although inherent in all earlier machines was accomplished by sets of independent multiplexers and a device address register within them. The switch was not yet identified as a major component of the computer system. The bus structures were also considered independently in the overall architecture.

CPU speed and capacity have increased dramatically (almost exponentially) over the last few decades. Size and power dissipation have also gone down

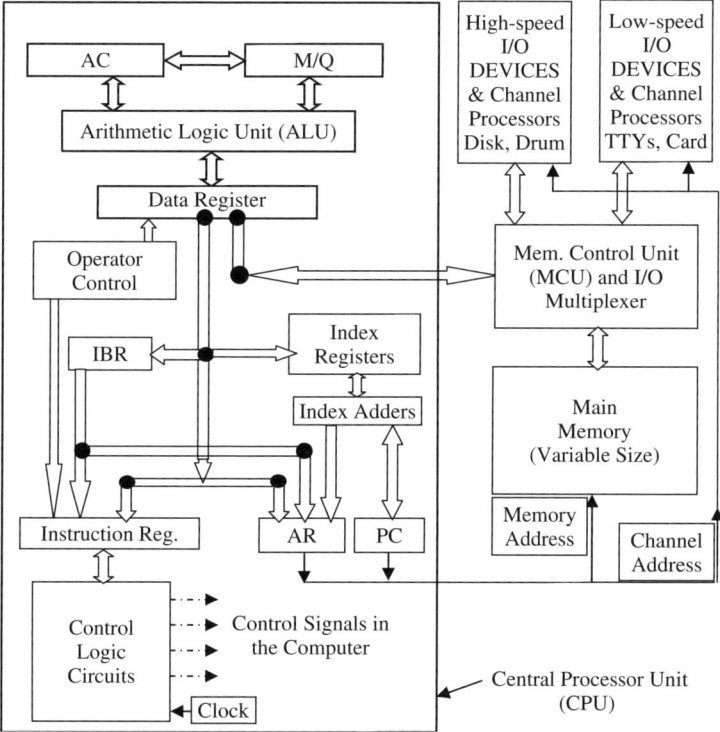

Figure 8.4 Simplified architecture of the second-generation machines of the late 1960s. The most successful of these machines were the 7090 and 7094. The 7094 arithmetic units could handle floating point and exponential numbers with mantissa and exponents in the range of −128 to +128 with 36-bit words.

dramatically. The clock speed of the Intel Pentium 4 670 CPU is 3.80 GHz and comes with 2 MB of L2 cache. The quad bus has a frequency of 800 MHz. The production of the device deploys 90-nm technology. An extended memory of 64 bits provides technology support. Possible applications include business, gaming, multimedia, or even Web browsing.

The hyper-threading technology permits multiple threads of instruction to pass through the processor independently (well almost). Multitasking and multi-threaded software applications are feasible. The processor also supports 4 GB of memory with a 64-bit architecture for 64-bit operating systems and applications. Security is also built in at the hardware level, which essentially prevents damage from malicious programs including Internet worms. The processor disables itself to prevent damage to the machine and from memory overflow conditions.

In more recent configurations, the switch and communications systems within the computer and in computer networks are treated in addition as essential components; They are generally designed and customized to suit the applications and environments. Conceptually, the layout of a modern computer with the elements

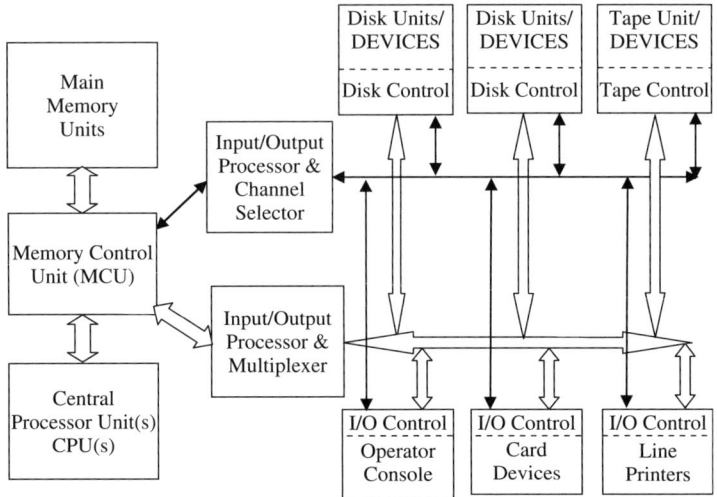

Figure 8.5 Architecture of the third-generation machines of the 1970s and 1980s. The central processor unit structure contained three registers (AR, IR, and DR), a program status word (with a built-in program counter), two sets of registers (16- to 32-bit general-purpose registers and 4- to 64-bit FP registers), three processor units (fixed point, floating point, and decimal arithmetic), and a memory control unit. For internal communication, a shared bus arrangement was also provided. Immense programming flexibility was thus a feature of the third-generation machines with five different types of instruction formats (RR, RX, RS, SI, and SS).

(CPU, main and secondary memories, I/O subsystem, buses, and an internal switch) is shown in Figure 8.6a. Most modern computer systems can be depicted by the five components. All the recent variations in CPU, memory, I/O devices, switch, and bus structures can be also accommodated. Various memory units, and direct memory access hardware units for high-speed data exchange between main memory units and large secondary memories are also depicted.

Over the last few decades, numerous CPU architectures (single-bus, multibus, hardwired, micro- and nanoprogrammed, multiple control memory-based systems) have come and gone. Some microprogrammable and reduced instruction set computer (RISC) architectures still exist. Efficient and optimal performance from the CPUs also requires combined SISD, SIMD, MISD, MIMD, and/or pipeline architectures. Combined CPU designs can use different clusters of architecture for their subfunctions. Some formats (e.g., array processors, matrix manipulators, etc.) are in active use. Two concepts that have survived many generations of CPUs are (1) the algebra of functions (i.e., *opc*) that is well delineated, accepted, and documented and (2) the operands that undergo dynamic changes as the *opcode* is executed in the CPU(s).

The ensuing conventional computer architectures with multiple processors, multiple memory units, secondary memories, I/O processors, and sophisticated operating systems depend on the efficacy and optimality of the CPU functions.

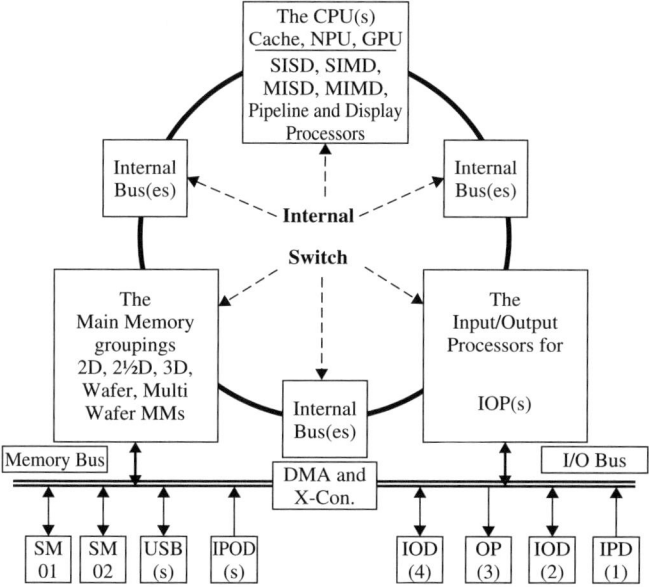

Figure 8.6a Typical architecture of a modern computer with five elements: central processor unit, main and secondary memories, input/output subsystem, buses, and an internal switch.

The CPU bears the brunt of action and of executing the operation code (*opc*) for mainstream programs. Distributing the traditional CPU functions to numerous subservient processors made the overall processing faster and more efficient.

The concept was projected to the storage and memory functions. Numerous memory architectures ($2\frac{1}{2}$-D, 3-D, wafer level, stick memories) were developed to make the storage capacity, efficiency, and speed suitable for Pentium-based processors. The secondary memory technology has kept pace by providing extremely fast store/retrieve instruction sets for the users of massive databases. Content-based (main and secondary) memories have yet to arrive at the common marketplace whose cost justifies the benefit they bring.

The I/O interface has seen explosive growth, from massive parallelism to a high-speed USB interface. The use of optical interfaces to transfer gigabytes of data in and out of computer systems is already on the horizon. Perhaps the economic incentives for deploying plastic optical fibers [4] for very-high-speed data will accelerate the progress of computer systems, transforming them into very-high-speed switches for local (and LAN) data transfer.

Three underlying innovations (still to come) will help streamline the computer and communication functions. The theme of each of the three innovations is addressed as follows:

1. The enhanced IP address space necessary to directly access global knowledge bases (KBs) from within the computer system enables it to gain access

372 KNOWLEDGE SYSTEM ARCHITECTURES

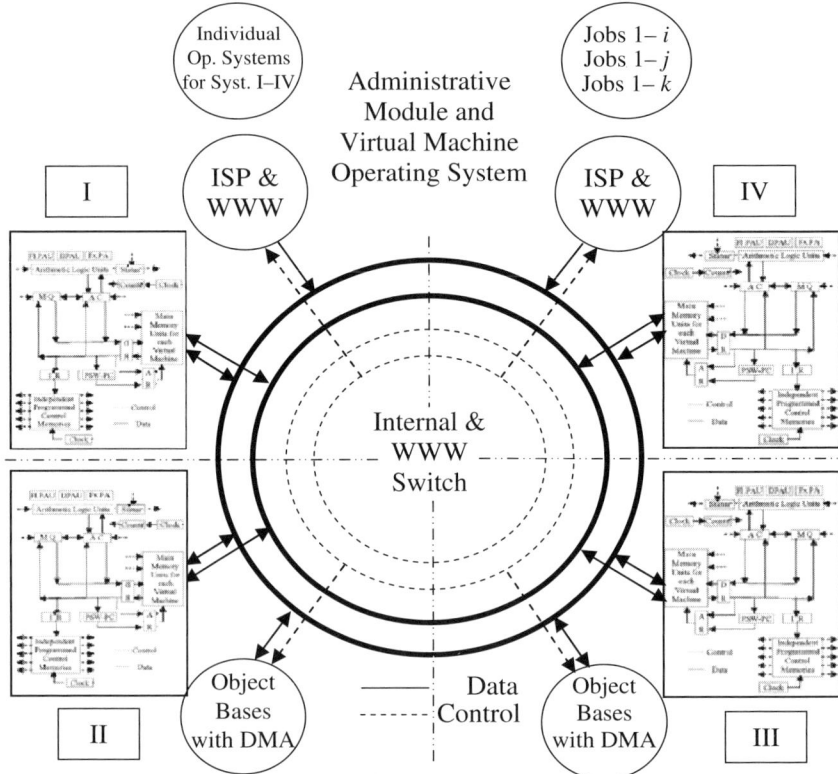

Figure 8.6b An architectural layout of a cluster of four computers (I–IV) based on conventional technology. Machines of this nature have been in existence since the mid-1980s. Such architectures can be tailored to specific applications, such as weather tracking, deep space, and genome pattern explorations. The job queues refer to computational jobs waiting to be executed by the machine cluster. Jobs specific to any particular computer (I–IV) are assigned by the main operating system of the cluster computer, and the machine tasks specific to each of the four machines are governed by their individual operating systems. The cluster data exchange takes place via the dual bus (concentric dark circles) and the control information flows via the dual-bus arrangement (concentric dashed circles). DMA = direct memory access, ISP = Internet service provider(s), WWW = World Wide Web or the Internet, AC = accumulator, MQ = multiply/quotient register, AR = address register, DR = data register, IR = instruction register, PSW = program status word used for multitasking, PC = program counter, FPAU = floating point arithmetic unit, DPAU = decimal point arithmetic unit, and Fx.PU = fixed point arithmetic unit.

to the Internet (TCP/IP) address space. A typical configuration is shown in Figure 8.6b.
2. The need to address the KBs is based on the contents of the information under process in the CPUs or KPUs with the Dewey Decimal System (DDS) or Library of Congress (LoC) classification. Such a classification will unify

and standardize content management systems around the world. The philosophy of standardizing intelligent network (IN) functions has provided developed nations with a global platform to provide IN services uniformly around the world.

3. CPU functions are enhanced to perform generic object-oriented functions at the hardware level to execute knowledge-oriented (economics [5], medical [6], education [7], electronic government [8], and financial [9]) instructions at giga instructions per second.

These three impending innovations are based on the limiting capacities of the central, knowledge, medical, educational, and electronic government processor units (the CPU, KPU, MPU, EPU, EG-PU, etc.) to address global knowledge bases and should unravel the barriers between very-high-capacity computers and the massive switches of future global systems.

In knowledge-processing systems the operands can be intelligent, process-dependent, assistive, neutral, or resistive to change. Their response can also be linear or nonlinear. When the relationship between the operation code and operands is dynamic, the operands can retain a memory/hystersis effect, thus affecting their future response. To maintain accuracy and dependability, the KM maintains a profile of the (intelligent) objects. The KPU architectures become increasing complex and they are explored further in Section 8.5.9.

8.5.2 Communications Processor and Electronic Switching Systems

The need for expanded and dependable telephone communications systems during the 1950s and 1960s mandated the need for error-free real-time switching. Typically, in plain old telephone service (POTS), with circuit-switched digital capability, digitized voice channels at 64 kilobits per second are actively allocated by electronic switching systems as they interconnect millions of telephone customers. The machines that provided enhanced capacity and faster switching in the 1960s were based on (integrated circuit) architectures and developed specifically for electronic switching systems (ESSs). The development of such machines was initiated by Bell Systems during the 1960s.

8.5.2.1 No. 1–3 Electronic Switching Systems in the United States

The No. 1 ESS is built much like a massive parallel-processor-based computer system. In 1965 the first software-driven (stored-program-controlled or SPC) central office was commercially introduced as the No. 1 ESS™ switch in Succasunna, New Jersey. The hardware for this central office was not entirely software-controlled. Numerous subsystems (such as a line scanner, signal distributor, and central pulse distributor) are combined into a basic hardware unit to interface with the line and trunk circuits and perform local functions. They are monitored by their own program control. The switching network inter-connects different segments of the communications path depending on need and availability. This switch was primarily electromechanical in nature. Because of software

374 KNOWLEDGE SYSTEM ARCHITECTURES

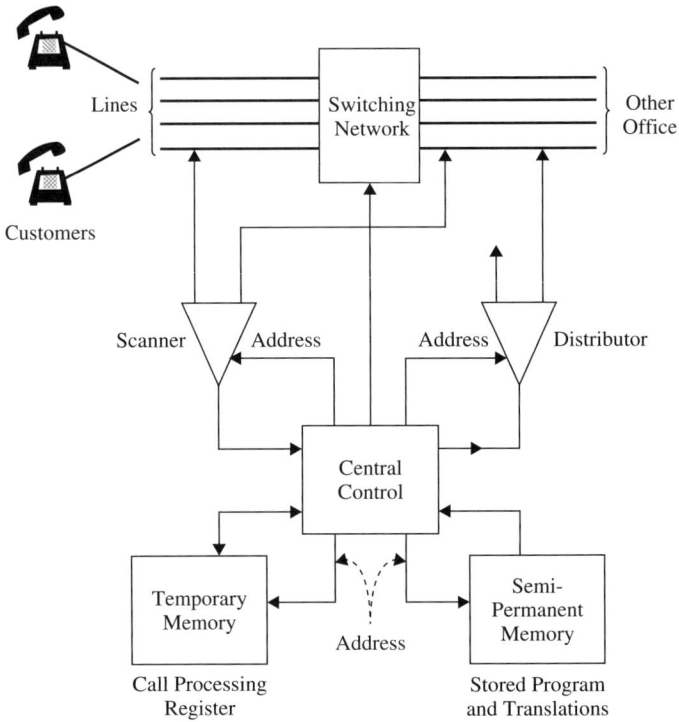

Figure 8.7a A Typical architecture of an earlier version of the No. 1 electronic switching system (No. 1 ESS).

considerations, the No. 1 ESS switch-based central office (see Figure 8.7a) was still designed specifically to handle communications functions in real time, with the dependability (about 2 hours of downtime) and life span (40 years) of typical central offices. The usual switching element was the ferreed switch to make or break a single connection at a time under high-speed centralized control. The centralized control for the No. 1 ESS is driven by one group of programs (about 90 different programs) with about 100,000 words. With the growth and number of specific services demanded by the No. 1 ESS, program size increased rapidly to 250,000 words of code. A program word for the No. 1 ESS is 44 bits (37 data bits and 7 bits for the single-error-correcting Hamming code and parity). The program storage size is 131,000 words, and each generic central office can have as many as 12 program stores for handling as many as 65,000 lines in parallel. The cycle time is 5.5 μs.

Other major developments that rapidly followed the introduction of No. 1 ESS architecture to switching systems included the No. 1A ESS switch. Integrated circuits and faster clock rates offered greater call-handling capacity. This development is similar to what happened when second-generation computer systems (see Section 8.5.1 and Figure 8.4) were introduced in the late 1950s and early 1960s. The translation of operational code and software code with the No. 1 ESS permited growth and expansion in the telecommunications industry that is typical of the late 1960s and 1970s.

The No. 1A ESS is capable of serving up to 128,000 lines and 32,000 trunks, with a total capacity of 10,000 erlangs. This number translates to 36 million (3, 600 × 10,000) call seconds per hour, since each erlang is equivalent to 3,600 call seconds per hour (centi calls per second or ccs). On average, 1 erlang (36 ccs) is a sufficient traffic load to keep one trunk busy for 1 hour because there are only 3,600 seconds in an hour.

In a sense, the No. 1 ESS initiated the impact of computerized control on local switching—something that computers had initiated in the data-processing field but 21 years earlier. Numerous other central offices (No. 2 ESS, No. 2B ESS, and No. 3 ESS switches) were also introduced during the late 1960s and early 1970s, but the impact of these switches was not as dramatic since the systems addressed the growth and speed aspects of switching systems. The microprogrammed control of the processor (a feature of third-generation computer systems) was introduced to the switching system in the No. 3A ESS switch with faster logic and automatic code conversion than the No. 2 ESS switch. Most of these switches provided sound economic justification for rapid expansion, and they provided a foundation for the eventual development of networked systems.

8.5.2.2 No. 4 and 5 Electronic Switching Systems in the United States

Switching in Toll Environments The electronic translator, consisting of the SPC processor No. 1, was introduced within the 4A Crossbar Toll System in Grand Rapids, Michigan, in 1969. This toll system was the first to include some of the same functions and features of computerized systems. For example, the system had a centralized message accounting system (CAMA) for local and toll billing. This system also included the peripheral bus computer (PBC) to interface with the 4A/Electronic Translator System (ETS) that monitored system performance in service. However, a major enhancement to the system included the ability to accommodate the common channel interoffice signaling (CCIS) between processor-equipped switching systems. With the introduction of the full-fledged 4A/ETS, the telephone system transformed itself into a computerized network for telephone systems with limited network capabilities but with great potential to grow.

No. 4 Electronic Switching System The next major step in toll switching was the No. 4 ESS toll-switching system. This machine was a logical (and managerial) extension of the No. 1 ESS switching system, it reduced fixed costs and running expense, and provide flexibility with the changing socioeconomic climate of the early 1970s. Processing capabilities for the toll, more sophisticated toll network resource and channel allocation, and pulse code modulation (PCM) transmission and switching capabilities were also included in the No. 4 ESS switch. The first No. 4 ESS switch was capable of handling about half a million calls per hour, with a trunk capacity of over 100,000 calls per hour. This switch became a benchmark for the capacity and reliability of trunk-switching systems.

To a degree, the No. 4 ESS switch, shown in Figure 8.7b, initiated the impact of computerized control on toll switching—what computers had initiated in the data-processing field 29 years earlier. This trend in the communications industry brought all the rewards to the digital processing and handling of telephonic information in the early stages. More recently, this same trend has evolved into networks that make any relevant information accessible to anyone authorized to receive it, at *any time*, at *any place*, and with *any degree of accuracy and resolution* within the framework of security and the legality of exchange of information.

As a telecommunications switching facility, the recent No. 4 ESS switch completes as many as 800,000 calls per hour as call attempts reach about 1,200,000 during peak calling periods. More recently, similar strides have been made in retaining compatibility between the No. 4 ESS switch and more modern communications systems and networks. Typically, the existing No. 4 ESS switch is capable of interfacing with the SS7 via its common network interface and the ISDN

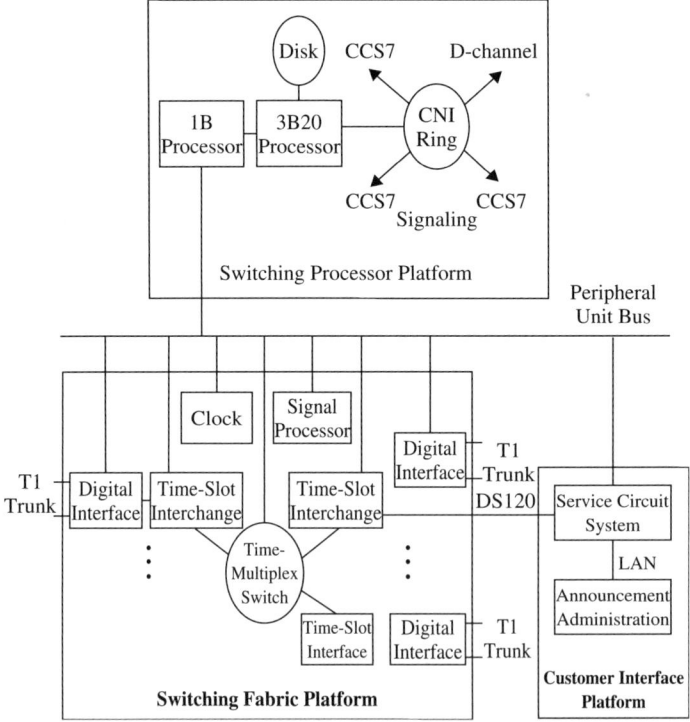

Figure 8.7b Architecture of the 1994 No. 4 ESS switch. LAN = local area network; IB processor = 1994 version of call processor for the No. 4 ESS switch; 3B20 processor = complementary process for overall call control; CNI = common network interface.

implementation with Consultative Committee of International Telephone & Telegraph CCITT (now International Telecommunications union ITU-T) Q.931 standards. The switching platform functions with a spatially separated time-multiplex switch capable of inter-connecting numerous time-slot inter-change units. "B" (i.e., the DS0 at 64 kbps) channel integrity is retained in this switch. Total synchronization between the switching clock and network clock is retained throughout the No.4 ESS switch. In its current design, the No. 4 ESS switch can also interface with DS3 carrier systems initially and then with the synechronous optinal network SONET-based transmission facilities.

The introduction of the new processor (1B) to replace the original processor (1A) in the No. 4 ESS switch has contributed to its increased call-handling capacity. The expanded volume that the switches handle is expected to increase at an approximate rate of 5% every year. The demand for an 800 number, cellular number, fax number, and video messaging services is likely to remain high. Rather than replacing the switch with the newer No. 5 ESS switch, a more economical and feasible alternative is to replace the processor. For this reason, the new processor is designed to function at a call-handling capacity 2.4 times greater than that of the original processor. The speed and memory access are more than doubled. Reliability is enhanced, and operations and maintenance are streamlined. The overall switch capacity (the No. 4 ESS) with the new processor is expected to be over 1 million calls an hour—a macro Pentium chip for a massive PC.

No. 5 Electronic Switching System   The introduction of the first single-module No. 5 ESS switch in 1982 (in Seneca, Illinois), and then the multimodule No. 5 ESS switch in 1983 (again in Sugar Grove, Illinois), started another major trend in central office capabilities. In essence, the integrated architecture and time division digital switching that are inherent in this machine provide for most of the distributed processing and multifunction capabilities of the expanding telecommunications market. This machine offers seven distinct features:

1. A single system can serve all operator, local, toll, and international calling.
2. It provides for digital switching between numerous subscriber loop-carrier systems and inter-exchange carrier circuits (the T or E family), making the transition to other digital networks easy and elegant. It also interfaces with the analogue inter-exchange systems.
3. It has a modular approach in hardware and software design, which permits adding newer devices as the technology evolves and which allows it to interface with older-generation switching systems (e.g., the No. 4 ESS switch or older No. 5 ESS systems).
4. Its hardware modularity permits gradual growth as the network and service demands increase. This permits a cost-effective deployment of central office resources.
5. Being mostly digital in its devices and control, it has a reliability comparable to, if not better than, most computer systems. Extra error correction

stages and parallel functionality of the CPUs permit the quick and dependable retractability of any call-processing step(s).

6. The computer orientation of the switch design permits local and centralized maintenance by running diagnostics in any module, combined modules, extended subsystems, or global systems. The operations, engineering, and maintenance (OE&M) functions become much simpler.

7. The No. 5 ESS operating system, being modular and encoded in a higher-level language (the C language), permits central office capabilities to be added easily. The system is portable and modular, and may be customized to particular central office needs. New digital services (such as integrated services digital network ISDN) and new intelligent network services within the service switching points (SSPs) can be accommodated by altering flow control within the call-processing environment; such network services can include the detection of the trigger condition, database lookup in a network information database (NID), invocation of an intelligent peripheral (IP), or exercise of a service logic program (SLP).

Numerous other digital switches also exist. In Europe, for example, Siemens' EWSD® and Ericsson's AXE® systems perform comparable functions. Equivalent systems in Japan have also been built and function with similar precision and dependability.

The No. 5 ESS-2000 switch (see Figure 8.8), in the American environment, contends with both ISDN and IN functionality. In addition, this newer No. 5 ESS platform permits enhanced processing capabilities, as they exist in the IN environment, and servicing capabilities with simpler operations, administration, maintenance, and provisioning (OAM&P). Servicing capabilities include handling as many as 200,000 lines, newer remote switching modules with line capacities of 20,000 lines, wireless services, etc. The provisioning aspect is unique in most of the recently designed ESS platforms in the United States and Europe. The reliability of the No. 5 ESS platform is better than that of most computer facilities in most No. 5 ESS installations around the world.

It should be noted that ATM technology has made circuit-switched digital systems almost obsolete. Packet, frame, and cell-switching methodologies offer superior cost benefits and performance over older conventional digital telephony techniques.

The newer wireless networks, SONET, independent packet networks (such as the Internet, knowledge nets, or medinets) are tied to the public domain networks. The No. 5 ESS-2000 environment permits it to perform as a totally software-driven versatile switch. The switching module (SM-2000) of this machine permits its interfacing with fiber-optic nets using the synchronous digital hierarchy (SDH), with OC-1 and OC-12 carrier capabilities. At the other extreme, the (mB + nD) broadband servicing features of ISDN make the provisioning of digital services quite easy and manageable with this No. 5 ESS-2000 platform. Such services can be just as easily offered with a SONET frame and ATM cell-relay systems.

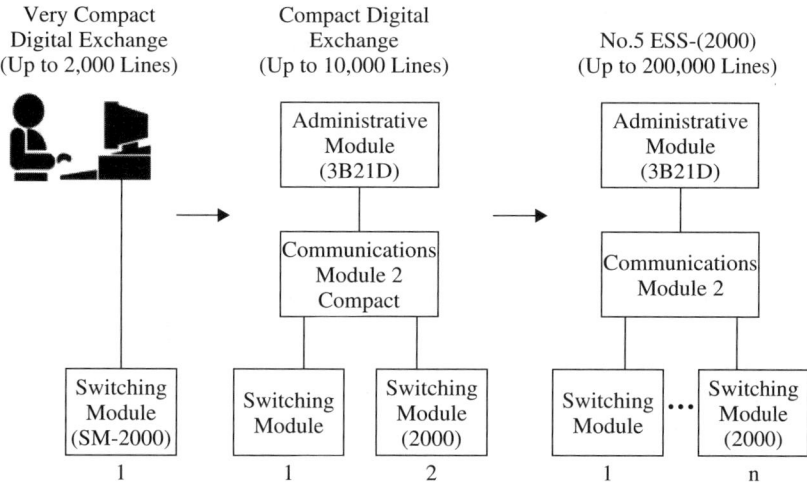

Figure 8.8 Three configurations of the No. 5 ESS switch. Most of the typical communications needs of individual customer and business needs are met by the versatile third-generation switching system. Siemens, Ericsson, and Alcatel switching systems compare well with the No. 5 ESS machines, and their performance is well regarded in the industry. The newer frame and cell relay systems function equally well at a fraction of the overall cost of the traditional electronic switching systems. More recently, routers and bridges have began to perform the switching of packets, frames, and cells in all digital networks.

8.5.3 Input/Output Processors

Input/output processors (IOPs) have been in vogue since the second-generation machines of the IBM 7094 era. Initially, these processors were designed to relieve data congestion at the CPU, making I/O functions partially independent of CPU functions. Processor speed was matched to device speed, with the direct memory access device cycle time reaching the main memory cycle time.

There are many possible designs for I/O processing systems, depending on application requirements and device speeds. The buffering of input/output data becomes necessary to scan/send and store/retrieve large amounts of high-speed data. Generally, the device transfers data in/out of the memory buffers and the buffers transfer data to/from the disk systems. The design of elaborate I/O processors can become as complicated as a tiny computer system with processor(s), memories, bus structure(s), and switches, with I/O interfaces.

If the CPU is to be spared of monitoring and handling the I/O functions, then the I/O processors become essential. Most efficient designs aim at making the CPU main-process-intensive, while attempting to make the I/O processors equally input/output-intensive. Such designs call for balancing the main processor capacity in the CPU with the I/O capacity in the I/O processors. Most normal microcomputer designs do not usually demand such stringent requirements.

Sequentially and/or interrupt driven I/O devices and DMA devices are generally available for various applications. The I/O interfaces can also be highly variable. The application requirement (such as for audio, video, graphics, medical, etc.) usually sets the primary constraints and limits on the design and integration of the overall system. Speed, buffer capacity, and bus structures with the I/O processors are the intermediate design variables.

8.5.4 Display/Graphics Processor Units

The earlier display devices (DPUs) for cathode ray tube (CRT) technology are obsolete. Cursor input in the CRTs is very outdated. The software was primitive and the display functions and colors needed considerable enhancement. Flat panel technology and the availability of very fine granularity and color options have rendered new venues for machines to carry visual impact deep into the minds and thoughts of scientists and researchers alike. The impact in every scientific and social direction of progress is monumental.

During their introduction, graphics processor units (GPUs) served as the intermediate processors between the CPU and graphic devices. GPUs were initially built to relieve the CPU of display functions and continue with application-based processing. Earlier designs (e.g., Intel 82786 graphics coprocessor) adapted a separate configuration, but later CPU architectures absorbed the basic GPU configurations within the larger VLSI chips that were offering larger capacity, dependability, and yield. The typical configuration of a CPU from the 1980s that could also be deployed as a GPU is shown in Figure 8.9.

The internal bus structure within the GPU is based on the word length and instruction set of the graphics/display processor unit. Typically, the older graphics processors and accelerators (the Intel 80860, i860, 1989 dating to the 1980s and early 1990s) delivered 13 MFLOPS or approximately a half-million vector transformations for graphics applications. The architecture deployed 10 bus structures, integer and floating point adders and multipliers, two cache memories, and numerous switches and multiplexer units. The algorithms for computer graphics functions are highly optimized to deliver the most desirable processor power consistent with the desired level of visual impact to the viewers based on application requirements.

Recent graphics processors are as powerful as CPUs. In fact, their VLSI technology is the same and some GPUs outperform traditional CPUs for specific applications. Processors for graphical inputs and outputs have specialized requirements. Most of the graphical and display processes are accomplished in real time and need to keep pace with the application-based computations. When the entire systems are designed for special applications (such as distance learning, remote surgery, MRI, stock market, games, etc.), the response time of the two (application- and graphics-oriented) subsystems requires consideration. Communication capacity influences channel capacity, and the three basic design parameters (computation, graphics, and communication) influence the accuracy, clarity, and quality of the output displays, movies, and images.

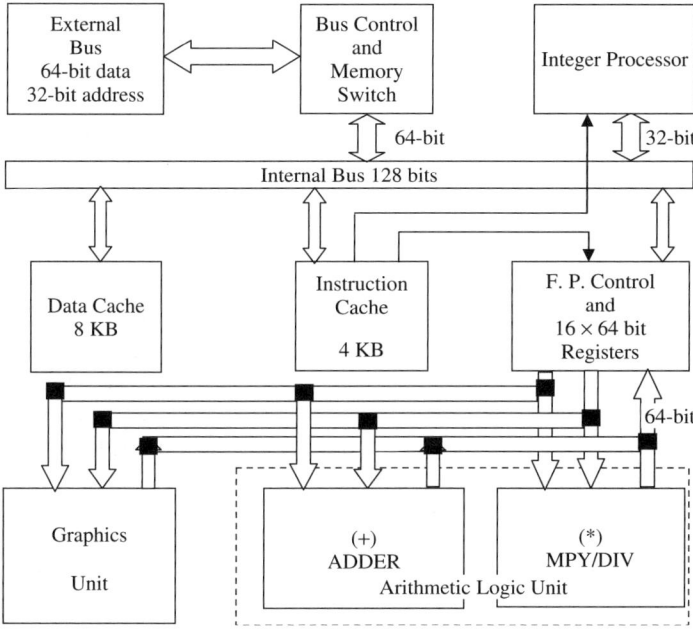

Figure 8.9 A Typical central processor unit configuration that was as the basis for a graphics processor unit unit from the late 1980s.

Graphical user interfaces (GUI) provide user input to the application programs that generate the resulting displays. Graphical devices are the main output devices for a computer to communicate with users. In certain applications such as computer games and movie making, the perception time of users and the viewer becomes crucial. When computer systems are dedicated to movies, graphics-processing systems can become as elaborate and intricate as mainframe computers.

The balance between processing power and graphics display capability becomes an option for most designers of standard desktop and laptop computers. In customized systems, processing powers are based on the choice of the CPU and graphics processor chips.

For intensive graphics application, independent graphics processor are deployed. Most modern computing systems have integrated GPUs built into the overall architecture. GPU computing has become a specialty in its own right. It is seen as a viable alternative to traditional microprocessor-based computing in high-performance computing environments. When used in conjunction with selected CPUs, an order-of-magnitude gain in performance can be expected in physics, games and in computational biophysics applications. The parallel architecture of the GPU is generally used to accelerate computational performance. These GPUs handle a broad range of process-intensive complex problems with visual insight into the nature of changes that occur as the computation progresses.

8.5.5 Numeric Processor Units and Array Processors

Much as GPUs were aligned with visual, gaming, and movie applications, numeric processor units (NPUs) were geared to research, scientific analysis, and the need for accuracy in user applications. The Intel coprocessors (8087, 80287, 80387, and even 80487 for a very short time) provided numeric support for their 8086 (1978–1990s), 80286 (1982–1990), 80386 (1985/6–2007), and 80486 (1989–1991), respectively, CPU chips. In the early versions, the NPU was used as an appendage to the CPU, to enhance the computer's overall performance, by forcing the NPU to perform most of the numeric functions. A series of NPU architectures were developed during the 1970s through 1989. In most cases, the NPU has been integrated within the main CPU architecture for almost two decades.

The more recent CPU designs incorporate the essential numeric and graphic functions. A variety of multimedia (audio, video, media creation, and multi-threaded gaming applications) functions can be accomplished in the Intel®, Pentium®, 900 sequence (extreme edition) processor. This particular processor has dual-core processing, with hyper-threading and virtualization technologies. It accommodates two 2-MB, level-2 caches and a front-side bus at 1,066 MHz for maximum bandwidth. An execute-disable bit provides additional hardware security, and the processor permits streaming for single-instruction multiple-data (SIMD) extensions to accelerate the performance of most applications, including multimedia, video, audio code/decode functionality. Three-dimensional graphics and image-processing functions can be performed in this processor environment.

Array processors (also known as vector processors) are an example of large numbers of SIMD architectures deployed in conjunction with numerous NPUs. Generally used for matrix and array operations, array processors perform large matrix operations (such as multiplications and inversions) at high speeds due to massive parallelism. Bus configurations and switches are designed accordingly to assure that most identical numeric functions (such as additions, multiplications, and divisions) can occur simultaneously to be able to reduce the overall execution time for matrix and vector functions.

8.5.6 Digital Signal Processors

Initially, digital signal processors (DSPs) were built as specially programmable processor units that convert external signals (mostly analogue) to digital formats and then process the signals to generate a new accurate and programmable set of output parameters. These external parameters can, in turn, be used to control the environment, feeding into other sensing and monitoring systems to control the stability or optimality of the overall systems.

DSP units have been used for many industrial, production, and robotic environments. For production-oriented and robotic lines, DSPs are firmly entrenched. The specific design of units is tailored to the particular industrial, production, or control environment. The nature of applications dictates the choice of a finite number of operations and $opc(s)$ that are embedded in DSP chip sets. The lifelong task of many digital designers is to optimize the performance and optimality of a particular DSP for a particular task.

The need to tailor a DSP design to any specific application brings about a wide variety of DSP architectures. From a conceptual perspective, it is possible to transform the simpler CPU design methodologies into DSP designs. VLSI chip design engineers incorporate the details to implement a DSP with a minimum number of gates and clock rate that offers a fast-enough response for applications. In the current state of technology, computer-aided VLSI design software offers ample freedom of architectures and allows chip manufactures to offer a marketable product in less than a month. Being highly specific to the application and less versatile, the DSP chips are generally less expensive than CPUs, if there is sufficient demand. For customized designs, the costs can be relatively high and the solution sometimes lies in using the less expensive DSP with necessary peripheral I/O devices.

The architecture of DSPs is highly application-dependent. Modern applications can range from DSPs for digital subscriber loop (DSL) transceivers to stream processors capable of delivering up to 512 GOPs [10]. The program code driving recent DSPs plays a vital role in realizing the highest overall performance. Modern processors are flexible enough to be used for a variety of applications needing signal, image, and video processing. Massive parallelism has been deployed in recent (2007–2008) DSP architectures. Most of the emerging process-intensive embedded applications (such as high-definition encoding, video processing, image processing, search operations, encrption, and wireless communications) stand to benefit from the new breed of DSP chips with highly specific code to drive them and realize data parallelism through the processors. Direct memory access capability is usually built into high-end processor chips.

8.5.7 Digital Object-Processing Environments

Object processing has been in vogue since the introduction of object-oriented programming and the days of computer graphics. However, the more powerful object-processing functions occur with object-oriented sophisticated programming language commands. Data structures that represent objects are processed in a mathematical sense as the objects undergo realistic/physical changes (e.g., when a ball bounces, when a bell is struck, when a wave rolls, etc.). Complex groups and collections of object entities may thus be processed quickly and accurately. The processed objects and their attributes represent the newly computed properties of objects or object groups. A complex chain may thus be processed sequentially or in parallel. A pipeline and parallel object processor can thus be readily built.

8.5.7.1 Object Processor Units Consistent with the architectural representation of typical central processor units (CPUs), the representation of a typical object processor unit (OPU) is shown in Figure 8.10. The design of the object operation code (*oopc*) will play an important role in the design of an object-oriented machine. This role is comparable to that of 8-bit operation code (*opc*) in the design of the IAS machine during the 1944–1945 timeframe. For this IAS machine, the *opc* length was 8 bits in the 20-bit instructions, and the 4,096-word, 40-bit memory corresponded to the address space of 12 binary bits. The design experience of the game processors and modern graphical processor units will serve as a platform for the design of OPUs and hardware-based object machines.

The intermediate generation of machines (such as the IBM 7094 and 360 series) provides a rich array of guidelines to derive the instruction sets for OPUs. If a set of object registers or an object cache is envisioned in the OPU, then the instructions corresponding to the register (R-series), register storage (RS-series), storage (SS), immediate operand (I-series), and I/O for the OPU can be initially designed. The instruction set will need to be expanded to suit the application. It is logical to foresee the need for control object memories to replace the control memories of microprogrammable computers.

The instruction set of the OPU is derived from the most frequent object functions such as (1) single-object instructions, (2) multiple-object instructions, (3) object to object memory instructions, (4) internal and external-object instructions, and (5) object relationship instructions. The separation of logical, numeric, semi-numeric, alphanumeric, and convolution functions (see Chapter 3) between objects will also be necessary. Hardware, firmware, or brute-force software (compiler power) can accomplish these functions. The need for the next-generation object and knowledge machines should provide an economic incentive to develop these architectural improvements beyond the basic OPU configuration shown in Figure 8.10.

As objects are more structured than logical and numerical operands objects need the operations to be directed toward any of their inherent and/or derived attributes. Some of the typical operations are presented in Sections 4.3 and 7.5.

Objects and their attributes are treated in unison and in synergy. Objects are configured for viability for synthesis from known objects when their attributes are manipulated (e.g., new drugs, new vehicles). The processes are checked to ascertain that the basic laws of science are not violated during the synthesis of new objects. Many other variations of the OPUs can be derived from this preliminary arrangement shown in Figure 8.10. The three decades of development between the IAS processor (Figure 8.2) and Intel i860 processor (Figure 8.9) can be radically shrunk by predictive human reasoning for the OPUs and KPUs of the next decade.

In the processing of objects, special situations can be foreseen. For example, intelligent objects may react to a *kopc* with time lag (e.g., jet lag in human beings), objects may display elastic and non-linear response (emotional response of teenagers), object groups may react differently from how the individual objects may react (student response vs. class response), etc. Under these circumstances, the KPU output (to the same *kopc*) will be intelligent and application dependent.

In the design of object processors or OPUs, the numeric/logical operands of the CPUs and GPUs should be treated as alphanumeric object matrices (see

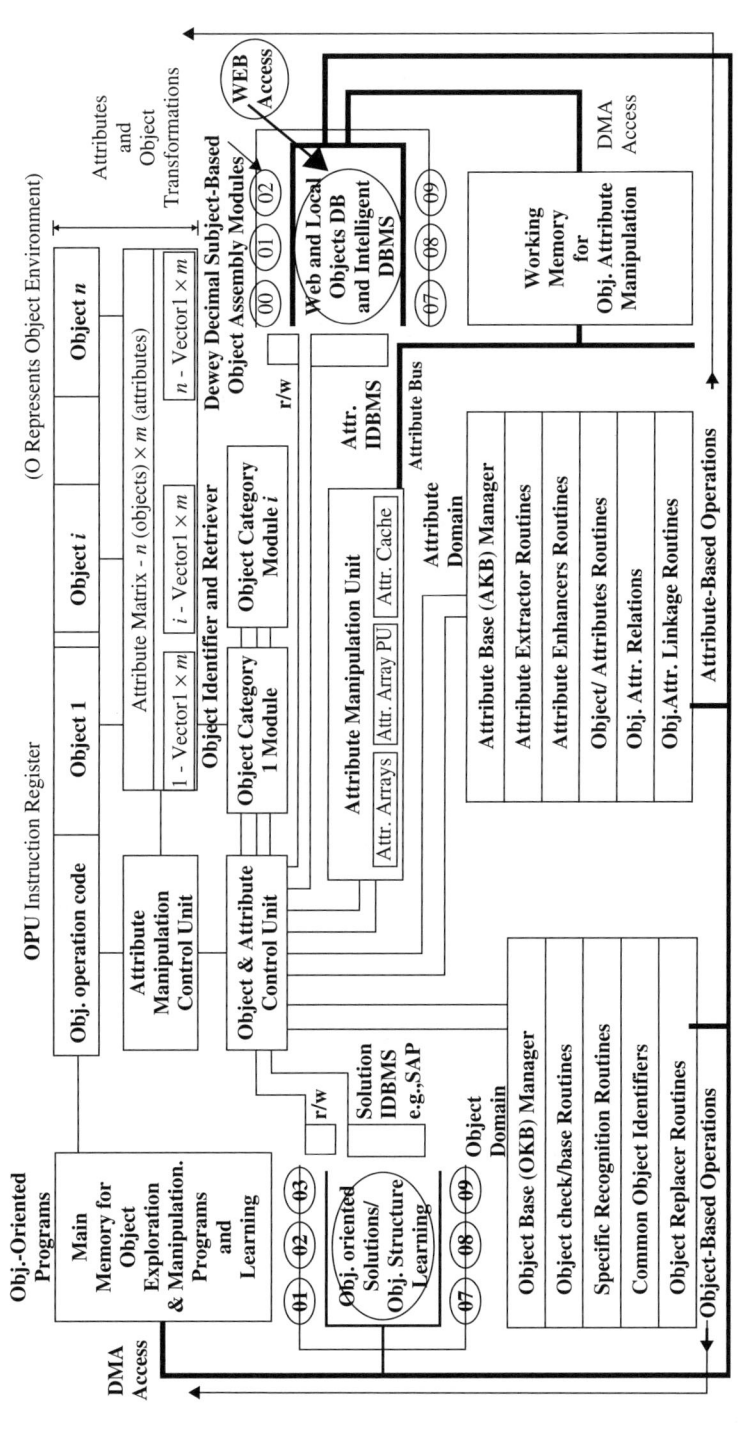

Figure 8.10 Schematic of a hardwired object processor unit. Processing n objects with m (maximum) attributes generates a $n \times m$ matrix. The common, interactive, and overlapping attributes are thus reconfigured to establish primary and secondary relationships between objects. DMA = direct memory access; IDBMS = intelligent, data, object, and attribute base(s) management system(s); KB = knowledge base(s). Many variations can be derived.

385

Chapters 3 and 10). The ALU of a typical computer/microprocessor needs the functionality of an object-convolution processing unit (OCPU) with an array of logical units (LUs) and numeric units (NUs). Some of the object functions are (entirely) logical from the humanistic domain, and some of the functions are numeric, probabilistic, and precise from the scientific domain. What becomes essential is that the judgmental functions be a weighted sum of the two (logical and numeric) outputs from the LUs and NUs. Essential feedback link between the two provides a microswitch for incremental judgment in a long chain of steps constituting a knowledge operation code or an entire knowledge program.

8.5.7.2 Object Machines The functionality of an object machine (OM) is greatly facilitated by the OO programming languages. The compilers for C++, LISP (LISt processing language), and PERL (a cross platform, general purpose, highlevel interpreted, dynamic programming language) offer excellent venues to force conventional computers to function as object/knowledge machines. From the perspective of the next-generation machines, these compilers may appear as the brute force of highly elegant algorithms and operating systems. To some extent, it appears like an effort to build an IBM 1620 computer of the 1960s with discrete logic gates.

In dealing with objects, access to external knowledge banks becomes essential. An automated search of the authorized, content-based, and related data and knowledge bases during execution can by allowed by the OM operating system. One such configuration is shown in Figure 8.11.

8.5.7.3 Object Networks Any generic backbone can exchange entire object blocks, that is, the data structure that defines the entire object. ATM and packet networks already communicate entire voice, video, and text messages. When complex objects or object clusters are to be exchanged, the hierarchical grouping of objects may become necessary. Such object groups are created routinely in computer graphics programs. The layering of objects permits another degree of freedom much like layering in graphic programs.

Object-oriented software needs greater flexibility than typical graphics-oriented software. The network commands for accomplishing object-oriented functions over the network can always be assembled by using the OSI layers. Typical interfaces that pass "objects" up and down the seven OSI layers are not streamlined, even though the application-layer programs can exchange and communicate objects from one location to the next.

In an ideal object network, a database lookup between an object's definition and its Web addresses will be desirable. Such database lookup in the intelligent network (IN) environment [5] is provided via the service control point [11].

8.5.8 Digital Knowledge Environments

Knowledge is focused around knowledge-bearing "objects" in society. These objects are referred to as "noun objects" that perform and undergo "actions."

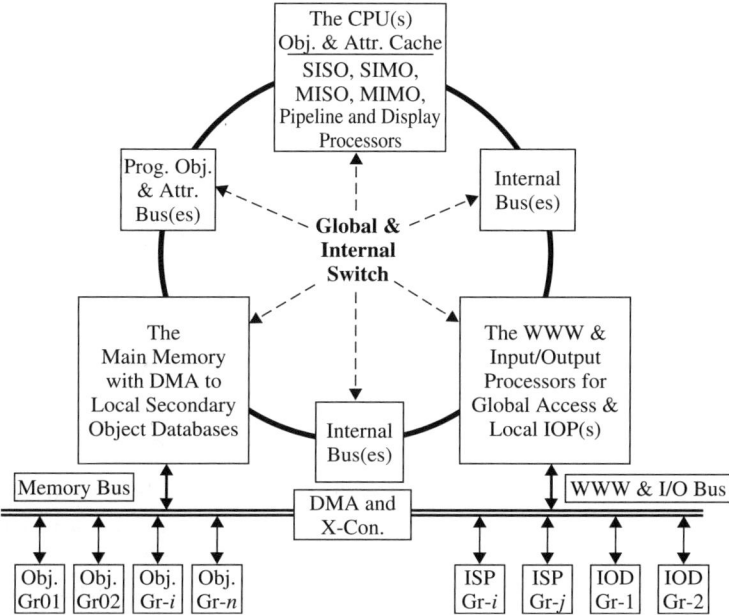

Figure 8.11 Architectural layout of an object computer with an enhanced object processor unit consistent with Figures 8.6a and b for the architectures of conventional computers. The main objection memories store and manipulate large chunks of object arrays. Each object can be a alphanumeric matrix in its own right. The hyper-dimensionality of the object space becomes apparent. The machine is thus able to process, objects, attributes, and their relationships.

These actions are referred to as "verb functions" that are performed on noun objects. The cycle brings a sense of reality and repeats itself indefinitely in society. Structured and orderly graphs appear for events that generate strong, stable, and scientific strands of knowledge. Such knowledge lingers on in society. Human beings weave a web of concepts and then collect dewdrops of wisdom along the strands of knowledge in the webs of concepts. A modern digital environment provides a precise handle on knowledge, concepts, and wisdom. It also provides the forward and backward pointers to noun objects and verb functions structured orderly in Web knowledge banks.

8.5.8.1 Knowledge Processor Units Knowledge is derived from objects, their nature and attributes, and their interactions. In fact, the entropy of knowledge is the graph of these four constituents upon which knowledge is built. Thus, the processing capability of knowledge entails processing objects, their attributes, and object interactions. Numerous designs for KPUs become evident, for the designs of the CPU's that serves and DSPs, GPU's, IOPs, etc. In fact, the KPUs can be derived initially from the many varieties of CPUs, then the GPUs that function as display, graphics and game processor units, and finally, the GPUs that can also sere as CPUs. The creativity of the individual KPU designer lies in matching hardware architecture to application needs. KPU chips, being more expensive and process-intensive than CPUs, are unlikely to become as numerous as CPUs that can be personalized to suit any whim or fancy of chip manufacturers. The function of KPUs depends on the capacity of the hardware to manipulate or juggle (global and local) objects, based on their own syntax, and environmental constraints in the semantics of the user objective.

The CPU's functionality depends on its capacity to execute stylized operation codes on arithmetic and logical operands (in highly specialized formats and date structures). The configuration of a simple KPU is shown in Figure 8.12. Other variations of KPU, based on SIMD, MISD, and MIMD[3] variations of the CPU architectures can be also built. Object processors that lie between CPUs and KPUs bring another degree of freedom because KPUs can deploy OPUs, much like CPUs can deploy ALUs and NPUs. Sequential, pipeline, and parallel execution of operations on objects in KPUs gives rise to at least eight (see footnote 3) possibilities: SKI-SO processors, SKI-MO processors, MKI-SO processors, and MKI-MO (see Figure 8.13) processors. Now if SO and MO processors have SOI-SO, SOI-MO, and MOI-SO (pipeline structure), and MOI-MO (pipeline and/or multiprocessor structure) variations embedded within them, then at least eight design variations become evident. The SKI-SOI-SO is the simplest to build (Figure 8.12), whereas the MKI-MOI-MO is the most cumbersome to build. From the first estimate, the hardware for the simplest KPUs should be an order of magnitude more complex than IBM 360 CPUs (even though these CPUs deployed microcode technology).

Knowledge processing is based on rudimentary knowledge theory [12]. Stated simply, human knowledge is clustered around objects and object groups. Such objects can be represented by data and information structures. Data have numerous representations and information has several forms of graphs and relationships that bring order and coherence to a collection of objects. Such a super-structure of data (at the leaf level), objects (at the twig level), the object clusters (at the branch level) can constitute a consolidated tree of knowledge. Specific graphs moreover, relationships that bind information into a cogent and coherent body of knowledge, bring about a nodal hierarchy (precedent, antecedent, and descendant) in a visual sense that corresponds to the reality of complex objects.

---

[3]A brief explanation of acronyms for this section is presented here. CPU = central processor unit; KPU = knowledge processor unit; ALU = arithmetic logic unit; NPU = numeric processor unit; SIMD = single-instruction multiple-data; MISD = multiple-instruction single-data; MIMD = multiple-instruction multiple-data; OPU = object processor unit; SKI-SO = single knowledge instruction, single object; SKI-MO = single knowledge instruction, multiple objects; MKI-SO = multiple knowledge instructions, single object; MKI-MO = multiple knowledge instructions, multiple objects; SO processor = single-object processor; MO processors = multiple-object processors; SOI-SO = single-object instruction, single object; SOI-MO = single-object instruction, multiple objects; MOI-SO = multiple-object instruction, single object; MOI-MO = multiple-object instruction, multiple objects; SKI-SOI-SO = single knowledge instruction, single-object instruction, single object; MKI-MOI-MO, multiple knowledge instruction, multiple-object instruction, multiple objects.

Knowledge processor units should be able to prune, build and shape, reshape, and optimally reconfigure knowledge trees, much as CPUs are able to perform arithmetic (and logic) functions on numbers and symbols and derive new numbers (and logical entities) from old numbers (and logical symbols). All the most frequently used knowledge functions need to be reviewed to ensure that the KPU can perform basic elementary and modular functions on objects.

In the design consideration of a CPU, the more elaborate AU functions are known to be decomposable into basic integer and floating point numeric (add, divide, etc.) operations. Similarly complex logical operations can be reconstituted as modular (AND, OR, EXOR, etc.) functions.

Knowledge-bearing objects can be arbitrarily complex. Numerous lower-level objects can constitute a more elaborate object entity. Like bacterial colonies, knowledge super-structures have a dynamic life cycle. The order and methodology in the construction and destruction of such knowledge super-structures lead to "laws of knowledge physics" in the knowledge domain under the Dewey Decimal System (DDS) classification 530–539 The traditional laws of Boolean algebra and binary arithmetic do not offer the tools for the calculus of dynamic bodies of knowledge undergoing social and technological forces in society.

However, if the new laws for the flow, dynamics, velocity, and acceleration of knowledge (see Chapter 3) can be based on a set of orderly, systematic, and realistic knowledge operation codes (*kopcs*), then these laws can be written as machine-executable routines that operate on knowledge-bearing objects. This approach is a bold digression from the approach in the classical sciences where new concepts enter the sciences as symbolic and mathematical equations. In present society, information is erupting as multimedia Web streams, rather than gracefully expanding in a coherent and cogent fashion. Time for extensive human contemplation is a rare luxury. Much as we needed digital data-scanning systems for DSPs in the past, we need machine-based, common-sense sensing systems to separate junk-level information (see Chapter 12, Figure 12.5 in [12]) from knowledge-bearing information. Knowledge filtering (Chapter 3) accomplishes this initial, robust, sensible, and necessary task.

The current scenario of science and innovation has given rise to the deployment of technology before it becomes obsolete. To accommodate this acceleration of knowledge, we propose a standard set of basic and modular *kopcs* (Chapters 5 and 6). The complex knowledge operations that encompass the newest concepts are then assembled from the basic set of internationally accepted standard *kopcs*. The proposed representation for the dynamics of knowledge paves the way between concepts that create new knowledge and the technology from old knowledge.

In effect, if the basis for technology resides in the DDS classification of pure sciences (500–599) and the applications of technology and innovation reside in the DDS classification of technology (600–699), then one of the functions of KMS is to explore the knowledge links (forward, backward, and lateral pointers) between the sciences and technology. Knowledge links between any two segments of knowledge can also be explored by KMs. In the current expanding roles of computers, networks, and knowledge-processing systems, KMs and their knowledge processors work on a continuous basis to explore, monitor, and deploy new linkages resulting from discoveries, inventions, and innovations. A typical computer serving as a KM is discussed in Section 2.5.2 and presented in Reference [12, see Chapter 8].

8.5.8.2 Knowledge Machines Knowledge computers and machines can be organized at a conceptual level, as in the diagram shown in Figure 8.14. The centralized KPU will itself be an individual unit or a collection of numerous subservient KPUs, OPUs, GPUs, and/or CPUs. The cache memories need special attention, and they can be individualized or shared to accommodate the and object clusters that undergo knowledge process(es). The design of the KPU can become as complex as that of a mainframe on a chip.

The high-level architecture of a KM for an educational environment is shown in Figure 8.15, and the detailed architecture of a general-purpose KM is illustrated in Figure 8.16. Numerous architectural variations are possible in designing KMs. In Figures 8.17, a more advanced KM is shown. Initially, the economic incentives for building these massive KMs are hard to justify in a purely business community. Public and state institutions that are interested in deploying such machines for monitoring social and human agencies are likely to serve as pioneers. To put things in perspective, it was the U.S. defense and NSF agencies that initiated the IAS machine and backbone networks.

The knowledge machines and systems can be designed to suit the application environment. Such design and architectural variations of mainframe, medical, educational, and sensing computers, or any number of electronic switching system configurations, already exist. The dilemma for the KM designers is not if a KM can be built, but which application can generate an income stream to substantiate the investment in building the first commercially viable knowledge machine. In looking at the history of computers, the demand for exactitude and precision has prevailed. In the history of magnetic resonance imaging, the demand for cancer detection has prevailed. In looking back at fiber optics, the need for global communication has prevailed, etc. However, the library and knowledge community appears less pervasive and demanding. The computer-aided scientific search for generic knowledge and wisdom did not begin as a technical venture. In terms national defense (e.g., strategic codes, their keys, etc.) and in certain highly profitable business situations (e.g., diamond cutting, polishing, etc.), the shear pursuit of knowledge and wisdom are both economic alternatives because of the social progress that follows.

8.5.8.3 Knowledge Networks Modern backbone networks can accommodate a large number of KMs. The size and configuration of these KMs depend on the demands imposed by each and network capacity to satisfy the data traffic requirement. The OSI layers permit most applications programs to exchange data effectively. It is feasible to build and dismantle knowledge networks as easily as the private virtual networks (PVNs) [5] of the 1990s.

However, knowledge networks have the unique requirement of communicating simple and complex noun objects and verb function consistently, cogently,

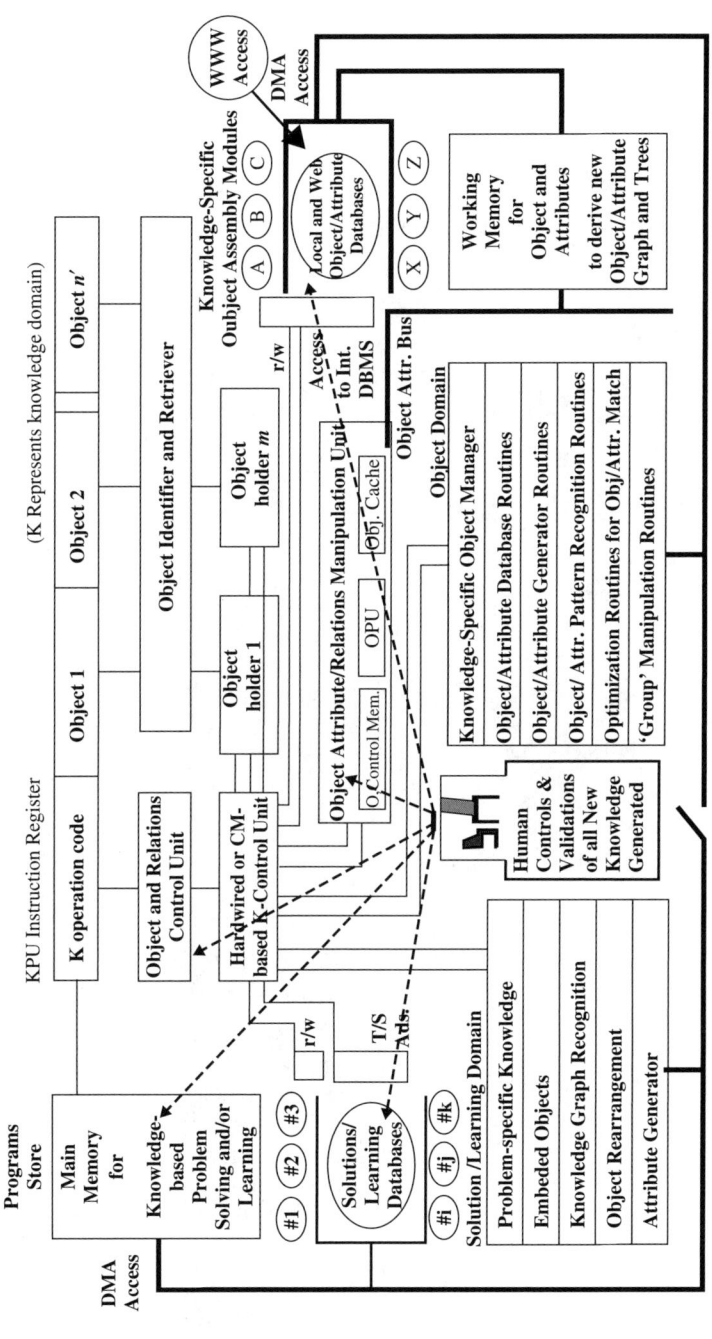

Figure 8.12 Switch S-1, open for the execution mode of knowledge domain problem solving and closed for the learning mode. The learning programs "process" existing solutions and are able to extract objects, groups, relationships, *opcodes*, group operators, modules, strategies, and optimization methodologies from existing solutions and store them in the object and corresponding databases. The architecture permits the knowledge-processing unit to catalog a new object in relation to existing objects and generate/modify existing pointers to and from new objects.

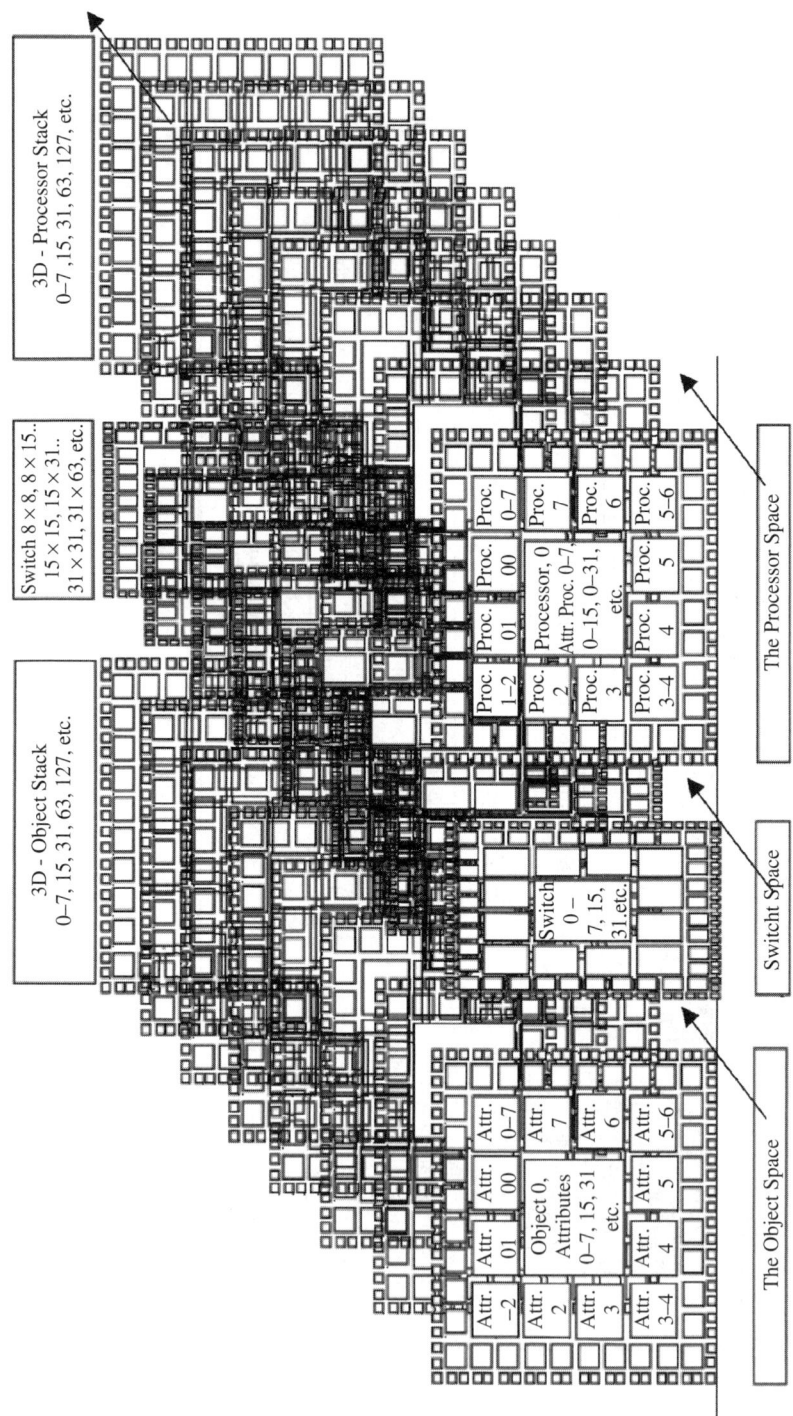

Figure 8.13 A three-dimensional knowledge-processing unit architecture to process multiobjects, each with multiple attributes. Multiple processors (processor space) each handle a specific knowledge operation code (*kopc*) and access objects and their attributes (in the object space) via the switching module (switch space). The processors can perform concurrently.

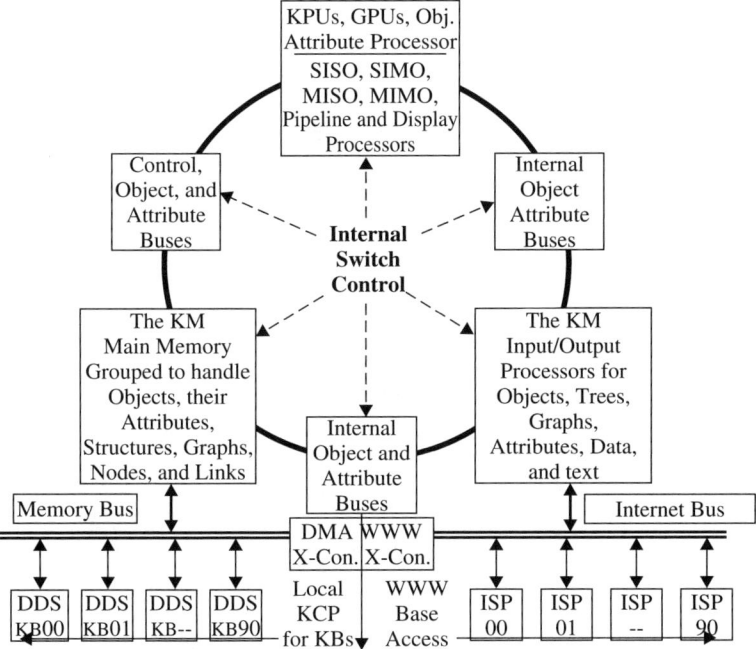

Figure 8.14 Initial configuration of a knowledge machine from a functional viewpoint.

coherently, and collectively. Attributes, relationships, and their structural relationships should retain their own contextual (i.e., syntactic and semantic) identity. A much more enhanced and elaborated application layer for handling knowledge may suffice, but if "search and add on" layers or hardware units are included in the bridges and routers, then the linked objects will be automatically retracted from Web knowledge bases and passed on to KMs for processing. Automatic forwarding of objects as "knowledge groups" from one network node to the next becomes feasible.

In the past, medical, educational, financial, and corporate networks have been suggested as independent networks. However, knowledge networks should encompass all application networks, as much as modern data networks encompass all digital networks (i.e., LANs, MANs, WANs, and the Internet). The integration of all application networks into one virtual knowledge network appears as distant as integrating all network computers into one virtual machine. KM operating systems require the enhancements of traditional operating systems, to become full-fledged global knowledge network operating systems. A unique blend of networking and computing environments that can communicate and process complex noun objects and their verb functions as well as the structured microcosmic entities will be desirable. Some of the commercial networks (such as integrated) billing systems have the features necessary to initiate the design of (integrated) knowledge communications and processing systems [12].

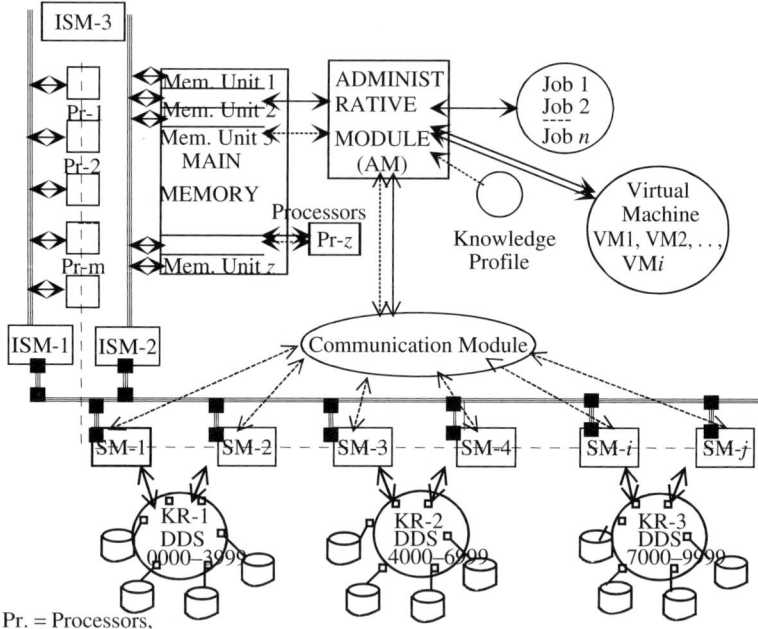

Figure 8.15 The typical architectural framework of a knowledge machine for educational and library functions. This architecture is feasible for local or globally distributed knowledge bases. The knowledge rings are classified based on their Dewey Decimal System classification, but these rings can be based on user preferences, such as classification of artists, countries, styles, etc. The knowledge bases may contain images, national data, various styles, etc. The switching, administrative, and communication addressing the numerous KR's modules can be small microprocessor-based machines. A lookup address in a knowledge control point database will render the knowledge machine intelligent.

8.5.9 Digital Medical-Processing Environments

8.5.9.1 Medical Processor Units
Medical processor units (MPUs) have not yet been built specifically for medical operation codes (*mopcs*). It is possible and viable that such processor chips be tailored and geared to medical functions. A variety of such generic functions can be readily identified. One such *mopc* is to instruct the MPU to compare bacteria strain i with all known bacteria ($j, j = 1, n$) from Web bacteria banks and identify any subtle patterns of change. A certain amount of object processing and pattern recognition procedures will become essential to complete this task. For computer scientists and VLSI designers, the modular tasks involved in executing this overall task have already been accomplished in other disciplines. The ordeal is to assemble an inter-disciplinary team of scientists.

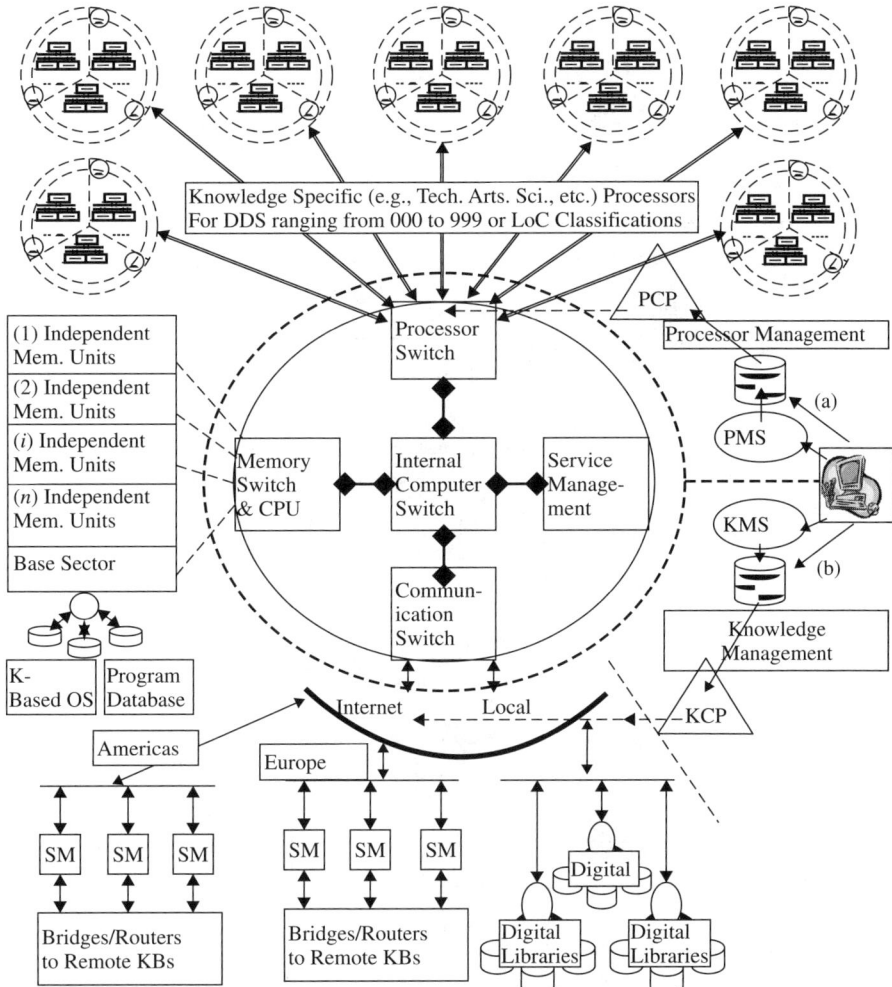

Figure 8.16 The architectural framework of an Internet, intelligent network-based intelligent content-based knowledge processing environment. KB = knowledge base; SM = switching module; OS = operating system; KCP = knowledge control point database; KMS = knowledge management system; PCP = processor control point; PMS = processor management system. Human interaction occurs via the control panel to control (a) the flow of knowledge programs and (b) the flow of knowledge being processed.

Other generic functions are the identification of ailments, patient complaints/cures, most effective treatments, and knowledge-processing units to bridge the various knowledge domains. Only a prolonged and consistent effort from medical staff can lead to generic operation codes for MPUs that can be the focal element of a medical computer. For an initial step, the knowledge operation codes (*kopcs*) should suffice as the *mopcs* discussed in this section.

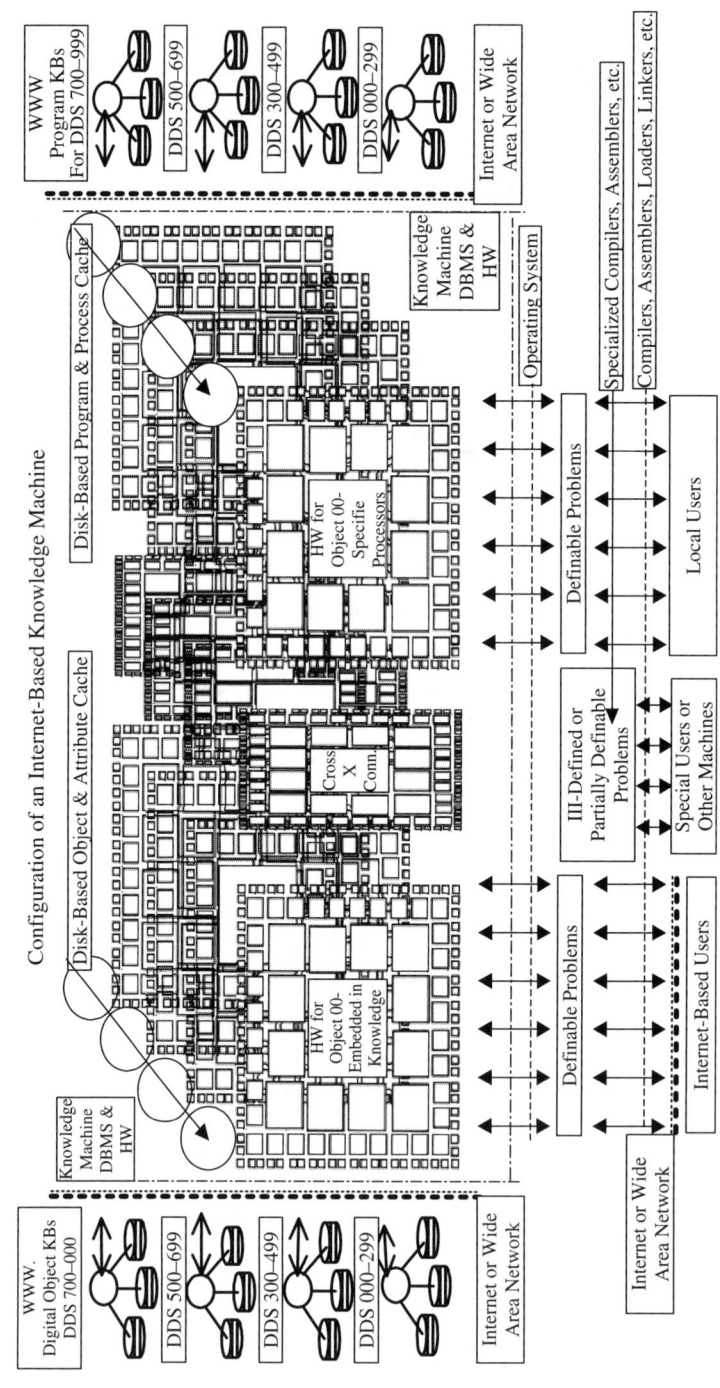

Figure 8.17 Architecture of a knowledge-processing system/machine with multiple wafer-level knowledge processor units.

Medical processors and medical data banks work in tandem to execute numerous instructions, retrieve/store, and process pertinent medical information. Numerous medical processor hardware units and modules (including conventional CPUs) coexist in the integrated system to track the confidence levels of medical functions, individually and collectively.

Medical computers constitute a special breed of cluster computers. Both have identifiable super-sets of traditional computers for routine functions. Although it is viable to force a cluster computer to act as full-fledged medical computer, it is desirable to build a medical computer on a distinctive track of its own *mopcs*. The demand for such medical computers is likely to expand and grow like specialized airborne computers for the aviation industry.

Patient databases, physician access points, patient access points, and service facilities are connected to medical data banks and a medical processor via several buses. In an alternative integrated medical computer system, numerous processors are included with their own memories and modules, and are linked together to establish a processor net unit. Such systems are amenable to hospital campus settings, where several buildings make up the hospital, or where several hospitals are inter-linked over local-area networks.

Figure 8.18 depicts the schematic of a medical processor unit (MPU) derived from the OPU shown in Figure 8.10. Some of the features for an automated learning procedure from the KPU shown in Figure 8.12 are included so the medical systems can activate AI procedures and suggest improvised procedures to the medical team.

Patients and their attributes are treated in unison and with one synergy. Patient records are dynamic and reflect the most recent state of health, and attributes are updated with a history of changes. Procedures are configured from known procedures on similar patients but with slightly different attributes. The procedures are then mapped against the attributes to ascertain their and other treatments' optimality and efficacy. The basic laws of medical science are not violated during the administration of treatments and procedures. Administering n subprocedures on a patient with m (maximum) attributes generates a $n \times m$ matrix. The common, conflicting, and overlapping attributes are thus reconfigured to establish primary and secondary safety rules in the practice of medicine.

8.5.9.2 Medical Machines Medical machines have all the architectural requirements of computers, objects and knowledge machines. Interaction with a medical teams becomes necessary to develop the medical codes., Only a close monitoring of MPU functions will generates consistent output during the early stages. The conceptual arrangement of a medical machine (MM) is shown in Figure 8.19.

Debugging of the MPUs and medical machines can be more complex than the debugging of IBM 360/370 CPUs (1967–1980s) or any defense surveillance machines. Much as customized computer systems are tailored to an application, medical machines may assume numerous customized profiles.

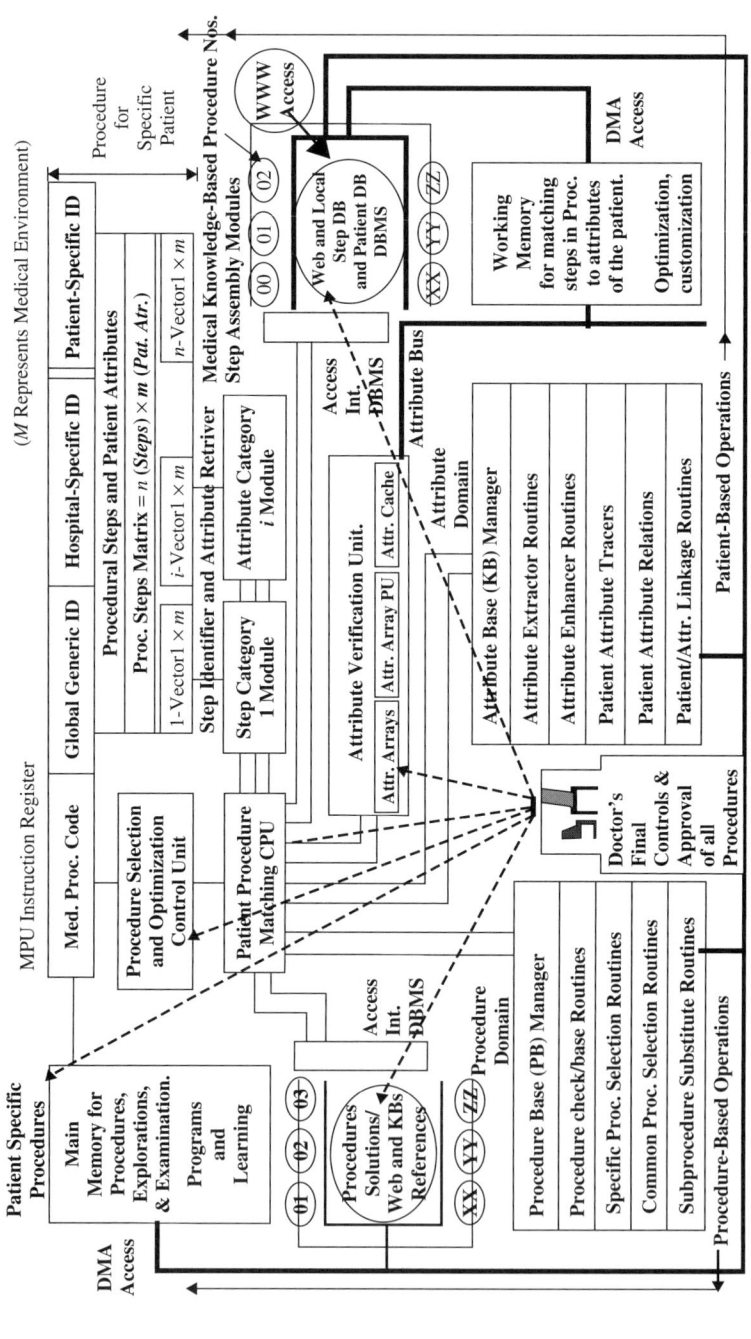

Figure 8.18 Architecture of a medical processor unit. DMA = direct memory access; DBMS = data, procedure, patient, and/or attribute base(s) management system(s); KB = knowledge base(s).

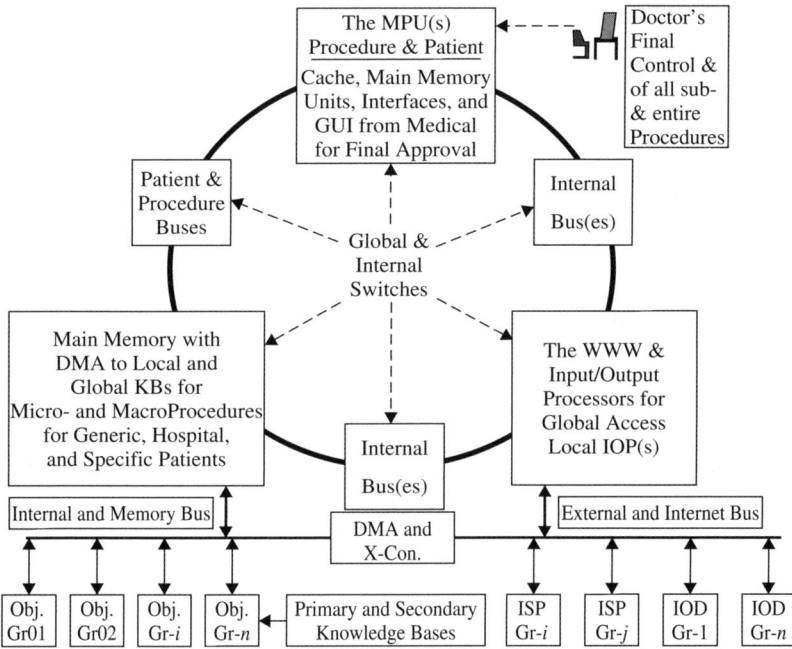

Figure 8.19 Initial configuration of a medical machine from a functional consideration.

Ranging from a simple AI-based MM for a doctor's office to global health systems to tackle world health issues, the machines can range from a simplistic PC to intelligent-network- (IN-) based, artificially intelligent (AI), intelligent-content-based clusters of analytical machines. The IT platform for MMs and medical networks (Section 8.5.9.3) can encompass the optical capacity of global fiber networks.

Figure 8.20 depicts a relatively simple hospital computer system with an extended I/O facility for Internet and remote access. The system offers a platform for medical services. Library, teaching, and research activities can be appended with distributed access and policy-based access control. The security of data and the system also needs to be addressed by appropriate I/O sensors and monitors.

This configuration is suitable for handling the traffic of a large community of hospitals or a local medical center. The capacity of the centralized main hub of the major hospital is made consistent knowledge module with expected services from the administrative module, knowledge module and the switching modules of a small electronic switching system (see Section 8.5.2). For small local clinics providing routine medical services, a standard server environment will suffice.

Hardware, firmware, and software constitute the main hub. The communications module (CM) also controls data flow between the other modules, the memory, and the medical processors. Effective communication from module to module is thus established. The hardware for the switching module(s) selects and switches between the various medical data banks for solving a particular problem. It also

400 KNOWLEDGE SYSTEM ARCHITECTURES

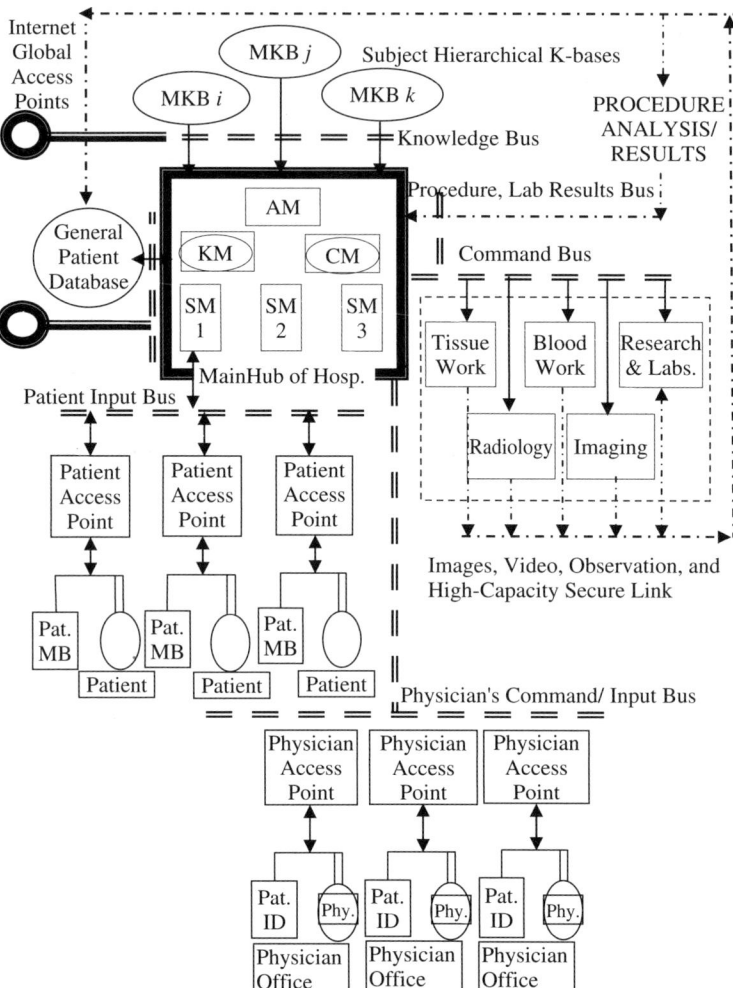

Figure 8.20 Architectural arrangement of a simple hospital or medical center computational/research facility. Pat. = patient; Phy. = physician; MB = local/personal medical base; MKB = specialty-based medical knowledge base; AM = administrative module for billing, insurance, medical codes, etc.; CM = communications module or intelligent hub/router/bridging router for security, monitoring, and/or records; KM = knowledge module; SM = switching module or simple routers.

enables the administrative module to perform housekeeping functions, including multitasking control with resource allocation, real-time multitasking, and scheduling of tasks. Within a KM, the hardware performs knowledge-processing functions and stores pertinent information in medical data banks.

General patient databases, physician-access points, patient-access points, and service facilities are connected to the medical data banks and medical processor via several buses. In an alternative integrated medical computer system, numerous processors are included with their own memories and modules are linked together to establish a processor net unit. This system can be used in a campus environment, where several buildings make up the hospital, or where several hospitals are linked via local-area networks. One such configuration is shown in Figure 8.20. The nature of components and their capacities are matched to the variety of services provided, and the sizes of links and routers are matched to the expected client base that is served by the medical facilities.

8.5.9.3 Medical Networks (Medinets) Numerous architectures for medinets are feasible. The customization of medinets is just as essential as the customization of computer networks. Ranging from a multicampus hospital network based on a LAN configuration, as shown in Figure 8.21, to a global medical services provider network, as shown in Figure 8.22, almost any medical service facility can be assembled based on commercial network components. Network, services, and artificial intelligence can be supplemented at any of the seven levels depicted in Figure 8.22.

8.6 THE TREND OF CONCEPTS

Social momentum is accelerating. The rate of change of social forces brings about increasing emotional pressure, demanding a changed outlook rather than simple adaptation. Intelligence becomes inadequate, needing the restructuring of social values. To fill the void, scientific contemplation, if it is possible, is necessary. Scientific contemplation initially investigated by knowledge machines and finally reiterated by human beings seems to be the frontier of machine processes and human ingenuity. Concepts become the "objects" at the interface that are both processed and contemplated.

8.6.1 Bridge between Science and Wisdom

CPUs are the center of numeric, arithmetic, and logical processes. Human beings have the focus for comprehension, intuitive search, and the farthest foresight. Space and time are both traversed freely during such a search. Concepts pave the illusive pathways between islands of universal knowledge. These initial pathways become major highways to well-proven wisdom if truth T, elegance B, and social values V are withheld, proven, and reinforced. Conversely, these initial pathways become lost in oblivion, if deception D, arrogance A, and hatred H are prized, practiced, and preached.

Modern (and AI-based) machines go a long way in helping the truth be investigated and validated, the elegance be consistent and beautiful, and social values be enhanced and strengthened. The methodology of machines lies in enforcement

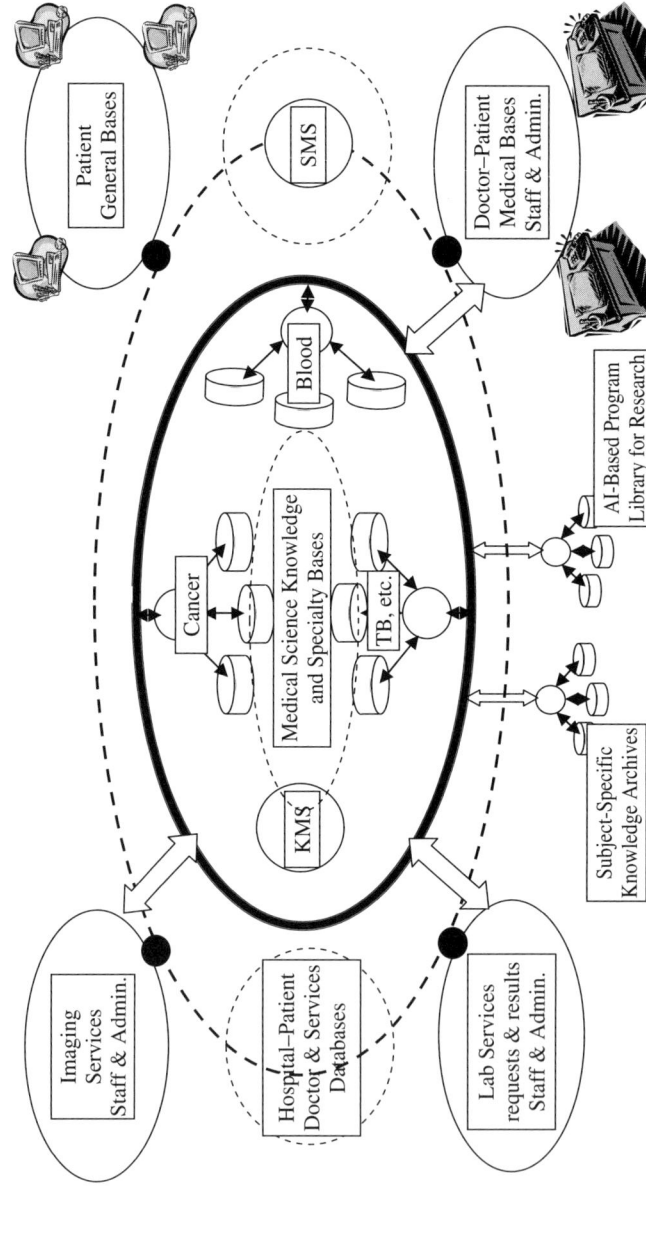

Figure 8.21 A local-area network implementing a core hospital-based medical network for doctors, administrators and medical staff. KMS-knowledge (medical) management systems; SMS-service/billing/administrative management services. Wide-area network and Internet-based configurations are presented in [12].

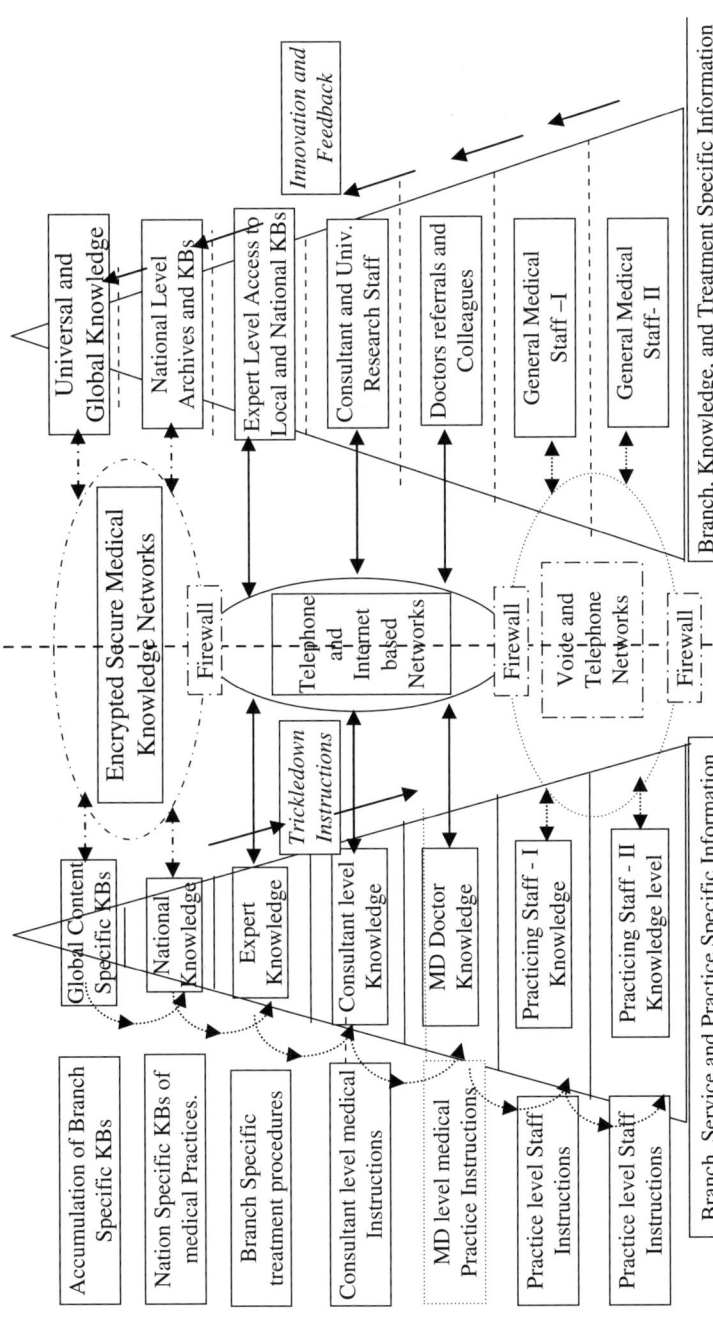

Figure 8.22 Seven-level intelligent medical environment. This arrangement responds to public medical needs via Web entry (level 7) into the hospitals. Routine physician–patient transactions are handled at the lower three levels (5–7). The facilities at the higher levels (1–4) provide venues for physicians and hospital staff to consult with local medical communities and interact with subject-matter specialists around the globe. Limited access to the public provides patients with the ability to check medications, procedures, and expert opinions. Network security is essential.

403

of the rigor of science, the precision of mathematics, and the optimality of algorithms. All three are used and apparent in the design of computers and information-processing systems. If knowledge and concept machines are thought of as predictable extensions of HW in computers and networks, SW in IT, AI in management information system (MIS) platforms, and lastly the adaptability of AI systems, then the HW, SW, and AI offer a much wider and diversified platform for migration toward the universality of TVB, which lies just beyond wisdom.

In this section, we propose the first few infantile steps in blending HW, SW, and AI with natural intelligence (NI) in charting the illusive path between diversified islands of knowledge. Figure 8.23 depicts the functions of a concept processor/machine. This hardware configuration evolves concepts from knowledge by disassembling knowledge into noun objects (NOs) and their inter-relating verb functions (VFs). The composite diagram represents graph(s) embedded in a module of knowledge. Many such graphs from different disciplines are processed to detect if an underlying concept can be derived from similar/different modules of closely/distantly inter-related knowledge modules.

The islands of knowledge are decomposed and mapped in numerous ways as graphs and neural nets of noun objects, verb functions, relationships, attributes, similarities, strengths, weakness, etc. The conditions for the stability of these "island objects" (e.g., the creatures of the Galapagos Islands) are investigated from feedback cycles in society, nature and project into the future by restructuring environmental and natural forces that shape the future societies. There is an enhanced probability that the new (noun) objects and (knowledge) islands so derived will have elegant survival qualities. Positive human inputs synergistic with the foundations of science, economics, and constructive technology can help the positive feedback process. Conversely, negative human inputs countering the basis of science, violating the laws of economics, and representing a destructive technology will not only cause the overall instability of the system but also rip it apart.

Examples abound as cycles of war, economic upheaval, and swings of technology become more severe and frequent. Concept machines (with positive human intelligence) can sense, monitor, and predict such conditions for collapse. This approach is quantitative and as quasi-mathematical as econometrics. Furthermore, a numerical breakdown of the various "economic costs" [13] for all the alternatives, including the doctrine of laissez-faire[4] that has come to express the philosophy of individualism and utilitarian ethics [14], now digs deeper into the hands of politicians.

---

[4]The French "physiocrats" of all the eighteenth century, who created this complete system of economics. This simply meant "leave it alone" with a sense of casual indifference to what happened. Unfortunately, there is no "invisible hand" (as defined by the physiocrats) that sets things right by the inaction or neglect of social balancing bodies. At present, it appears that collective wisdom (machine, artificial, and human) wisdom is not only desirable but also necessary for national and global issues such as global warming, world hunger, social injustice, even child abuse and international terrorism.

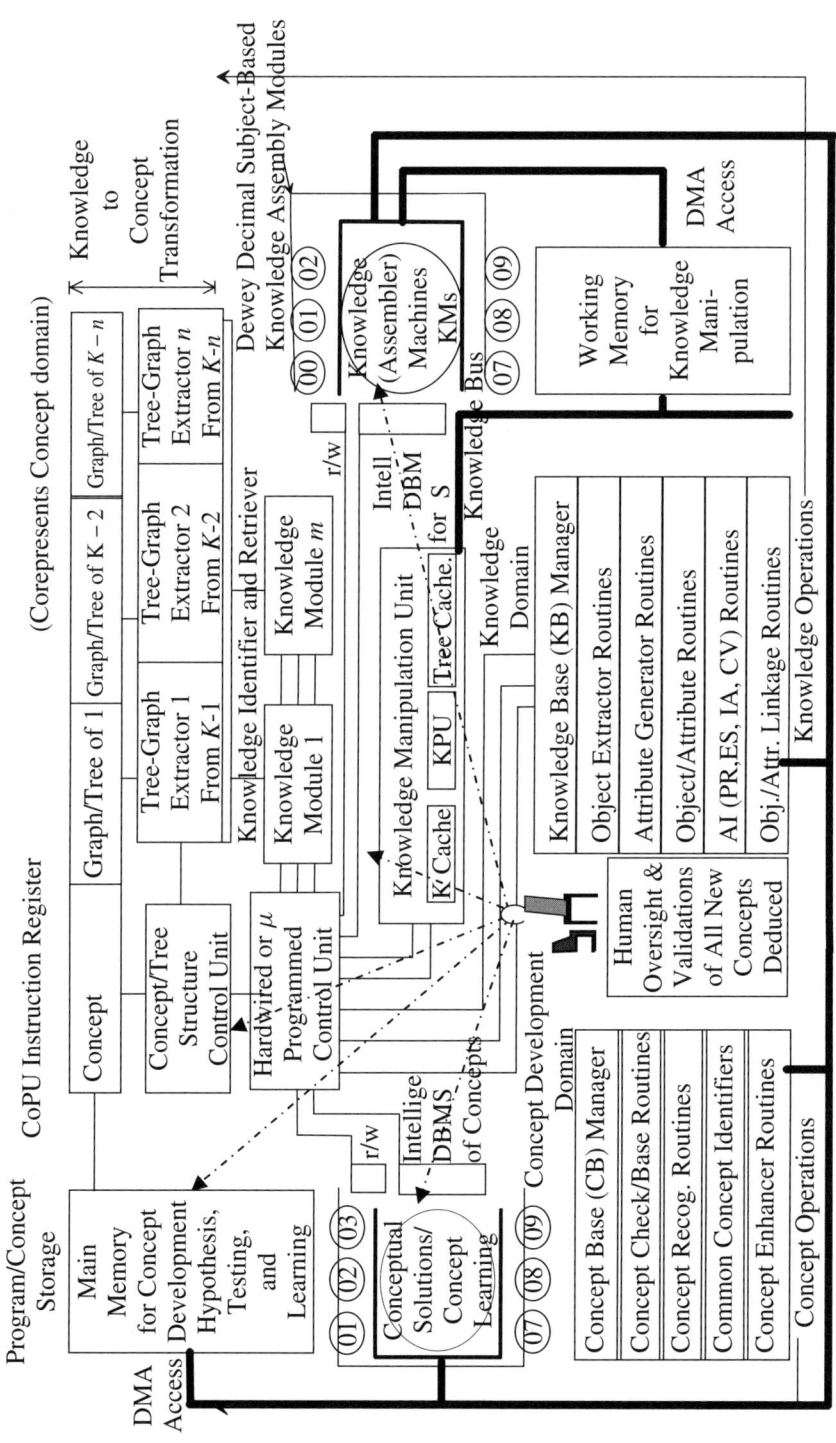

Figure 8.23 The hardware configuration of a concept machine that helps concepts from knowledge evolve by disassembling knowledge into noun objects and their relationships (verb functions).

8.6.2 Concept Processing

Modules of knowledge contain graphs of related objects. This composite diagram of mutually attractive "noun objects" that gravitate toward each other by the linkages of "verb functions" gives the modules their structural stability. In a sense, knowledge is such a graph. When such graphs are drawn on different layers (of thought) of object-oriented graphics programs (such as Designer, CorelDraw, or even common photo editors) and examined with semi-transparent (thinking) pattern-recognition algorithms, then newer images evolve. However, more than that, the gravitation between the vast array of objects starts to exhibit commonality. The commonality and contrasts between the attractive or repulsive forces between noun objects lead to the causes of harmony or disarray of the larger body of knowledge that contains the modules of knowledge. Concept processing is then a derivation of the Newtonian laws for the attraction or dispersion of real noun objects.

Many such graphs from different disciplines are processed to detect if an underlying concept can be derived from similar/different modules of closely/distantly inter-related knowledge modules. The human insight for such processes far exceeds machine intelligence. We present a diagram in Figure 8.23 for a concept processor to help human thought rather than to build a VLSI chip. In the clearest thinking, a human can be an intriguing concept processor.

8.6.3 Human Concept Machines

It is premature to try and build a concept machine since the time, resources, and energy involved are monumental. Instead, it is more human to derive/build the concepts. Concepts unlike wisdom are less clearly defined and highly individualistic during their infancy and nascence.

In a sense, the blueprint for a concept machine can be deduced from the figures for knowledge machines (Figures 8.14–8.17) and medical machines (Figures 8.21 and 8.22). For this reason, we do not present the architectures of any concept machines.

8.7 RECENT TRENDS TOWARD WISDOM

During the last century, computational venues have profoundly influenced human thinking. Alongside human thought, the notion of wisdom has evolved in three discernible directions. The focus of knowledge has also changed accordingly. These effects are summarized as follows.

First, consider absolute wisdom. Modern society has nurtured a few instances where absolute wisdom remains alive and well in the minds of individuals and isolated communities. The authenticity of absolute wisdom is a heritage retained by older generations and painstakingly passed down to younger generations. If culture is itself treated as a complex "object" or entity, then time alters the entropy of culture and altering it means being aligned with modern concepts and values. From archival times, absolute wisdom may be found in Aristotle's notions of truth, virtue, and beauty. Hence, for those who retain this orientation toward wisdom, wisdom remains absolute, even though there is no definition for absolute.

Second, consider materialistic wisdom. Modern society has squeezed reflection time out of human lives. Socioeconomic pressures tend to exaggerate the pursuit of wealth rather than the pursuit of wisdom. The pathways to traditional wisdom are quickly cluttered with materialistic objects rather than abstract and virtual objects. Knowledge is soon abandoned as a lofty goal. To accommodate the modern trend toward materialism, the trend of traditional wisdom has been curved down to "*bizdom*," which includes the business of acquiring materialistic wealth in a form of neo-wisdom. The orientation of such bizdom becomes closer to the SET orientation of wisdom introduced in Chapter 3.

Third, consider political wisdom. Modern society has also squeezed values out of human lives. The passion for wealth also transmutes itself into greed for power. Historically, there has been huge overlap between power and politics in pseudo-democracies around the world. In an ironic sense, the pathways to wisdom reappear as ruthlessness to reach the goals of power and politics. Knowledge is diverted toward a collection of numerous byways (without regard to ethics and social values) that lead to the satisfaction of political and power-hungry ambitions. If such wisdom can be coined as "*polidom*," then its orientation become closer to the DAH orientation of wisdom also introduced in Chapter 3.

8.7.1 Artificial Wisdom

CPUs have been the center of numeric, arithmetic, and logic operation. In an extended mode, the CPU-based machine can function well as a software-driven object machine that can process graphic objects, blocks of text, tables, and spreadsheets. Such machines are unable to generate axioms of generic wisdom. The derivation of wisdom calls for scientific validation, representation of universal truth, an economic basis, and social value. Machines can penetrate deeply into the scientific, universal, and economic domains and create solutions consistently based on a given confidence level. The highly unpredictable task of evaluating social value can be accomplished by appending a "societal machine" that has yet to be built.

Machines can however scan social and cultural knowledge bases to determine if the proposed axiom (or similar axioms) of wisdom has any social value. Machines can also extrapolate if an axiom of processed wisdom is likely to have future social value. Human insight can only navigate such machines.

In the same vein as we presented the concept processor unit in Figure 8.23, we present a wisdom processor unit in Figure 8.24. The architecture exists to help human thought processes and their visualizations in guiding the functions of a wisdom machine. If concepts are a pathway to wisdom, then a clear cataloguing of related concepts can steer through such concepts to find an integrated theme on which a new axiom could be based. Conversely, a catalogue of unrelated concepts leads to the fundamental properties of dissimilar noun objects and their

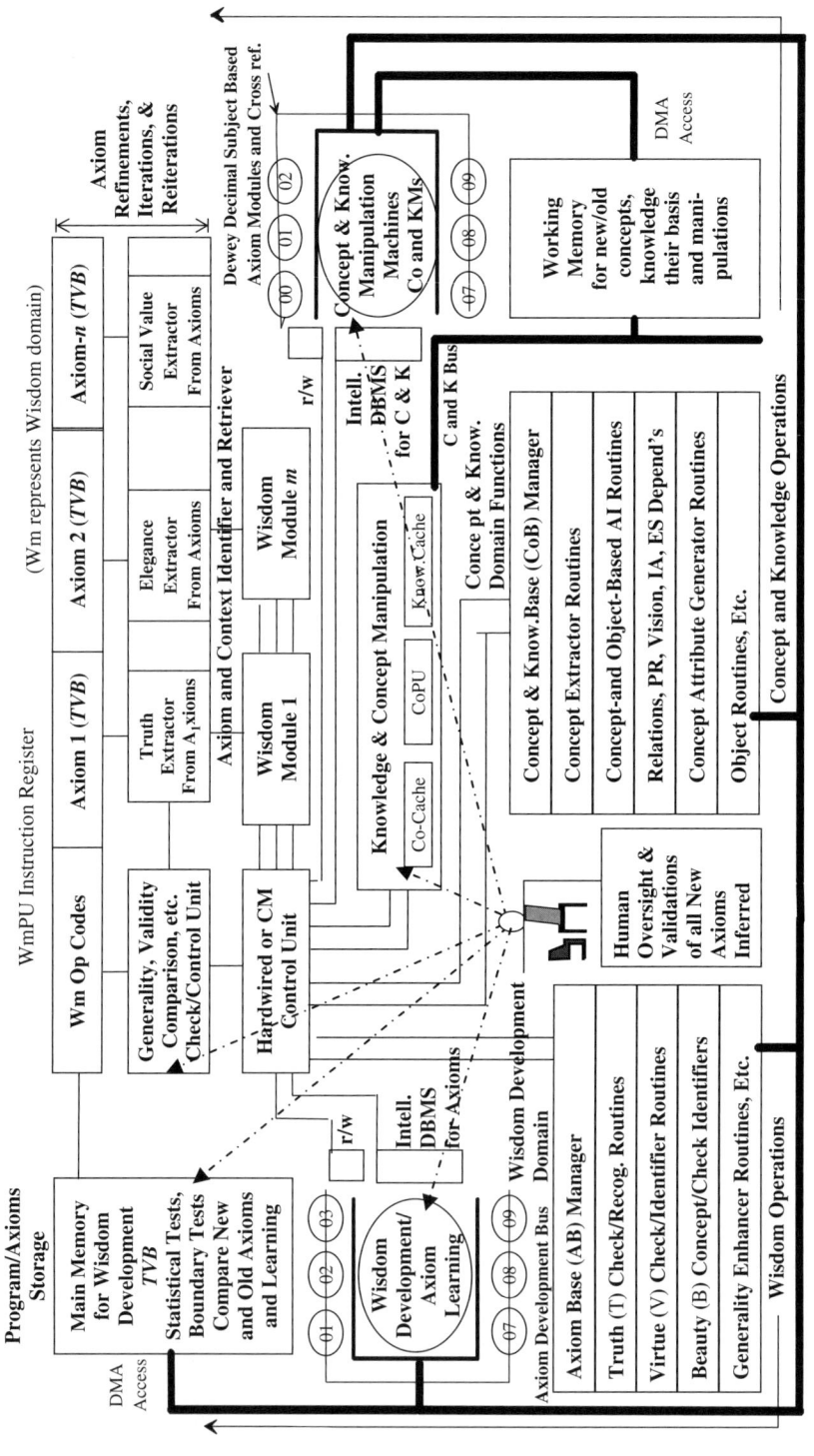

Figure 8.24 Configuration of a wisdom processor unit (WmPU) to facilitate TVB composition (if any) from the derived concepts and then refine the axiom(s) to suit the social and cultural setting.

verb functions etc. The wisdom machine leaves no room for nature to hide in obscurity.

The hardware configuration of a wisdom processor unit (WmPU) creates wisdom from concepts and knowledge by verifying universality, social value, and elegance, which are (statistically) likely to have the most knowledge and concepts embedded in a statement, equation, theorem, invention, etc. The machine can be commanded to find the "best" piece of literature, art, or movie. It is capable of creating a masterpiece, if the requirements of a masterpiece are embedded in the objective function (OF).

8.7.2 Old and New Wisdom

Wisdom like music and art does not have natural, intellectual, or aesthetic boundaries. The partitioning in classical (wisdom), neoclassical (bizdom), and modern (polidom) forms of wisdoms is merely an artificial venue for forcing the pre-selected role of a wisdom machine. From the history of human thought, the strides of wisdom appear almost random, but the footprints statistically align themselves in these three directions. As much as music or art is partitioned into classical neoclassical, and modern groups, wisdom can also be artificially confounded in classical neoclassical, and power-(or politics-) based wisdom.

Wisdom machines can be built as a cluster of machines supervised by humans or by groups of humans assisted by machines. The blending of old (classic) wisdom and new (neoclassic) wisdom can be controlled by the user. This choice offers another degree of freedom between the TVB-based or DAH-based orientations of wisdom (see Chapter 3). In fact, the absolute and TVB-based wisdoms exhibit overlapping directionalities, as do the power- and DAH-based wisdoms and materialistic and SET-based wisdoms. In general, any strain of wisdom can be synthesized by machines to blend the old with the new: the TVB with SET, the SET with DAH, and inwardly DAH with outwardly TVB wisdoms.

Figure 8.25 depicts a configuration in which the old and new breeds of wisdom can be artificially blended by the user. It is possible to build a general-purpose wisdom machine that can offer different shades or hues of wisdom. The wisdom that the machine generates lies in the wisdom of the user. In a sense, the axiom of wisdom (beauty lies in the eye of the beholder) is generalized as a series of axioms, such as truth lies in the honesty of the interpreter, wisdom lies in the virtue of the user, etc. Such axioms can be rewritten as the beginning and end of wisdom (in the limited context of available knowledge) to exist in the wisdom base of the machine and that is personalized according to the sense of values of the user.

A wisdom machine with inter-leaved computational and human elements generates new axioms (see the top right, of Figure 8.25) with a variable blending of "old" and "new" wisdoms to suit the tastes of the user. It deploys all disciplines based on semantic Web access. Data bus structures are inter-leaved with human communications channels to mix, merge, and synergize age-old wisdom with the

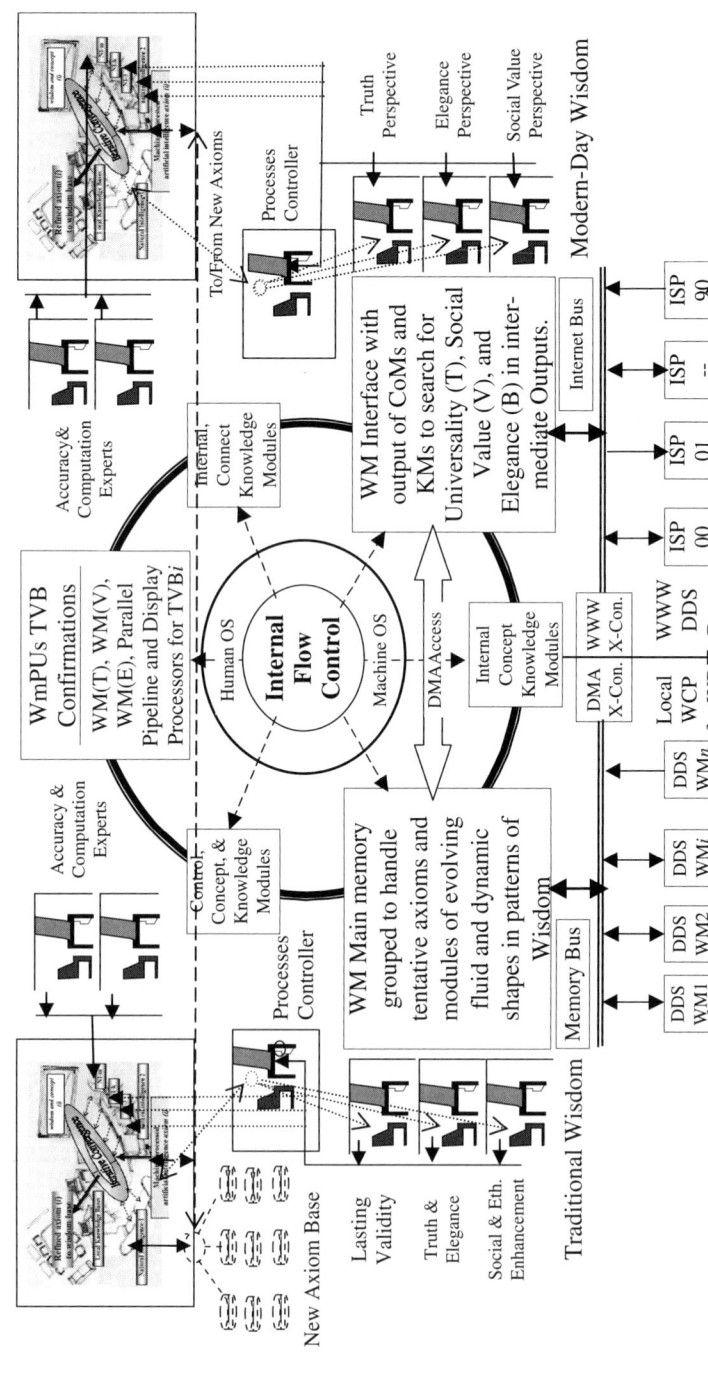

Figure 8.25 A wisdom machine with inter-leaved computational and human elements to generate new axioms (see top right). It deploys all disciplines based on semantic Web access. Data bus structures are inter-leaved with human communications channels to mix, merge, and synergize age-old wisdom with modern-day concepts.

Note for Figure 8.25: B = elegance, within the new axiom, device, concept, book, commodity; CoM = concept machine (see previous architectures); DMA = direct memory access; DDS = Dewey Decimal System; KM = knowledge machine; ISP = Internet service provider; T = truth in the statement of wisdom; Machine OS = operating system for the wisdom machine; Human OS = operating system for the human beings operating the wisdom machine; V = virtue and social benefit from the new invention, device, concept, book, commodity, etc.; X-Con = cross-connect; and finally, WCP = wisdom Control point. The functions of the entire WM is equivalent to the standard operating procedures for executives in corporations.

modern-day concept interpretations of wisdom. Music and art are also amenable to such blends in their own respective realms.

8.7.3 Wisdom in Machines and Instinct in Humans

Machines can retain the basis of wisdom and they can scrutinize local human moves and compare them with optimal/expert-system-based moves. Based on expert-system rules, machines can block certain harmful moves in society by other human beings. In society, such examples exist when police stop the movements of suspected terrorists, or when ATM transactions are aborted when fraud or a bank over draft is detected.

When machines detect corruption or an illegal transaction, the (knowledge bases of) legal framework needs to be interrogated for a partial/entire illegality. When localized social and ethical laws are violated, a knowledge base of society-based standard operating procedures (SOPs) needs to be interrogated. In corporations and financial transactions, such a procedure of checks and balances tends to keep corporate crime under (some) control. When the arrogance of political leaders [Nixon's Watergate scandal, Clinton's indiscretions with Monica Lewinsky, Bush's non existant weapons of mass destruction (WMD) deception, etc.] or the indulgence of corporate executives (at Enron, Arthur Andersen, Global Crossing, etc.) becomes an ongoing trait, then a knowledge base for a set of rules of conduct and ethics needs to be investigated.

The evolution of modern society has failed to create such knowledge-bases for the rules of conduct, ethics, and behavior for leaders and executives. In a sense, human arrogance is the impediment. As much as a computer programmer will not like being instructed by a machine about how to code, the wealthy and politically powerful will not like being blocked by a machine on how to deploy the wealth and power to which they have access for a while. Some of the most damaging moves by the politicians (such as invasion of Iraq despite UN warning about the absence of weapons of mass-destruction (WMD)) or by the corporate executives (such as the efforts to cover up the impending Global Crossing bankruptcy) can be averted or exposed on the media.

When this logic is pushed to an extreme, the conflict between the unethical use of wealth and responsibility toward society can be partially resolved by machines, as much as labor–management conflicts (see Section 5.B.2 in Chapter 5) can be

412 KNOWLEDGE SYSTEM ARCHITECTURES

resolved on an economic basis. The role of wisdom (artificial and/or human) is to find an intricate and unbiased balance. As a judicial system resolves conflicts, the cumulative and weighted average of truth, virtue, and elegance in society is implanted as encoded social expert-system laws, in wisdom machines.

8.8 THE TREND OF VALUES AND ETHICS

Along the knowledge trail (Figures 2.5a and b), values and ethics remain deep in the human domain. However, close to the philosophic attitude toward Aristotelian notions of truth, virtue, and beauty (TVB), values and ethics are social derivatives. Transient societies have inculcated cruel and fragile ethics and its converse. Stable and long-lasting societies reflect the theme of ancient Romans favoring TVB disposition over DAH disposition, wisdom over ignorance. Ironically, Romans favored themselves over their slaves, Romans over Israelites, and men over women.

Barring their intolerant social attitudes, what the Romans preferred for them had and still has a lasting impact that can be realigned with the knowledge society and information technology. One of the functions of the wisdom machine is to distill knowledge and purify it from the cruel concepts of deadly societies. Hence, the customization of wisdom to suit modern times can be partially a machine function and human ingenuity. The machine functions establish scientific validity and human ingenuity enforces practicality. The SET (Chapter 3) orientation of wisdom can only strengthen the TVB wisdom of modern days, leading to stable societies. If the SET orientation is indifferent to TVB or DAH orientations, this can also strengthen the DAH wisdom of modern days, leading to cruel and unstable societies.

The newer CPUs refuse to execute self-destructive commands. The role of social organizations is much more passive. Subversive instructions from political leaders do not automatically initiate whistle blowing or the exposition of corruption. DAH wisdom can thus survive to see another human disaster in the name of political freedom.

8.8.1 Ethical Issues and Role of Machines

The elite have commanded machines. In a very insidious fashion, humans have blocked machines from being corrective or even suggestive when they commanded machines to carry out unethical and definitely harmful instructions. This mold was recently broken when CPUs (e.g., the recent Intel®64 ± processors with an execute disable bit) refused to execute self-destructive actions and network virus' propagation command.

In general, the options for traditional machines are limited by the choice of operating systems and system software resident in memory. In short, machines can only provide computing power and humans have to provide direction and creativity. When machines are equipped with concept and knowledge processors to execute knowledge, concept, and wisdom instructions, the realm of freedom for the machine is its ability to compute solutions that optimally fit knowledge, concepts, and wisdom given the current socioeconomic situation. Machines may execute Marshall's laws of economics and tune them according to Web knowledge bases for universality, elegance, and social values; but the ethical framework of future generations can only be extrapolated by machines as much as it can be fuzzily guessed by human beings. However, the prediction is based on a quantitative basis derived from authenticated data. It is our contention that the two components (prediction and fuzzy guess) can only reinforce and strengthen each other. In this unexplored space of union between thought and computation, the ingenuity of humans based on the wisdom to be gained from machines can find an Athena of creativity.

8.8.2 Conflictive Roles of Humans and Machines

The elite (wealthy and powerful rather than wise and socially benevolent) command and prime the next-generation machines. When the founding members of elitists become depleted of the qualities that gave them wealth and power, a reversal process starts to take shape. Societies, cultures, and nations are drawn into the relentless cycle of swings from every perspective, ranging from selfish economics to self propagating ethics.

Networks in the information age and machines in the knowledge society both enhance the intensity and frequency of such a cyclic swing. Conflict is intensified and wars emanate. In a sense, traditional computer and network technologies can give rise to human conflicts (e.g., Israel and Palestine, both deploying, lopsided IT tools and expert-system approaches) and misconstrued wars (e.g., the U.S. invasion of Iraq based on unfounded reports of WMD).

The false instinctual drives of humans can hack into layers of network and machine security. Abuse of wealth and power has no bounds. In conflict situation humans will win, since code is written based on experience favoring wealth and power and self interest creativity gets projected into the future.

Social acceptability in the balance (based on social expert-system rules) implanted in the machine and the TVB human component implanted in human counterparts with veto power over each other can offer the viable solution in a peace institute or research center. The role of the United Nations will become unbiased and uncorrupted. Unfortunately, the converse scenario is equally possible, whereby the wealthy and powerful nations can not only influence the expert-system rules in the machines but also bias and corrupt the negotiators. The power to be unbiased is not to be controlled by wealth and power. Machines can offer immense protection from power hackers and corruptors, if their TVB-based operating systems are preserved in tact.

8.9 CONCLUSIONS

This chapter provides architecture for the hardware configurations of object processor units (OPUs) and object machines (OMs). Knowledge and medical processors are presented and discussed in detail. Knowledge and medical machines that can be derived from these configurations are also presented. We take the bold step of presenting concept and wisdom processors to introduce the methodologies to derive concepts from knowledge and wisdom from concepts and knowledge. The purpose is to highlight that human perseverance can enter the realm of being concept-driven and wise. Innovations and creativity become by-products of this systematic and perseverance-based methodology. It must be appreciated that a flow diagram of human activity can be deduced from the architectures of concept and wisdom processors. The major functional blocks of human activity to become concept-oriented and wise can be inferred from the architectures of knowledge and wisdom machines.

Machines by themselves cannot invent. However, being driven by human intuition, machines can perform scientific, validity, economic, and social acceptance checks for inventions with their incremental innovations. Machines can also provide a basis for the enhancement and modification of a nascent concept or its link to other disciplines. When real-time feedback is provided by knowledge and wisdom machines, human ingenuity and intuition are enhanced. A higher level of scientific nirvana is reached. This effect is similar to that of computer-aided design methodology in the optimizing engineering designs in any given society.

Machines cannot provide the solution to all social ills and, the solutions provided are robust and quantitative. The alternatives are examined based on the economic costs of each step in a string of possible solutions. The graph of the final solution is an optimal selection of the numerous steps in the proposed solution. Various strategies—(minimization (expected risk, cost, exposure, etc.), maximization (expected gain, profit, reward, etc.), minimax (minimize maximum risk, loss, exposure, etc.), maximin (maximize minimum gain, reward, profit, etc.)—become immediately feasible using standard DSS, MIS, and/or IT software. The methodology is borrowed from economic analysis, decision support systems, and operations research.

The nature and extent of solutions can be user alternatives. Blending and administering the nature of wisdom in the machine-proposed solution depend on access to and use of Web knowledge bases. The local knowledge bases of standard operating procedures can provide very rapid results for the crisis management of social problems. For a global solution to pervasive problems such as world hunger, terrorism, global warming, and extinction of species, machines provide robust and well-articulated solutions with quantitative data and finite confidence of accuracy to support its conclusions and expected benefits and costs of implementing such solutions.

REFERENCES

1. J. von Neumann, "First Draft of a Report on the EDVAC," Contract No. W-670-ORD 4926, U.S. Army Ordinance Department and University of Pennsylvania, Moore

School of Electrical Engineering, June 30, 1945. Also see A. W. Burks, H. H. Goldstine, and J. von Neumann, *Report of Ordinance Department*, U.S. Army, Washington, DC, 1946.

2. G. Estrin, "The Electronic Computer at the Institute of Advanced Studies," in *Mathematical Tables and Other Aids to Computation*, Vol. 7, IAS, Princeton, NJ, 1953, pp. 108–114.

3. A. Clements, *The Principles of Computer Hardware*, Oxford University Press, New York, 2006.

4. Gale Reference Team, Infineon Introduces POF Transceiver, *Fiber Optics Weekly Update*, 26, Issue 20, May 19, 2006.

5. S. V. Ahamed and V. B. Lawrence, *Intelligent Broadband Multimedia Networks*, Kluwer Academic Publishers, Boston, 1998.

6. N. Waraporon, "Intelligent Medical Databases for Global Access," Ph.D. dissertation, City University of New York, 2006. See also N. Waraporn and S. V. Ahamed, Intelligent Medical Search Engine by Knowledge Machine, in *Proceedings of 3rd International Conference on Information Technology: New Generations*, IEEE Computer Society, Los Alamitos, CA, 2006.

7. M. S. Pinto, "Performance Modeling and Knowledge Processing in High-Speed Interconnected Intelligent Educational Networks," Ph.D. dissertation, City University of New York, May 2007.

8. N. Mollah, "Design and Simulation of International Intelligent Medical Networks," Ph.D. dissertation, City University of New York, 2005. Also see S. V. Ahamed and V. B. Lawrence, Evolving Network Architectures for Medicine, Education and Government Usage, in *Proceedings of Pacific Telecommunications Council*, PTC, Honolulu, HI, 2002.

9. N. Goldburt, "Design Considerations for Financial Institution Intelligent Networks," Ph.D. dissertation, City University of New York, 2004.

10. B. Khailany, T. Williams, J. Lin, E. Long, M. Rygh, D. Tovey, and W. J. Daly, A Programmable 512 GOPS Stream Processor for Signal, Image, and Video Processing, *ISSCC Digest of Technical Papers, ISSCC*, 2007: 272–602.

11. J. O. Boese and R. B. Robrock, Service Control Point, the Brains behind the Intelligent Networks, *Bellcore Exchange, Bellcore, Navesink, NJ*, December 1987.

12. S. V. Ahamed, *Intelligent Internet Broadband Networks*, Wiley-Interscience, Hoboken, NJ, 2007.

13. J. P. Van Leeuwen and H. J. P. Timmermans (Eds.), *Innovations in Design & Decision Support Systems in Architecture and Urban Planning*," Springer, New York, 2006. Also see M. Krol and D. L. Reich, Development of a Decision Support System to Assist Anesthesiologists in the Operating room, *Journal of Medical Systems, Kluwer Academic Publishers Boston*, 24, 2000: 141–146.

14. S. Gordon "The London Economist and the High Tide of Laissez Faire", *Journal of Political Economy, University of Chicago Press, Chicago, IL.*, 63(6), 1955: 461–488.

CHAPTER 9

HUMANS, MACHINES, AND NETWORKS

CHAPTER SUMMARY

Human networks existed long before computer and machine networks. The social, and application aspects of human networks are merged into the hardware, software, speed, and accuracy of machines and networks in this chapter. The basis for the existence of human networks is derived for the needs of individuals, corporations, and social entities. The innovations for machine networks are derived from their economic value. When these four aspects (*needs, innovations, machines, and networks*) are portrayed as the four sides of an open pentagon, then ever-present social progress, constitutes the fifth side. The inter-dependence of these five aspects in a rapidly evolving pentagon is explored in this chapter.

The conceptual bond between humans, machines, needs, innovations, and social progress is the manifestation of natural intelligence innate to human nature. The sense of balance necessary to maintain a dynamic equilibrium is explored in this chapter. The knowledge society has materially altered this balance through glossy information highways. Silicon and glass have altered the rhythm of the mind that works as alpha, beta, and gamma wavelengths in the neural nets of the brain.

If harmony is restored while not being downtrodden by global treachery of junk/deceptive information, machines need to more human rather than for humans to be more robotic. The middle ground for unison is further delineated and charted

---

Computational Framework for Knowledge. By Syed V. Ahamed
Copyright © 2009 John Wiley & Sons, Inc.

in this chapter by making the innermost software layers of machines tamper-proof against deception, greed, arrogance, and hate that surface in scum of the society. Humanistic machines (Chapter 5) acquire the behavioral traits of humans rather than the mechanical traits of robotic arms. The most modern innovations that have propelled machines and their networks are now diverted toward building the bridges between human needs and their elegant and noninvasive solutions. These customized solutions that humans and wisdom machines (Chapter 2) generate appear to be longer-lasting than the primitive slash-and-burn approach that is characteristic of the quest for knowledge.

9.1 ETHICS AND VALUES IN HIGH-TECH SOCIETY

Positive values and prudent ethics in society make it enduring. Much of the financial and social transactions stems from business and commercial activities. These activities offer stability to and the expansion of the economic base that sustains society. The strength of the economic fabric within society reinforces its intellectual and cultural growth, thus contributing to values and ethics in society. Peace and justice prevail and harmony endures. This idealistic and positive full cycle is most often interrupted by unrest and wars on a macroscopic basis and by deception and greed on an microcosmic basis. In its degenerate mode, segments of this full cycle foster viral hot beds (e.g., in Vietnam, Korea, Afghanistan, Iraq, Palestine, etc.) of hate and violence, fear and agony (e.g., in Burma, Darfur, Kashmir, the Occupied Territories, rural city blocks, etc.), within local (global) society.

The role of human beings who are pivotal in stabilizing the full positive cycle of an enduring society can fluctuate precipitously, Ranging from the deeply entrenched positive foundations of ethics and tolerance in Buddhism, Christianity, Islam, etc., to numerous negative offshoots in political scandals, corporate thievery, white-collar crime, etc., have emerged. Traditionally, society has deployed "positive (good)" human agents to block "negative (bad)" human agents. In the earlier era, machines were plain with traditional operating systems, hardware, software, firmware, sensors, I/O devices, and their interfaces driven by the thought processes of software designers. As society moves deeper and deeper into the information age, computer systems programmed with AI programs, machines, and networks have become necessary, playing an important corrective and stabilizing role. In recent times, human vision and wisdom integrated deep into the core (hardware and firmware) and crux (software and application) of the machines toward a peaceful new world need to supplement the arithmetic and logical functions of the older machines.

The ethics and values of a nation are based on the wisdom in the constitution and on the legal system to be able to enforce it. The wisdom of a nation consists of refined and derived axioms (from knowledge and concepts) and that have evolved over a period of time, accepted by society at large and documented in national archives of established knowledge (and wisdom) bases. The knowledge and concepts that are the precursors to these axioms also need to be stored in the event that an enhancement or a modification of an axiom is necessary. An insidious stabilization of the (knowledge-concept-wisdom-ethics) cycle is achieved. Like the national archives of science, the national archives of wisdom provide a foundation for ethics and values that guide the legal arm and judiciary of a nation. The evolutionary steps of the wisdom-cycle of any nation is accumulated and iteratively refined over the history of its growth. The chain relies on judgment and rationality to derive successive axioms of wisdom. The wisdom trail is transparent to later generations as a clear deduction, as logical (and if possible) as a mathematical derivation. The language of concepts and wisdom has not evolved sufficiently to be able to compile knowledge and concepts to derive a (wisdom) machine-executable code. However, refined lexical analysis techniques along the knowledge trail (see Section 2.5) will facilitate due processes in deriving the axioms for the wisdom of nations.

The constitution and concurrence behind the collective will of society in any democratic nation carry through the embryo of wisdom to the evolution of the nation. Wisdom learned during the growth of the nation provides the blueprint for future growth. A similar procedure occurs at the inception of corporations and human organizations. A documented set of ethics and code of justice leads to the knowledge trail (Figure 2.5) that instills wisdom, concepts, knowledge, information, and data to guide the activities of the nation. Prior to the information revolution and knowledge networks, the documentation and reinforcement of evolutionary concepts for nations were manual and dexterous.

Over many decades, the human knowledge trail (Figure 2.5) becomes ingrained and the seven nodes are implied in academia. It converges into the ethical foundations of the very society that hosts the constituents, who will become a nation. The nation generally follows the path of convergence led by the charismatic and social appeal of the leaders. As the nation ages, the trail gets worn and leaders become stale. Charisma gets infused with a strain of the greed and selfishness of leaders who violate the notion of "selfless wealth." Both these traits are evident in the leaders of many nations (especially those of aging nations). The wealth of nations fills the bank accounts of corrupt politicians full of "selfish wealth". Stagnation in the wealth of wisdom becomes a precursor to the decay of a nation. Innovative concepts and constructive knowledge decline. Senseless information in networks and worthless data streams fill the fiber-optic pipelines. The wisdom of nations is soon crucified on the Cross of personal wealth.

In this book, the theme is for machines to think as a nation but act as well as (if not better than) the human being. Collective and global wisdom is the driving force behind these new machines. It is our contention the if such machines are networked into knowledge and wisdom bases [1] that collect and distribute validated standard operating procedures [2] of human society and networked machines, then these machines will prevent human beings from falling into a downward spiral of social bugs and viruses. These new machines will prevent other machines from being infected by social bugs and viruses. Such machines can also track their human[1] counterparts who are infected by "societal" bugs who go on to damage a healthy and working society. Such steps are routinely practiced in the field of medicine and enforcement of computer security. The human–machine team will be a part of the wisdom machine [3], in which a group of independent and autonomous leaders (like Supreme Court justices nonpolitical United Nations professionals, etc.) would play the lead role and machines would provide global access to world-wide knowledge banks based on intelligent Internet knowledge networks [1]. With an internal machine-based on audit system, the chances of another Hiroshima, Watergate scandal, or Chernobyl disaster will be averted. The past axioms of wisdom [3] in computers are built into the "humanware" of these brand-new machines.[2]

9.2 NEEDS THAT DRIVE

Individuals and organization have needs. An organism without needs soon finds itself unneeded. The balance of interaction and communication hinges on perceived values of assets exchanged that satisfy reciprocal needs. There is room for perceptual differences, negotiations, expectations and generosity. The human variations make the circumstances different for humans. Whereas humans guess, machines compute, weighing and considering circumstances precisely, and explore choices based on logic embedded in the decision support systems, their reason and wisdom. Machines also compute expected losses to avoid any impending dangers because of reckless moves.

Organisms can range from an individual to a developed nation. Order within the nation is the basis of its existence. Organisms become aware of their self-existence and the conditions for survival. The inherent need to maintain and conserve themselves becomes a major concern for social and biological organisms.

Complexity of order within the organism is amplified in two ways. *First*, as the level of needs varies from the rudimentary (physiological) to the abstract (realization [4] or unification [5]), the gratification of the need becomes increasingly sophisticated. *Second*, as the organism varies from a rudimentary form (e.g., a cell, an individual, etc.) to an evolved social structure (e.g., an organization, a nation, etc.), the dynamic relationship between the needs and their gratification also becomes complex. It becomes clear that a generalized theory of needs satisfaction for every need, for every organization becomes impossible; instead, it becomes feasible to optimize the needs gratification diagram or a graph for

---

[1] Physical security procedures in airports and public places are well accepted. Canines track drugs. Automated devices track explosives. The trend is already established for machines to track socially destructive individuals and organizations before prolonged damage is done.

[2] We propose that humanware refer to the group of individuals (and their documented code for) enforcing social values and ethical norms in the daily functions of machines and networks. The existing context of peopleware who inject technical and managerial personnel (and their code for) into enhancing and smoothing the activities of organizations is left intact.

420 HUMANS, MACHINES, AND NETWORKS

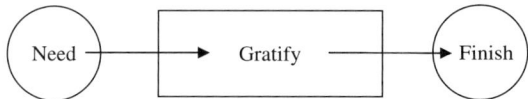

(a) Three-step gratification of primitive needs.

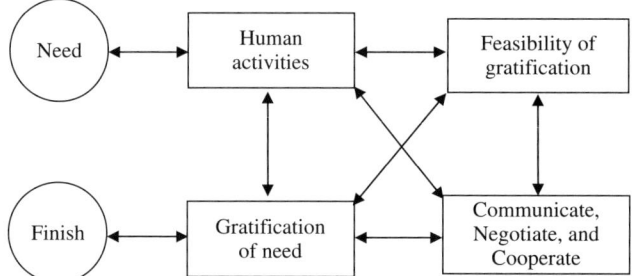

(b) Contemplated human approach to resolution of complex needs invoking communication and cooperation within society (see Figures 9.2 and 9.3).

Figure 9.1 Evolution of the pattern for resolving primitive and complex needs.

dependability, repeatability, and learning to strategize on an individual basis. The search for solutions and strategies (the best design for a spacecraft, fuel-efficient cars, etc.) involves the combined best efforts of humans and machines over a long time. For this reason, we start from the needs that drive human beings who design and build systems that serve others, as depicted in Figure 9.1, and then offer machines that complement human functions.

9.3 NETWORKS THAT TRANSPORT

The fabric of society rests on the backbone of communication. The essence of social transactions is conveyed by human interactions. In the age of multigigabit transport capacity of networks, human interactions and communiction have become synonymous. The need to communicate, cooperate, mitigate, and negotiate while resolving complex needs becomes an integral part of the gratification of needs. The core component is communcation, and the most commonly used methodologies are based on an Open System Interconnect (OSI) model with frame relay techniques and the asynchronous transfer mode (ATM) of communication.

Modern switch systems and routers handle conventional digital traffic, computer communication, and Internet communication with the utmost flexibility. Almost all digital networks provide data transport with a high degree of reliability. The configuration of a typical network segment is shown in Figure 9.2. The immense flexibility required for various networks to be connected and inter-connected, and also serve a broad spectrum of network and communication services, is evident. The key component in the figure is the add-drop multiplexer (ADM) [6].

NETWORKS THAT TRANSPORT 421

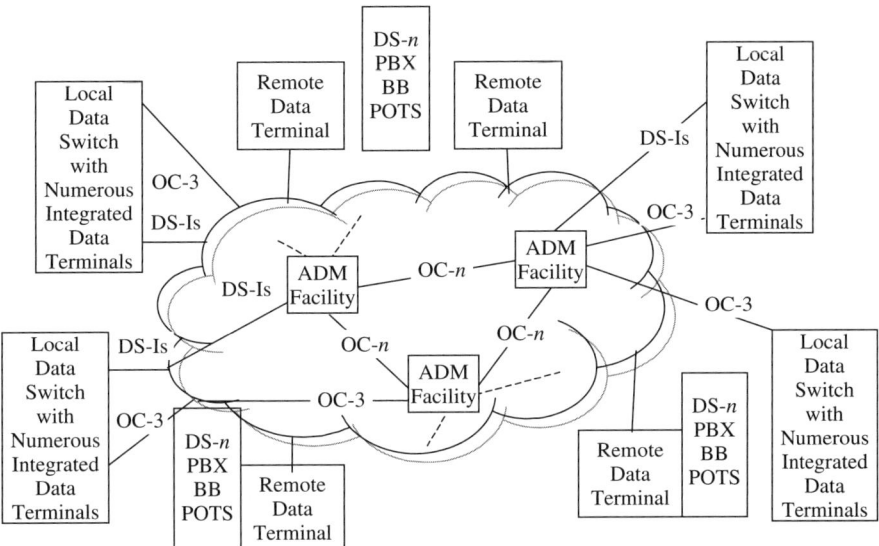

Figure 9.2 Interconnectivity for optical carriers with other digital carriers (such as DS-1, DS-3, etc.) and services (such as plain old telephone or BAE-band services, etc.). ADM = add-drop multiplexers; PBX = private branch exchange; POTS = plain old telephone services; DS-1,3, and 4 = digital signals 1, 3, and 4.

Digital cross-connect switches (DCS) inter-connect with SONET-based optical carriers (OC-n) with traditional circuit-switched network digital signal (DS-m) carrier systems to offer the seamless backbone network of any nation or community. The age of fiber merges smoothly into the age-old networks of the 1950s and 1960s. The introduction, refinement, and acceptance of the OSI model shown in Figure 9.3 resulted in local and global communication. A more compact version of this model is illustrated in Figure 9.4.

The functions of modern society get through the fabric of modern networks implemented as data streams for networked computers driven by application programs. Recent advances in perfecting the graphical user interface (GUI) between users and computers provide the link between users with their individual needs and the data carried in the backbone networks via a hierarchy of application programs.

To a limited extent and in targeted applications, such as business, graphics, computer-aided design (CAD), or management information systems (MIS), etc., specialized application programs bridge the gap between human beings with their targeted needs and the data streams in networks. It is our contention that specialized (knowledge and concept) machines with their unique software directed toward the human needs, social needs, and global needs of society can serve human beings and social organizations directly. Layers of specialized software (knowledgeware and conceptware) executed in traditional or specialized hardware (knowledge and concept machines) will provide the mechanized link between human thought and its manifestation as data streams in the networks.

422 HUMANS, MACHINES, AND NETWORKS

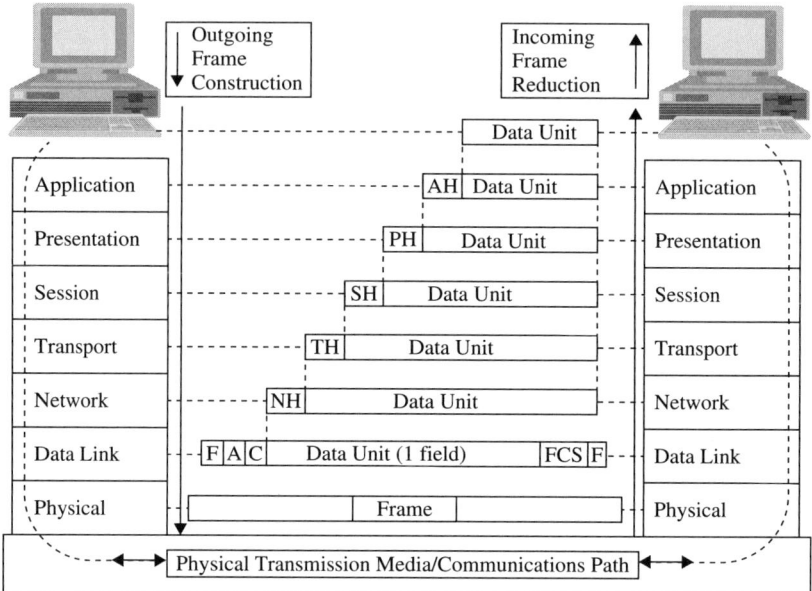

Figure 9.3 The functional representation of the seven layers of the open-system inter-connect reference model. A = address field (one octet); C = control field (one octet); F = flag sequence (one octet); FCS = frame check sequence (two octets); xH = header block at layer x.

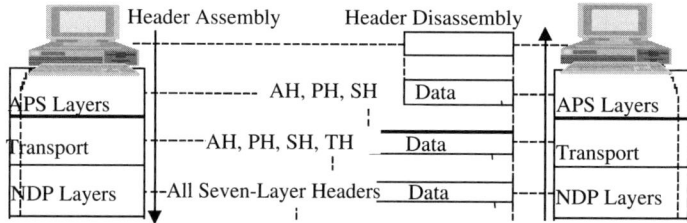

Figure 9.4 The simplification of the established seven-layer open-system inter-connect model into three layers. APS = application, presentation, and session layers; NDP = network, data link, and physical layers of the traditional OSI model; H = header information.

Reassembled at the receiver end, knowledge and concepts will be integrated, quickly and effectively, in human minds via customized graphs, graphics, movies, or any other multimedia techniques. The source and target human beings will be spared of knowledge and concept functions that machines can perform with greater ease and accuracy.

These new machines cannot satisfy needs in a physical or materialistic way unless they are connected to the robotic arms of mechanical devices. Instead, these machines provide an information profile and a conceptual linkage to/from knowledge banks or their complied and personalized information that will satisfy the particular need in any given particular socioeconomic setting. Currently, human beings such as doctors, lawyers, consultants, etc., provide such informational/consultative services. Furthermore, these machines being connected to knowledge, concept, wisdom, and ethical conduct banks also serve better than their average human counterparts. In the medical field, AI- and expert-system-based [7] programs outperform an average physician, much like a computer can beat an average chess player, or create a verse or poem better than the average human being could compose. In such situations, machines outperform the average human but rarely the exceptional individual.

9.4 OVERLAP OF NEEDS AND NETWORKS

In a nutshell, humans needs are satisfied as networks carry out the procedures and strategies for the gratification of such needs. Optimal strategies for solution are essential to resolving the needs arising from most complex needs. The artificial intelligence (AI) algorithms incorporated in Figure 9.1b, permit the recognition of patterns of needs and their solution strategies. Sensing and learning algorithms permit the design of optimality in solution strategies. However, as needs and societies become complex and inter-dependent, the simple enhancement of Figure 9.1b will not suffice. A new hardware platform and algorithms become necessary to implement these new solutions. For example, if the problem is generic and global, the search for strategies becomes crucial to the solution, and inputs from numerous cultural variants, economic settings, war and peace, etc., also need to be included in the overall solution. In addition, the problems are sometimes important enough to require specialized knowledge and its evolution. For these reasons, we broadly classify organizations (Sections 9.4.1, 9.4.3, and 9.4.5), the levels of need that drive them, and some possible generic hardware platforms (Sections 9.4.2, 9.4.4, and 9.4.6) that behave with the collective wisdom of the population and social benovolence as the goal. Such hardware platforms can simplify knowledge, concept, and wisdom machines based on the embedded intelligence of the recent knowledge society, or such machines can be humanistic, based on the values and ethics of the entire human race.

9.4.1 The Human and Social Aspects

Needs drive individuals and societies. For machines to respond to such needs, algorithms, strategy, and implementable code become essential. Human needs may thus be categorized for the machines to respond accordingly. Maslow introduced a well-documented five-level needs pyramid [4] for traditional office and production workers in the 1940s [8]. Individual motivation and personality were linked throughout the 1960s [9, 10] and 1970s.

Simon and March [11, 12] approached the behavioral aspects of individuals and organizations from a rational and economic basis. Other significant

424 HUMANS, MACHINES, AND NETWORKS

contributions from social scientists such as Fromm [12], Mead [13], Chomsky [14], etc., also exist for individuals, firms, and societies before the age of information. Ahamed [1, 5] has enhanced the five-level needs pyramid of Maslow to a seven-layer open structure for the information age human who "searches" (at the sixth level) for the best or optimal solutions to his or her needs. At the pinnacle of needs resides the human longing to unify (at the seventh level) knowledge and wisdom into one universal order. All other needs, machines, and networks become subservient to this illusive and exhaustive need.

9.4.2 Individual Needs

An open-ended needs structure (at the top) is presented in Figures 9.5. For an individual, the traditional needs pyramid of Maslow [4] is constituted by the lower five layers of Figure 9.5 and is the basis for designing the artificially intelligent platform of knowledge machines that provide customized information leading to the solution of individual and customized needs.

9.4.3 Individuals, Machines, and Networks

Machines and networks operate in synergy and unison to serve humans. As humans need[3] cooperation and communication within society, machines and

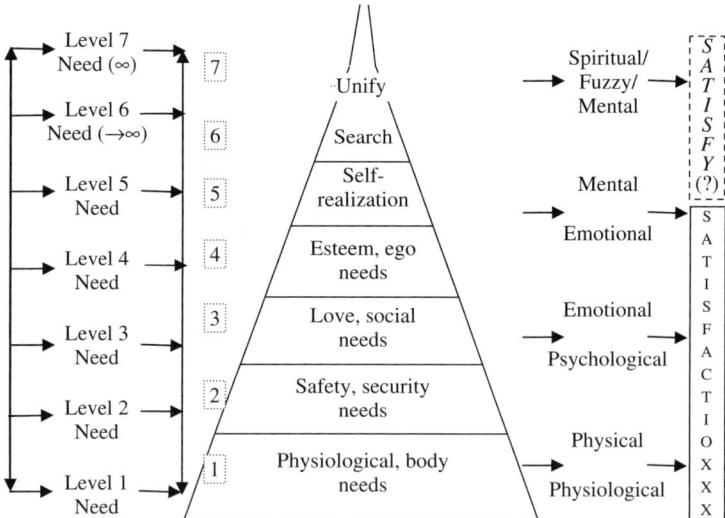

Figure 9.5 Individual human needs and activities. Levels 1–6: computers, networks, Internet tools. Level 7 is mostly human activity.

---

[3]To will distinguish the word "need" in its use as a noun from its as a verb, we use spell the verb form as "need" and the noun form as "Need."

networks need embedded intelligence to provide cogent and error-free communication. At the current stage of computer and network evolution, embedded intelligence serves the scientific, business, and financial communities well. The extent of service provided starts to decrease as the level of human need increases. Current systems do not serve emotional needs as effectively as they serve intellectual needs.

Machine capacity to serve human spiritual needs is nonexistent. When needs change gradually from mental, through emotional, to spiritual, the capacity of the machine declines rapidly. In tracking the decline, current machines offer two major stumbling blocks:

1. Traditional computers and networks fail sharply to process knowledge sufficiently to extract the concepts that constitute the basis of wisdom.
2. When social values have to monitor the operations of modern machines, operating systems do not offer sufficient room for flexibility.

In order to attire machines with more human tendencies, two categories of machines have been proposed as follows:

(i) In the first category, the machine imitates the human thought process and attempts to perform knowledge derivation, concept compilation, and wisdom extraction, based on embedded AI techniques and algorithms.
(ii) In the second category, the machine monitors its own functions by amplifying and enhancing the positive[4] values embedded in the "humanware" of the machine.

This humanware enhances the machine options toward[5] truth, virtue, and beauty (T, V, and B used as noun objects) and is elaborated as follows:

Truth and Universality: "Truth" defined as a noun: reality, accuracy, fidelity, correctness, veracity, exactitude, exactness, verity, freedom from deceit or falseness—*American Heritage Dictionary*; fact, information, reality, actuality, authenticity, certainty, veracity, honesty, sincerity, integrity, frankness, candor, openness—a modern dictionary; universality, authenticity, reality, certainty, exactness, precision—an older Dictionary. Alternate definitions are also feasible.

---

[4]Classification (ii) has two aspects. The positive, more favored philosophic and peaceful aspect is considered first. The negative and less desirable aspect, is machine options set toward the undesirable, harmful, and disruptive, with its considered next. In reality, there are a large number of gray aspects that tend to make machines biased and compromising. Like human beings, machines can become corrupt and political agents in the hands of exploitive leaders. The human aspect of the humanware becomes venerable.

[5]All the synonyms and the synonyms of synonyms are included for the machine processes. The circular definitions are eliminated to prevent the machine from becoming trapped in the execution phase.

"Truth" has this set of synonyms: (fact, reality, certainty, genuineness, precision, legitimacy, veracity, and honesty). Each of the synonyms has its own synonyms: (information and detail); (actuality, authenticity, truth, certainty, and veracity); (confidence, conviction, faith, belief, and assurance); (authenticity, reality, substance, legitimacy, and validity); (accuracy, exactitude, exactness, case, correctness, and meticulousness); (legality, authority, and authenticity); (reality, actuality, genuineness, sincerity, and truth); (candor, integrity, dedication, loyalty, devotion, and uprightness), respectively. Removing the words that contribute to circular definitions and words that have a meaning very close to that of neighboring words, we get a set of eight words (noun objects or concepts) for "truth" as follows:

> Certainty, fact, genuineness, honesty, legitimacy, precision, reality, veracity

Virtue and Social Benevolence "Virtue" defined as a noun:: highest (value, quality, merit, worth, stature, caliber), purity, morality, goodness, righteousness, probity, rectitude—*American Heritage Dictionary*; asset, good quality, good feature, desirable quality, good worth, high merit, high caliber—a modern dictionary. Alternate definitions are also feasible.

"Virtue" has this set of synonyms: (asset, quality, caliber, merit, value, and worth). Each of the synonyms has its own synonyms: (benefit, advantage, feature, quality, skill, and talent); (excellence, superiority, exceptionality, selection, and high quality); (righteousness, integrity, honesty, morality, and uprightness); (quality, ability, completeness, and level); (value, advantage, good point, worth, and plus); (price, cost, charge, rate, assessment); (value, merit, appeal, significance, attractiveness, importance), respectively. Removing the words that contribute to circular definitions and words that have a meaning very close to that of neighboring words, we get a set of seven words (noun objects or concepts) for "virtue" as follows:

> Asset, caliber, goodness, merit, quality, value, worth

Beauty and Elegance "Beauty" defined as a noun: distinction, virtue, merit, excellence, perfection—*American Heritage Dictionary*; loveliness, attractiveness, exquisiteness, splendor, magnificence—a modern dictionary. Alternate definitions are also feasible.

"Beauty" has a set of distinct synonyms—namely, loveliness, attractiveness, goodness, prettiness, exquisiteness, gorgeous, splendor, and magnificence. Without actually writing the synonyms of the synonyms, we present a set of seven words (noun objects or concepts) for "virtue," after removing the words that contribute to circular definitions and words that have a meaning very close to that of neighboring words, as follows:

> Loveliness, splendor, attractiveness, good looks, prettiness, exquisiteness, gorgeous, magnificence

While considering nouns that are virtual and abstract (such as truth, virtue, and beauty), the definitions tend to get fuzzy. For this reason, the center of focus of the cluster of definitions is suggested to carry the meaning of the term. The context of use defines the central focus. For example, if the context of truth is used in a scientific sense, the focus shifts toward universality, merit, worth, integrity, caliber, and exactitude rather than good looks, legality, candor, etc. Similar centers of foci exist for virtue and beauty and the context of their usage.

The mapping of individual human needs and the development of computer systems are depicted in Figure 9.6. The machines that serve the lower needs of human beings are in place as computer systems, intelligent networks, and internets. The room at the top is vacant for next-generation machines and humanistic internets. Conceptually, such machines can be designed, constructed, and built. The functionality of existing super-computers can be restructured to serve as the proposed knowledge, concept, wisdom, and humanistic machines. However, it would be desirable to configure a new hardware platform for each machine as proposed in Chapters 4 and 5.

9.4.4 Corporate Needs

Figure 9.7 is also an open-ended diagram depicting the needs of a corporation. This representation is based on expert-systems concepts from an inter-corporate negotiations model [15]. The corporate needs of commercial institutions are akin to those of individual human beings.

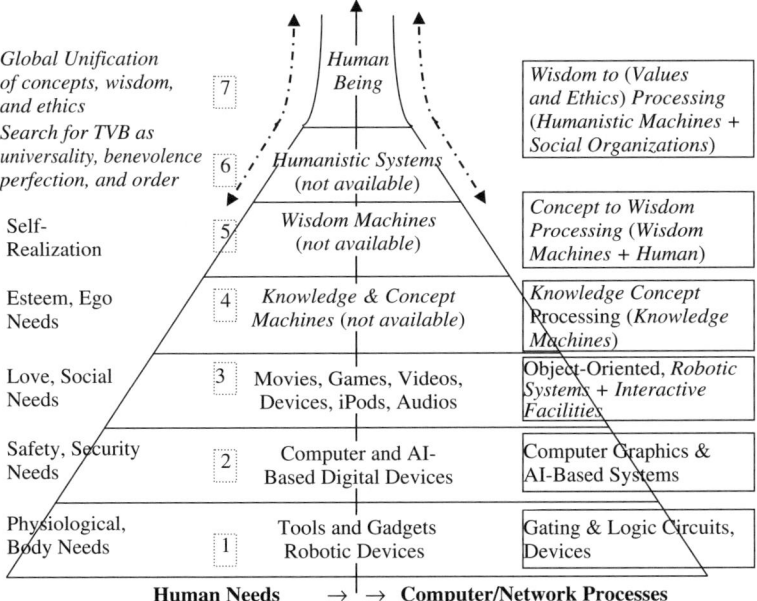

Figure 9.6 Mapping of individual needs on the development of computer systems.

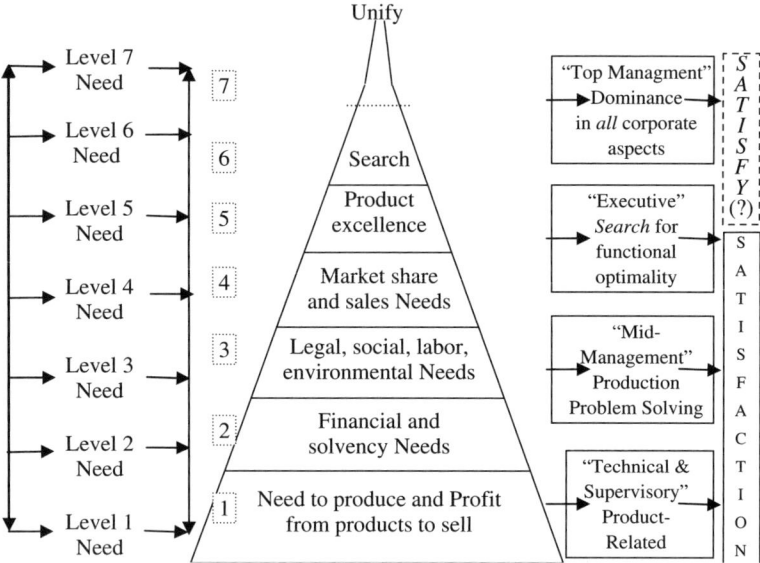

Figure 9.7 Corporate needs and activities. Levels 1–5: computers, networks, Internet tools. Level 5–7: human activity.

Corporate needs motivate the staff. At the lowest level when the future of the corporation is at stake, production needs dominate. Technical and supervisory staff use rudimentary computers and networks to meet the production goal. Robotic devices and control systems may be networked to secure the targeted goals. Sophisticated MIS software is generally available for middle and upper management to achieve short-term and mid-range goals. For top management, a blend of personal style, job-acquired wisdom, and teamwork supplements the few long-range guidelines that any advanced MIS software may provide. Full-fledged tools and methodologies for precise control of the corporation are not available. A similar scenario occurs in the Federal Reserve system when attempting to control the national economy by adjusting the interest rate for the upcoming three-month period.

9.4.5 Corporations, Computers, and Networks

Management information systems have grown progressively intelligent. Multinational corporations thrive on the capabilities of the network deployed by international firms. The role of human beings is supplemented by intelligent decision-support systems that use local and global knowledge bases and AI algorithms to reduce the margin of error. The dependability of the solutions and strategies proposed by modern executive information systems (SAP, PeopleSoft, etc.) [16, 17] is comparable to the diagnosis, prognosis, and procedures supplied by Mycin [18], NeoMycin [19], etc., in the medical field.

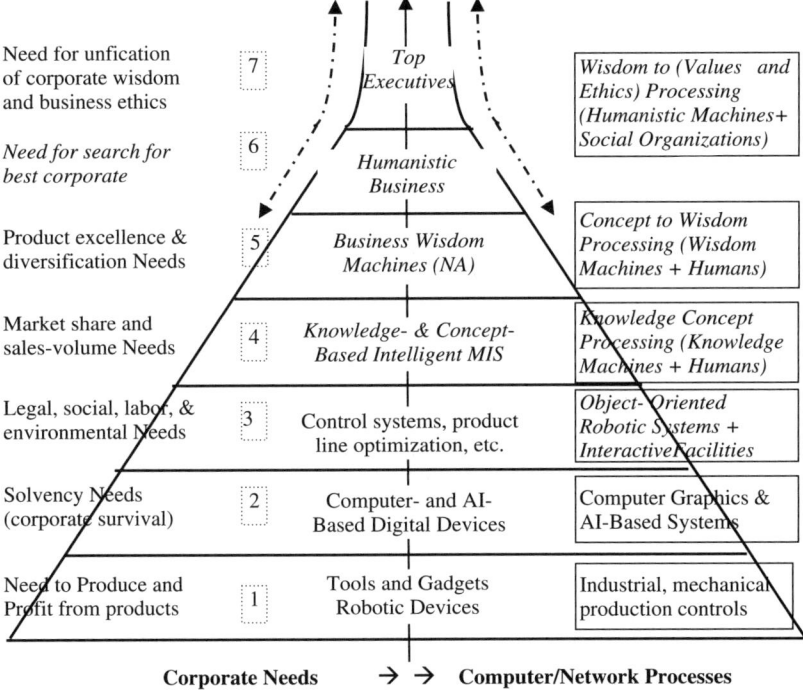

Figure 9.8 Mapping of corporate needs on the machines and networks. MIS = management information systems, NA = not available.

From the existing platforms of intelligent MIS systems, progress toward wisdom in business is inevitable in the long run. Figure 9.8 depicts the overlap between corporate Needs and the machines that serve the business communities.

9.4.6 Societal Needs

In Figure 9.9, the needs pyramid of a society or nation is depicted. The methodology for the deployment of more sophisticated machines, networks, AI tools, and software has not yet systematically evolved.

The effects of events and the actions of politicians and officials in an entire society generally go unchecked. A wide loophole exists for mass deception and lack of accountability in most nations and societies. In certain applications, the room for medical malpractice and errors is restrained by AI programs that validate events and actions to some extent. A similar scenario exists in the diagnostic field. Distance learning and electronic government platforms are taking over some of database management's achievement and networking capabilities. In the majority of applications, specialized algorithms for the upper two levels of needs in society (level 6, the search for optimality, and level 7, the unification of global

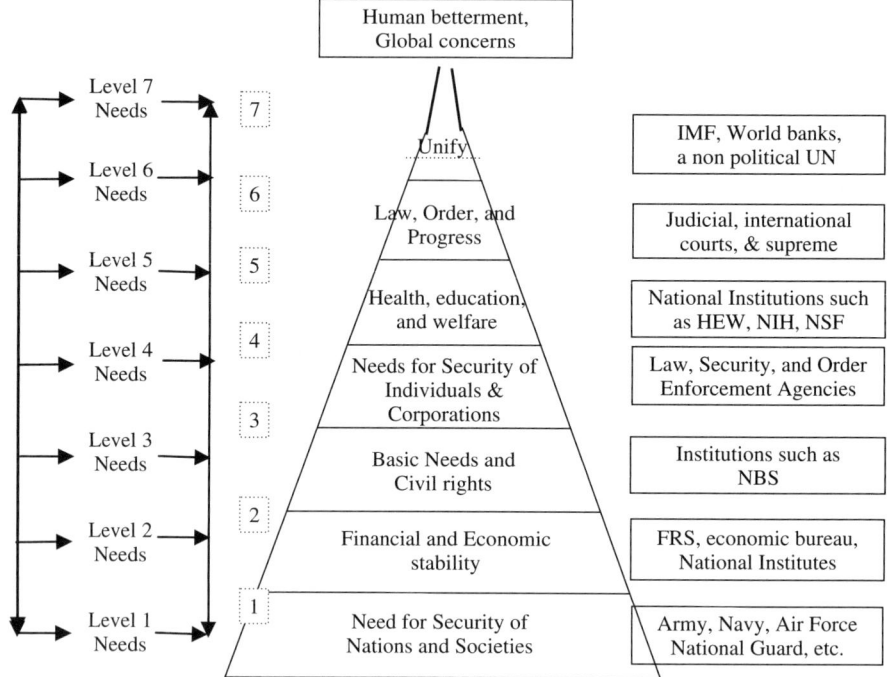

Figure 9.9 Social/humanistic need of nations. Levels 1–5: computers, networks, and Internet-assisted human activity in most national agencies. Specialized systems exist for levels 1–6. Level 7 is mostly human activity. IMF = International Monetary Fund; UN = United Nations, HEW = Health, Education, and Welfare; NIH = National Institute of Health; NSF = National Science Foundation; NBS = National Bureau of Standards.

solutions) are left unaddressed and unresolved. Specialized knowledge banks, and the analysis and derivation of concepts to deal with levels 4–7 require algorithms and machines that are akin to those of Federal Reserve system (FRS) machines that monitor the economy of a nation.

9.4.7 Societies, Machines, and Networks

Societies exist because of their inherent order and structure. The nature, stability, and robustness of the structure have contributed to the acceptance and longevity of that particular society. The internal forces that enhance nature, stability, and robustness are historically based on values and ethics. In a sense, a long-term stable loop is generated, one that traverses the triad of values and ethics, continued social justice, and internal reinforcement of the conditions for contentment that support values and ethics.

From a sinister side, it is possible to conceive of a society that thrives on a doctrine of mistrust, exploitation, and opportunism. Such societies have lasted for a while until internal forces (injustice, discontent, and oppression) within destroyed their very fabric. The proponents of this kind of exploitation and opportunism within society are but a few who stand to gain at the cost of social chaos and inequities. Imperialism and colonialism have faced a painful demise. Whereas populations exhibit an ever-expanding nature, for justice, exploiters number few in comparison and are likely to lose the edge and become victims of social unrest. The French Revolution and the rise of native nationalism within British, Portuguese, and French colonies were startling examples of the death and decay of power by oppression rather than by reason. The social awareness that accompanies education and knowledge, the role of the mass media, and a sense of justice are likely to deploy machines that reason and think rather than machines that bomb and destroy.

Seven levels of societal achievement are depicted on the left side of Figure 9.10. In stable Western societies, these well-structured entities exist in some form or another. The differences in emphasis and rigor of the levels are inconsequential to the major themes on which the society rests.

If, in putting things in perspective, an insidious structure of social order can be justified, then a network of machines and information technology platforms is necessary to support such social and cultural systems. These systems can be designed, built, and streamlined much like data networks but with social needs addressed. The structures of the machines are linked to the functionalities within the seven levels presented in Figure 9.10.

9.5 RATIONALITY OF HUMANISTIC MACHINES

Humanistic machine are processing systems that respond with human traits. The "interface" is a humanistic nature and face. Far from menu-driven answering systems, these new machines "think" (process) like a well-informed, knowledgeable, astute, intelligent, and wise human being and respond in a cogent, coherent, concise, and convincing way. The information bases of the machines are updated with current events from near and far. Local and global events will result in an update. The capacity of the machine to process information to derive pertinent and relevant knowledge becomes a factor in solving a local problem. This astuteness results from the knowledge-processing capabilities of the machine. More than just being astute, the machine has "tunable" astuteness and assumes a flavor of intelligence to suit the problem and/or the user.

The methodology to solve local problems is derived by the use of machines and networks in conjunction with the organizations of collaborative human beings, that is, groups of human beings who can work together to solve immediate social problems. This scenario is depicted in Figures 9.11 and 9.12.

In Figure 9.11, a pyramidal view (on the left-hand side) is shown with seven levels. A corresponding table with seven rows is also depicted. The four panels of the pyramid signify human needs (A), the information age (B), computers and networks (C), and the humanistic machine age (D). The A, B, C, and D panels correspond to the four columns in the table on the right.

432 HUMANS, MACHINES, AND NETWORKS

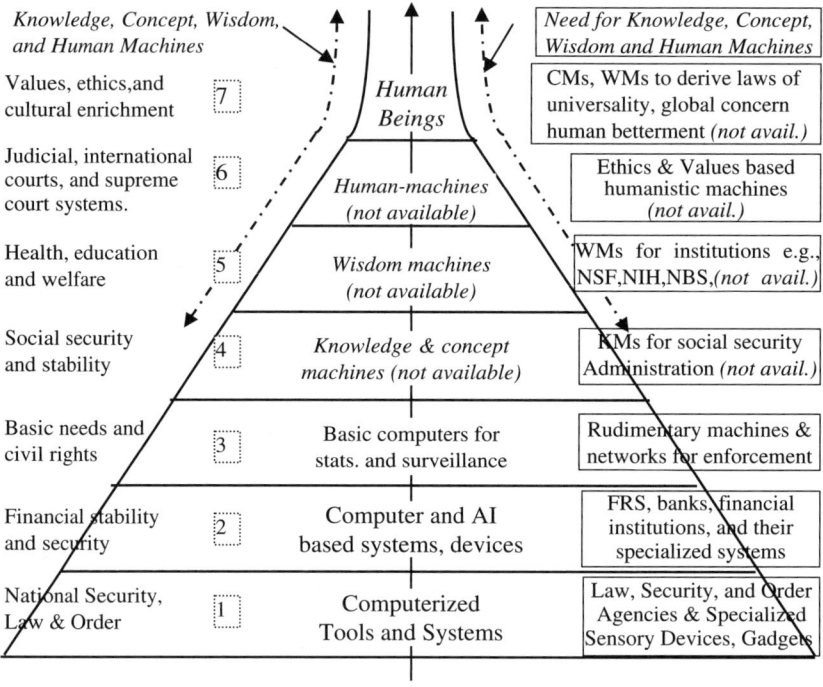

Figure 9.10 Social/Humanistic Needs. Levels (1–5) computers, networks, and Internet assist human activity in most national agencies. In addition, specialized systems exist for (6). At level (7) most of the activity is human. FRS – Federal Reserve System, IMF – International Monetary Fund, UN – United Nations, HEW – Health, Education and Welfare, NIH – National Institute of Health, NSF – National Science Foundation, NBS – National Bureau of Standards.

If human progress is a vertical ascent along the axis of the pyramid, then the final goal would be convergence atop the pyramid. It lies along the main arrow through the table, and movement from left to right and a region of union will occur at all seven levels (rows) and four columns of social need. The ultimate region of convergence lies at column (D) and row (7). Knowledge, concepts, and wisdom machines will facilitate the mass migration of societies. Internet and broadband networks are already paving the path for nations and societies to follow toward the (D, 7) coordinates of the table or the apex of the pyramid.

Figure 9.12 is an alternate representation of the human struggle in the current knowledge society. Computers, broadband backbone networks (see the lower-left box in Figure 9.12), and human organizations (see the right side of Figure 9.12) are closely inter-twined. Intelligent internets ($\underline{I} - \underline{I}$); intelligent-network-based intelligent internets [20] ($\underline{I}^2 - \underline{I}$); then artificially intelligent, intelligent networks, and Intelligent internets [20] ($\underline{I}^3 - \underline{I}$) will become the top layers on current networks and Broadband backbone systems. The current platform of six

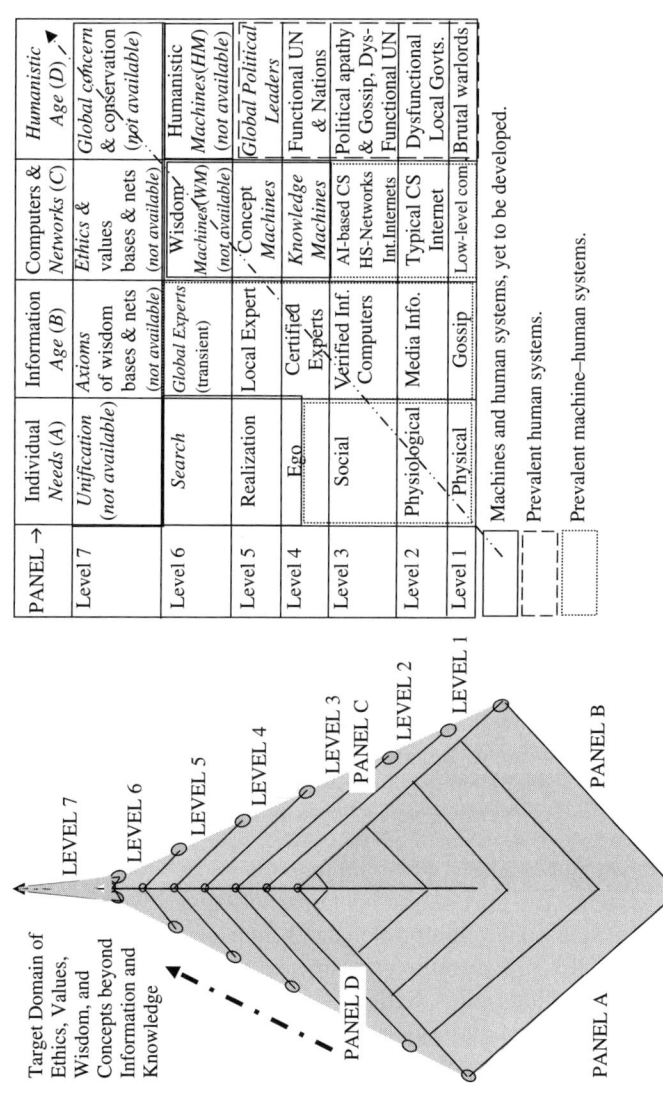

Figure 9.11 Confluence of human needs, computers, networks, and the information age. The target domain lies in the concepts, wisdom, and ethics beyond current information and knowledge in the Internet age.

434 HUMANS, MACHINES, AND NETWORKS

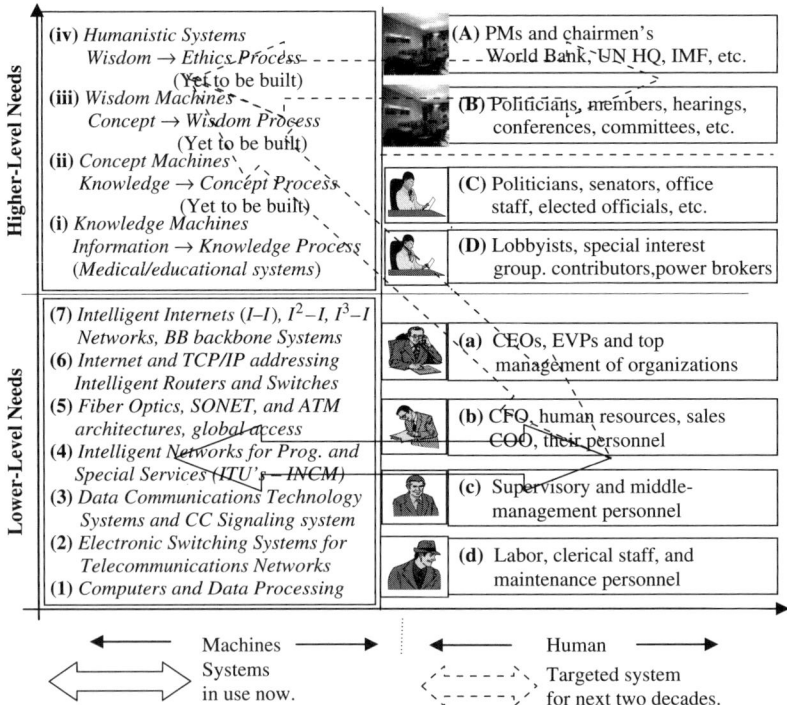

Figure 9.12 Society and communication are inseparably linked. Fruitful and positive links exist currently within a large segment of society via computers and broadband backbone networks. CFO = chief financial officer; CEO = chief executive officer; COO = chief operations officer; EVP = executive vice presidents; ITU = International Telecommunications Union; INCM = intelligent network conceptual model of ITU; FRS = Federal Reserve system; IMF = International Monetary Fund; PM = prime minister of a nation or the president of a corporation or a nation; UN = United Nations. Intelligent internets (I − I); intelligent network based $I - I$; intelligent knowledge processing ($I^2 - I$); and ($I^3 - I$) systems serve society with mixed results. See Ref. [1].

technological achievements namely, (1) computers and data processing, (2) electronic switching systems for telecommunications networks, (3) data communications technology systems and common-channel inter-office signaling systems, (4) intelligent networks for programming and special services standards or (ITUs, INCM) services [21], (5) fiber optics, SONET and ATM architectures, and global access, and finally, (6) Internet and TCP/IP addressing via intelligent routers and switches developed during the last six decades and depicted in Figure 9.12, provides a strong platform for a new breed of machines.

Current status and future trends are evident. In the lower half of Figure 9.12, the current scenario is represented. Futuristic goals are shown in the right-top quadrant of the figure, where social and political leaders of developed nations can deploy knowledge, wisdom machines, and humanistic systems to solve major issues (such as global warming, world hunger, the political arrogance of emerging nations, the energy crisis, etc.) that face humanity. It is our contention that social machines can tackle human problems as well as computers can handle scientific problems.

When next-generation systems and networks emerge, politicians and social agents stand to benefit from the capabilities of the new breed of machines. With appropriate humanware, machines will help humans to progress toward a more advanced and ethical society and its converse. These new machines will also expose the actions and unethical moves of politicians who run counter to the social norms of a more civil society.

Such a trend in the knowledge society is comparable to the time when accuracy in weights and balances was introduced in the commerce-driven societies of the early twentieth century or when the accounting systems were introduced in the business community. A sense of honesty and integrity was forced on the humans in commerce and business. The proposed breed of new machines, that is, knowledge, concept, and wisdom machines and humanistic systems, will inject a degree of ethical and social standards on politicians and social agencies. Their corruptions and deceptions will be exposed more quickly and their weapons of mass deception rendered inoperative. Illegal and unethical practices will be curtailed much like crime and violence are curtailed although not completely eliminated. The machines also act as sensors of negative inputs from their human counterparts who practice deception, exploit mass hysteria, and deplete human dignity from society.

9.5.1 The Human Element

The main reason for uncertainty is the human attitude in social leadership. Whereas machines may seek idealistic goals, human operators can command machines to generate self-fulfilling goals. Such examples abound even in recent history. In most instances, the survival of political power is based on deception (e.g., weapons of mass destruction (WMD) deception during the early twenty-first century, the nonexistent national pride that followed the invasions of Iraq and Afghanistan, continued congressional support for the massive drain of human and monetary resources, persistent conflict in the Middle East, etc.). This uncertainty can cause serious ripple effects in the path of social benevolence and justice. In fact, a few days of irresponsible actions on the part of leadership can set back decades of progress.

Such effects can be forcibly duplicated in knowledge and wisdom machines by altering the nature of wisdom being pursued. This scenario is depicted in Figure 9.13. The figure is divided into eight octets with four quadrants and two leadership modes in each quadrant. The two lower quadrants indicate the state of society when lower-level human and social needs are addressed and vice versa. Eight logical scenarios[6] can be detected. The flow graph is shown in Figure 9.14. The four general modes are as follows:

---

[6]The eight (2^3) scenarios occur with (lower-level/upper-level) needs, the satisfaction of such needs (yes/no), and (+ or TVB/− or DAH) societal dominance.

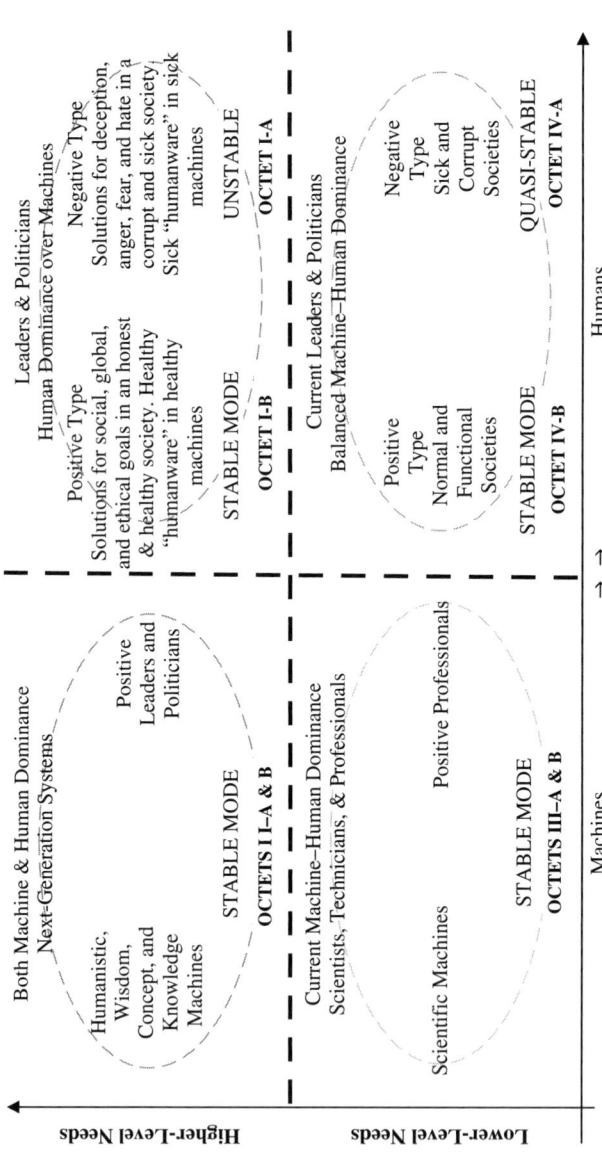

Figure 9.13 Superposition of three machines (knowledge, wisdom, and humanistic) and their networks within the existing framework of the Internet and frame-relay systems. The new machines integrate human values and ethics in current human activities. Global social concerns such as pollution, world hunger, imminent wars, occupation, etc., can be addressed.

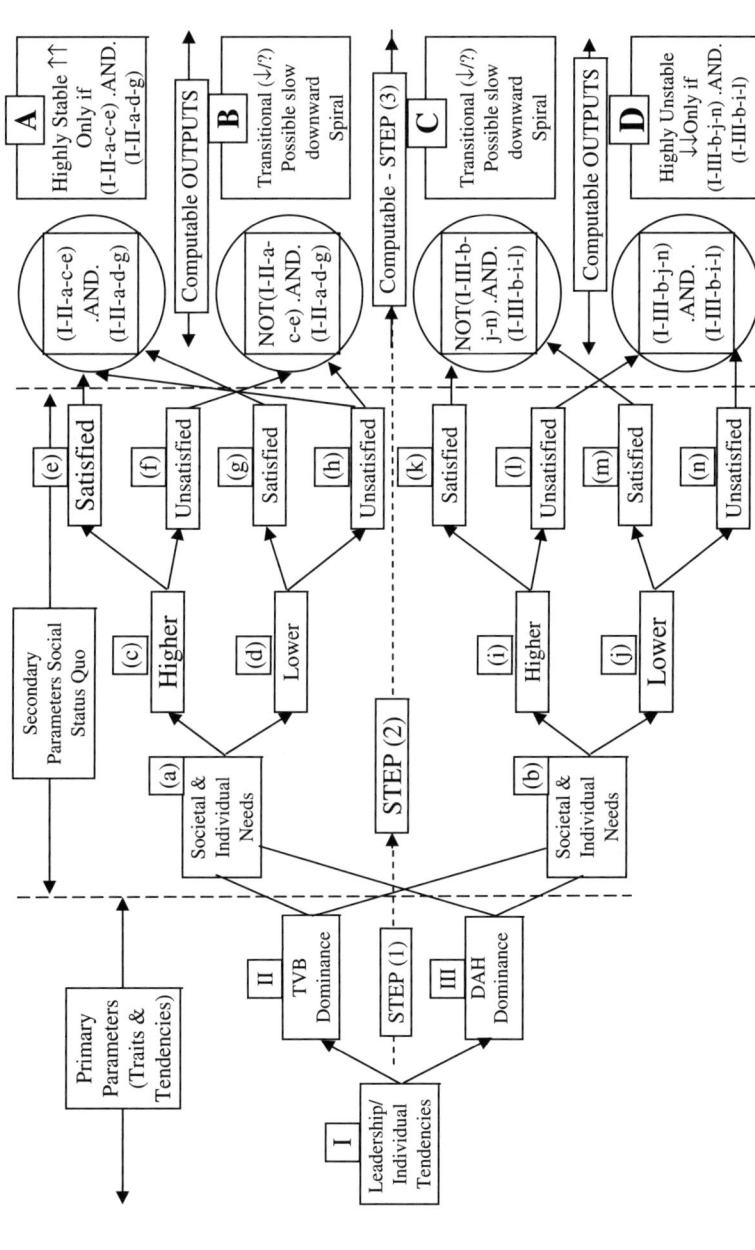

Figure 9.14 Flowchart for a simplified three-step, social domain predictive system that simulates the combined effects of leadership, needs of society, and future need satisfaction.

438   HUMANS, MACHINES, AND NETWORKS

(a) A utopian society (with positive human/leadership styles, positively primed machines, and a well-satisfied society, see Box A in Figure 9.14).

(b) A downwardly mobile, decaying utopian society due to the (more positive than negative) style of human/leadership, and/or corrupted machines, and/or dissatisfied lower/higher needs, (see Box B in Figure 9.14),

(c) An upwardly mobile, struggling society with (more negative than positive) style of human/leadership, indifferent scientific machines, and partially satisfied society (see Box C in Figure 9.14), and

(d) A totally degenerate society (negative human/leadership styles, negatively primed machines, and a totally dissatisfied society, see Box D in Figure 9.14)

Stability exists over an extended period of time, under (a), with positive human/leadership styles and a set of positively primed machines and mostly satisfied society. Functional societies result with scientific and social machines when both the lower- and higher-level needs of constituents are addressed by positive styles of leadership. This would be the ideal situation where by all humans and *(all)* their machines work in a synergistic way to fulfill national/corporate goals.

Human deprivation and moral degradation start to surface as the machine is corrupted by corrupt leadership goals (as occurs in some developing countries or nations where the leadership has gradually grown corrupt). This condition, designated as (d), fosters pockets of positivism in isolated packets and initiates change (e.g., the French Revolution against the French aristocracy, or Quit India Bill against the British, etc.). Instability can be sensed by humanistic/social machines in mode (d) covering ironic conditions in society. Unstable and quasi-stable states can be sensed and computed by these machines to regain their status as ethical and just societies. Venues and cost/benefit ratios are computed by such new machines as exact procedural steps during their operation.

9.5.2   The Open Human Inter-Connect Model

Human and corporate models exist [2]. Such social organizations have needs and resources and abide by Marshall's laws of microeconomics in routine activities and more global laws of macroeconomics to deal with conditions of corporate and organizational stability, financial security, etc. When human values, needs, and resources are governed by humanistic (rather than economic) considerations, then the needs structure of the corporation can be superposed on the computer and communications model of the organization. This model is shown in Figure 9.15.

Seven layers are depicted in Figure 9.15, with traditional OSI layers compacted into the three lowest layers of the figure. The methodology and technology are both well established in the communications industry. The top three layers of this model (again, Figure 9.15) are for the global/long-term activities of the corporate board, implementation of wisdom/values within the corporation, and knowledge functions within the corporate MIS. The fourth layer consists of routine human activity at the executive and senior management level.

RATIONALITY OF HUMANISTIC MACHINES 439

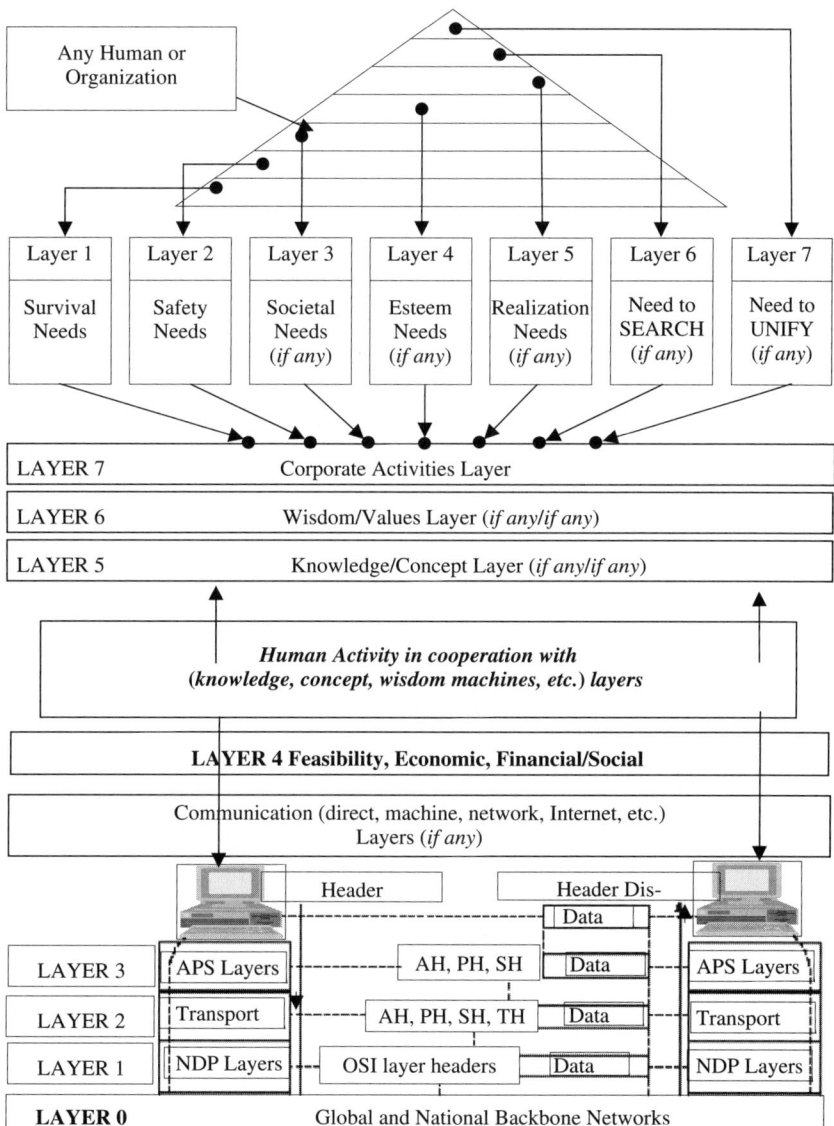

Figure 9.15 The seven layers of the open human inter-connect model. The functions of the seven layers correspond to the functions at the seven nodes of the knowledge trail (Figures 2.5a and b). APS = applications, presentation, and session layers; NDP = network, data link, and physical layers; H = header.

A conceptual parallelism exists between this model and the knowledge trail. If the knowledge trail illustrated in Figures 2.5a and b is drawn along a vertical line, the functions in the top three layers of the model correspond to the functions at the K, W, and E nodes of the trail. The corporation or human organization serves to progress along the knowledge trail.

When computational functions in the trail are to be executed in a network environment, then the lower three layers of the model will become essential. The model (Figure 9.15) now represents the seven layers of an integrated human organization (IHO) rather than an open system inter-connect (OSI) model. An open human inter-connect model (OHI) model appears more appropriate than the OSI model for the knowledge society.

The seven-layer OHI model is derived as a superposition of three machines [the knowledge machine (KM), wisdom machine (WM), and humanistic machine (HM)] and their networks with atop the existing framework of the Internet and frame-relay systems. The new machines integrate human values and ethics in current human activities. Global social concerns such as pollution, world hunger, imminent wars, occupation, etc., can be addressed by using the OHI framework of computation and communications model.

9.5.3 Role of Open Human Inter-Connect in National/Social Settings

The OHI model can be made as generic as the OSI model. The top three/four layers are adjusted to suit the macroscopic application (corporate activities, wisdom/values, and/or knowledge/concepts) for any corporation, nation and society. When the model is deployed for solving national human and social problems, the functionality of the seven layers is as depicted in Figure 9.16. The integration of old (traditional OSI) and new machine (knowledge and wisdom) systems becomes necessary to the integrate the lower three layers with upper four layers. Machines for these upper layers still need to be built. The methodologies and technologies already exist for the lower three layers. Massive standardization and software effort for the entire realization of the OHI model appears as distant the OSI model appeared in the late 1970s.

9.6 WISDOM DOMAIN AND KNOWLEDGE RUSH

If the domain of wisdom encompasses universality, endurance, and elegance, then wisdom has no beginning and no end. The horizons of wisdom span many universes, globes, and countries. The dawn of wisdom spans many millions of years, centuries, and lifetimes. The impact of tools, gadgets, and machines is relatively constricted and short. The computerized rush for pearls of wisdom and diamonds of knowledge is even more constricted and shorter. However, technological strides are long and offer wings for inter-galactic travels.

The search for infinite wisdom and eternal beauty needs to be toned down, much like the search for infinity and eternity. Intellectual space still lies open to

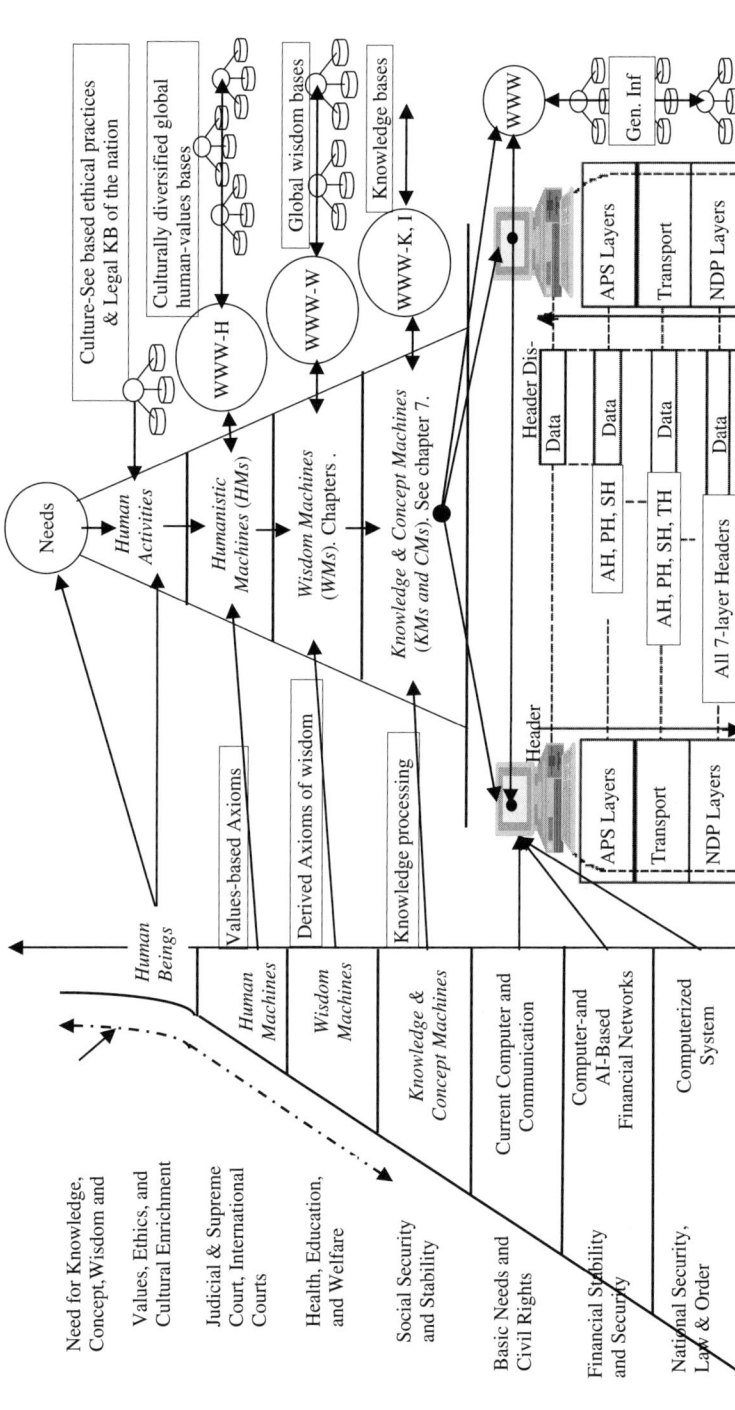

Figure 9.16 The superposition of three machines (knowledge-, wisdom, and humanistic-) and their networks within the existing framework of the Internet and frame-relay systems. The new machines integrate human values and ethics in current human activities. Global social concerns such as pollution, world hunger, imminent wars, occupation, etc., can be addressed.

be explored with the help of machines that can indulge in the multiprocessing of concepts many layers deep and sensors that can sense beyond many visual cues. The tools and techniques for the exploration of this intellectual space are founded in human partnership with humanistic machines and concept processors. The rewards are indeed the pearls of (relatively longer-lived) wisdom and diamonds of (relatively more intensified) knowledge.

The prospectors for wealth and fame need to fear deception, arrogance, and hate (DAH) that can lie hidden in their agenda. For those who are trapped in the golden valley of money, power, and greed, infinity can appear deceptively close and knowledge can become CPU-deep. Values and ethics are generally brushed aside in the humdrum of number crunching. Life becomes an end unto itself rather than a door into society, a distant infinity or a prolonged eternity.

Modern machines and computer systems serve corporate needs well, just as the well as Internet serves a large segment of society. Specialized machines and interfaces serve the medical, educational, and financial communities well. Search engines and ISPs serve relatively elite communities well. Knowledge machines and concept processors should be able to serve thinkers and leaders to make a conscious effort to extend the search beyond finding immediate problem solutions for individuals [4] and corporations [5].

Greater needs (such as global warming, world hunger, poverty, etc.) of nations and cultures require the assertion of power from social leaders and their networks of "wise" machines and unbiased intellectual "software." The machines separate the needs of nations and cultures from the passing whims and fancies of their leaders. It is evident from history that longer-term solutions toward selfless wealth (green money) need shelter from the fire and fury of selfish wealth (gray money). For a select few, the pearls and diamonds of the mind are the thrill and glory of searching intellectual space in the knowledge society, that is more monumental than the mid-American Incas empires for the Conquistadors. Realistically, exploration must be a team effort between a group of persistent humans and many scores of tireless machines. The rewards are likely to include an enduring society that will outlast the global threats on any day in any age.

9.6.1 Pitfalls in a Knowledge Society

The dark side of an ideal society is the shadow it casts in the minds of a few for whom lust and greed are more important than values and ethics. Legal and judicial systems can only block the downward slide in a few instances. However, the efforts of socially destructive individuals and organizations can be blocked by machines far more effectively. Humanistic machine refuse to implement negative moves. In information age, an individual without machines is a carriage without horses.

Negative knowledge macrocommands are nonexecutable in such machines. The security of knowledge machines becomes many times more important and stringent than the security of plain old computer systems. The security of information and knowledge is a practiced priority of most financial [22] and telecommunications industries [21] in this digital age.

Historically, greed, apathy, arrogance, and hate have actively initiated the downfall of societies. National volatility is generally ignited by the selfish underpinnings of leaders, and insecure leaders become easy victims to personal ills. Largely, the downward spiral of any social cycle originates in the selfishness of leaders.[7] When deception prevails at the highest social, corporate and political levels, waves of disasters follow in families, organizations, and countries. Policies are diverted to personal agendas. Honesty and candor become negotiable, and integrity gets murky. Public opinion is manipulated for personal gains.

It is possible nonetheless to conceive of selfless leaders (Gandhi, Lincoln, King, Carter, etc.) it becomes viable to track their movements toward a powerful knowledge-oriented society. Social goals of a knowledge society and information age become as realistic as architectural goals were of Roman society and the granite age. Though painstakingly long, such achievement in the knowledge and concept domains is comparable to nanotechnology in semiconductors and to dense wave division multiplexing (DWDM) in fiber-optics domains. The intellectual age is embedded in truth, virtue and beauty. Wise social goals become the prime movers of nations and societies.

9.6.2 Humanistic Systems That Refuse

Machines primed with basic values and an ethical framework start to turn humanistic. In computer systems, an operating system prevents users from encroaching on protected memory locations (typically, the zero sector, OS memory space, otherwise allocated memory space, etc.) or from executing illegal operations (divide by zero, square root negative real numbers, etc.). In security systems, servers refuse to communicate harmful data (e-mail viruses, access blocked addresses, etc.). The Internet refuses certain information (spam, porno, etc.). When the basic rules of operation of a computer and network systems are implemented to adhere to the legal and ethical code of a nation, only positive social activities are enforced and negative activities are resisted. The initial level of resistance from machines blocks at least some human beings from becoming thugs, even though it may not block professional wrongdoers from becoming criminals. A degree of initial entry-level resistance to social and ethical misconduct is thus enforced by humanistic machines.

9.7 NEEDS PYRAMID OF A SOCIETY

Dominant civilizations have pursued values and ethics to benefit their peoples. A sustained and happy state of existence was the result for a majority of people. Both natural circumstances and a well-planned strategy resulted in

---

[7]For example, see Scott McClellan's *What Happened Inside the Bush White House: Washington's Culture of Deception*, released May 2008, or read about "Bush's poodle" and the lack of evidence of WMD.

sustained growth. To consider, weigh, and provide natural venues for satisfying social needs, a set of well-planned physical and architectural strategies has been deployed in the past. From a historic perspective, when the lower levels of need are adequately met for the majority, then the elite have the time and energy to address a higher level of needs (such as arts and literature, higher education, cultural growth, affluence, leisure, etc.) for individuals and society. Such societies progress in many significant directions, for example, food and shelter for the poor, individual security, economic development, institutional research, etc. When extensive social discontent becomes apparent, changes become imminent.

Change can be chaotic and catastrophic at one extreme, or gentle and pervasive at the other extreme. Historically, radical changes toward social change (e.g., the abolition of slavery) or social destruction (e.g., the Chinese occupation of Tibet) have occurred. Generally, democratic societies experience slow and well-articulated change stemming from the free choice of a majority of their constituents.

More recently, another scenario has been in the making. Broadband networks, the high-speed internet, and knowledge society are bringing three significant changes to the status quo. *First*, public awareness of society and any recent transformations is publicized quickly, due to the diversity and dispersion of news media banks around the world. *Second*, the infrastructures of broadband networks deployed by news, media, and TV networks make it possible to convey the news accurately and graphically. *Third*, the public response to the greedy and selfish actions of political leaders also is exposed effectively and convincingly. When society grows weary of such changes, a certain amount of public sentiment, emotion, and even unrest can follow. Unstable societies suffer more, and many times revolution becomes more attractive than mediated and moderated change. Broadband media networks can convey emotion and charisma as well as they can transmit reason and restraint.

In a sense, broadband networks can be disruptive when the downward spiral has already begun by opening the wounds of controversies and conflicts within society's constituents. Conversely, such networks can accelerate growth when the dissemination of educational and cultural information to the people experiences an upward spiral. In the information age, the deception and selfishness of leaders are quickly exposed. Broadband networks can have the insidious effect of amplifying, quickening, and catalyzing social change.

Examples abound of situations in which the broadband media has ignited public reaction. When restraint in society is worn down due to the unpopular and harsh actions of leaders (e.g., during the Vietnam War, Watergate scandal, recent troop deployments to Iraq, etc.), high-speed networks bring severe changes to fragile leadership. Both the dissemination of sensitive information and ensuing public reaction are accelerated by backbone broadband networks and the Internet. Graphical images of violent crime and war scenes evoke emotional reactions rather than well-planned rational change. The Internet becomes a catalyst in drawing a quick response from the population, thus making the role of the media crucial in conveying truth/deception and events in an ethical/unethical format. Being unbiased, the processing of information and knowledge by knowledge machines will thus become instrumental by the undesirable and unworthy human intervention.

In most developed countries, social changes are contemplated, articulated, and slowly implemented by the majority of constituents. The population is generally convinced that the proposed change will be positive for the majority. When leaders become corrupt and unpopular [14], the corrective reactions can be radical, unpopular, and sometime cruel.

In than Internet- and knowledge-based society, such measures of change by corporate entities are in the best interests of owners and executives. Generally, such short-lived cover-ups do backfire (e.g., Enron, Arthur Andersen, Global Crossing, etc.). Deception also is quickly exposed over high-speed media and news networks. In the end, the modern networks pose a threat for the deceptive. For a short while, the networks may provide a venue to this group to become even more deceptive to cover up an earlier deception (Watergate Scandal, Monica Lewinsky affair, etc.). For societies that learn, broadband networks encourage honesty and truthfulness.

9.7.1 Individuals as Reputatuble Writers

From the dawn of civilizations, reality within a society is documented in the writings of (unbiased and politically independent) intellectuals. In most cases, information and knowledge in society have been processed in the minds of people who provide a snapshot of reality. The reputation of writers was based on their honesty in portraying reality (truth), their choice of words for consistency of thought, and their composition of information in the elegance of expression. In a sense without being explicit statements, the concepts of (Aristotelian) truth, virtue, and beauty were conveyed. Without being artificially wise, the axioms of wisdom were written for thousands of years.

9.7.2 Machines as Processors of Information

In modern settings, the basic information that leads to knowledge and inference is processed by machines and stored in knowledge bases. The process is greatly automated and unchecked. The reputation of a keypunch operator or data-entry clerk does not match the thoughtfulness of writers. In the same vein, the professional responsibility of software writers to process data to derive information and knowledge is less than the contemplations of those who consider the ideas that produced the data in the first place. The accuracy of the data node (D) to knowledge node (K) becomes questionable without a well-derived information node (I) in Figure 2.5. However, machines can be queried to generate quick responses rather than await human interpretation.

The verifications and deductions are still a human responsibility. For this reason, the wisdom, values, and ethics derived in these modern times are likely to be more cogent, coherent, and convincing if knowledge bases are primed by authoritative sources [23] and being verified by knowledge and wisdom machines rather than by human beings or plain old computer systems. Knowledge-processing systems have a direct impact in distilling wisdom and evolving the framework of ethics and values for generations to come. Danger exists when the temporary gain of a few individuals along the knowledge trail [1] starts to outweigh the concerns for a long-term benefit to society.

Intellectual corruption can become just as damaging as terrorism. The proposed machines have the ability to process knowledge, concepts, and wisdom rather than bits, bytes, and dollars as typical computer systems do. A high degree of statistical confidence can thus be evolved for the any new deduction of knowledge from information, for the distillation of wisdom from knowledge, and for the reinforcement of ethics from wisdom.

One of the chief responsibilities of human operators is to maintain a pristine environment for knowledge processing, knowledge bases, and embedded AI algorithms for social betterment. A level of inter-dependence is necessary between those who run knowledge machines and those who prime the knowledge bases. In current all-human endeavors, such a cooperative inter-dependence (between the judicial branch and law enforcement branch, between MRI teams and physician teams, etc.) has been successfully inculcated.

It is still to be resolved if a command center for authenticated knowledge domain functions needs the oversight of a federal agency such as the National Bureau of Standards (NBS) that oversees standards for the measurement and documentation of scientific entities (such as electrical and physical parameters, etc.). At this early stage of building a global knowledge environment, it appears illogical to permit wealth-oriented commercial institutions to command the national or global network of knowledge machines.

The accuracy of processes and dependability of circuits and systems from the perspective of these new machines can propel society forward. Freedom from bias and corruption from a human perspective is ingrained in knowledge domain processes by safeguarding software and implantable hardware. A converse scenario is equally true. Deviation from perfection exists in the synchrony of mind and machine that offers infinite hues of deception. Clever manipulators find infinite variations for deceiving the unwary by exploiting the disharmony between the thoughts of humans and VLSI processes in machines. The dynamics of society thus swings in a bilateral cone of uneasy perceptions for humans. However, a numerical and statistically weighted average of good (with three deterministic flavors of truth, virtue, and beauty within individuals and society) is likely to dominate the bad (with its own three deterministic flavors of deception, arrogance, and hate within individuals and society).

The installation of key checks and balances in proven software modules and operating systems can retard greedy advances from corrupt human beings. The values and ethics embedded in the "humanware" of such machines can be made as immune and impenetrable as the honesty and integrity of the judicial system in a nation. Although slightly imperfect, the pros for such machines will prevent the widespread abuse of power and prevailing corruption in some developing countries. In developed Nations, the system is likely to be at least as valuable as the Supreme Court in blocking gross social injustices.

9.8 SELF-PERPETUATING POWER LOOPS

Power loops are a by-product of the greed for power (and/or wealth). Numerous manifestations of these power loops exist in almost every nation and corporation, or any social entity. From a distant perspective, the feuds for power and/or wealth are pasttime for those trapped in the lower needs levels [4, 5]. The enhancement and preservation of power and wealth occur within these groups. Any sense of truth, honesty, and justice is conveniently swept under the rug by blanket statements such as "all is fair in love and war". There is no T, V, or B in such statements; they are unlikely to bring any scientific validity or universal justification. However, such (civilized) feuds occur on a daily basis in most (civilized) societies.

If politicians have power, lobbyists have the money to influence politicians and thus the power wielded by them. The social climate is thus altered by those who can buy politicians, who in turn buy votes in a democratic system by misinforming the average citizen. The bond between power and wealth becomes tighter and tighter to squeeze out ethics in society. Absolute wisdom plunges deeper and deeper into opportunistic wisdom.

A diagram for the power cycles within society and the factors that influence self-perpetuating power loops is shown in Figure 9.17. The diagram has three major players: (1) self-interest groups (SIGs) like individuals, corporation, unions, tribal leaders, etc.; (2) social leaders like politicians, industry proponents, landowners, etc., who influence the social and legal framework and can alter functions within society; and (3) lobbyists and power brokers. The economic game resembles those described by von Neumann and Morgenstren [24] but with a triad of players. The goals are known to each: The SIG group tries to maximize the local objective by buying a politician's influence, the politician tries to maximize his or her personal wealth by "selling" influence, and the lobbyist/broker tries to maximize his or her own wealth and power base from the transaction(s) brokered.

The dynamics of transactions between special-interest groups (SIGs), lobbyists (power brokers), and public leaders in a corrupt society are shown as one loop configuration:

$$a - b - c - d - e - f - g - h, \quad \text{back to } a \quad \text{(Figure 9.17)}$$

Traversing this loop permits an enhancement of the private wealth of every party involved, and the net effect is corruption in society and the contamination of ethics. Society bears the brunt. Both sets of wisdom (SET and DAH) tolerate corruption and the contamination of ethics.

An elaborate version of such interactions is shown in Figure 9.18 when the effects of positive wisdom (i.e., a sense of honesty and integrity, or predisposition to TVB) are included for the SIGs and politicians. It is interesting to see that lobbyists and power brokers lose out completely when SIGs and politicians

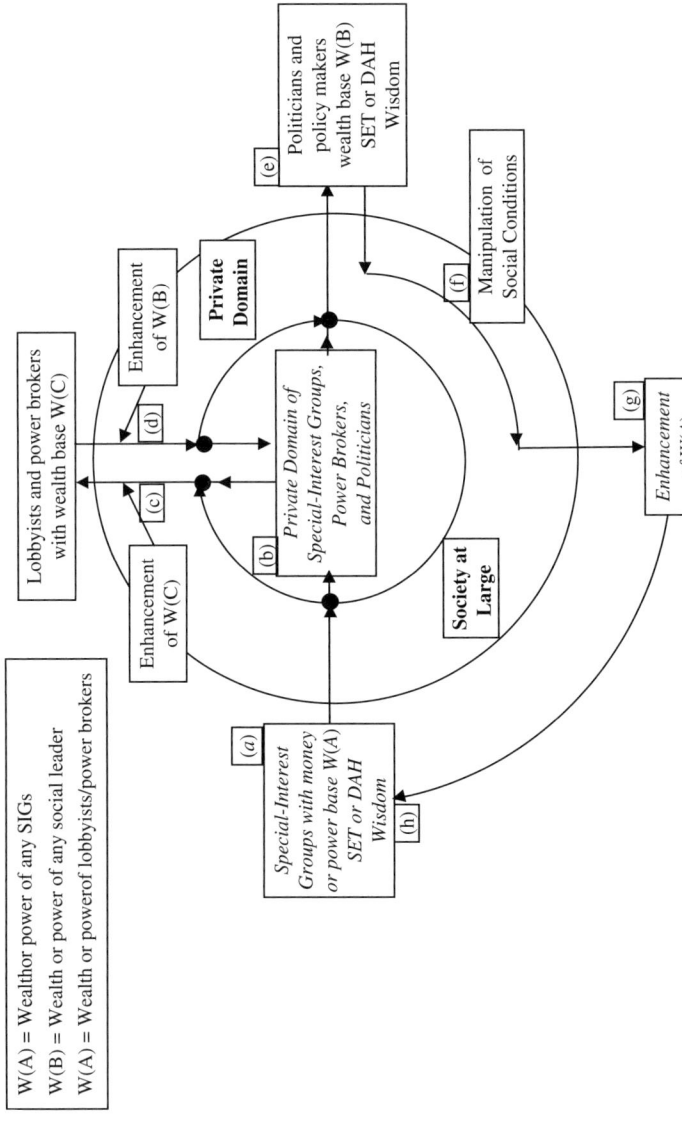

Figure 9.17 Dynamics of transactions between special interest groups, lobbyists (power brokers), and public leaders in a corrupt society. The loop a-b-c-d-e-f-g-h permits an enhancement of the private wealth of every party involved, and the net effect is corruption in society and the contamination of ethics. Society bears the brunt. The Science, economics, and technology (SET) and deception, arrogance, and hate (DAH) wisdom sets tolerate both corruption and the contamination of ethics.

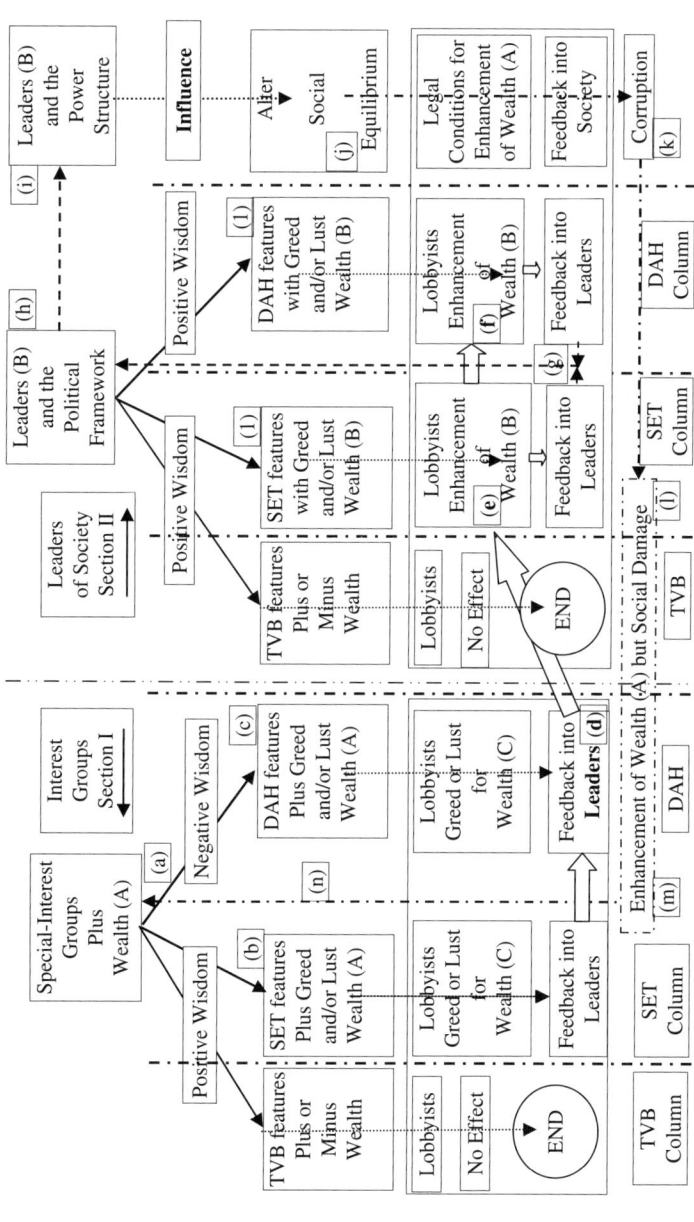

Figure 9.18 Social setting for the growth and stability of self-perpetuating power loops in any society. There is only one loop in this diagram the shown as a-b(or c)-d-e(or f)-g-h-i-j-k-l-m-n, back to a. There are three types of wealth; it is or power involved: (A) the wealth of self-interest groups, (B) wealth of leaders, and (C) wealth of lobbyists.

both have a predisposition to TVB and Figure 9.18 terminates in two ENDs. In other cases, the goals of SIGs (A), and the wealth of politicians (B) and lobbyists (C) are enhanced. The self-perpetuating loop now has this configuration:

$$a\text{-}b(\text{or } c)\text{-}d\text{-}e(\text{or } f)\text{-}g\text{-}h\text{-}i\text{-}j\text{-}k\text{-}l\text{-}m\text{-}n, \qquad \text{back to } a \quad \text{(Figure 9.18)}.$$

The cycle of corruption continues and the damage to the entire society starts to become evident as box m is reached. The source and reason for the enhancement of the three types of wealth—(A), (B), and (C)—are the depletion of wealth of the entire society and the resulting damage. From an econometric perspective, traversing this loop more than once can have a "multiplier" effect and the sale of political power continues at a higher power than during the prior sale and the politician get more corrupt as the power of Mafia had grown past. It causes long and sustained damage to the entire society.

9.9 CONVERGENCE OF KNOWLEDGE AND MOTIVATION HIERARCHIES

Motivation (M) occurs to satisfy human needs. Knowledge (K) serves to guide the motivation to achieve concrete results that satisfy such needs. Motivation and knowledge are seminally linked in the human mind. Human curiosity intuitively forces the paradigm that to know a little is the path to knowing a little more. Since the days of Aristotle, humans have pursued knowledge in its own right. More than that, humans have moved beyond knowledge and pursued perfection and wisdom. It is an insatiable quest that brightens the chasm between the created and nature, the creator of two scenarios (reality and perception). In this gap lie buried the limits of understanding. Yet knowledge is pushed beyond one frontier only to find another universe[8] hidden below the fallen tides of time.

The development of any society lies embedded in human minds or, more recently, the knowledge banks that supplement trails of knowledge. An organized and archived collection of knowledge bases provides a digital snapshot of the extent of social development. When partitioned according to the disciplines in the DDS or LoC classifications, the many trails all converge into one knowledge trail for all of humanity. This abstraction is also explored in Chapter 10.

All of humanity has one schema in common: human needs. From the individual to whole nations, needs structures exist and provide the motivation for societies to move forward. Embedded in the drives lie hidden their needs structures, reflecting the agenda of individuals, corporations, societies, and cultures.

One of the lingering goals of this book is to uncover any underlying basis for the convergence of the two hierarchies (of needs and of knowledge) that have been in existence since the beginning of human civilization. Genetic codes

---

[8]In the words of an Indian poet who implores: "There are many worlds to explore beyond the stars, There are many victories to be won beyond the wars."

CONVERGENCE OF KNOWLEDGE AND MOTIVATION HIERARCHIES 451

and the knowledge embedded in them are essential for initiating the process of life that has its own needs structure. Codes are embedded knowledge, and the enhancement of knowledge gives rise to an understanding that the codes for survival mean satisfying the needs to live. In essence *knowledge, life and needs* form a triadic relation, each one being affected by the other two. Thus, it appears logical that any hierarchy in each one depends on the hierarchies of the other two. In this section, we unravel the link between needs and knowledge hierarchies, even though the two links may exist in their own right but soon become inter-dependent via the hierarchy of human behavior.

This rather complex scenario is depicted in Figure 9.19. The central core of the figure is taken from Figure 2.2. The movements for any society are jagged and appear almost random. Closer examination reveals the computable socioeconomic forces and leadership styles that initiate change and preserve positive or the negative social momentum.

The equations governing the dynamics of social entities are discussed in Part IV of [1]. Social inertia, and the nature and magnitude (power, money, and charisma) of the net forces in each direction of incremental movement enter into the extent of social change.

Although too complex for human guesswork, these parameters are well defined and computable by knowledge and wisdom machines (see Chapters 2 and 5). These machines train themselves by examining the response to similar social forces and computing the extent of social change. Multivariate analysis and estimation techniques are used. The results have a probability number and confidence window at each stage of the knowledge functions. The approach is similar to that used by medical diagnostic systems (e.g., Intern, Mycin, NeoMycin, etc.), and cascaded knowledge processes reduce the accuracy of the results generated, but are precise and exact numbers projected far into the future or under highly variable social conditions. The response is akin to that of human beings when they are exposed to totally new situations and settings.

The two extreme orientations (TVB and DAH) of society are depicted to the left and right of the central core of Figure 9.19. The upward and downward forces of the two extremes keep society dynamic and adaptive. For the business segment of society, the economic gains to be made during the swing of social tastes and attitudes become important; for the scientific community and elite, the progress toward greater truth and knowledge becomes important; and for the corrupt, the fulfillment of greed and self-interest becomes pronounced. The limits of vacillations within society are controlled by the extent of satisfaction/dissatisfaction of the social needs shown at the left and right edges of the figure.

9.9.1 The Hierarchy of Knowledge

The seven-node knowledge trail was first introduced in 2006 [1]. This knowledge trail (see Figure 2.5) is also a hierarchy since the initial nodes (B, D, and I) nodes are essential to build the latter nodes (K, C, W, and E). In fact, every prior node is essential to construct the current node. Humans and machines both ascend the

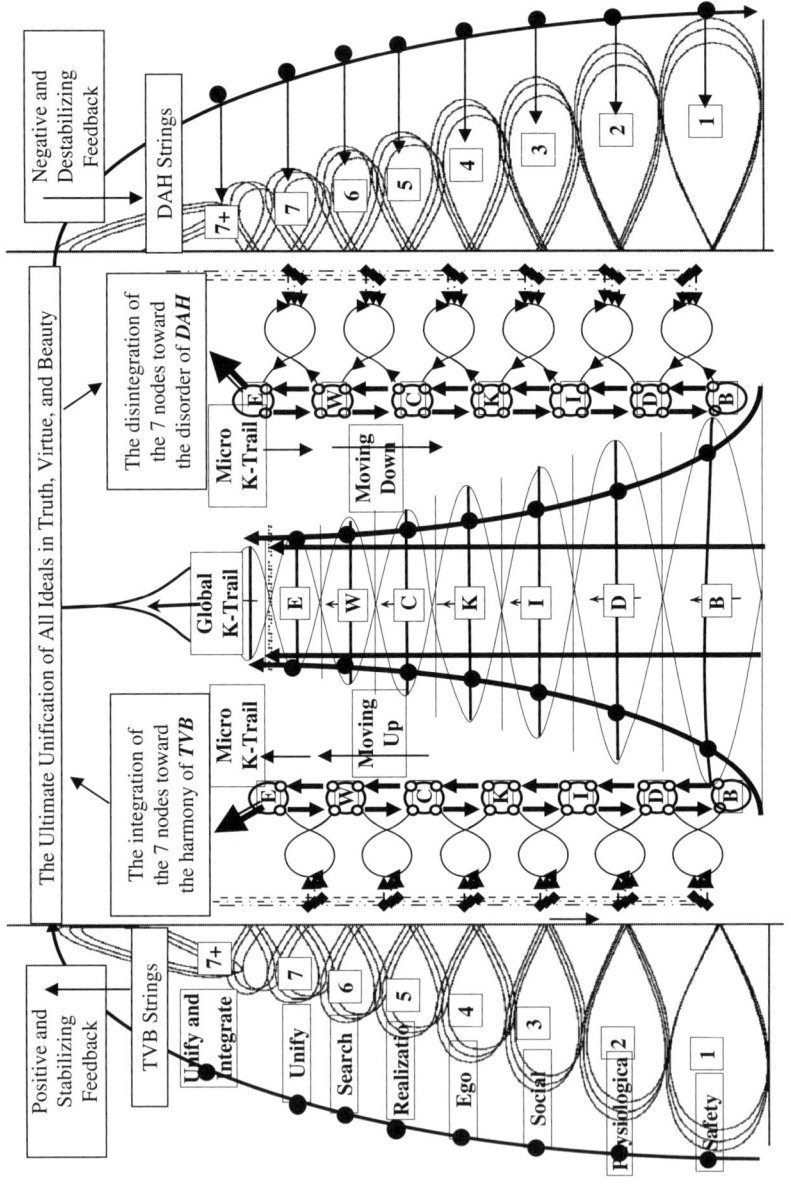

Figure 9.19 Positive feedback loop from truth, virtue, and beauty social forces upto the knowledge trail and vice versa.

hierarchy, and differences between conventional machines and wisdom machines exist. Robots and humanistic machines encompass the architectures, software, and firmware for systems in between.

Banks can hold monies from any origin and for any purpose. Data can be encoded in any wavelength of light traversing a fiber-optic link from any laser source and to any photo-sensitive device. In the same vein, knowledge banks can hold information of any origin and for any purpose. Three extremities of flavors exist:

1. Pristine information (from and for TVB purposes)
2. Green information (from and for SET orientation)
3. Corrupted information (from and for DAH orientation)

When the TVB, SET, and DAH orientations are equated to the three original colors [red, yellow, and blue (RYB)], then innumerable shades (hues) of information can exist in the knowledge bases. Unless information is maintained in a pristine state, derived knowledge can be corrupted. The filtering and cleansing of information (Chapter 3) become essential.

Free and unabated thought in a TVB orientation of wisdom is generally the result of highly humanistic processes. The three fundamental phases appear to be (1) the perception and perpetuation of truth, (2) the incessant struggle toward perfection of virtue, and (3) the realization and personification of beauty and elegance. Free and unabated spirit in a SET orientation of wisdom is generally the result of highly rational and logistic processes. The three stages appear to be (i) the pursuit of each of the sciences, (ii) search for the basis and laws of economics, and (iii) the search for breakthroughs in technology. In contrast to the earlier two orientations of wisdom, the DAH orientation displays three fundamental traits: (a) grandiose deception, (b) the globalization of arrogance, and (c) the grossness of hate.

Through the vastness of options, the detailed prediction of individual human behavior appears unrealistic. However, a cultural and social norm can be predicted by knowledge and wisdom machines. Such a norm will facilitate the direction and velocity of social change that can be expected in individuals, societies, and nations.

9.9.2 The Hierarchy of Motivation

If knowledge and motivation can be correlated, then the search for knowledge ranks with the motivation to resolve the outstanding needs that must be gratified. However, once satisfied, the motivation to resolve the next immediate need becomes dominant. The level of motivation may remain the same, but the sophistication and methodology to resolve successive higher needs also get higher, thus making the required efforts harder and results scarcer. Personal challenge and creativity take over in the few who dare to go anywhere equipped with science, economics and technology.

The five-layer motivational hierarchy of needs was introduced by Abraham Maslow in 1943. It forms a basis for correlating the behavior and actions of successful (rather than ethical and righteous, socially oriented) people to their drives and motivations. This needs structure was enhanced [1] to include aspects of social responsibility and the well-being of the community and culture, rather than expending all the intellectual energies of successful people toward the "*self-actualization*" needs at the fifth level of Maslow's needs pyramid.

In Figure 9.20, the migration from unbiased human needs (on the left side) and their associated motivations are linked to the TVB orientation of wisdom (right side). The intermediate stage is the deployment of computational/machine-based systems (center of the figure).

It is possible to chart a similar migration from unbiased human needs and their associated motivations to the DAH orientation. In Figure 9.21, the use of machines is shown to be quite limited due to the limited development of negatively oriented machines and the associated software. In the post-9/11 era, the funding for destructive systems (as a response to virtual threats) has increased enormously, and it is quite apparent that numerous nations are building field robots, drone planes, and killer machines. A vast change in the nature of human beings and nations can make the DAH counterpart (Figure 9.21) of TVB wisdom (Figure 9.20) realizable in hardware, software, firmware, and humanware, much to the detriment of the human society that has evolved since the ancient days of Aristotle.

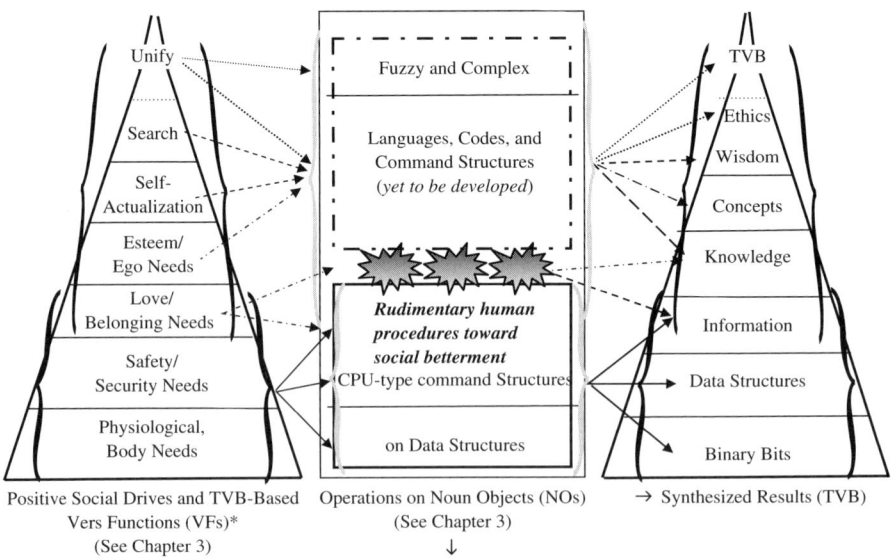

Figure 9.20 Positive Driving forces that move society forward along the knowledge trail (Figure 2.5) and enhance the wealth of knowledge, reinforcement of concepts, enhancement of wisdom, and establishment of ethics in society. The model is based on the ideals that truth and honesty are the favored modes of social functions.

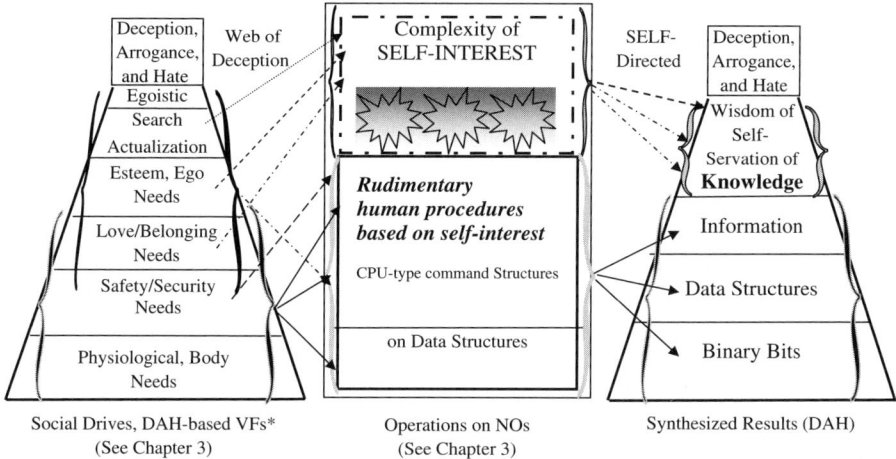

Figure 9.21 Negative Driving forces that vascillate society along the knowledge trail (Figure 2.5) and deplete the wealth of knowledge, distorts concepts, challenges wisdom, and disintegrates ethics in society. The model is based on the ideals that deception and hate are the favored modes of social functions.

9.10 KNOWLEDGE MACHINES ON KNOWLEDGE

It is possible to instruct a knowledge machine to cleanse (filter) the knowledge that it generates. The functionalities of machines are based on the commands discussed in Chapter 3. In any specific area of discipline, two knowledge machines are used.

1. The first machine operates under its global command:

    ENHANCE Knowledge (Global VF = ENHANCE;
      Global NO = Old Knowledge)

2. The second machine operates under its global command:

    VERIFY Knowledge (Global VF = VERIFY;
      Global NO = Nascent Knowledge)

When the two machines are identical and have the same capacities, the VERIFY process becomes as through as the ENHANCE process. All the VERIFY operatives will become superposed on the ENHANCE operatives. In a sense, a grid of processes will be superposed on old and nascent knowledge(s) together. The result is likely to be robust new knowledge. The situation is very similar to one subject-matter specialist checking out the output of another equally proficient expert.

Even when the two machines are not identical, then all (hardware, software, firmware, and hardware) aspects of the ENHANCE machine will be reviewed by the VERIFY machine. If the role of the two machines is now turned around, all aspects of both machines will be deployed in both global processes. The resulting new knowledge at the output is likely to achieve a very high degree of confidence.

Such a procedure is adapted in telecommunications switching systems where two ESS [25] generate the same output commands to the network to establish a telephone connection. Extremely dependable service is provided to customers. In accounting procedures, the audits are conducted by numerous accountants to ascertain the accuracy of corporate balance sheets.

Figure 9.22 depicts the role of two knowledge machines that operate in a cooperative mode to generate a high grade of knowledge, inventions, processes, etc. The mechanization of the creative and innovative process becomes feasible in this mode of operation of the two machines.

9.11 CONCLUSIONS

In this chapter, the fabric of social organizations is inter-woven in the processes of machine functions. The basis of the existence of human networks is derived for the needs of individuals, corporations, and social entities. The basis for computer networks derives from their accuracy and connectivity. Needs, social norms for human activity, computers and their accuracy, and finally networks and their connectivity are the four cornerstones of this chapter.

The conceptual bond between humans, their needs, their machines, their innovations, and social progress is compiled. Both the positive and negative movements of society are investigated and traced back to the predisposition of its constituents to truth vs. deception, social betterment vs. self-interest, and finally, optimality and beauty vs. aggression and hate.

The sense of balance necessary to maintain dynamic equilibrium within society is explored in this chapter. The knowledge society has materially altered this balance through sub micron wavelengths in Silicon and glass. Yet, data and information at this wavelength influence the mind that operates at sustained long wavelengths neural. No doubt, Silicon and glass have altered the rhythm of the mind that works at alpha, beta, and gamma wavelengths in the neural nets in the brain. It is our contention that the new knowledge and wisdom machines can restore the harmony between binary data and human ethics.

If harmony is restored for not being downtrodden by global disasters, machines need to become more human rather than for humans to become more robotic. The middle ground for unison is further delineated and charted in this chapter by making the innermost software layers of machines tamper-proof against deception, greed, arrogance, and hate. Humanistic machines (Chapter 5) acquire the behavioral traits of humans rather than the mechanical traits of robotic arms. The most modern innovations that have propelled machines and their networks are now

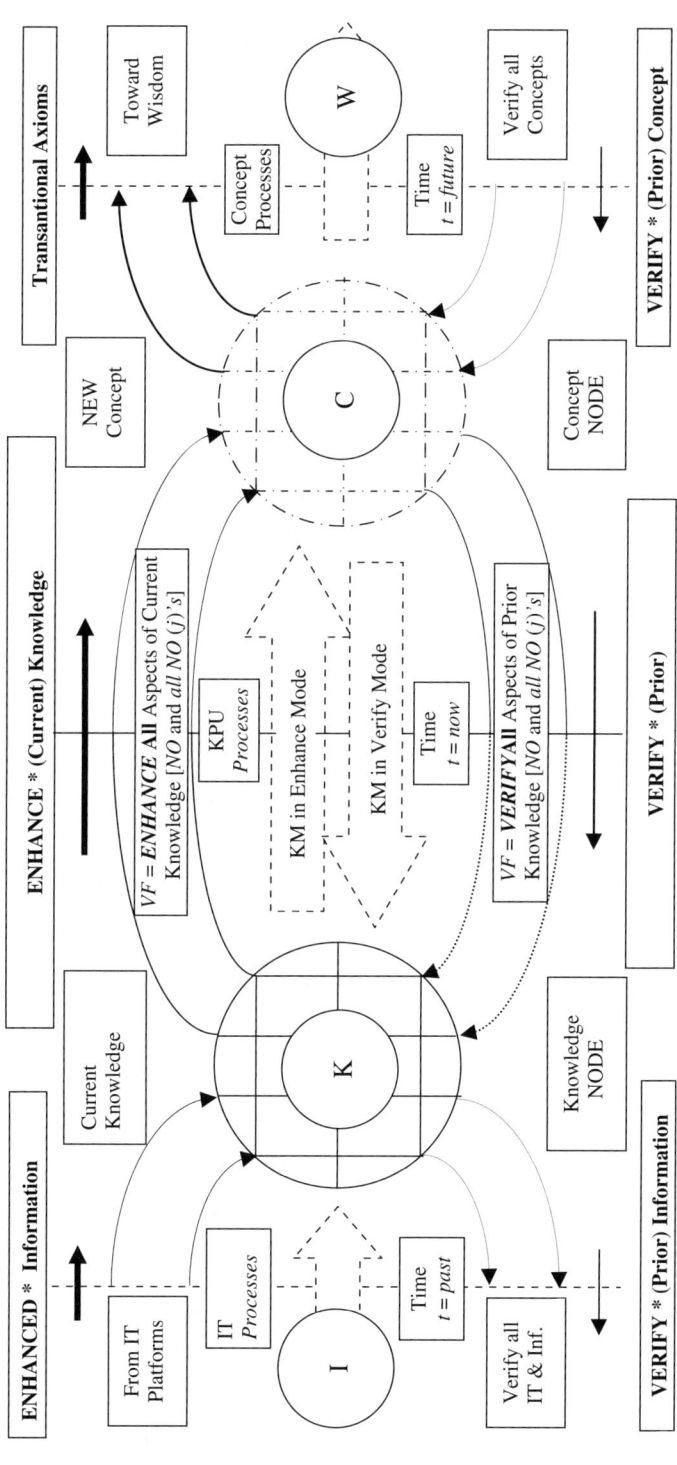

Figure 9.22 The superposition of knowledge machine principles for movement along the knowledge trail (Figure 2.1) at the I-, K-, C-, and W-nodes. The machines can be primed with slightly different knowledge bases to generate flavors of knowledge, concepts & wisdoms to suit different social, economic, and cultural settings.

diverted toward building bridges between human needs and their elegant, noninvasive solutions. These customized solutions that humans and wisdom machines (Chapter 2) generate are perhaps longer-lasting than the primitive slash-and-burn approach in the quest for knowledge and the ensuing disharmony with nature.

REFERENCES

1. S. V. Ahamed, *Intelligent Internet Knowledge Networks*, Wiley-Interscience, Hoboken, NJ, 2006.
2. R. Cyert and J. G. March, *A Behavioral Theory of the Firm*, Prentice-Hall, Englewood Cliffs, NJ, 1963.
3. S. V. Ahamed, "The Architecture of a Wisdom Machine," *International Journal of Smart Engineering Systems Design*, 5(4), 2003: 537–549.
4. A. Maslow, *Motivation and Personality,* Harper & Row New York, 1964.
5. See Chapter 11, "Human Needs and Machines," in [1], *Intelligent Internet Knowledge Networks,* Wiley-Interscience, Hoboken, NJ, 2006. Also refer to S. V. Ahamed, "An Enhanced Need Pyramid of the Information Age Human Being," International Society of Political Psychology (ISSP) Scientific Meeting, Toronto, Canada, July 3–6, 2005.
6. S. V. Ahamed and V. B. Lawrence, *Design and Engineering of Intelligent Communication Systems*, Kluwer Academic Publishers, Boston, 1997.
7. C. M. Bishop and C. Bishop, *Neural Networks for Pattern Recognition,* Oxford University Press, Oxford, UK, 1996. Also see J. R. Parker, *Algorithms for Image Processing and Computer Vision*, John Wiley & Sons, Hoboken, NJ, 1996.
8. A. H. Maslow, "A Theory of Human Motivation," *Psychology Review*, 50, 1943: 370–396.
9. A. H. Maslow, *Motivation and Personality*, Harper & Row, New York, 1970.
10. A. Maslow, *Farther Reaches of Human Nature*, Viking Press, New York, 1971.
11. H. A Simon, A Behavioral Model of Rational Choice, *Quarterly Journal of Economics*, 69, 1955: 99–118. Also see H. A March and J. G. Simon, *Organizations*, John Wiley & Sons, Hoboken, NJ, 1958.
12. E. Fromm, *Anatomy of Human Destruction*, Holt, Rinehart and Winston, Austin, TX, 1975. See also E. Fromm, *The Art of Loving*, Harper & Row, New York, 1954.
13. G. H. Mead, *Mind, Self and Society*, University of Chicago Press, Chicago, 1934.
14. N. Chomsky *Distorted Morality*, Production Koch, Studio Wea Corp., DVD release, March 2003.
15. E. G. Roman and S. V. Ahamed, "An Expert System for Labor-Management Negotiations," in *Proceedings of 1984 Summer Computer Simulation Conference*, North Holland Publishing Co.
16. *SAP Solutions and eLearning*, SAP (UK) Limited, Feltham, Middlesex, England, 2002.
17. R. Rowley, *PeopleSoft v.8: A Look under the Hood of a Major Software Upgrade*, White Paper, PeopleSoft Knowledge Base, 2005. Also see R. Gillespie and J. Gillespie, *PeopleSoft Developer's Handbook*, McGraw-Hill, New York, 1999.

18. N. S. Stuart, *Computer-Aided Decision Analysis,* Quorum Books, Westport, CT, 1993. For decisions in medical diagnostics, see E. H. Shortliffe, *Mycin: Computer-Based Medical Consultation,* Elsevier, New York, 1976. See also B. G. Buchanan and E. H. Shortcliffe, *Rule-Based Expert System: The Mycin Experiment at Stanford Heuristic Programming Project,* Addison-Wesley, Reading, MA, 1984.
19. W. Kintsch et al., *About NeoMycin: Methods and Tactics in Cognitive Science*, Lawrence Erlbaum, Mahwah, NJ, 1984.
20. See Chapter 11, "Knowledge Environments" in [1], *Intelligent Internet Knowledge Networks,* Wiley-Interscience, Hoboken, NJ, 2006.
21. S. V. Ahamed and V. B. Lawrence, *Intelligent Broadband Multimedia Networks*, Kluwer Academic Publishers, Boston.
22. N. Goldburt, "Design Considerations for Financial Institution Intelligent Networks," Ph.D. dissertation, City University of New York, 2004.
23. See Chapter 12, "Wealth of Knowledge" in [1], *Intelligent Internet Knowledge Networks,* Wiley-Interscience, Hoboken, NJ, 2006.
24. J. Neumann and O. Morgenstern, *Theory of Games and Economic Behavior*, Princeton University Press, Princeton, NJ, 2004.
25. See Chapter 4, "Current Digital Networks," in [6], *Design and Engineering of Intelligent Communication Systems*, Kluwer Academic Publishers, Boston, 1998.

CHAPTER 10

ARCHITECTURE OF KNOWLEDGE

CHAPTER SUMMARY

In this chapter, we extend the notion that knowledge evolves continuously. It is derived ceaselessly from information and leads to concepts that humans conceive of and machines distill. The continuum of knowledge has prevailed since the beginning of human communication. Bits are transmitted and received, data structures are assembled and enjoined, information is retrieved, but knowledge is communicated. Knowledge is the bond of civilization. It traverses the highway of thought that enjoins mind with mind, soul with soul, and compassion with love. In the twenty-first century, the widest gaps between the past, present, and future are linked through the knowledge bases of human civilizations and global Internet works. Knowledge, human activity, and high-speed networks constitute an expansionary triangle, with its vertices spread far and wide around the globe, if not nature itself.

The main theme of this chapter is time as a dimension in the knowledge space wherein nature, humans, and machines influence noun objects. Machines serve a two-fold purpose: They tracks the changes in objects, and they (if possible) extrapolate the circumstances for the change in objects to be desirable. The knowledge space of all times encompassing all civilizations is too vast even for super-computers. However, confined knowledge spaces have accrued over time and been classified by major subject classifiers, such as the Dewey Decimal

Computational Framework for Knowledge. By Syed V. Ahamed
Copyright © 2009 John Wiley & Sons, Inc.

System (DDS) or Library of Congress (LoC). It can be seen that current Internet search techniques will create knowledge bases specific to any noun object or verb functions that are associated with any known noun object. Thus, the DDS or LoC systems are initial mechanisms to enter into selected regions of the generic knowledge space, but the regions can be stretched far and wide, depending on the whims and fancies of a knowledge worker in the modern information age.

In this chapter, we also introduce a virtual object "knowbula." This word is derived from and merges two words: "knowledge" and "Nebula".[1] Ideally, a knowbula is an infinitely large three-dimensional envelope of all activities in every field of human endeavor whatsoever. Such an assertion has two prerequisites: (1) that human activity deals with objects (real, abstract, virtual, or just about anythings) which bear coordinates in a more encompassing universal knowledge space, and (2) that there is a time constraint (i.e., begin, middle, and end identifiers) associated with such human activity. Thus, knowbula starts to become a finite three-dimensional virtual object, with the (noun) objects (objective, subjective, or virtual) along the X-first axis, the correlated (verb) functions (active, passive, or hypothetical) along the Y-second axis, and time along the Z-third axis for a subset of human knowledge. The extreme limit for the volume of knowbula is infinite and occurs when all known objects from all knowledge bases around the world are placed along the X-axis, all known and correlated verbs from the same knowledge bases are placed along the Y-axis, and an eternal semi-infinite interval of time (T) is allocated for the convolution (see Chapter 3) of the verb functions with the noun objects. Inconceivable as it sounds, the convolution is indeed the evolution of the planet Earth from the even more encompassing knowledge space of the universe. However, by limiting objects, verbs, and time, an artificial constraint can be imposed on the maximum volume of knowbula. The ratio of the volume of a particular knowbula to its maximum volume is thus an indication of the success of that human activity in that particular society. We discuss the implications of this ratio in Section 10.17. When civilizations experience their demise, the shape and volume of their particular knowbula are X, Y and T dimensions are frozen.

10.1 A NEW BREED OF KNOWLEDGE

Future knowledge is likely to be more structured, robust, and coherent than past knowledge. The increasing role of computers and knowledge machines will assure the validity and precision of every fragment of information. The next generation

---

[1] In the discipline of astronomy, "nebula" refers to "a diffused mass of interstellar dust or gas or both, visible as luminous patches or areas of darkness depending on the way the mass absorbs or reflects incident radiation" from the *American Heritage Dictionary*, 4th ed., copyright © 2000 by Houghton Mifflin Company. In the knowledge domain, "Knowbula" entails areas of understanding that appear as patches of intelligible information which becomes clearer as new understanding emerges.

of scientists will interject human creativity and intelligence into every concept conveyed by any body of knowledge (BOK discussed in Chapter 6). Validity and accuracy are the main stays of science and mathematics and overlap with the broader overview of truth. Quick learning and predictive thinking are the main strengths of the new breed of knowledge workers. In synergy, humans and machine will offer structure and accuracy in knowledge that can now be viewed as a scientific entity.

Over the evolution of knowledge, scientists have not built "social" machines and almost completely ignored "elegance" systems. Engineers and scientists have ignored the disciplines that would have enforced human integrity and bonds. Perhaps such machines could have instilled a sense of social justice and prevented wars based on deception and the corruption in political systems.

Based on current computational and network facilities, we can only address the question of validity and accuracy (or "truth") in every sentence[2] in a paragraph, every paragraph on a page, every page in a chapter, every chapter in a book, every book in a discipline, and every discipline in the global classification of knowledge. Our ability to climb such a structured methodology in softer disciplines (such as arts and literature) declines rapidly. In the deepest folds of mechanization emerge the viability and dependability of the central processor unit (CPU) to execute every machine instruction with scientific validity and mathematical precision. Silicon pathways and fiber highways do not respond to human whims and fancies as easily as they can respond to clock cycles and binary bits.

The lack of "behavioral processors" forces social scientists to emulate behavior on general-purpose computers. Such an exercise is as naïve as grinding stones to get water. Yet, such attempts have been partially successful with extensive humanware to clothe a bare silicon machine. Perhaps a general-purpose computer with silicon chips and a behavioral code will generate the binary responses according to the intellectual rhythm of a human mind. It is not inconceivable that the firmware in the control memories of such processors will carry the notions of Freud, Maslow, or Jung into the next generation of knowledge machines. Personalized behavioral machines stand a good chance of becoming neutered partners for humans as much as the spacecraft is for astronauts.

10.2 THE KNOWLEDGE LOOP AND ITS STABILITY

The feedback effects of the knowledge trail pursued by human activities in any society occur through the environment or nature that sustains that society. In a sense, the trail becomes a loop (see Figure 3.14, Chapter 3) because

---

[2] From the laws of grammar in any language, every sentence has some form of noun object(s), and one or more verb function(s) and tense(s) involved in its structure. We use this requirement to characterize the incremental three-dimensional knowledge space where every sentence will fit as an increment in the (semi-infinite) global space of knowledge.

the cumulative effects of social, scientific, or economic actions in a society come back to reward or haunt it. This loop is defined as the K-loop, since the knowledge accumulated in the past plays a critical role (Sections 10.2–10.4) in shaping the future. Over a long time, such loops can indeed alter the social trend.

The evolution of knowledge is a systematic process. From a computer science perspective, binary bits, data structures, and information are precursors to cogent, coherent, and consistent knowledge. In Figure 2.5, the movement from left to right indicates the scientific processes that accompany the processing of data in routine computational environments at nodes B and D. The processing of information at node I leads to knowledge that is more universal but restricted to the particular environment from where the data are derived. The processing of knowledge at the K node offers a concept that is more universal and generic in most settings which are similar to the particular environment. The derivation of wisdom and ethics at the W- and E-nodes carries a social impact from one environmental setting to another.

The knowledge trail shown in Figure 3.4a was used for the development of the general theory of knowledge presented in Chapter 3. A node for nature (N) was prefixed to the binary node in Section 3.9.1. Three short-term destinations (TVB, SET, and DAH) are feasible since human activity is influenced by the nature of wisdom that drives humans and the machines that process the inputs to the D-, I-, K-, C-, W-, and E-nodes. In essence, humans and their machines are seriously impacted by the motivations and objectives of the activities that are pursued.

The circulation of noun objects (NOs) and verb functions (VFs) along the knowledge trail over sustained long periods of time is depicted in Figure 10.1. The three orientations of wisdom are pulled out to the left, even though their individualized effects become apparent at every microscopic movement (every arrowhead) in the figure. In a computational environment, the stabilizing, neutral, or destabilizing effects of the feedback processes involved can be tracked accurately, and the incremental effect of each social attitude traced. Similar approaches in econometrics offer a precise picture of every financial and social policy in a nation.

In the long run, the cumulative activities (verb functions) of nature, humans, and machines tend to stabilize or destabilize the environment that hosts societies. This effect[3] is authenticated by the rise and fall of most civilizations. In recent human endeavors, nature has become a means to an (good, neutral, or evil) end. The harmony between humans and nature prevailed in Eskimo society while the

---

[3]The roots of truth, virtue, and beauty (TVB) reach beyond humans into nature. Science, economics, and technology (SET) are unaffected by emotions and natural consequences. Finally, deception, arrogance, and hate (DAH) engulf the minds and destroy the nature that hosts societies. The rationality and precision of thought usually drown in deception, arrogance, and hate. The Freudian super-ego, ego, and id assume a distorted manifestation in Aristotle's TVB, DAH, and SET that lies in between the two.

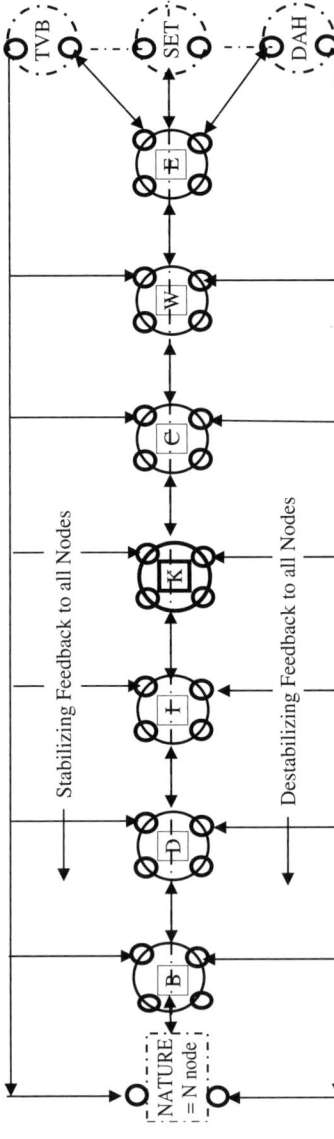

Figure 10.1 The framework for truth, virtue, and beauty is ingrained in most societies if the environment is self-preserving, beneficial, and natural. The cycle is completed from nature, which gives rise to the environment where the knowledge cycle exists in the society that is hosted by nature, thus providing a feedback signal for the stability of the loop. In nature, numerous examples exist. The life cycles of the fig wasp and fig tree, ants and plants, bees and flowers, etc.— have mutually beneficial dependencies. These cycles of stability (involving strong mutually beneficial relationships) have existed for millennia before human societies evolved.

N-Nature or environment that hosts a society; B-binary data; D-data structures; I-information; K-knowledge; C-concepts; W-wisdom; E-ethics and values in society. TVB represents truth, virtue, and beauty. DAH represents deception, arrogance, and hate. SET represents science, economics, and technology that can be deployed in both the TVB and DAH environments. In some cases, inter-nodal feedback is also dominant. The effect can be that of a multiplier (from traditional macroeconomics). Such "multipliers" can accelerate or decelerate the feedback processes. By suitable control of the magnitude and direction of the feedback (generally by well-intentioned news media), the characteristics of a tapped delay line filter (from traditional filter theory) can be imposed, maximizing the desirable effects on society. Conversely, undesirable effects can be minimized.

Portuguese and British were trading slaves and waging wars. Over time, nature thus is regenerated, depleted, or even destroyed due to the cyclic feedback effects of human endeavors. Global warming is a tell-tale sign of the human exploitation of nature.

The representation of the trial can be enhanced dramatically to incorporate the impact of science and technology over the centuries. In order to depict movement over decades, six zones between the B- and E-nodes are shown in Figure 3.4b. The rules for localized navigation through these zones depend on the (TVB, SET, or DAH) orientations of wisdom. For example, the manipulation of bits at the B- and D-nodes so they become deceptive information contradicts the laws for enhancing the objects for truth at the I- and K-nodes, etc.

10.3 CONTINUITY OF KNOWLEDGE

Knowledge is a dynamic entity. Nature, humans, and machines continuously mold the size, and shape, and alter the entropy of knowledge. Like an image on the retina, the entropy of knowledge is a virtual object in neural space. Machines processes overwhelm human efforts in tracking such rapid changes in the entropy of knowledge in any given subject matter or classification. When the knowledge trail is automated (by knowledge and wisdom machines), any significant change at the K-, C-, W-, and E-nodes (Figure 10.1) will update entropy and be specific and precise about the changes and their causes, deviations, and derivatives.

Such accuracy is captured by re-plotting the K-loop (Figure 10.1 in Section 10.2) for each discipline or subject as a lamniscate[4] (Figure 10.2) with two focal points. Conceptually, the two focal points are the CPU and knowledge processor unit (KPU) processes. Each of the two sets of processes has a "multiplier" effect on the other. The locus of the point tracing the lamniscate is that the product of the distance from each focus be constant. When the CPU and KPU processes are constructive and balanced, the lamniscate assumes its preferred shape though not its prefect shape. When the two processes get uncoupled, the product (and hence the spread of the locus) can be oscillatory between infinitely large numbers leading to instability. The numerical representation and the concepteral representation of knowledge gets decoupled.

Nature provides numerous instances where such balances have been successfully reached for many thousands of years. The life cycles of the fig wasp and fig tree, and survival of leaf-cutter ant colonies and plants—both have mutually beneficial dependencies. These cycles of stability (involving strong

---

[4]The lamniscate (also spelled lemniscate) of Bernoulli is written out as $[(x - a)^2 + y^2)(x + a)^2 + y^2)] = a^4$ or more generically as the Cassini oval, written as, $[(x - a)^2 + y^2)(x + a)^2 + y^2)] = b^4$ in Cartesian coordinates.

466 ARCHITECTURE OF KNOWLEDGE

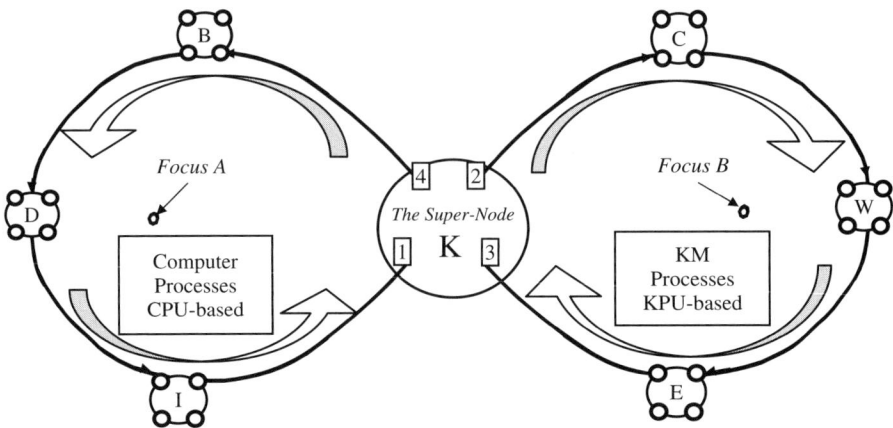

Figure 10.2 The rearrangement of the BDIKCWE chain as a BDIK1-K2CWEK3-K4B loop that is continuous in the current knowledge society. The focus of this (regenerative) loop is the knowledge node. The directionalities of movement, K1-K2 and K3-K4, are important if the effect of time is to be considered. The cycle time for this loop can be a few hours or minutes (as in stock market environments) or it could be centuries (as in Chinese and Buddhist cultures).

mutually beneficial relationships) have existed for millennia before human societies evolved. The methodology proposed in this chapter finds the fine balance between modern societies (that depend extensively on binary data, data structures, and the infrastructure of computers and networks) and human wisdom (that depends on well-founded ethics) with nature as an integral part of survival.

Computational and conceptual processes (within the framework of knowledge processing) get inter-twined in modern society, such that every computational process has a conceptual process as a counterpart (to conceive and predict it). Conversely, every conceptual process has a computational process as a counterpart (to confirm and validate it). Knowledge serves as a fulcrum to balance the two processes.

Knowledge plays a crucial key role in the economic cycles in societies: social cycles in nations: national and political cycles in the world: and power and control cycles, around the globe. This aspect is explored further in Figures 10.4–10.8. Computer processes are CPU-based and knowledge machine (KM) processes are KPU-based. The super-node K serves as the space of union between the CPU and KPU groups of processes.

10.4 THREE ORIENTATIONS OF WISDOM

Knowledge precedes the wisdom that will follow and past wisdom permeates current knowledge. The constant and irresistible flow of time is implied. The

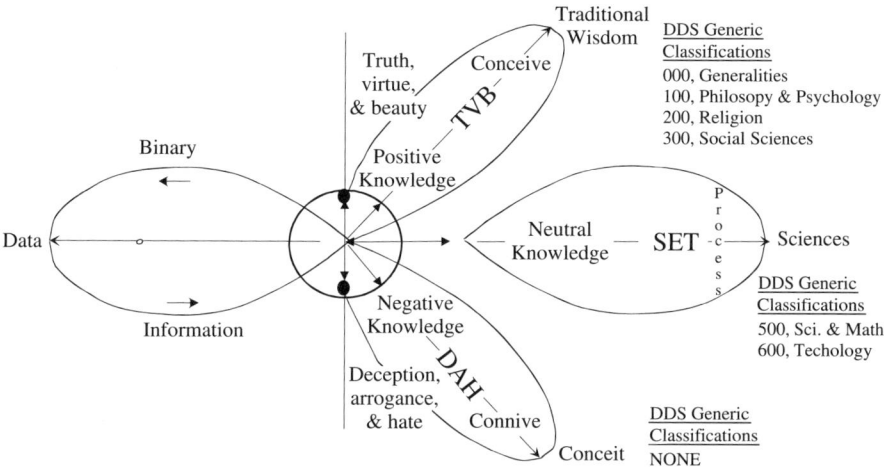

Figure 10.3 Three lobes of the knowledge loop (or orientations of wisdom) are shown on the right side and correspond to the three orientations of wisdom presented in Chapter 2: truth, virtue, and beauty (TVB); sciences, economics, and technology (SET); and deception, arrogance, and hate (DAH). The forward paths from I-node or Information penetrate the K-node or knowledge node and become the concepts oriented toward TVB, SET, or DAH. For DDS classification see footnote in Section 4.3.4.

cycle is pervasive. Knowledge being more abundant forms the foundation for wisdom, and wisdom determines the shape and entropy of future knowledge. In studying prior societies, wisdom can have infinite variations, but three orientations are discernible. The three right lobes of the knowledge loop in Figure 10.3 correspond to the three orientations of wisdom presented in Chapter 2: TVB, SET, and DAH. The forward paths from node I or information penetrate the K-node or knowledge and yield concepts that are oriented toward TVB (based on DDS 000–300), SET (based on DDS 500 and 600), or DAH (no DDS numeric assignments; See footnote in Section 4.3.4). In addition, the K-node plays a crucial role in all cases since it is penetrated twice and exited twice during routine human activity in almost all computational and conceptual domains.

It is evident that there are infinite locations for the right-side lobe in Figure 10.3. In fact, the lobe can have as many variations as humans can conceive of interms of the orientation of wisdom: what it can or should be. In order to simplify machine processes, we broadly align the TVB, DAH, and SET orientations on three radial lines $\theta = +60°, -60°$, and $0°$. Being variable, human nature can make the alignments swing incrementally or dramatically over time. For instance, events in society (such as wars, economic booms and busts, successes and failures, etc.) can alter the human psyche. Events in nature (such as calamities, epidemics, droughts, etc.) can alter social norms.

468 ARCHITECTURE OF KNOWLEDGE

10.5 LONG-TERM MOVEMENTS WITHIN THE K-LOOP

Over the short term and during routine activities within a social environment, the movement of societies along the K-loop is small and well articulated. In fact, it could be so minuscule that it is hardly noticeable at the knowledge base level. However, a protracted effort in society leaves behind an impact. For example, the research at NASA is more appropriately depicted by a K-loop, as in Figure 10.4 rather than in Figure 10.2.

Space science had a well-defined beginning at the START node and has not yet reached the FINISH node. But, it is likely that the science will reach its end. Routine activities in a typical society are confined to the BDI($K_1 - K_2$) CWE($K_3 - K_4$) B-loop. The movement of the knowledge loop from the N-node (nature) to the positive wisdom of TVB occurs in very few societies. The point of entry is the B-node. The final breakaway point occurs at the E-node. Major shifts to break with routine are quite rare but they bring about fairly drastic changes in the social norms. During this phase, the entire value system is scrutinized and changed for the better or worse.

Typical examples from recent history are the French Revolution, and the abolition of slavery and Equal Rights Amendment in the United States. Though computer systems have evolved recently, the human estimation of the overall effects of drastic changes has preceded any computer models. Other changes in

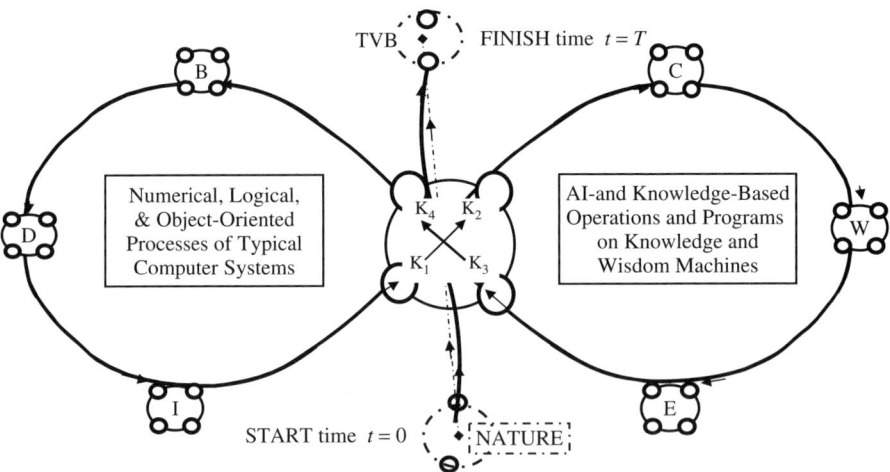

Figure 10.4 Movement of the knowledge loop from the N-node (nature node) to the positive wisdom of TVB, twice through the K-node or knowledge node. The entry point is the B-node. The final break-away point occurs at the E-node. Major shifts that break with the routine BDI($K_1 - K_2$) CWE($K_3 - K_4$) B-loop occur rarely in society. During this phase, the entire value system is scrutinized and changed for the better. Two examples from recent history are the deliberations over the U.S. Constitution and the abolition of slavery in the United States.

society have also occurred due to natural events, such as tsunamis, earthquakes, and the ice ages. Such natural events leave a trail of knowledge behind.

10.6 CONFLICTIVE ROLES OF TVB<>DAH ORIENTATIONS

A sustained social attitude leaning toward either side can be more easily tracked by machines rather than human beings. The moving window average of the collective record of achievements that favor one orientation over the other will indicate the trends and shifts.

Such an approach is adapted in economic studies to determine stock market trends, public confidence, social preferences, etc. Social accounting of this nature becomes crucial in bringing about desirable change and blocking undesirable change in small and large societies. In enhancing the progress of most societies, ethics and value systems will not be sacrificed by the machines. Over any given duration, the group velocity of indicators along the forward path in Figure 10.1 will achieve the desired goals of a society or nation.

Human nature can be volatile, ranging from the saintly to the savage. In a more controlled social environment, truth becomes science, virtue becomes generosity, and elegance becomes art. Conversely, in an adverse environment, deception becomes politics, arrogance become intolerance, and hate becomes war. Fortunately, moderation prevails. And society offers a tolerable refuge for most human beings ranging from philosophers to things. The conflicts between these two extremes of orientation does exist, and machines that "sense" the insidious movements of K-loops can "respond" in an intelligent and stabilizing fashion.

Figure 10.5 depicts continuous movements through the knowledge loops and includes the effect of time. The K-loops lie in a plane perpendicular to that of the paper or page. These loops although not plotted and tracked do prevail in societies. For example, Dr. King's efforts to bring about social justice, scientific efforts to expand and stabilize the Internet, and federal efforts to curtail the activities of sex offenders would fall in the upper half of this figure.

By the same token, Nixon's Watergate scandal and his efforts to cover it up, Clinton's impeachment for lying about a sexual indiscretion, Bush's deceptions regarding weapons of mass destruction (WMD), and the subsequent invasion/occupation of Iraq fal in the lower half of Figure 10.5. Such incidents are examples that point to the insidious social tendencies which can creep up in any healthy society.

10.7 TRACKING OF THE K-LOOPS

The movement between the numerous K-loops shown in Figure 10.5 is also a continuous process in society. In some cases, the shift between one K-loop to the adjoining loop is a gradual process and follows an orderly legal process (e.g., the constitutional amendment to abolish slavery). In other cases, it could be a

470 ARCHITECTURE OF KNOWLEDGE

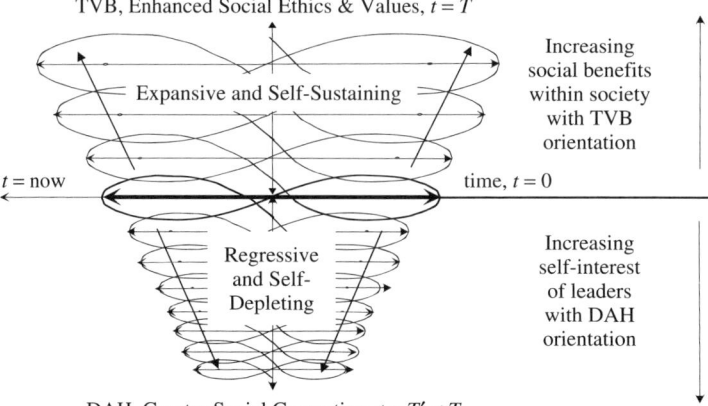

Figure 10.5 The two opposite orientations of wisdom (truth, value, and beauty, and deception, arrogance, and hate) and their insidious role in enhancing (upper K-loops) or depleting (lower K-loops) society are shown. The TVB orientation brings about social benefits and the DAH orientation serves the self-interests of leaders. In this Figure, the time axis is the vertical axis in the plane of the paper or page. Each of the K-loops is in a plane perpendicular to that of the paper.

chaotic and sudden event (e.g., the French revolution or the fall of Aztecs). When the shift follows an orderly process, then the discontent of society is moderated by the ethics and wisdom in the current loop. Instances such as the current corporate scandals in the wake of greed for wealth, or the injustices delivered to the African American community during the mid-nineteenth century provide a window to explore higher goals and aspirations for society. Two variations exist; the loop can (incrementally) move up or move down. Its movement is discussed in Sections 10.7.1–10.7.2.

10.7.1 Expansionary Upward Movement (TVB)

In the expansive mode, the passage from levels 1 and 2 is as depicted in Figure 10.6. Since the feedback path enhances values for society in an environment or natural setting that supports society, society perpetuates the environment that hosts it and vice versa.

This symbiosis between society and nature assures the continued growth of a value system, and the level 2 knowledge loop (K-loop) will be stronger and more stable than the level 1 K-loop. Over time, this symbiotic growth continues until the DAH orientation of wisdom corrupts society, the nation, or culture and the K-loop starts moving downward. Like the wealth of nations, knowledge and wisdom also experience cycles of expansion and contraction. The bankruptcy of knowledge based on truth, virtue, and beauty is the abundance of deception, arrogance, and hate that is written in the dark pages of history. It appears almost as a zero-sum game that society plays again and again on humans. Despite the

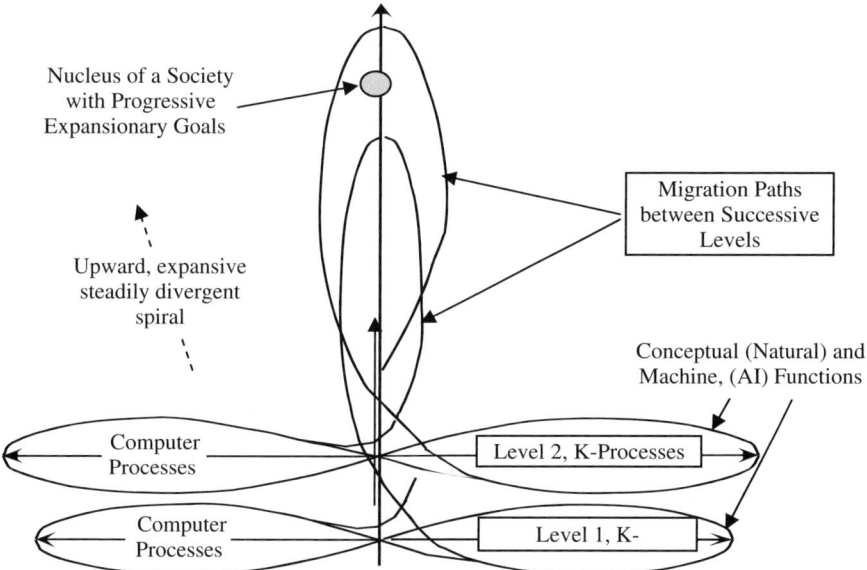

Upward and Expansionary Path of Migration of the Knowledge Loop (K-loop) from Level 1 to Level 2 with traditional and symbiotic TVB Orientation of Wisdom. NI = Natural.

Figure 10.6 Migration between level 1 of the knowledge loop to the next higher loops based on truth, virtue, and beauty (benevolence).

negative pressures, peace and prosperity have prevailed many times in many cultures.

10.7.2 Recessive Downward Movement (DAH)

In the recessive mode, the feedback path of the K-loop drains society of its moral and ethical values. In a very deceptive way, the downward migration, shown in Figure 10.7, depletes society in the environment (nature) that supports it, and society in turn depletes the very environment (nature) that hosts it.

This destructive relationship between society and nature soon depletes the respect and integrity of the social system, and the level 2 knowledge loop will be weaker and less stable than the level 1 K-loop. The DAH orientation corrupts society, the nation, or culture as the K-loop starts moving downward in the time dimension.

Over time, this mutually destructive relation continues until society is morally bankrupt. Nature runs into environmental disasters such as global warming and the depletion of the ozone layer. Like the wealth of nations, knowledge and wisdom also experience cycles of boom and bust. Indulgence in DAH orientation is the bankruptcy of accumulated knowledge based on truth, virtue, and beauty. The death and destruction of societies have occurred many times in the dark pages in the history of human civilizations.

472 ARCHITECTURE OF KNOWLEDGE

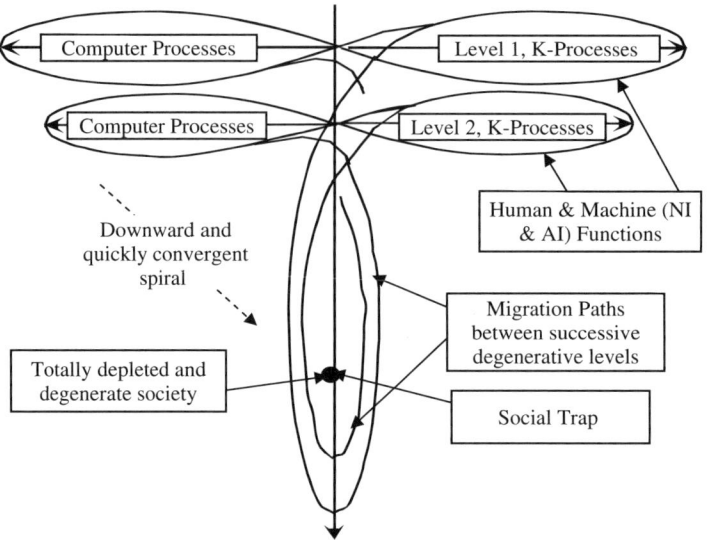

Figure 10.7 Migration between level 1 of the knowledge loop to the next lower loops based on deception, arrogance (aggression), and hate.

10.8 DETAILS OF UPWARD MIGRATION

The bounds, frontiers, and entropy of any body of knowledge (BOK) are dynamic and volatile. Every act of nature, humans, or the processes of the machine modifies some aspect of knowledge. Like the ocean, knowledge has a dynamic profile. Ripples and waves persist. In the vast majority of cases, the microscopic contour is unimportant, but in some cases the impact is monumental.

On the upward movement of K-loops, a relatively simple statement such as the ratio of the circumference of a circle to its diameter gives rise to π, whose value is held fixed. This has profound implications in mathematics. Statement such as $(P = E \times H)$ is extremely significant in all electromagnetic field propagation, etc. Other less important and profound manipulations of knowledge occur on a continuous basis.

When the contour of knowledge is shifted up, the sense of truth, social benevolence, or elegance is reinforced. Human beings themselves (e.g., those in the judicial system) make it impossible for hypocrisy to survive, political and corporate leaders to cover up their wrong doings, and criminals to be pardoned (e.g., the series of scandalous pardons that rocked the end of Clinton's presidency, Ford's pardon of Nixon after he was faced to resign the presidency in disgrace, etc.).

When the processing of knowledge becomes as common as the processing of data, machines will sense and track illegalities early on and block the spread of rampant political and corporate crime. Sex offenders, drug dealers, and porno distributors will be stopped before any crimes occur and long before crime becomes a way of life.

10.9 DETAILS OF DOWNWARD MIGRATION

On the downward movement of K-loops, the process is reversed. The illegitimate and unethical manipulation of knowledge has profound effects. When information networks and the mass media carry falsified and corrupted facts (e.g., Bush's assertion that weapons of mass destruction existed in Iraq, Nixon's role in the coverup of the Watergate break-in, Clinton's continued denial of improper sexual behavior with Monica Lewinsky, etc.), the media becomes a weapon of mass deception for the public. In a sense, negatively primed knowledge machines assure the authenticity of falsified information and – hypocrisy and any coverup to protect those in power.

On the contrary, the movement from the K-loop to a higher level is a well-planned procedure. Much as in a surgical process, numerous steps are collected, assembled, graphed, and optimized. A series of these steps is depicted in Figure 10.8, and the migratory path can be smooth as the rationality in science rather than the crimes committed during war.

The entry into and exit from the loops at the K-node are also elaborate. The bulk of human activity is inconsequential to the structure of knowledge during any given interval of time. The effect is portrayed in Figure 10.9 as the knowledge loop traverses its mundane path, that is, f (from the E-node) to g, (a location in the K-node), to b (toward the B-node). The entire K-loop (i.e., BDIKCWEB in Figure 10.8) is thus uneventfully completed. When major expansive or cataclysmic social events occur, the entire loop is shifted up or down, respectively. For upward movement, the break-away point to K_{i+1} from the K_i loop occurs hear the E-node along the path k, and the come-back point from K_{i-1} to the K_i loop occurs along the a path. Thus, only monumental events in society mold the directions of future knowledge.

10.10 KNOWLEDGE LOOPS IN THE X-AND Y-DIMENSIONS

Knowledge is a virtual entity. However, it is centered around noun objects (NOs) of varying complexity that perform and experience verb functions (VFs) and processes. Objects can be grouped as easily as processes can be programmed. Groups of objects undergoing sequences of valid and realistic processes constitute knowledge programs. The progression along the knowledge loop offers a methodology for building knowledge banks and axioms of wisdom that contribute to the progress of society per se.

The knowledge super-node K_i, shown in Figure 10.9, has at least two entry and two exit points. Through each of these four points (K_1 to K_4), a whole array

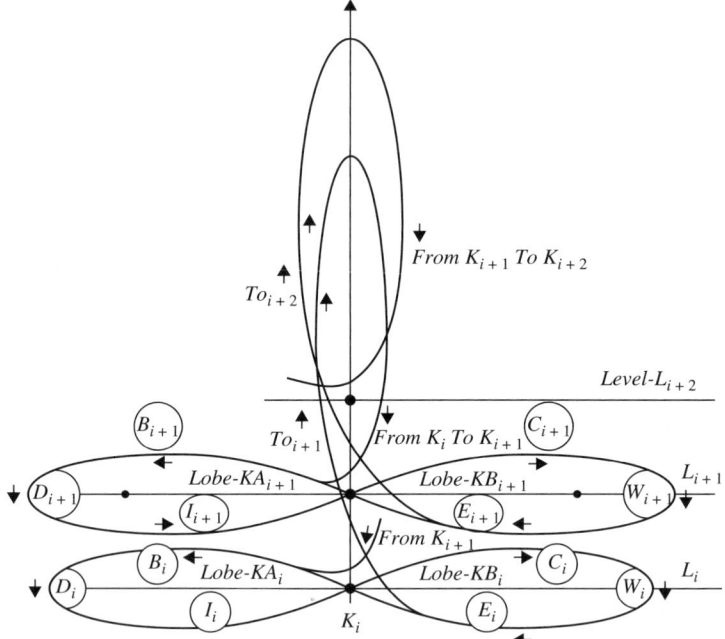

Figure 10.8 Migration between level L_i of the knowledge loop to the next higher loops L_{i+1}, and L_{i+2}, based on a truth, value, and beauty orientation of wisdom. All the nodes along the knowledge loop (K-loop) are shown. The path starts from the K-node K_{i-1} at level $i-1$ and traces a K-loop at the ith level. Most of the routine social functions occur when traversing the K-loop at any of the three levels shown. At the inception of a major change in society (such as a knowledge revolution, freedom of the nation, space travel, etc.), the path shifts from level L_{i-1} and moves up to the K-node K_i. The same process repeats itself at the ith level to continue the movement to the K_{i+1} node.

of related NOs and a whole string of corresponding VFs get (computationally) threaded. Circulation along the K-loop continues. In order to arrange an orderly passage through these points, the implicit NOs and their corresponding VFs can be aligned as the rows and columns of a matrix (see Figure 3.3). If such a matrix is shrunk to a point, then this Euclidian point traverses the knowledge loop. Totally disoriented and disordered human activity and/or execution of error ridden knowledge programs does not constitute a knowledge loop.

However, scientific, systematic, orderly, and consistent effort brings about knowledge from information, concepts from knowledge, wisdom from concepts, and then values and ethics to make society richer and more civilized.

Historically, knowledge loops are poorly organized. When integrated over decades, hundreds of such major loops (such as the inception, maturity, and decay and demise of a major corporation, social programs, institutions, etc.) can be detected. A snapshot of an organized K-loop can now be projected on an XY-plane by collapsing the NOs (as they pertain to any given discipline or

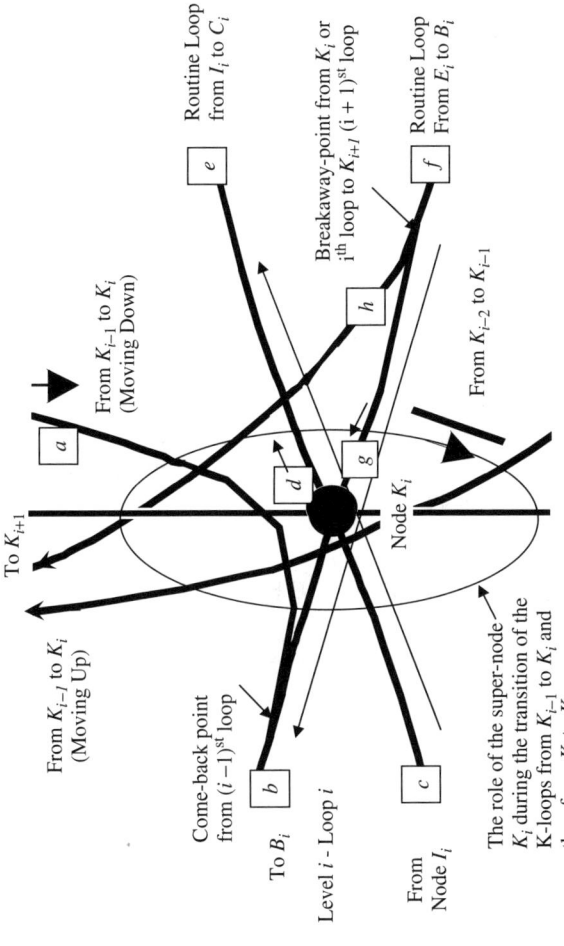

Figure 10.9 Routine loop ($bcdefgb$, also see Figure 10.8) is shown. The break-in path from K_{i-1} and onto the routine loop is (a-$bcdefgb$), and the break-out path to K_{i+1} from the K_i loop is (h-$abcdefgb$) of the $(i+1)$st loop are also shown.

476 ARCHITECTURE OF KNOWLEDGE

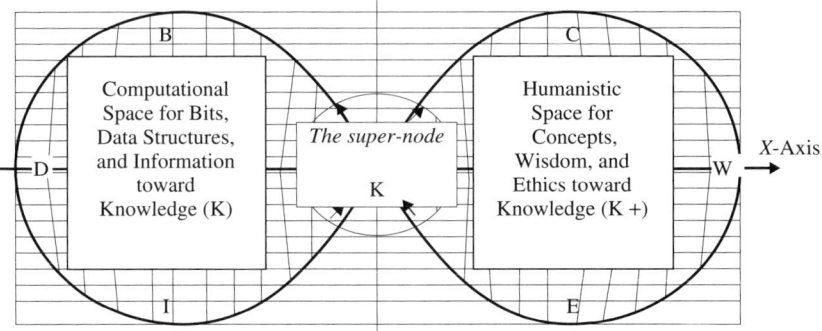

Figure 10.10 In routine instances, the activities of humans and machines are mapped as incremental movements within the BDIKCWEB loop. Traversing the K-loop alters the entropy of noun objects in society during that interval of time. A time sequence is thus generated due to the execution of the processes (verb functions) on the noun objects. A time series analysis depicts the migration of knowledge in the time domain.

organized entity) on the X-axis and VFs on the Y-axis. The process is depicted in Figure 10.10.

The purpose of this representation of the K-loop is to facilitate the architecture for a knowledge plane and then a knowledge space. An ordered knowledge space will provide the mathematical means to penetrate the zones of unknown knowledge coherently and cogently, just as it has become possible to penetrate outerspace coherently and cogently.

10.11 FOUR PORTS OF SUPER-NODE K

Positively accumulated knowledge loops in society provide a platform for forward social mobility for sustained growth. Such loops serve as catch basins to avoid the downward social drift that depletes society. Social enhancement and regression are the two phases that are triggered by the use and abuse of current knowledge bases and the incessant turmoil within them. Whereas the deployment of knowledge bases (e.g., authenticated information bases of cancer research centers) is desirable, the deployment of knowledge bases with negative intent (e.g., porno and prostitution Internet sites of sex abusers) is deplorable.

The two entry points to the K-node from the I- and E-nodes and the two exit points to the C- and B-nodes are shown in Figure 10.10. The procedures at these four points require a special protocol for the passage of appropriate information in and out of super-node K. The I to C path via K needs to ascertain that the content and quality of information from the I-node is suitable for knowledge processing in the K-node, and that output generated at the K-node will be appropriate for the C-node to examine new concepts (if any) in it.

The convolution of VFs and NOs (Figure 3.3) and its effects on the knowledge trail (Figure 2.5) are shown in Figure 10.11. Iterative movement along the trail and along the loop becomes the constant incremental change in entropy. Major convolutions lead to new objects and notable inventions, and innovations may thus be generated anywhere throughout this loop, but the overall continuity of the major disciplines remains a relentless process (of traversing the knowledge loops) through the ages.

10.12 CONSTRUCTION OF A KNOWLEDGE PLANE

When the NOs and VFs are collapsed in the X- and Y-dimensions, a plane of reference for knowledge starts to emerge. The two discrete steps necessary to construct the knowledge plane are presented below in Sections 10.12.1–10.12.2.

10.12.1 Embedded Noun Objects and Verb Functions

The source for such NOs and their associated VFs is embedded in the nodes of any knowledge loop or K-loop. The process is depicted in Figure 10.11.

Whereas the current K-loop gives rise to NOs and VFs, the convolution between the two sets of NOs and VFs gives rise to new NOs and VFs appropriate to the context of the K-loop traversed. Properly navigated, the K-loops occurring in nature, the human mind, and KPUs provide the group velocity for NOs and VFs as they circulate through these loops, offering a stable and steady pace for social progress. Conversely, disorganized and jittery movements along the loop only result in noise in the knowledge, which leads to confusion and chaos.

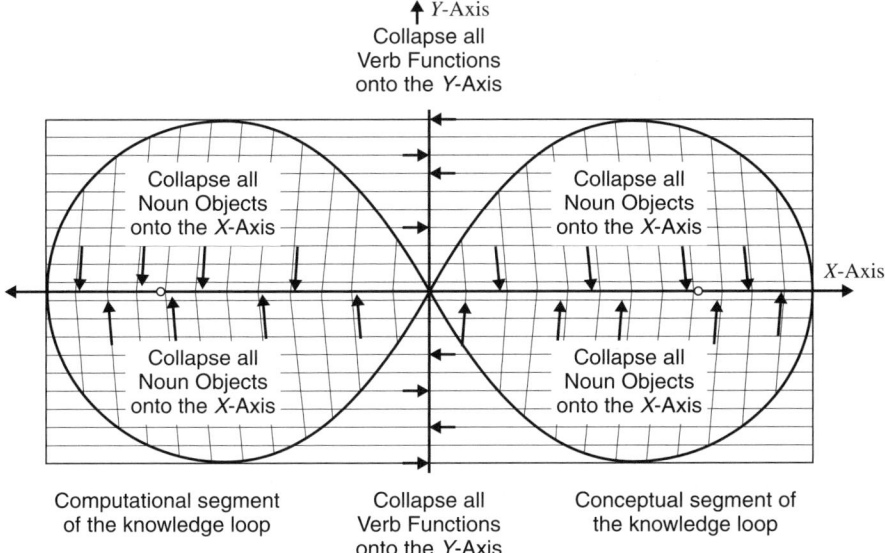

Figure 10.11 Collapsing the noun objects onto the X-axis and the verb functions onto the Y-axis to generate an interim two-dimensional plane for the knowledge loop.

478 ARCHITECTURE OF KNOWLEDGE

In a computational environment, collapsing the NOs and VFs is a routine data/object base operations/management function. The entire process simply consists of building two data/object-oriented bases: XNounObjectBase and YVerbFuctionBase. In practice, such object bases are routinely built into almost all application environments, such as patient, drug, procedure, physical therapy, etc., databases in medicine, or human resources, inventory, sales, revenue, and profit, etc., databases in corporations.

10.12.2 Primary K-Plane in the *XY*-Dimensions

The convolution of NOs and VFs was introduced in Chapter 3. From the mathematical perspective of implementing the process, the NOs and VFs were arranged as a row and column of a matrix (see Figure 3.3). From a computational perspective, we considered the NOs and VFs as two data/object bases in Section 10.12.1. From a conceptual perspective, we construct a virtual knowledge plane of NOs (X-axis) and VFs (Y-axis), as depicted in Figure 10.12. There are two major reasons for constructing the K-plane: to accommodate (1) the three[5] orientations

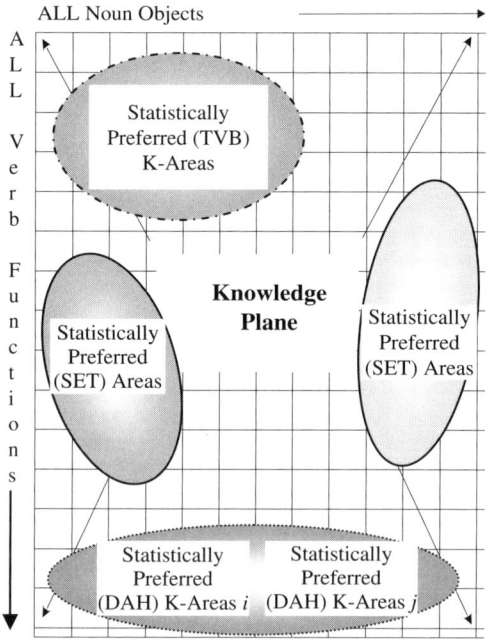

Figure 10.12 The enlargement of the knowledge node to a knowledge plane to indicate that entry at the exit points can occupy different levels of knowledge in the node. The three paths (TVB, SET, and DAH) on the right side of the lobe (Figure 10.3) will alter the return path, thus resulting in an impact on the knowledge loop.

---

[5] In Chapter 3, the three orientations of wisdom, (TVB, SET, and DAH) were introduced. For indicating the numerous dimensions, the variables *tvb*, *set*, and *dah* are used in this section.

(*tvb, set*, and *dah*) of wisdom and (2) the effect of time T that are instrumental in circulating NOs and VFs in the K-loop. The construction of the K-plane leads to the building of a *XYT*-knowledge space or to many four- or five-dimensional spaces (*XYT-t/s*; *XYT-t/e*; *XYT-t/t*; *XYT-v/s*; *XYT-v/e*; *XYT-v/t*, etc.).

10.13 KNOWLEDGE BASES IN THE PRIMARY K-PLANE

Knowledge has evolved randomly across cultures and civilizations. At the outset, the evolution of knowledge appears to be as complex as the evolution of species. Rather than tracking the evolution of knowledge, it is more practical to track the NOs and VFs already evolved and embedded in existing Internet knowledge bases (KBs). When the KBs are scanned based on the DDS or LoC classifications, the shadows of NOs and their associated VFs appear haphazardly on the knowledge plane (K-plane), as shown in Figure 10.13.

In more recent subclassifications such as fiber-optic communications, VLSI, cancer research, etc., the area projected on the K-plane becomes discernible and the data/object bases manageable. In particular, when the NOs and VFs are limited to a specific timeframe and/or from certain local/Web geographical bases, the number of entries in the cache bases becomes specific. The K-programs at the K-node perform specific tests to validate or repudiate any localized information in order to verify the truth, virtue, or elegance in the newly found local NOs/VFs. This methodology is common to all generic topics, even if they are not classified by the DDS or LoC numbering systems.

10.14 SUCCESSIVE K-LOOPS IN THE TIME DIMENSION

Much as space and time form an incessant bond, knowledge, its spatial coordinates, and the instant of change in entropy constitute an incessantly spinning triple helix. The *XY*-space and time (*T*-) coordinates are firmly connected by the information highways on the Internet. However, barring the noise in knowledge, major historic knowledge events have well-defined coordinates. For example, the U.S. Constitution, Bill of Rights, Marshall's marginal utility theory, Maxwell's equations, Einstein's $E = mc^2$, etc., all have definite knowledge (i.e., DDS or LoC classification), geographical location, as well as *XY*-space and *T*-coordinates. Events within society also occur in the hyper-dimensional knowledge space. Mapping the movement of NOs along the K-loop thus becomes conformal in the "object space" of a knowledge machine (KM).

10.14.1 Stacking of K-Loops

In order to build knowledge space, if the numerous K-loops in any definite discipline are stacked in the plane of the paper (or page) and time is orthogonal to the K-loops, then a three-dimensional virtual book with *XYT*-dimensions can

480 ARCHITECTURE OF KNOWLEDGE

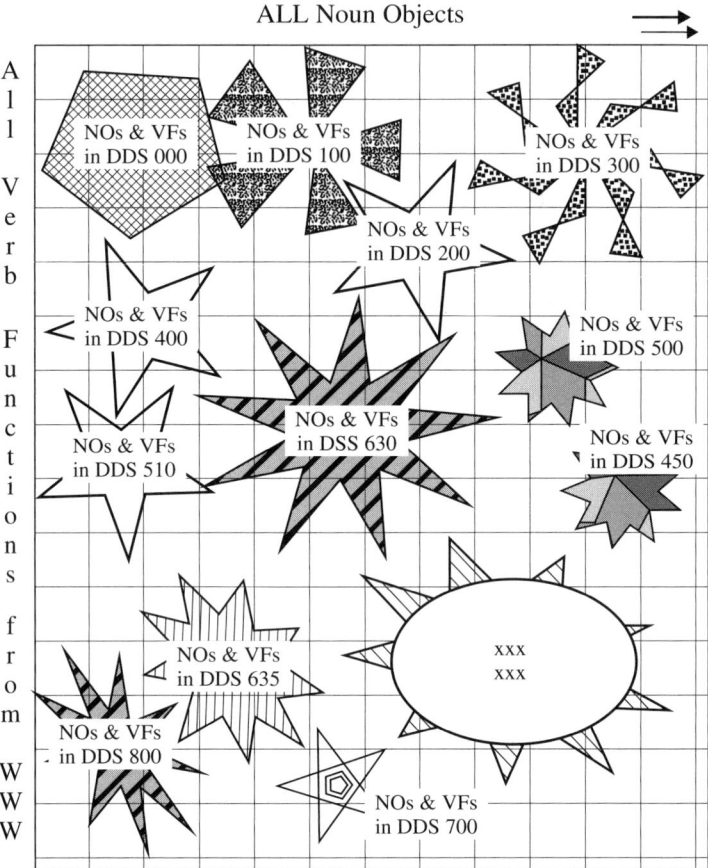

Figure 10.13 Grouping of noun objects and verb functions that exist in knowledge bases around the world for each of the Dewey Decimal System (DDS) classifications. For example, all the noun objects and verb functions that exist only for fiber-optic communications systems are grouped in a FO knowledge base and become a subset of the DDS classification 621.3. When the verb functions in FO (e.g., Raman effects) overlap with the noun objects in other groups (e.g., theoretical and mathematical physics, with a DDS classification of 530.1), then forward and backward linkages provide the web of knowledge with commonalities of noun objects and verb functions as they have been studied in the past.

be envisioned. The structure of such a three-dimensional knowledge space is shown in Figure 10.14. All the NOs and VFs at any instant of time fall along the K-loops. In scientific and idealized settings, the migration from one loop to the next loop occurs as a well-designed transition, as shown in Figure 10.9, but it can also be disorderly and chaotic driven by natural calamities, or by acts of war and the vengeance of humans.

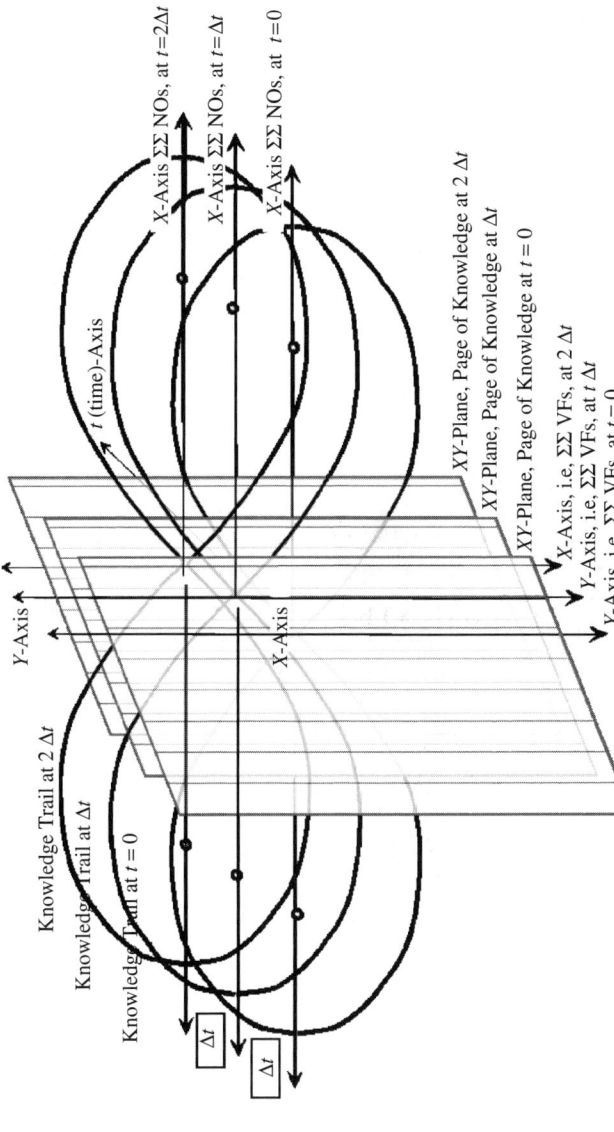

Figure 10.14 Depiction of the insidious nature of change in the entropy of knowledge with time. Three instants are chosen at $t = 0$, Δt, and $2\Delta t$ along the t-axis. The XY-plane carries a cumulative record of all $\sum\sum$ noun objects and all $\sum\sum$ verb functions. These may be convolved to get the desired changes in noun objects. In this figure, the time axis (t-axis) is perpendicular to the plane of the paper.

481

In the past, the transition of knowledge from one generation to the next has not been smooth. The major monuments of knowledge (K-loops) of many civilizations have been squandered, disjoined, disintegrated, and in some cases lost forever. With the tools and techniques of the modern information age, an orderly *XYT*-coordinate system will prevent such knowledge disasters. More than that, the progress of any civilization that has been frozen in time can be recovered and reshaped to suit modern or future times. When the three coordinates are combined to generate a knowledge space, the idealized cube of knowledge is shown in Figure 10.15. The continuum of NOs, their correlated VFs, and time is evident in that illustration. It establishes a framework: that future NOs and VFs can be extrapolated from current and prior NOs and VFs, and the innovations of prior generations become a leading indicator of future creativity. Such a knowledge space almost resembles a curtailed non-Euclidian space with appropriate rules for the geometry of knowledge. Extrapolation and prediction become complex but viable mathematical operations.

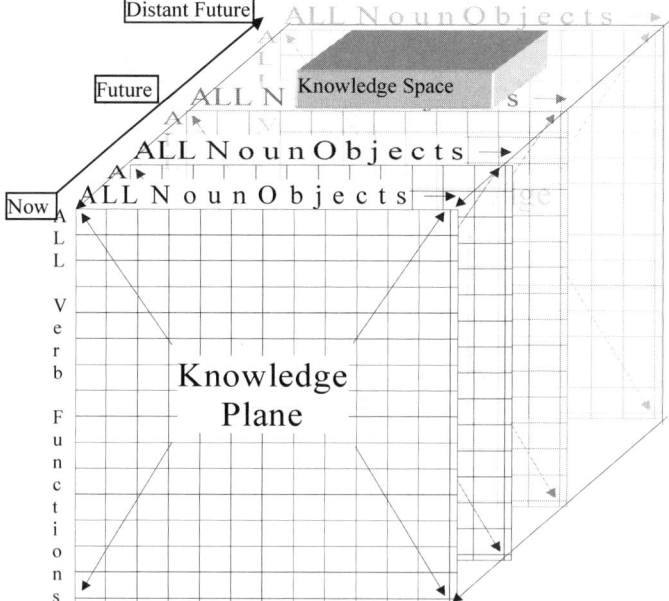

Figure 10.15 The effect of time on processing knowledge is shown by adding the time dimension (perpendicular to the plane of the paper). Time moves away from the instant $t = $ now and follows the t-dimension in the right-handed system of coordinates. Time lapses independently of the paths of the K-loops. Only major shifts of the K-loops bear significant impact on the t-axis and become islands of human achievement in a positive sense, or scars of hurt in a negative sense.

10.14.2 Lessons from Knowledge Lost

The achievements of the past generation have left behind scores of physical splendors. The path of their intellectual processes can hardly be captured from their monuments. In fact, the greed of some[6] has even destroyed the remains of tell-tale signs of civilizations. The reconstructed K-loops from earlier generations shed disordered imagery in the current knowledge space. In fact, no sufficient proof exists that their K-loops were robustly designed or that circumstances provided enough stability such that major concepts (Aristotle's democracy, Newton's gravity, the pharaoh's pyramids, Aztec and Mayan temples, etc.) could be conceived. If such major and conclusive K-loops are pushed into a virtual K-space, the picture appears disorganized and disconnected. Such a cluttered knowledge space is portrayed in Figure 10.16.

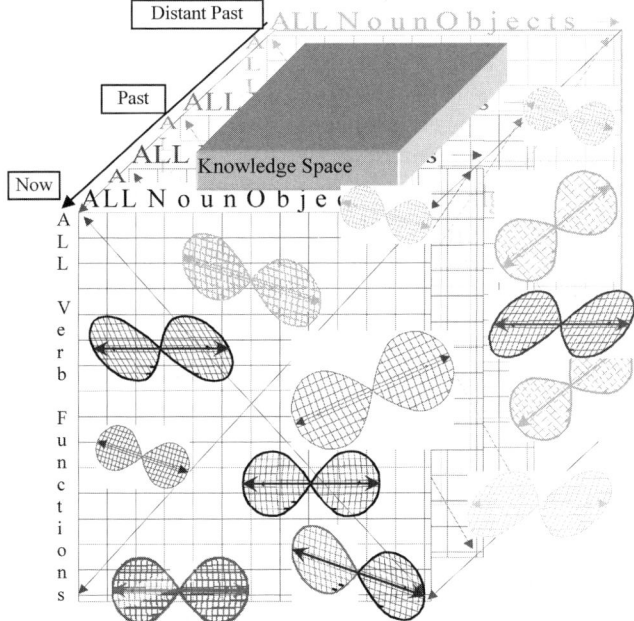

Figure 10.16 A disordered collection of knowledge loops in the knowledge space over time, leading to uneven and unpredictable path of knowledge into the future. It appears that if a dictionary of knowledge loops is uploaded in libraries of arts, literature, sciences, technology, etc., then the progress and paths for knowledge can be arranged so they have order and structure. A well-defined course for the development of knowledge is thus feasible.

---

[6]Raiders of the Egyptian pyramids and their treasures; the Spanish explorers Hernando Cortez, and Francisco Pizarro who plundered Native American heritage and treasures; the British Clive who seized (stole) gems and diamonds during the occupation of India, etc.

484 ARCHITECTURE OF KNOWLEDGE

10.14.3 Cosmic Space, Knowledge Space, and Intellectual Space

Cosmic space with celestial NOs and colossal VFs forge their own script of knowledge. For philosophers and scientists, their knowledge is a subset of the knowledge derived from cosmic and subcosmic laws of nature. To bring the universal picture within human comprehension, scaling down the cosmic and scaling up the microscopic knowledge spaces become necessary.

Human understanding of the entire universal space of knowledge and time is limited at best. Cognizance of cosmic space simply provides an outer bound for the knowledge of all human civilizations. Human knowledge is the tiny segment shown in Figures 10.17a and b that is carved out of an irregular (and vastly ununderstood) cosmic space.[7]

The laws of physical sciences and mathematics start to fall short, beyond the documented knowledge of human civilizations shown in Figure 10.17b. For this reason, we suggest that the cosmic space be so finely subdivided that within its own elemental cube the laws of science and mathematics become applicable.

Currently, human knowledge has two major means of classification: the DDS numeric and the LoC alphanumeric systems. To push human understanding beyond these systems and maintain a scientific methodology, we transport this smaller cube of knowledge into a bigger knowledge space that is the XYT-space for *all* NOs, *all* VFs, and *all* times. Even such a bigger knowledge space is too vast for super-computer systems and elitist human comprehension.

The intellectual space is well structured for the sciences and mathematics, with DDS classifications of 100–600 (see Figure 10.3). When a smaller, elemental cube is carved out for the science of knowledge for a particular discipline, then the cube has a configuration as shown in Figure 10.17c, and the dimensions are $(\Delta X, \Delta Y, \Delta T)$. To represent the Address look-up points for Web KBs NOs in a discipline, VFs of that discipline, time cycle for selected K-loop(s).

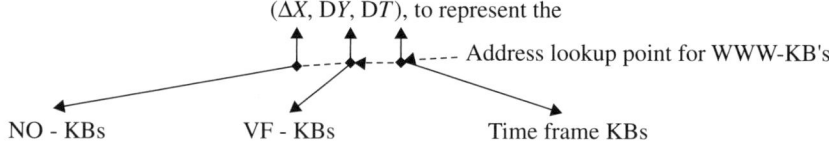

The volume within the elemental cube (Figure 10.17c) of knowledge is thus organized to follow the guidelines of convolution of VFs on appropriate NOs (see Chapter 3, Figure 3.3) that are specific to the discipline. The connectivity between the numerous disciplines is established between the (known) domains of knowledge, and it will be possible to modify, enhance, and translate the rules of convolution from one subject matter to the next. Such a translation of rules and the transportation of elemental cubes of knowledge from one knowledge space to another add a new dimen-

---

[7]In reality, such segmentation occurs many times. Eskimos carve out blocks of ice from vast chunks to build their igloos, mathematicians have caved out a sequential numbering system from an infinitely large sets of numbers, coding theorists select a subset of binary sequences for error detection and correction.

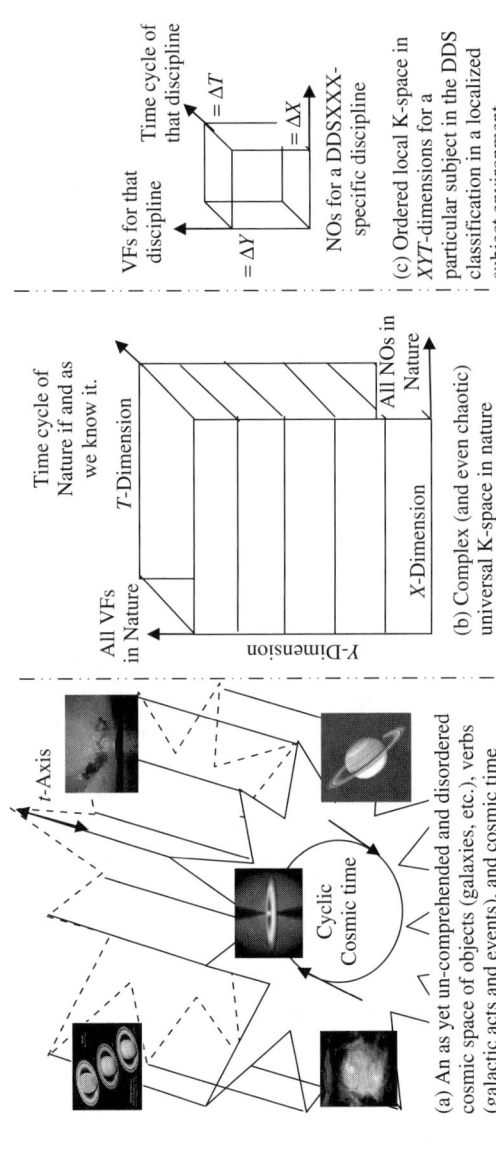

Figure 10.17 Two-step degeneration of universal space and time in (a) to an ordered elemental cube of ΔX, ΔY, and ΔT in (c), to facilitate the processing of knowledge. In this tightly enclosed and ordered k-space occurs the subject-matter classification of the noun objects, that is, ΔX, their associated verb functions that is, ΔY, and an increment of time ΔT. The processing of knowledge can be relaxed or tightened by the user. Processing over a larger cube brings objects and functions from neighboring disciplines and makes the search for new objects and processing more generic. This element of time represents the time cycle of the elemental cube of knowledge that is constantly being altered by human understanding and time. Order, knowledge, and time facilitate the isolation of this elemental cube (c) from the more complex and universal K-space in nature (b).

sion to being able to program at least one aspect of automated machine-based creativity.

The graphical superposition of the three images of cosmic, universal, and subject-specific knowledge spaces is shown in Figure 10.18. The composition of the three spaces presents possible techniques for navigating the surfaces between adjoining spaces. The relaxation of the rules for one discipline and simultaneous tightening of the rules for another can now be forced to become dependent on the XYT-coordinates of the rules for dealing with nonlinear transformations of NOs, *s and VFs.

Enormous flexibility in the creative manipulation of NOs and VFs in almost any discipline is gained. In fact, it is possible to go back in time and perform a forensic study of the development of knowledge during any given era in hindsight with the computerized vision of current objects and technology and foresight of knowledge predictability. Knowledge, although a virtual entity, now becomes a super-object in the object space of knowledge machines.

The representation of an ordered knowledge space (K-space) in the XYT-dimensions shown in Figure 10.18 facilitates an orderly transition of structured knowledge from one discipline to the next. The subject-specific elemental K-space is a block within the generic infinite space of knowledge. A block may then be identified in the generic K-space by limiting the X-dimension (ΔX) to consist of all the NOs in a particular discipline (such as biology, cancer research, medicine, etc., initially, with similar disciplines combined later on for broader research). The Y-dimension can also be limited (ΔY) to all the VFs that occur in that particular subdiscipline or the combined disciplines. Now, adjoining elements correspond to the subclassifications in the overall DDS subject classification of knowledge. Thus, the knowledge machine (KM) performs its knowledge domain (KD) functions in a limited or an expanded K-space. Such an expansion or contraction is accomplished by expanding or contracting the Web addresses for the KBs accessed by the KM. The T-dimension enforces the timing constraint on knowledge about the era or window of time (ΔT) during which any K-loop was traced in that discipline. All aspects of knowledge,—including its changes (caused by nature, humans, and machines), its entropy, and its processes—are stored and processed in this XYT-knowledge space.

10.15 THE NEBULA OF KNOWLEDGE: KNOWBULA

The three-dimensional knowledge object "knowbula" is an envelope of all knowledge loops during any given interval of time and is depicted in Figure 10.19. Knowbula can be as microscopic as a small computer program or it could be macroscopic, encompassing all human knowledge. The time span can also be pico seconds or span eons. However to be realistic, the size of knowbula is selected to suit a particular segment of knowledge that is of current interest and the time span that will likely be operative for the investigation of that topic(s). Typically

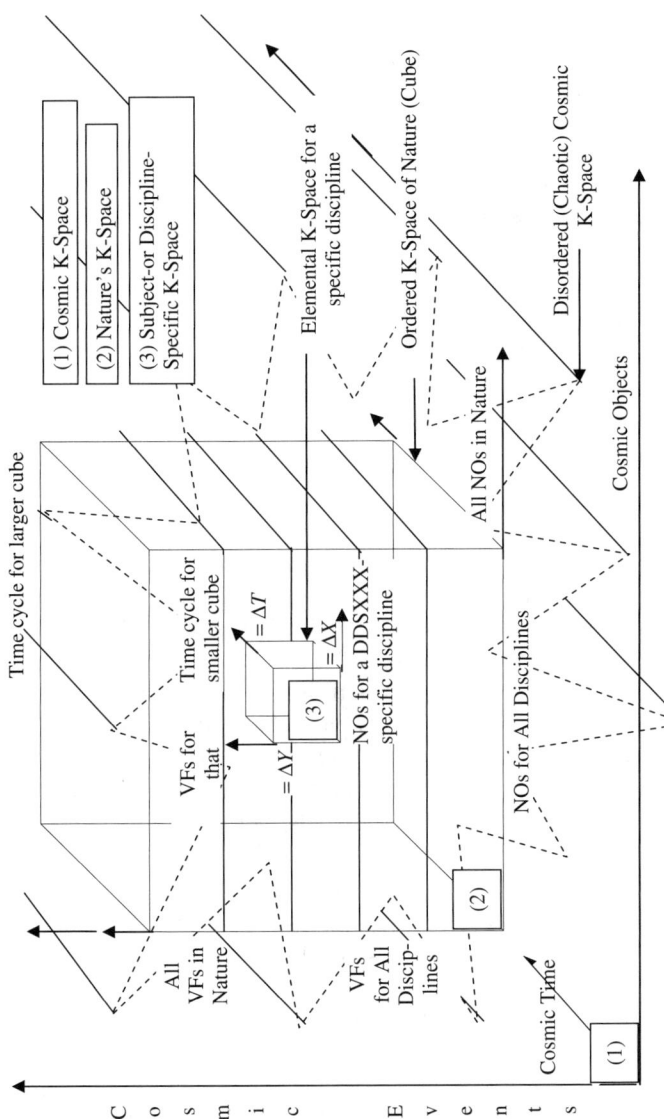

Figure 10.18 The superposition of the three knowledge spaces to derive an ordered knowledge space (3) from an overall cosmic Knowledge space (1) in the *XYT*-dimensions.

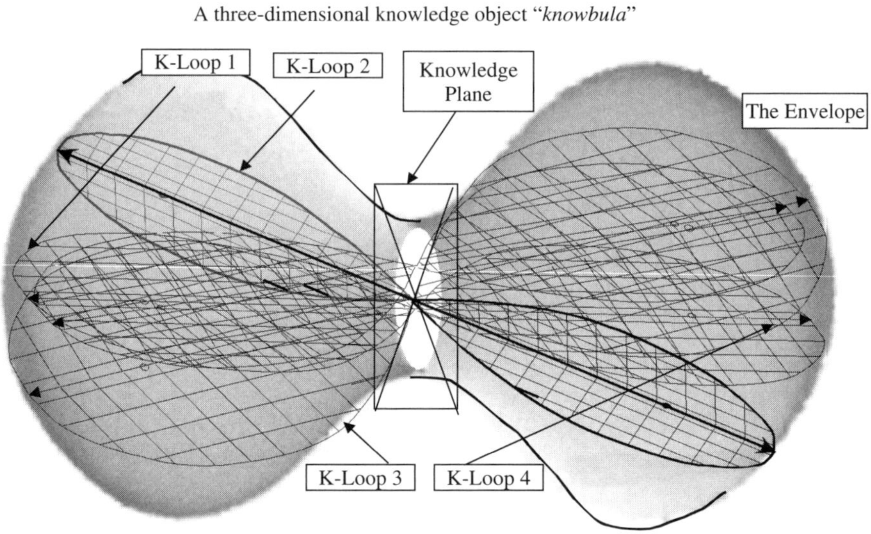

A three-dimensional knowledge object *"knowbula"*

Figure 10.19 This three-dimensional knowledge object "knowbula" is the envelope of all knowledge loops during a given interval of time.

lasting for 24 hours for stock mark daily fluctuations or a decade for trend analysis, the window of time is moderated by accessing the knowledge machine that accesses the appropriate time-sliced knowledge bases on the Web.

In the knowledge space, the convolution of VFs with appropriate NOs leads to their modification, enhancement, or restructuring to suit a particular need of society, a corporation, or any larger entity that engulfs this particular set of NOs. This fuzzy object defined as the *"knowbula,"* (Figure 10.19) is the knowledge in numerous forms that surrounds the K-node along the knowledge trails or K-loops. In essence, it is the envelope of all the knowledge loops during a predefined interval of time in a specific discipline. In a constrained sense, it has been referred as $(I \ll \gg K)$ in [1] and in Chapters 3 and 4. In this chapter, the knowledge around the K-node has at least four characteristics, as indicated by the four ports in the K-node as shown in Figures 10.2 and 10.10. In the present context, knowbula may be treated as a fuzzy mathematical entity that can be processed and its entropy modified by the execution of knowledge operation codes (*kpocs*) in the KPUs of knowledge machines. If it is frozen in time, knowbula is an instantaneous snapshot of the matrix shown in Figure 3.3.

Knowbula's constant change reflects the profile of knowledge of every size group of NOs and VFs, particularly those that are important to the society. Knowbula is essentially a three-dimensional envelope of all the knowledge loops that are and have been active in any environment during a finite interval of time (such as a four-year presidential term, the rise and fall of the Roman empire, or the campus life of a college student, etc.).

The envelope has X-, Y-, and T-dimensions because of the time required to traverse these k-loops when they are active during any finite duration of time. In practical situations, these various knowledge loops can be any function or social program, such as the writing of the U.S. Constitution, Lincoln's Emancipation Proclamation, Gandhi's pursuit of nonviolence, Kofi Annan's Middle East peace efforts, or even a PhD student's dissertation project, etc.

On the one hand, wisdom and worthy knowledge are derived from traversing these K-loops. On the other, instances (such as the Spanish Inquest, British slave trade, Clinton's Impeachment hearing, Bush's war plans in Iraq, the Blair–Bush "poodle" drama during the invasion of Iraq, etc.), these knowledge loops offers pain and suffering through society. Recorded as knowledge loops, they are preserved as indignations of humanity. In a true sense, the evolution of negative knowledge also involves its integration over the forgotten millennia. Major civilizations have left their imprint, sometimes bloody footprints, in the pages of time. The earlier knowledge methodologies to study and document the embedded VFs, NOs, and their interactions, were frail and error-prone. In the modern world, numerous digital techniques are developed. These are application specific and provide the intellectual highways to traverse and integrate many disciplines and events into one coherent pyramid of knowledge or the misdeeds of human beings and cultures.

10.16 HINDSIGHT AND FORESIGHT

A survey of knowledge offers rich hindsight. A scan of current knowledge methodologies also offers substantial and predictive foresight. In this section, we survey the old and scan knowledge domains to build a platform in a horizontal XY-plane and conceptual coherence along the vertical axis.

10.16.1 Coarse Survey of Human Struggle

Every significant struggle within a society has a tell-tale knowbula that freezes the achievement in time. It is thus appropriate to envision a cube containing a knowbula with its own knowledge plane and knowledge space (see Figures 10.13–10.16). This knowbula cube configuration that scans history is shown in Figure 10.20. The knowbula does not entirely fill the knowledge space since only a few NOs and VFs are deployed in traversing the K-loop. For example, when pharaohs were building the pyramids, mathematics and language were both sufficiently developed, yet the architects did not leave behind any significant information[8] or strategies on how they constructed their structures. As an another

---

[8] As far as we know, the pharaohs, Aztecs, and early scientists did not have binary representations (B) and data structures (D) at their disposal; however, they had the skills for communicating, storing, and retrieving information (I). Their K-loops appear to have been shortened as N-IKCWE-N, providing them with the capacity to learn directly from nature (N) and return to it as necessary to retune their skills sets.

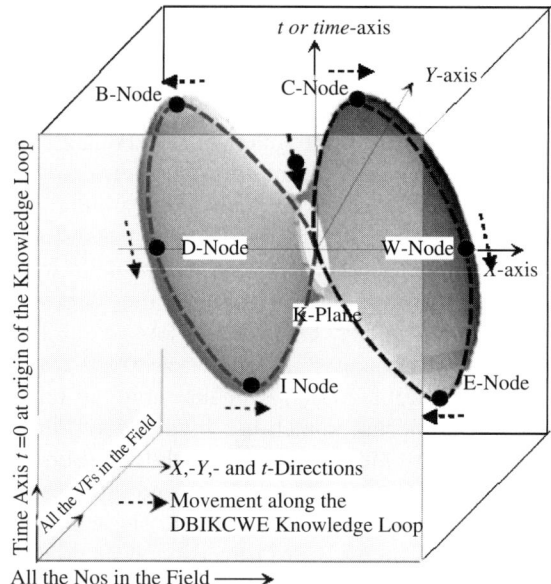

Figure 10.20 The insidious movement of society along the knowledge loop constantly alters the entropy of knowledge and exact location of the BDIKCWE nodes.

example, when the first steam engine was assembled by Thomas Newcomen in 1792, the thermal efficiency or heat loss was not calculated; this did not happen until a much later window of time. Other significant examples exist even today as the effectiveness and cost–benefit ratios of most major social reforms are neither computed nor monitored. Tremendous inefficiencies, corruption, and losses still continue.

In hindsight, if decades and even centuries are aligned on a vertical line and many hundreds of knowbulas are spread on a horizontal plane, then a heap of human activity results. Arranging each knowbula in its own time slot creates this three-dimensional structure. Since there has been no mastermind in arranging the continuity of knowledge over many civilizations and centuries, the newly created three-dimensional structure is a heap haphazardly thrown on a horizontal plane. Such a haphazard collection of knowbula with their time axis along a vertical axis is shown in Figure 10.21.

10.16.2 Scan of Modern Methodologies

Current methodologies of scanning are based on extensive computing systems and AI techniques. During the last few decades, pattern recognition, expert systems, intelligent agents and systems, etc., have facilitated processing and network technologies have facilitated (almost) instantaneous communications. By coordinated deployment, the structure, shape, and entropy of knowledge can be molded, shaped, and manipulated. On the positive side, social reform and progress accumulate, and on the negative side, deception, corruption, and social depletion accrue. Human beings remain firmly in control to drive society up or down. In the current knowledge environment, machines only accelerate the process.

The continuous movement of society (see Figure 10.20) along the knowledge loop constantly alters the entropy of knowledge and the exact location of the BDIKCWE nodes. In most instances, it is not necessary to track knowledge to such a level of fine detail. However, in some instances (such as wars, space missions, emergency room procedures, etc.) the shape and configuration of the K-loop can be monitored and adjusted in very small increments of time. For this reason, the knowbula can only be a three-dimensional object since there is no (noun) object that is immune to time and there is no (verb) function that is absolutely still.

In order to facilitate the representation, the XY-plane (NOs and VFs) is oriented to be perpendicular to the plane of the paper, and the time axis falls along the vertical line. The representation permits the partitioning of K-space such that the passage of time occurs along the Y-axis and human and machine activities take place on a horizontal plane. The positive struggles of humans (such as Newton's formulation of classical mechanics, Einstein's concept of relativity, Poynting's P-vectors, etc.) and the machines that have been invented move up the knowbula. The negative struggles of human (such as the Vietnam War, the search for nonexistent weapons of mass destruction, and flagrant disregard of global warming, etc.) move down the knowbula.

10.16.3 Structured and Predictive Knowledge

The Dewey Decimal System (DDS) and Library of Congress (LoC) have classified the subject material in numerous disciplines around the globe. However, knowledge that is the gateway to concepts still lies as an ill-organized heap of individual knowbulas trapped in cubes of NOs, VFs, and time. The convolution between VFs and NOs (see Chapter 3) leads to inventions that push society forward. The methodology for organizing any heap into a pyramid was been an achievement of the pharaohs and Mayan culture. The base is wide and the peak is high. In the knowledge domain, the base is the subject matter for the subjects included in the DDS and LoC classifications and the peak is the concepts that unify the knowledge distributed in the knowledge bases on the Internet. The reinforcements for the knowledge pyramid are the forward and backward pointers between the NOs, VFs, their convolutions, and underlying concepts that make various disciplines distinctive segments of global wisdom.

When a number of knowbula cubes are organized by the knowledge sciences of the twenty-first century, the disorderly heap of knowledge from the K-loops in the cubes (Figure 10.21) assumes the form of a knowledge pyramid. Such a pyramid is shown in Figure 10.22. The pyramid is also a virtual entity but totally reinforced with forward and backward pointers in all three X-, Y-, and T-dimensions. Any knowledge-processing machine (KM) will have the capability

492 ARCHITECTURE OF KNOWLEDGE

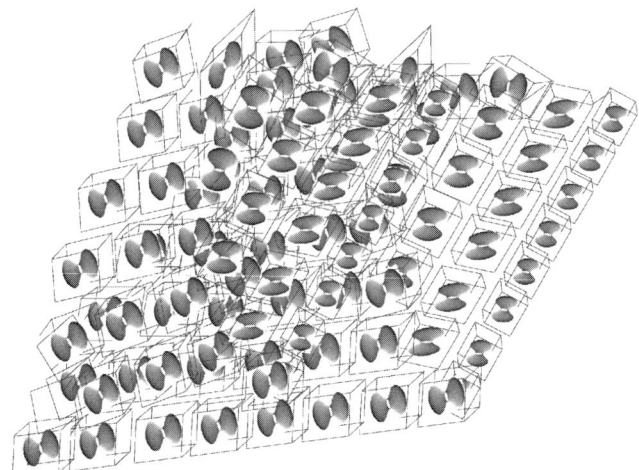

Figure 10.21 The Dewey Decimal System and Library of Congress have classified the subject material in numerous disciplines around the globe. When knowbulas are not organized in K-space, the K-cubes appear as heaps rather than as architectures or pyramids.

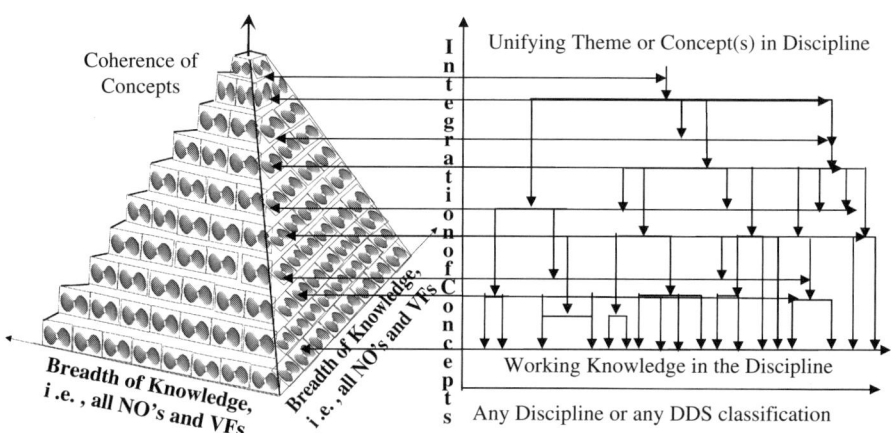

Figure 10.22 Movement from the universal knowledge level to a universal concept level is based on the unique human tendency for most cost structures to have a sense of order and elegance, even though such may be initially overlooked. However, if knowledge is based on scientific principles, then concepts unify such principles into a tight circle that is the focus of a broader wisdom.

to navigate in and out of the K-pyramid, and penetrate it, just as a spacecraft can navigate its way through cosmic space.

Two aspects that are crucial in Figure 10.22 are *(1)* that the vertical axis of the pyramid represents the path between the K-node and C-node of greater knowledge that encompasses the current discipline, and *(2)* that the X- and Y-axes at any horizontal plane through the pyramid contain all the NOs and VFs of the current discipline. In order to hop between one K-pyramid to the next, and from discipline to the adjoining/sister discipline, the wisdom from the W-node plays a crucial role. It affirms the parallelism between many adjoining disciplines and then draws on them to find a common thread of underlying wisdom, if any. Human beings are far more adept at this type of K-pyramid hopping and the role of human beings in wisdom machines (see Chapter 2) becomes essential to monitoring then. In this figure, the focus of wisdom lies just beyond the highest block of coherent concepts.

10.17 KNOWLEDGE, FREEDOM, AND CREATIVITY

Well-founded scientific knowledge and intellectual freedom provide a good basis for creativity. In the context of the K-loops embedded in knowbulas in a structured pyramid of concepts (Figure 10.22), the entry and exit points at the K-node (see Figure 10.2) play an important role in the future direction of traverses along the K-loop. When a knowbula is enclosed in an elemental cube of knowledge, maneuverability along the path of traverse of the loop can influence the creative deviation of the current loop from earlier predecessor paths.

Human ingenuity attempts to alter the pathways between nodes in the K-loop so as to be optimal and most direct. In essence, creativity results and inventions are made. The size and shape of a knowbula that fits the elementary cube of knowledge offer three possibilities: (1) The knowbula is such a tight fit that no movement or adaptation is feasible, (2) the knowbula is a good fit that restrains violent swings in its orientation and structure but offers some room for negotiations with the environment to find a more optimal fit within the cube (innovations and inventions result), and (3) the knowbula becomes a line joining the computational/mechanical intelligence of a machine and the natural intelligence of a human being (see Figure 10.2), that is, the volume of the knowbula collapses to zero. These three cases are depicted in Figure 10.23.

10.17.1 Knowbula Almost Fills the Cube of Knowledge

This condition occurs when all the NOs and their associated VFs in a given window of time are completely deployed in the K-loop. The freedom to readjust the shape and structure of the loop for even minor reasons is not possible. Under such conditions, the knowledge plane at the K-node is the entire XY-plane, as shown in Figure 10.23a. The typical loop that is generally close to a lemniscate (see Section 10.4) assumes the shape of Cassini's oval. In the limiting case, the volume and shape of the lemniscate approach those of the cube of knowledge itself. The cube of knowledge can be artificially compressed by oppression and dictatorships. When access to NOs (such as books and libraries) is denied and VFs (such as research and development) are made illegal, knowbulas assume the shape of hardened clay.

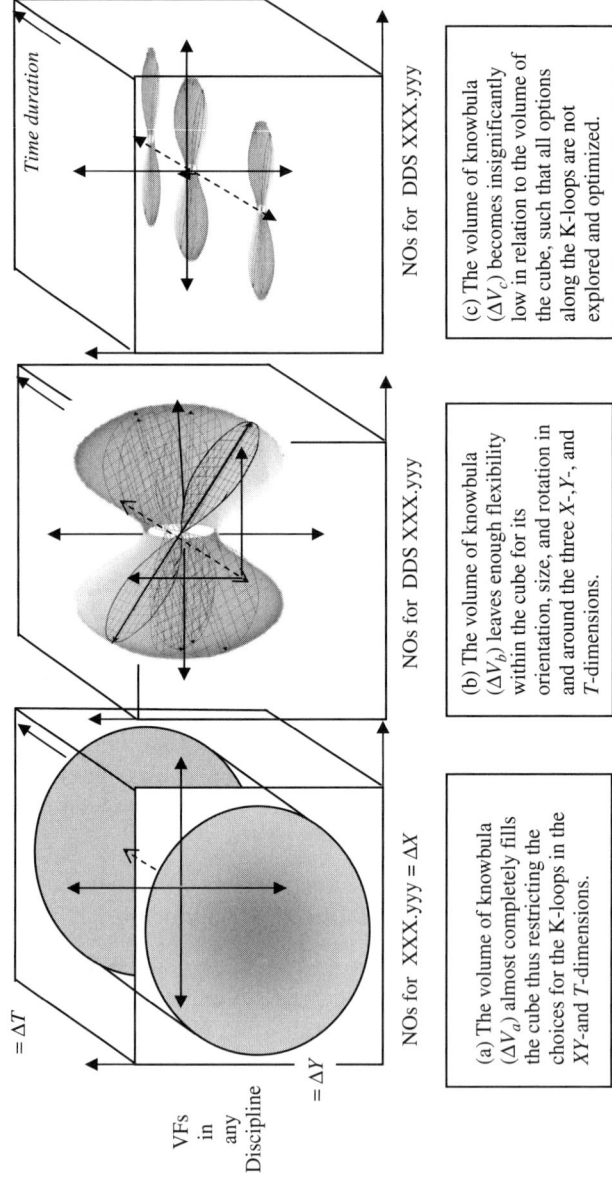

Figure 10.23 Knowledge, freedom, and creativity: the effect of size, shape, orientation of knowbula in exploring and inventing new objects in the knowledge domain. Three instances (a) oppressive society, (b) dynamic society and (c) state society, are presented. The effects of the size and shape of the knowbulas is discussed in the text.

In this window, scientific creativity comes to a stop; societies and nations stagnate. Artistic and literary talent is likely to survive in isolated pockets that are unaffected by science, economics, and technology, which usually bring about a profusion of new NOs and VFs. Closed societies and the dictatorships of the past offer excellent conditions for the knowledge loop to become trapped in its own (selfishly) generated wisdom (W) and values (E) dictated by political leaders. The human spirit that is born free becomes enslaved by brutal forces of occupation, whims of dictators and their single-minded goal of self-glory.

10.17.2 Knowbula Loosely Fits the Cube of Knowledge

The loose fit of knowbula offers the possibility of stretching K-loops (within the knowbula) creatively and making inventions to suit the application, situation, and particular phase of social adaptation. In a sense, this is a self-perpetuating syndrome. The convolution of NOs and VFs during any interval of time, gives rise to new NOs and VFs for the next interval. However, the prevalent freedom for the convolutions of NOs and VFs can yield the three (TVB, DAH, and SET) full-fledged orientations of wisdom. Social progress or social regression follow the attitudes, the deeds of leaders and their hidden agenda.

When the freedom for extensive convolutions is taken to an extreme, the human reaction is purely economic; the maximization of derived utility (or wealth) becomes most appealing. At one extreme, when freedom is derived from integrity, society stands to gain. Data faking, mass deception, and fear mongering at the hands of the media will gradually come to a stop. It appears that once the tides of society have started to move in a downward spiral, social resistance in genuine democracies becomes more deeply ingrained for any reversal to occur. The expenditure of energy required for this social reversal (from DAH to TVB) becomes far greater than that needed for reversing the trend (from TVB to DAH) from an upward social track to a downward path.

At the other extreme, when society seeks freedom but becomes tolerant of dishonesty, individuals lean toward stealing rather than earning, deceiving rather than convincing, especially if the facts are distorted to suit the purpose, war becomes easy choice for the military industrial complex resident in a nation.

These conditions are amplified many times over even today due to the self-centered reasoning of politicians, executives, and state officials, let alone the pressure tactics of thugs and criminals. Science, economics, and technology (SET) become weapons of mass deception in the hands of the fraudulent and conniving. When deception becomes a tactical commodity in perpetuating self-interest, the know-not pay the price without ever knowing it.

Directional friction effect occurs in nature when the fear (smoke) of destruction strikes a bee colony. Smoke alerts the bees to a forest fire and they abandon the hive after gorging themselves on the honey. This event comes after many months of the slow and painstaking work of collecting honey is wasted. In a similar vein, humans in threatening situations tend to slide toward the chaos of DAH much more easily than continuing to protect the hard-earned goals of wisdom, ethics, and values based on TVB.

10.17.3 Knowbula Becomes Insignificant in the Cube of Knowledge

This condition prevailed at the dawn of human civilization. It existed after major calamities in nature when the history of human achievement seemed all but lost. There is no residual scientific knowledge to support human intuition and imagination, and there is little systematic creativity to grow and extend scientific findings. However, human creativity and values survive in pockets and civilization finds a new beginning at a much slower pace. On an individual basis, this condition occurs for a new-born infant while it learns to cope with a new environment or for a new employee in a corporation.

Total conceptual freedom in the absence of all scientific methodology is comparable to the feathers of flight for an arctic tern without the genetic navigational capability. Reinventing the wheel would be an essential exercise for humans. Learning an entire social and ethical framework would be a necessary step. With an essential component of scientific methodology, the path toward a new digital-age society and civilization can become more direct without the dark ages of prolonged ignorance and witchcraft.

Even though an experiment with the entire human race is impractical, a set of machines may be forced to operate in a knowledge vacuum. If enough AI rules (analysis, learning, and adaptation) are primed, such machine will invent new NOs and VFs yet unknown to the human species. In another practical application, if a group of untrained Einsteins were to dapple with the mysteries of cellular biology, perhaps the theory of relativistic biological sciences would evolve. In a more realistic setting, if a raw human intellect was confined to the mind and self, to trace a K-loop without any recourse to local or Web knowledge bases, then the psyche of Hitler becomes just as likely as the spirituality of Gandhi.

When break-through approaches are sought, partial isolation from society becomes a necessity. History provides many examples of those (Buddha, Christ, Moses, Rumi, Schweitzer, and even some recent sufis) who sought retreat in their lives. In the knowledge society (where total isolation is impossible), this condition is achieved by balancing the size and shape of an individual knowbula in relation to the volume of the cube of knowledge (i.e., $\Delta X \times \Delta Y \times \Delta T$). This delicate balance is a self-achieved equilibrium condition that maximizes the achievement of a personal goal in any social setting. It holds the promise of being a stable state of equilibrium through which machines can help human beings and nations to achieve.

10.18 KNOWLEDGE AND MONEY

Knowledge and money are both resources. Invented wisely, both offer rewards. When the velocity of money is zero, economic activity comes to a halt. Severe recessions follow. Conversely, when velocity is too high, economic activity becomes rampant and severe inflation follows. The Federal Reserve Commission monitors economic activity by controlling the availability of funds to the public by controlling prime lending rates. A healthy economy prevails for the long term to make conditions suitable for the constituents of a nation to enjoy a stable life in a stable society.

When the velocity of knowledge tends to zero, an intellectual freeze occurs and massive ignorance about dealing with critical social issues takes over. A state of apathy prevails. Many strains of intellectual starvation follow. Fortunately, human memory retains enough low-level knowledge to maintain subsistence living. Conversely, when the velocity of knowledge become too high, two effects are evident: *First*, the media carries junk information of no intellectual value, and *second*, if such information is mostly educational, then the economy cannot offer employment to match the level of education and understanding of the general population. Many strains of local and national social unrest follow. Between the two extremes (intellectual starvation and junk information highways) of social fear, media opportunism floods information highways with a palatable cocktail of gossip-level information. A vast majority of the public consumes this zero value information as entertainment.

Unfortunately, there is no centralized commission to regulate the velocity of knowledge that can instill social content with intellectual growth and also curtail the flow of tainted information on the knowledge highways throughout the nation. This status quo suits some leaders just fine; it enables them to cover up their own incompetencies. The mass media thus perpetuates the self-interest of many, and the seeds of mass deception are planted in those who monitor the velocity of the flow of knowledge. The result is apathy and lethargy, thus paralyzing the social/intellectual growth of an entire nation.

In this section, we propose a mathematical strategy that permits the moderation of the flow of knowledge, neither starving the mind with too little knowledge, nor corrupting it with a massive infusion of deception and tainted knowledge. The following two strategies are proposed:

1. Monitoring the size of knowbula in relation to the *relevant* elemental cube of knowledge over a finite interval (such as a four year presidential term), reveals the balance between scientific knowledge and a glut of junk knowledge. Irrelevant media information, bias, and gossip are systematically filtered out. Science and ethics are leading indicators of social progress.

2. Matching the rate and relevance of information feedback from intelligent knowledge bases keeps the curiosity and creativity of the user at a level consistent with his or her interest and intellectual profile. This strategy requires the smart design of intelligent knowledge bases so they are predictive and adaptive. Inappropriate responses only annoy the user. Some of the design considerations for intelligent educational systems are presented in [1]. Much as a lending institution performs a credit check on borrowers, knowledge banks should maintain an intellectual profile of every user. Much as an intelligent medical network [1] keeps a medical profile for a patient, knowledge providers should keep an

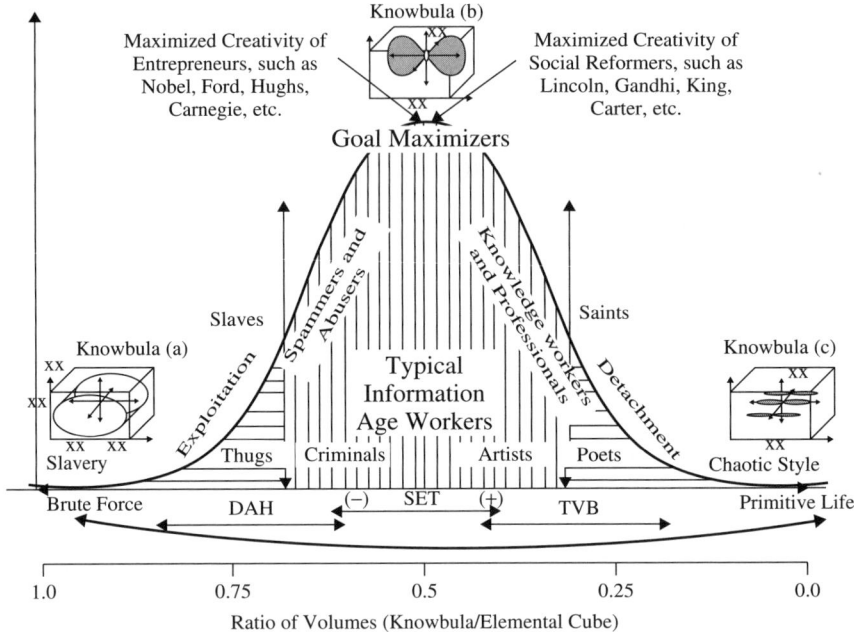

Figure 10.24 The influence of the ratio of volumes of knowbula to elemental cube on the freedom and choice of creativity depending on the TVB, SET, or DAH orientations of wisdom. Ratios are depicted in the (a), (b), and (c) inserts. (Also see Figure 10.23)

intellectual profile on each of its users and dynamically update it from time to time. Much as a corporation maintains an employee profile in a human resources database, a central knowledge agency (CKA) should compile a profile of those who can alter the destiny of a nation or the genes of futuristic knowledge.

The methodology for migration into well-structured knowledge for the future is depicted in Figure 10.24. A utopian society will exist at the peak of a bell-shaped curve. Most constituents will gravitate to the peak rather than drift down to the skirt of the bell-shaped curve. Such upward mobility in the knowledge space will occur in the pursuit of genuine and beneficial knowledge, concepts, and wisdom in an infinite cube of knowledge, where both the semi-infinite X- and Y-axes are primed with all NOs, and VFs from all of nature as we know it and eternity along the T-(time) axis exist. This ideal goal is perhaps a forbidden zone for scientists, philosophers, and poets and brings us back to the confined reality of $\Delta V = \Delta X \times \Delta Y \times \Delta T$.

10.19 PINNACLE OF MIND AND INFINITY OF THOUGHT

In the utopian space of infinite objects, their convolutions, and eternal time, the continuum exists with strings of thought. Such strings of thought are duly represented as the K-loops of human activity. When an infinite number of such

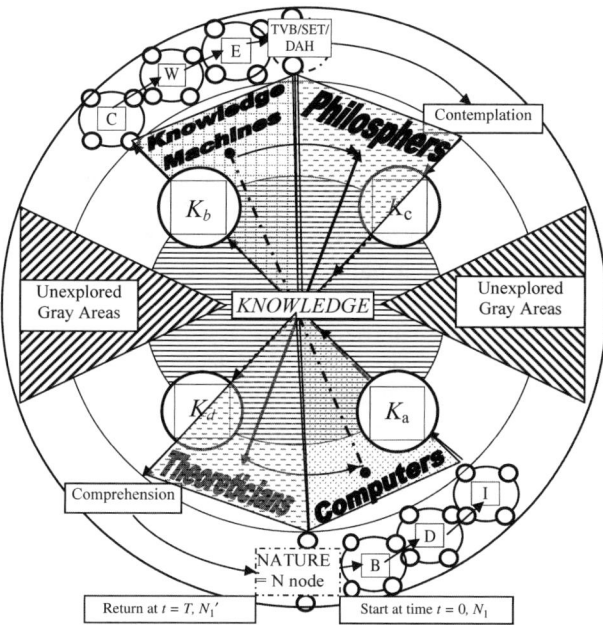

Figure 10.25 The secondary plane of knowledge with a protracted K-loop N-BDIKCWE (TVB)-N1, from nature back to nature. This loop incorporates human intervention in the way TVB knowledge evolves. Similar loops for SET or DAH orientations can also be deduced from this figure. Computers and knowledge machines remain, but philosophers and theoreticians are replaced by humans of a mixed disposition, ranging from saints and the virtuous to thugs and criminals.

strings are huddled in a knowbula, a sense of order starts to appear in this super-space. Being a virtual entity, knowbula needs the human mind to be perceived. Human beings would be lost in this virtual space without computers and knowledge machines to serve as navigation tools. Machines would be impotent without philosophers and theoreticians to jump between one state of nature to the next. A state of symbiotic coexistence becomes essential for survival. If nature created human beings, then human beings created the mind to conceive of nature. We present this stable condition in Figure 10.25. Stability is assured until nature destroys the human species or the human species destroys nature.

The crucial node in the forward path, '$N_1 - BDI (K_a - K_b) CWE - N'_1$, is knowledge—in the forward and return paths of philosophers and mathematicians. Computers and knowledge machines become subservient to mathematicians and philosophers in regaining a sense of reality in this rather prolonged excursion of the K-loop. This is a synergistic and symbiotic relationship that occurs in the background as society reverses its momentum toward one orientation to the other.

The roles of humans (as philosophers and theoreticians) and machines (as computers and knowledge machines) are depicted. The time-cycle of this loop can be centuries long be or quite short. The knowledge revolution over the last few decades has hastened the cyclic processes that drive this K-loop in every discipline. The effects on human beings are rampant.

The top node in Figure 10.25 has three (TVB SET and DAH) orientations of wisdom embedded in it. Being neutral, the SET mode of wisdom can work equally well with the TVB or DAH mode. Different flavors of wisdom become evident in the political posture of a nation or society. When arranged in the plane of the paper, the impact on society can be derived as being expansionary or regressive. Thus, the present political posture becomes a leading indicator of social migration. The role of computers, knowledge machines, philosophers, and theoreticians becomes almost subservient to the political posture of politicians within society. This posture can reconstruct a deteriorating society or deteriorate a constructive society. The mathematical framework for the outer limit of the machines is generally constrained by national priorities. Such priorities come and go from fragmented windows of time over the cycles of human evolution. On many occasions, human effort drowns in a slice of cosmic time but eternal knowledge lives on.

10.20 SOCIAL MIGRATION AND POLITICAL AGENDA

The long-term implications of political pressure on society are evident on every page of history. In the recent past, defense spending has shifted the types of jobs available, tourist revenues have changed peoples' attitudes, lack of medical insurance has prompted better and healthier lifestyles, junk foods have made people obese, liberal gun laws have increased crime rates, etc. These isolated cause–effect relations have a tell-tale story. In essence, wise national policy breeds a healthy society.

The three orientations of wisdom (TVB, SET, and DAH) and their numerous permutations (such as *tvb, set*, and *dah*) were discussed in Chapter 2 (Section 2.6.4, also see Figures 2.12 and 2.14) and bear a long-term effect on cohesion, attitudes, and the preferences of society. On the positive side, the Industrial Revolution brought an end to forced labor, farm mechanization brought relief from famines, the internal combustion engine revolutionized transportation, the Wright Brothers brought about the beginning of the airline industry, etc. On the negative side, Chinese society suffered stagnation during the monarchy, British and Portuguese traders found the freedom to traffic slaves, Spaniards raided Inca temples, Southern states practiced cruel slavery—all these events endured for countless decades.

When the effects of TVB, SET, and DAH orientations are perceived, political posture does not always stand out. However, when machines perform a detailed analysis of cause–effect relationships, rather than a coarse guess on the part of humans, the tell-tale effects will be more definitive and precise.

Policies leaning toward DAH will be investigated in greater detail by knowledge-processing machines. Deception in the media and the falsified

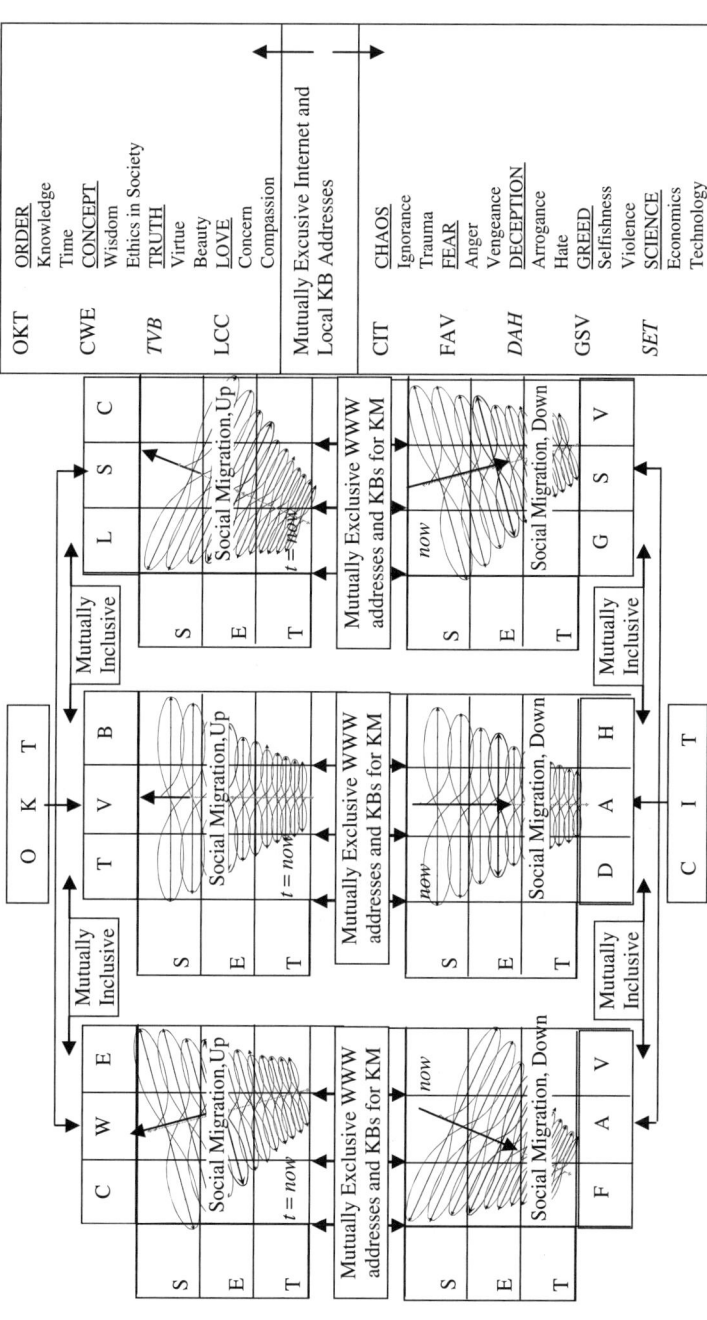

Figure 10.26 The last node in Figure 10.25 has three orientations of wisdom (TVB, SET, and DAH) embedded in it. Being neutral, the SET mode of wisdom can work well with the TVB or DAH mode. The social migration up or down is shown by arrows. In reality, the effects may be incremental and a knowledge machine can accurately track them.

501

statements of leaders will be intercepted early, before serious damage is done. Combinatorial policies such as (high-level) deception (from DAH) with (artificially sweetened) virtue (from TVB) will soon be trapped by machines operating on a scientific basis and through statistical analysis. Uncorrupted machines will gravitate toward truth because science is based on truth and universality.

We depict this entire scenario in Figure 10.26. The SET mode of wisdom, shown as neutral, can work well with the TVB or DAH mode. Different flavors of "wisdom" become evident in the posture of a nation or society. In Figure 10.26, two major variations (OKT for order, knowledge, and time; CIT for chaos, ignorance, and trauma) of wisdom are also shown. The OKT orientation is further divided into three: CWE for concepts, wisdom, and ethics; TVB for truth, virtue, and beauty; and LCC for love, concern, and compassion. The CIT is also subdivided into three: FAV for fear, anger, and vengeance; DAH for deception, aggression, and hate; and GSV for greed, selfishness, and violence. When these eight variations of wisdom are arranged in the plane of the page or paper, the impact on society can be derived as being expansionary (upward movements) or regressive (downward movements).

A mathematical framework for the outer limit of machines becomes enveloped in the political posture, which can itself be shaped by machines, provided politicians have the capacity and insight to deploy such machines for social reform rather than personal gain.

10.21 CONCLUSIONS

This chapter culminates the general framework for knowledge, the processing of knowledge, and a systematic integration of the past and future domains of knowledge. It provides both the conceptual highways and byways between old and new knowledge domains by building systematic forward and backward pointers to NOs and VFs that populate the past and current knowledge spaces. The chapter also summarizes the growth and evolution of knowledge. Human beings, being logical and infused with economic behavior in modifying the environment, follow the movement toward deriving knowledge from information, concepts from knowledge, and finally wisdom from concepts; they will mold the environment wisely and hopefully in an ethical sense. Knowledge is visited many times over, with only a little more wisdom gained each time.

In this chapter, a mathematical framework for traversing the knowledge loop is delineated and made re-entrant. Knowledge is thus constantly updated and replenished. The factors that make the general population more creative in any given situation are embedded in constructing knowbula as a virtual object that envelopes human and individual effort in learning from the past and realigning for the future.

Neither human beings nor machines by themselves can reverse well-established trends in any knowledge society especially if such trends are based on deception, arrogance, and hate. The symbiosis and synergy of scientific facts based on truth, virtue, and beauty lead to rapid change and the quick reversal of any faulty trends that may be entrenched or evolve. Sudden and disruptive changes are relatively few because humans retain the basic powers of (mental) processing, (biological) memory, and genetic intelligence. In the same vein, machines retain quick data- and information-processing capability, massive knowledge banks and memories, and generic algorithms of artificial intelligence. Even superposition and a division of tasks yield to the genesis of wisdom machines.

An elemental module of knowledge "knowbula" is introduced in this chapter. It acts a basic building block in the knowledge domain. Knowbula is evolved from the NOs embedded within it and reinforced by the VFs that provide a structure and coherency within the element. Major linkages to the human, corporate, social, etc., needs that this element serves and the orientation of the embedded NOs and VFs provide the hierarchical address of the element in relation to other knowbulae (a cloud or multiplicity of knowbula) that dominate the knowledge space. The newly formed knowbula serves as a body of new knowledge created from old knowledge and bears the proper credentials to be scientifically deduced from older and well-established bodies of scientific knowledge.

This chapter introduces the concept of predictive knowledge that is capable of being artificially creative rather than passively intelligent, thus paving the path for the design of creative machines rather than intelligent systems. When one knowledge machine is primed with knowledge ware to function in a scientific deduction fashion and is coupled to another knowledge machine (knowledge ware) that is also primed with to verify and validate, then the output from that combination provides refined and validated axioms of wisdom to be scrutinized by human beings. Intelligent Internet knowledge networks provide the inputs to both machines, thus imparting a scientific basis for the derivation of axioms of wisdom from Web bases by knowledge-processing systems.

Intelligent Internet knowledge banks match the novelty and rate of delivery of knowledge to users. Such knowledge banks provide entire frameworks and creative methodologies for knowledge. Such platforms offer solution beyond the immediate needs of the users thus projecting artificial creativity of the knowledge machines and knowledge processing system onto the users. The quality of knowledge (QoK) is thus personalized to each user or user group.

REFERENCE

1. E. Hubble, *The Realm of Nebulae (Sillman Memorial Lectures)*, Yale University Press, New Haven, CT, 1982.
2. S. V. Ahamed, *Intelligent Internet Knowledge Networks*, Wiley Interscience, Hoboken, NJ, 2006.
3. C. Peylo, et. al., Adaptive and Intelligent Web-Based Education Systems, *International Journal of Artificial Intelligence in Education (2000)*, IAIED Society, Edenburgh, v.11, 2000, 1050–1059.

4. C. Pearce, et. al., Doctor, patient and computer—A framework for the new consultation, *International Journal of Medical Informatics*, Edenburgh, v.78, Jan. 2009, 32–38. also see N. Waraporn and S. V. Ahamed, Intelligent medical search engine by knowledge machine, *Proc. 3rd, Int. Conf. on Inf. Tech.: New Generations*, IEEE Computer Society, Los Alamitos, CA. 2006 and N. Waraporon, *Intelligent Medical Databases for Global Access*, Ph.D. Dissertation at the Graduate Center of the City University of New York, 2006.

ACRONYMS

$(I « » K)$	Information transiting into knowledge
$(I « » K)$ Design	Design of *computer based* $(I « » K)$ systems
$(I « » K)$ Filters	Filters that permit the *filtering* of Information to Knowledge
$(I « » K)$ Systems	Systems that facilitate the *flow and transition* of Information to Knowledge
(DAH, CIT, FAV, GSV)	Three triplet features of (Deception, Arrogance, Hate); Chaos, Ignorance, Trauma; Fear, Anger, Vengeance; Greed, Selfishness, Violence. (see Figure 10.28)
(TVB, OKT, CWE, LCC)	Three triplet features of (Truth, Virtue, Beauty); Order, Knowledge, Time; Concept, and Wisdom, Ethics; and Love, Concern, Compassion. (see Figure 10.28)
(SET)	Science, Economics, Technology; An indifferent platform that can interface with TVB or with DAH and carried into either of the two conflicting orientations of wisdom. (see Figure 10.28)

1

1A processor	Special communication processor for call-processing applications in electronic switching systems.

2

2B1Q code	two binary to one quartinary code
2-D	trellis coding, two dimensional $(x + iy)$ code algorithm for data communication
2G, 2G+	coding standards by ITU cordless telephone (CT) usage.

3

3G,	Third generation wireless technologies for CT applications
3XX, 4XX, 6XX,	Numbering System for special services in the CSDC

4

411, 911	
48 octets	payload in the ATM cell
4ESS	
4ESS, 4ESS™	electronic switch toll center Part of the ESS developed by Bell
4ESS™	switch toll center

5

53 octets,	ATM cell size
5ESS,	A generic electronic switch for handling a wide variety of telephone switching services
7XX, 8XX, 9XX,	Special services numbering systems

8

800, 900, 700,	numbers reserved for special services
800-network	first generation Intelligent network IN/1 toll free calling
802.x,	ITU standards for packet switched networks
899 or 898	numbers and numbers for specialized services to the customers

9

900 services,	Reversed charges special services network services
9-row/13-row	SONET frame configuration debates and deliberations

A

ABS -	alternate billing service
ABS,	alternate billing service
ADMs,	Add-Drop Multiplexers for the new and older generation networks

ACRONYMS 507

ADSL,	Numerous versions and configurations of the asymmetric\
ADSL/ADSL ADSL+	Asymmetric digital subscriber line (ADSL)
A-I	Nine variables in the Knowledge Convolution Matrix between VFs and NOs
AI,	Artificial Intelligence
AIKN	advanced intelligent knowledge network
AIN,	advanced intelligent networks proposed by Bellcore.
AKSP,	Adjunct knowledge processor for knowledge functions in knowledge environment
ALKP	Application Level (Knowledge) Programs
ALUs,	Arithmetic Logic Units in the CPU environment of a computer
AM	Administrative Module
AM, CM, and SM	Administrative communications and switching modules in a ESS facility
ANSI,	American National Standards Institute
AP,	adjunct processor
APs	Applications programs
AT&T,	American Telephone and Telegraph Company, no longer in existence
Atlantis-2,	One of the many Atlantic crossings for voice and data communication
ATM,	Asynchronous transfer mode for cell relay systems
ATTr	attribute or attributes of NO's
AUs,	Arithmetic units embedded in the ALU and CPU environments.
AVPU,	Audiovisual processing unit
AW	Artificial Wisdom
A-WAB,	Wisdom axiom base for the type A human being (philosopher)
AXE,	A generic European switching environment for data and voice communications by Ericsson.

B

B	Bearer channel of the ISDN
BAL,	Basic assembly language
BALK Instructions	Basic assembly language knowledge Instructions
BB	broadband networks
BDI-K-CWE,	A chain or a trail the permits linkage of binary, data, information, to the knowledge, and then to the concept, wisdom and ethics of human beings

508 ACRONYMS

Bellcore,	Bell Communications Research, no longer in existence
B-H loop	Graphical representation of the cause-effect relations with memory effects
BI,	behavioral intelligence evolved to solve routine human problems at a behavioral bevel.
BOK-R	Body of knowledge with rhyme R-Zero Filtering
BRISDN,	basic rate integrated services digital network (ISDN)
B-values,	Maslow's concepts of the values of human being (see page 383)
B-WAB	Wisdom axiom base for the type B human being (scientist)
BX.25,	version of X.25 protocol/network developed by Bell system

C

C language	A computer language evolved with the Unix operating system.
C, C+ and C++	computer languages for generic programming applications
CAD	computer aided design
CAMEL	Intelligent network services for mobile users to select specific service.
CANTAT-3,	one of the numerous Trans Atlantic Cables for data/voice services
CATV,	cable TV
CCIS,	Common Channel Interoffice Signaling
CCISS,	Common Channel Interoffice Signaling System
CCITT,	consultative committee of the ITT
CCS	Common Channel Signaling
CCS7,	Common Channel Signaling #7
CCSN	Common Channel Signaling network
CDRs,	call data records
CEOs criminal	CEOs of defamed organizations the operate under the DAH type of perverted wisdom.
CEPT	central European post and telegraph
Change of Operators ($+ \times / $ and $=$)	
CIT	Chaos, Ignorance, Trauma
CLASS	special group of IN/2 services provided by second generation INs
clergy	
CM,	communication module or concept machine
CN	Campus networks
COA	Cost of Agreement
COD	Cost of Disagreement

COMNET	An event based computer simulation program to evaluate network performance.
CPM,	critical path method
CSDC	circuit-switched digital capability
CT2,	cordless telephone 2nd generation
CT3,	enhanced CT-2 telephone services towards the 3rd generation
C-WAB	wisdom axiom base for the DAH human being
CWE	concept, wisdom, ethics in Society

D

D channel banks	CO equipment for multiplexing and demultiplexing the DS0 and DS1 signals through the Circuit switched networks.
D channel	A signaling and low speed digital channel proposed for use with ISDN type of networks and services.
D I K W and E	Data, Information, knowledge, wisdom and ethics along the knowledge trail.
D to D	Data node to data node transfer of data structures, and modules of information.
D, I, K, W, and E,	Data, information, knowledge, wisdom and ethics levels of progression.
DACS	digital access cross connect systems
DAH Filter	A knowledge based processing environment that prevent flow of deceptive, (falsified), arrogant, (aggressive), hateful (degenerate) information, knowledge and data through it and does not propagate such DAH attitudes of humans, corporations and nations.
DAH	Deception, Arrogance, Hate
DAH	Deception, arrogance and hate that takes over perverted humans, leaders of corporations and nations.
DAH,	deception, arrogance and hate for the (C1-C5) and (D1-D5) type of human beings as identified by type II personalities identified as Aristotle and as "B-values" as identified by Maslow.
DBM,	database management
DBMSs,	database management systems
DCS,	digital cross connect systems in the SONET environment
D–D traits	Characteristic correspondence between DAH personalities in individuals,. Corporate officials and leaders of nations.

DDS,	Dewey decimal system (in context to classification of knowledge)
DDS,	digital data services (in context to digital transmission systems)
DI	Demand Intensity
DIL & DIM	Demand Intensities of Labor and Management
DL	distance learning
DM,	Data management/data manager
DMA	direct memory access (in traditional computer systems)
DMT,	discrete multi tone (for DSL transmission)
DPUs/GPUs	Display/Graphics Processor Units
DQDB,	distributed queue dual bus
DS	D-system digital hierarchy in Americas
DS0,	digital signal 0 is 64kb/s channel at the lowest level of digital hierarchy
DS1,	digital signal 1 is 1.544 Mb/s signal with 24 DS0 channels + signaling bits, also known as T1 signal
DS1C,	digital signal 1C is at 3.152 Mb/s
DS2,	digital signal 2 is at 6.312 Mb/s also known as T2 signal
DS3,	digital signal 3 is at 44.736 Mb/s
DSL,	digital subscriber line
DSLAM,	uplink multiplexer typically converts the DSL data and signal onto the Layer 2 ATM format
DSSs	decision support systems

E

E system	European digital hierarchy for data transmission
E1,	CEPT digital hierarchy signals at 2.048 Mb/s
E2, E3,	CEPT digital hierarchy signals
EC	echo cancellation
EG	electronic government
EG-IT	IT platform and numbering system for EG operations
EG-IT	IT platform for EGs of Nations
ES	Expert Systems
ESS,	electronic switching systems
EW1000,	Education and welfare wing of electronic governement systems
EWSD	a series of ESS facilities built by Siemens

F

FAV	Fear, Anger, Vengeance
FCs,	functional components of the intelligent network conceptual model

FDC,	falsehood, distaste and corruption of type DAH individuals
FEPs,	front end processors of the SCP facility
FLAG,	one of the numerous transoceanic fiber optic networks
flip-flop	attitude of human beings in doubt or under stress
FO	Fiber Optic
FTTC,	fiber to the curb distribution data/voice/video networks
FTTH,	fiber to the home distribution data/voice/video networks
fuzzy NPU	A numerical processor unit working with imprecise rule for manipulating the attributes of objects.

G

G.992.2,	ADSL standards for ADSL Lite discrete multi-tone
G.DMT,	ADSL standards for ADSL discrete multi-tone
Gb/s	gigabits per second optical rates in FO networks
GNP,	Gross National Product of any nation
GPRS,	Global packet radio services
GPU,	graphical processor unit
GSV	Greed, Selfishness, Violence
GUI,	graphical user interface

H

HD	Hamming distance
HDSL,	high-speed digital subscriber line
HDTV,	High Definition TV
HLLs,	higher level languages

I

I K C W and E	Nodes in the Knowledge Trail
$(I \ll \gg K)$,	information under process for refinement to the knowledge level
I^2-KPS,	intelligent and Internet-based; Intelligent and IN-based; or Internet-based and IN-based knowledge processing system
IAS,	Institute of Advanced Study known for von Neumann's SPC machine
IC,	integrated circuit
IKN,	intelligent knowledge network
IKPS, I-KPS	intelligent, Internet-based, or IN-based knowledge processing system
IKW	intelligent-knowledgeware
IKWV chain,	information, knowledge, wisdom and (human) values chain
IMF	International Monetary Fund
IN,	intelligent network

IN/1,	first generation intelligent network
IN/1+,	enhanced version of the IN/1
IN/2,	second generation IN
INCAD	intelligent computer aided design
INCM,	intelligent network conceptual model
INKW	intelligent knowledgeware
INOs	intelligent Noun Objects
INTERN	An AI program for medical diagnostics
IOPs	Input/output Processors
IP,	intelligent peripheral or Internet protocol
ISDN,	integrated services digital network
ISP,	Internet service provider
IT	information technology
ITU	International Telecommunications Union

J

JCL	job control language

K

KAND	Knowledge based AND gate
kb/s	kilobits per second
KBMS,	knowledge base management system
KBs,	knowledge bases
KCOs	Knowledge Centric Objects
KCP,	knowledge control point
KCPU,	hardware unit to accomplish knowledge functions and CPU functions
KCS	knowledge communication system
KCSs,	knowledge communication systems
KD	knowledge domain
KEE,	knowledge enhancement environments
KI	knowledge or content based intelligence
KIP,	knowledge intelligent peripheral
KIR,	knowledge instruction register
KM,	knowledge machine
KM for LM	KM for Labor Management (LM) negotiation
KMM,	knowledge management module
KMs and WMs	Knowledge machines and Wisdom machines
KN	knowledge network
KNID	knowledge network information database
KNOT	Knowledge based NOT gate
Knowbula	A three-dimensional knowledge space to contain the knowledge loops in a particular discipline for a given span of time for a particular culture or society.

Kompilation,	compilation of knowledge programs
kopc,	knowledge operation code
kopcode	Operation code within BAL knowledge instructions
KOR	Knowledge Based OR Gate
KP	Knowledge processing
KPE,	knowledge processing environment
K-Plane	Knowledge plane in XY Dimensions
KPS	A knowledge processing system
KPU,	knowledge processing unit
KS,	knowledge systems
KSCP,	knowledge services control point
KSMS,	knowledge system management system
KSP(i)	i-th switching point in a series of knowledge/content base switches
KSP,	knowledge-switching point used for switching in knowledge-based intelligent (KI) systems based on LoC classification of content knowledge. The KBs are distributed throughout the world. The content of knowledge carried in the channel determines the Web address of the KB to be accessed. This address is similar to the WWW address but the content determines the appropriate LoC classified KB address that gets translated to the exact WWW address
KSW,	knowledge software or knowledge switch
KTP,	knowledge transfer point

L

LA	Lexical analyzers
LCC	Love, Concern, Compassion
Lite,ADSL2+,	One of the numerous varieties of ADSL
LM	Labor-Management
LoC,	Library of Congress
LSI	large-scale systems
LUs,	logical unit

M

M13, M34	Group of digital multiplexers used in the American digital hierarchy
M1C, M12,	
MAN	Metropolitan Area Networks
Mb/s	megabits per second
MIMD,	multiple instruction multiple data CPU architecture
MIN	medical intelligent network
MISD,	multiple instruction single data CPU architecture

MKCP	medical knowledge control point
MKNs,	medical knowledge networks
MKTP	medical knowledge transfer point
MP	Medical Processors
MP-MCKO-MDO-MA	one of a Class-kopc Instruction
MP-MCKO-MDO-NA	one of a Class-kopc Instruction
MP-MCKO-MDO-SA	one of a Class-kopc Instruction
MPMO	Multiple process multiple object architectures of the KPU
MPSO,	multiple process single object KPU architecture
MRI,	magnetic resonance imaging
MSMS,	medical services management system
MSPs,	medical service providers
MSs	knowledge management systems
MWMM	Multiple Wisdom Multiple Machine Systems
MWSM	Multiple Wisdom Single Machine Systems
MYCIN	An artificially intelligent SW system for medical diagnostics

N

N	Nature Node as the Origin/Termination of Knowledge Trail
NA	node administration
NBES,	Network-based educational system
NCOM,	network communication unit
NKSPs,	network knowledge service providers
NOs and VFs	Noun Objects and Verb Functions
NPU,	numerical processor unit in the CPU or KPU environment
NT-1 and NT-2,	network terminations 1 and 2 for ISDN serices

O

OA&M	
OA&M,	operations, administration and maintenance
OAM&P,	operations, administration, maintenance and provisioning
OC	Optical Carrier designation
OC-n,	optical carrier systems in the SONET environment
OHI	Open Human Interconnect Model
OKT	order, knowledge, time
OKT	Order knowledge and time

OOL	Object Oriented Languages
OOP,	object oriented programs
OPNET	a discrete event networks simulation package
OSI model	open system interconnect model

P

PBXs,	private branch exchanges
PCM,	pulse code modulation
PCP	processor control point
PCP	processor control point
PCU,	program control unit
PIN,	personal intelligent network
PLCP,	physical layer convergence protocol
PM	Prime Minister's Office
PMPC code,	machine instruction code to drive the different stages in the PMPC
PMPC,	patient medical process cycle
PMR,	private mobile radio
POTS,	Plain old telephone service
PPNs and EPNs,	processor port network and expansion port network
PR,	pattern recognition
PROLOG,	one of the numerous OOLs
PS	packet-switched

Q

QAM	quadrature amplitude modulation coding,
QoK,	quality of knowledge
QoS,	quality of service

R

$R(D_x)$, and $R(D_y)$, $R(F_x)$, $R(F_y)$,	resistance factors in social dynamics of intelligent objects
RACE,	research and development projects for advanced communication in Europe
R-DAH	Recessive Downward Movement in DAH environment
RDTs,	remote data terminals
RGB	red green and blue to symbolize truth, virtue and beauty
ROM	read only memory

S

SAFE,	spanning the globe, a global network
SANs,	storage area networks
SAP,	one of a series of intelligent MIS and EIS programming systems
SAT-3/WASC	one of the newer fiber optic transoceanic network
SCE,	service creation environment
SCP,	service control point
SDSL,	symmetric DSL
SDTV,	standard TV
SEA-ME-WE 2,	segments of the newer fiber optic transoceanic network
SEA-ME-WE 3,	spanning the globe
SET Filter	SET filter to verify the validity John Tyndall's statements
SET	Science, Economics, Technology
SET	Science, Economics and Technology Platform of wisdom
SIMD,	single instruction single data CPU architecture
SIMO,	single instruction multiple data CPU architecture
SIP,	service intelligent peripheral
SM,	switching module of an ESS facility
SMDS,	switched multi-megabit digital service
SMS,	service management system (in conventional INs)
SNI,	Service network interface
SNMP,	simple network management protocol
SNR,	signal to noise ratio
SONET,	synchronous optical network
SONET/SDH,	SONET/SONET digital hierarchy
SOPs,	standard operating procedures
SPC,	service control point (in conventional INs)
SPSO,	single process single object KPU architecture
SS7	signaling system 7
SSI,	small scale integration
SSKO	Single Simple Knowledge Object
SSKO-SA	Single Simple Objects with Single Attribute
SSLI,	special services logical interface
SSPs,	service switching points (in conventional INs)
SST,	spread spectrum technique
STM-i,	STM-i designation, with $i(=n/3)$ for the STS-n designation
STPs,	signal transfer points (in conventional INs)
STS-n	synchronous transport signal at level n
SWMM	Single Wisdom Multiple Machine Systems
SWSM	Single Wisdom Single Machine Systems
S-Zero Filtering	A knowledge filter to validate the scientific validity of information passing through the filter

T

T P and S	truth, philosophic and scientific validations
T P S E	truth, philosophic scientific and ethical validations
T V and B	truth, virtue and beauty
T, P, S, E,	true, philosophic scientific and economic validations
T1,	Signal at DS1 rate of 1.544 Mb/s
T1/E1 lines,	T1 or E1 lines
TAT-8 and 9	transatlantic telecommunication fiber optic systems
TCP/IP,	transmission control protocol, connection oriented transport protocol of the Internet architecture; reliable byte stream delivery service. IP Internet protocol that provides a connectionless best effort delivery service across the Internet
TINA,	TMN + IN, (telecommunication management network + intelligent network)
TMM,	transaction management machine
TPC-5,	Trans Pacific cables
TR Filter	Truth and Rhyme filter for Henry Higgins (Henry Higgins)
TR-Zero	Zero impedance filter for Truth and Rhyme filter (Henry Higgins)
TVB Filter	A knowledge processing system that validates the truth, virtue and beauty in the information and knowledge flowing through the filter.
TVB	Truth, Virtue, Beauty
TVB,	truth, virtue and beauty borrowed from Aristotle's writings
TVB <> DAH	Conflictive Roles of two universal orientations of positive and negative wisdom
TWS	Tsunami warning system
T-Zero	A filter that offers zero impedance to the passage of truthful information

U

U, T and S	interfaces for ISDN
UMTS,	universal mobile telephone system
UN	United Nations
UNIX,	a powerful operating systems developed by Bell System to operate ESS type of telecommunication nodes and network functions. It is generic enough to operate most computing environments.
USIN,	universal services intelligent network proposed by AT&T during late 1980s, but never finally implemented.

V

VCs,	virtual circuits through the network environments
VDSL,	Variable rate DSL, one of the numerous versions of the digital subscriber lines
VHDSL,	very high-speed DSL
VLSI,	very large scale integration for IC chips
VoIP,	voice over Internet deploying established IPs
VT	Virtual terminal

W

WAN	wide area network
WC	wire center of a typical switching node where the subscriber line terminate at the central office
WDM	wave-length division multiplex for FO transmission
WHO	World Health Organization
WMs	Wisdom Machines
WWW	world wide web

X

X.25	ITU recommendation that specifies the interface between users data equipment (DTE) and the packet-switching data circuit terminating equipment (DCE)
Z_0	characteristic impedance of transmission media

INDEX

(I ≪≫ K), 253, 258, 266, 272, 275, 295, 505
(I ≪≫ K) Systems, 266, 505
[A-I] matrix, 507

1940s, 2, 9, 423
1950s, 2, 5, 9, 12–13, 17, 82, 346, 368, 373–374, 421
1960s, 5, 13–14, 17, 163, 310, 346, 362, 369, 373, 375, 421, 423
1970s, 4–5, 14, 17, 82, 93, 206n9, 310, 345, 370, 375, 382, 423, 440
1980s, xxvi, 4, 13–15, 23, 59, 203, 276, 309, 316, 345, 370, 372, 380–381, 397, 517
1990s, xxvi, 5, 13–15, 17, 59, 345–346, 380, 382, 390
1A processor, 505

2B1Q code, 506
2-D trellis coding, 506
2G, 506
2G+, 506

3G, 506

4ESS, 506
4ESS™ switch toll center, 506

53 octets, 506
5ESS, 506

800, 162n5, 506
800-network, 506
802.x, 162n5, 506

900 services, 162n5, 506

ABS, *See* alternate billing service
absolute wisdom, 68, 74n5, 88, 90
abstractions, 113, 359
abuse, 90, 404n4, 413
abuse of knowledge, xxvi–xxvii, xxvin1, 476
accumulation of information knowledge and wisdom, 111–112
actions and entropy, 321
active filters, 251, 260
active kopcs, 341, 343–346
active kopcs for intelligent KCOs, 345–346
active kopcs for passive KCOs, 344–345
active kopcs for reactive KCOs, 345
Add-drop multiplexer (ADMs), 420–421
adjunct knowledge service processor (AKSP), 59–60, 507
adjunct processor, 507
Adjustment and relationships, 360, 363–364
ADMs, *See* Add-drop multiplexer
ADSL Lite, 511
ADSL, *See* Asymmetric digital subscriber line
advanced intelligent network (AIN), 58, 507
Advanced Research Projects Agency (ARPA), 14

Computational Framework for Knowledge. By Syed V. Ahamed
Copyright © 2009 John Wiley & Sons, Inc.

AI concepts, 54
AI programming, 82, 84, 123, 127–128, 133–134, 184, 417, 429, 512
AI Response, 199
AI tools, 71, 76, 92, 94, 187, 429
A-I, 282, 286, 507
AI, *See* Artificial intelligence
AIDS SARS, 98
Ailment, 52, 111, 114, 158, 168, 271, 395
AIN, *See* advanced intelligent network
AKSP, *See* adjunct knowledge service processor
algorithmic adaptation, 192
Algorithmic representation, 193–197
ALKP, *See* Application Level (Knowledge) Programs
alternate billing service (ABS), 506
ALUs, *See* Arithmetic logic units
AM administrative module, 372, 379, 399–400, 507
AM CM and SM, 507
American National Standards Institute (ANSI), 507
American Telephone and Telegraph Company (AT&T), 149n1, 507, 517
Ampere, 29, 97
Anderson, Arthur, 73, 411, 445
ANSI, *See* American National Standards Institute
Anti-Terrorism Networks, 55–58
antivirus and antispy programs, 323, 343
Application Constraints, 266
Application for Filtering and Comparing, 270–271
Application Level (Knowledge) Programs (ALKP), 165–168, 172, 174, 507
application to three negative noun objects, 319
Architecture of Knowledge, 460–503, *See also* Tracking of the K-loops
  conflictive roles of TVB <> DAH orientations, 46
  construction of a knowledge plane, 477–479
  continuity of knowledge, 465–466
  downward migration, 473
  foresight, 489–493
  four ports of super-node K, 476–477
  hindsight, 489–493
  knowledge and money, 496–498
  knowledge bases in primary K-plane, 479
  knowledge loop and its stability, 462–465
  knowledge loops in the X-and Y-dimensions, 473–476
  knowledge, freedom, and creativity, 489–496
  long-term movements within the k-loop, 468–469
  nebula of knowledge, 486–489
  new breed of knowledge, 461–462
  pinnacle of mind and infinity of thought, 498–500
  political agenda, 500–502
  social migration, 500–502
  successive K-loops in the time dimension, 479–486
  three orientations of wisdom, 466–467
  upward migration, 472–473
architecture of the KM, 265
Architecture, 348, 355–413, 460–503
archives, 402–403, 417–418
Aristotle, 30, 64–65, 96, 122, 130, 161, 276, 357–359, 358n1, 359n2, 407, 450, 454, 463n3, 483, 509, 517
Arithmetic logic units (ALUs), 322, 388, 507
ARPA, *See* Advanced Research Projects Agency
ARPANET, 14
Arrogance, 67, 94, 120–121, 298n10, 301, 319, 448, 455, 463n3, 467, 505, 509
Artificial intelligence (AI), 6, 35, 55, 61, 71, 148, 219, 248, 250, 361, 399, 423
Artificial Wisdom (AW), 407–409, 507
Aspects of knowledge, 33–34, 486
Assistive neutral and resistive objects, 152–153
Asymmetric digital subscriber line (ADSL), 507
asynchronous transfer mode (ATM), xxvi, 16, 51–52, 378, 386, 411, 420, 507
AT&T, *See* American Telephone and Telegraph Company
Atlantis-, 507
ATM, *See* asynchronous transfer mode
attribute, 314, 326, 328–330, 332, 334, 507
Audiovisual processing unit (AVPU), 507
Aus, 507
AVPU, *See* Audiovisual processing unit
AW, *See* Artificial Wisdom
A-WAB, 507
Awareness and Knowledge, 359–361
AXE, 378, 507
Axioms, 66n3
axioms of wisdom, 73–75, 77–78, 86, 91, 96–97
Aztec leaders, 470, 483, 489n8

Background of Economics Based Interactions, 230–231
Background of LM Negotiations, 222–223
BAL, *See* Basic assembly language
balanced transactions, 155, 233
Bargaining and Negotiation, 363
base cells, 35, 361
Basic assembly language (BAL), 174, 507, 513
basic building blocks of intelligent networks, 16–17

basic rate integrated services digital network (BRISDN), 508
Basis for Theory Of Knowledge, 102
BDI-K-CWE, 507
beauty/elegance/grace, 200–201
behavioral intelligence (BI), 362, 364, 508
behavioral models, 70
Bell, 11
Bellcore, 507–508
Berne, E., 206, 218, 240
best solutions, 61
BI, See behavioral intelligence
Bilateral Human Relations, 205–206
billing system, 333, 393
Black money, xxvi
Block diagram of the KPS, 43
Boole, 112–114, 389
Bose, 29
Boundaries, 30, 117, 168, 348, 409
breeds of intelligence, 11
Bridal choice, 292
bridal dilemma, 292–293
Bridge between science and wisdom, 401–405
BRISDN, See basic rate integrated services digital network
broadband networks, xxvi, 35, 40, 64, 83, 98, 359, 432, 444–445, 507
Buddhism, 417
building blocks, 5, 16–17, 40, 71, 75, 82, 86, 172, 251
Bus Configurations and Switch Locations, 84–86
Bush, 131n5, 411, 443n7, 469, 473, 489
business records, 7, 17, 72, 379, 407
B-values, 508–509
B-WAB, 508
BX, 508
B-Zero filter, 261

C language, 13, 378, 508
C&C checker, 153
C++, 13, 113, 386, 508
CAD environment, 15, 75, 96, 249, 330, 421, 508
calculated risk, 292–293, 414
call data records (CDRs), 333, 508
call monitoring, 374–375
Call Processing, 374, 378, 505
CAMEL, 508
CATV, 508
CCIS, See common channel inter-office signaling
CCISS, See common channel inter-office signaling system

CCITT, See Consultative Committee of International Telephone & Telegraph
CCS, See Common Channel Signaling
CCSN, See Common Channel Signaling (CCS) network
CDRs, See call data records
cell format, 378
cell relay, 16, 378–379
Central Processing, 108–111
Central Processing and Knowledge Processing, 108–111
Central Processor Units and Generic Computers, 366–373
centralized sensing interface, 55, 81, 83, 87, 357, 367
CEPT, 508, 510
change in entropy of objects, 102, 108, 342
change of entropy, 20, 142–143, 320–353
Change of Operators ($+ \times /$ and $=$), 283–285, 508
characteristic impedance, 277–278
chicken flu ebola epidemic or malaria, 347
chromatic dispersion, 310, 406, 444
circuit-switched, 4, 180, 373, 378, 421
circularity, 27, 154, 313, 425–426, 425n5
CLASS, 508
Classical Migration Path of Knowledge, 175–179
Classification of knowledge-centric objects (KCOs), 326–333
  SSKO, 327–328
  SSKO, multiple attributes $(1 - n)$, 328–329
  SSKO-SA, 328
  SSKO, single dependent object, 329
  SSKO, single dependent object, multiple attributes, 330–331
  SSKO, single dependent object, single attribute, 330
Clergy, 508
client(j), 42–43, 47, 59–60
client/agent(j), 42–43, 47, 59–60
Clinton, 411, 469, 472–473, 489
Clusters of Complex KCOs, 333–336
CM, See communication module or concept machine
COA, See Cost of Agreement
Coarse Survey of Human Struggle, 489–490
COD, See Cost of Disagreement
coefficient of corruption, 248, 412
common channel inter-office signaling (CCIS), 375, 508
common channel inter-office signaling system (CCISS), 59, 508
Common Channel Signaling (CCS), 376, 508

522 INDEX

Common Channel Signaling (CCSN) network, 508
communicating computing and switching, 15–17
communicating knowledge, 15–16, 25, 76
Communication Module of the KM, 78–79
communication module or concept machine (CM), 7, 160, 394, 399–400, 408, 508
Communication Processor and ESS, 373–379
Communications and Interactions, 362
COMNET, 509
Compilation of knowledge, 156–163
  knowledge machine code, 163
  lexical analysis for knowledge, 157
  parsing bodies of knowledge, 157–161
  semantic analysis for knowledge, 161–163
  syntactic analysis of knowledge, 161
Compilers, 9, 28n6, 162, 174, 344, 386, 396
complex objects, 157, 168–169, 331–332, 339
Complexity theory, 26–27, 350
Components of Knowledge, 21–22, 165
compound semiconductor, 12, 15–16, 443
Comprehension Nature and Knowledge, 102–108
Compromise during transactions, 363
Computational aspects in negotiations, 233–235
Computing, 15, 99, 151, 187, 202, 322
Concept, 5, 54, 77, 87, 94, 113, 141, 155n3, 175, 177, 179, 287, 366, 405, 411, 464
Concept Processing, 406
concept-based switching, 5, 15–17, 373
Conclusions for LM Negotiation, 228
conditions for stability, 240–241
confidence level, 30–31, 33, 45, 116, 272–273, 305, 332
Configuration of $(I \ll\gg K)$ systems, 266–273
  application constraints, 266
  application for filtering and comparing, 270–271
  design concepts for $(I \ll\gg K)$ systems, 266–270
  design methodology, 272–273
  design steps for $(I \ll\gg K)$ filters, 272
  low-pass and high-pass information filters, 271–272
Configuration of a Two Port Knowledge Network, 279–283
Conflictive Roles of Humans and Machines, 413
Conflictive Roles of TVB <> DAH Orientations, 469
Connection oriented data, 27, 42, 55, 203, 517
Connectionless services, 517
Constituents of Model for LM Negotiations, 223–224
Construction of a Knowledge Plane, 477–479

Consultative Committee of International Telephone & Telegraph (CCITT), 377, 508
Contamination of $(I \ll\gg K)$, 295–297
contamination of knowledge, 297–300
Continuity of Knowledge, 33, 465–466
Contours of Knowledge, 155
control communication computation, 16, 38–39
control signals, 47–48, 201, 328, 362, 369, 394
control units, 47, 49, 408
Convergence of K & M Hierarchies, 450–455
Convolution, 105–106, 108, 111, 180, 263n4, 285, 296
coordinated coherent cogent and consistent, 59, 112, 200, 388, 431, 445, 463, 476
cordless telephone (CT), 509
Corning Glass, 149n1
corporate crime, 72, 411, 473
corporate interaction, 229–243
Corporate Needs, 233–234, 427–429, 442
Corporate Negotiations, 238, 243–244
Corporations Computers And Networks, 428–429
corrective strategy, 84
Cosmic Space Knowledge Space and Intellectual Space, 484–486
Cost of Agreement (COA), 226, 229, 235–237, 242–243, 363, 508
Cost of Disagreement (COD), 226, 229, 235–237, 241, 243, 363
CPM, *See* Critical Path Method
Creativity, 493–496
criminals, 98, 201, 319, 443, 472, 495, 498–499
criminal CEOs, 508
Critical Path Method (CPM), 237, 509
cross-talk and impulse noise, 250, 297
CSMA/CD, 14
CT, *See* cordless telephone
Culture, 56, 115, 126, 135, 137, 257, 299, 358n1, 359n2, 367, 441, 470
customer care, 456

D channel banks, 509
D I K W and E, 509
D to D, 509
DACS, *See* digital access cross connect systems
DACU, *See* data acquisition and control units
DAH Filter, 509
DAH, *See* deception, arrogance, and hate
data acquisition and control units (DACU), 47–49
data link-layer, 422, 439
Data management/data manager (DM), 510
data manager, 58, 510
data transmission system, 510

INDEX

database management (DBM), 405, 509
database management systems (DBMSs), 509
DBM, *See* database management
DBMSs, *See* database management systems
DCS, *See* Digital cross-connect switches
DDS or LoC, *See* Dewey Decimal System
DDS, *See* digital data services
deception, arrogance, and hate (DAH), 67, 69, 74n5, 88, 90–92, 94, 126–127, 133–134, 138, 312, 358, 345n6, 463n3, 467, 469–470, 478n5
Decision, 42–45, 73, 91, 296
decision support systems (DSSs), 73–74, 96
decontamination of knowledge, 297–300
Demand Intensities of Labor and Management (DIL&DIM), 224–225, 510
Demand Intensity (DI), 235–237, 510
Department of Defense (DOD), 14
Derivation of Knowledge, 163–171
Derived Axioms, 73–74, 417, 441
Derived Wisdom, 441
Design Concepts For $(I \ll \gg K)$ Systems, 266–270
Design Methodology, 183, 254, 272–273, 328, 414
Design of $(I \ll \gg K)$ Filters, 258–266
Design of Conventional Signal Filters, 249–252
design of networks, 52
Design Steps For $(I \ll \gg K)$ Filters, 272
destructive goals, 98
Dewey Decimal System (DDS or LoC), 50, 59, 104–106, 131, 141, 153, 162n5, 173, 175, 301, 323, 372, 389, 461, 480, 491, 510
DI, *See* Demand Intensity
digital access cross connect systems (DACS), 509
Digital cross-connect switches (DCS), 421, 509
digital data services (DDS), 510
Digital Filters, 251–252, 255
Digital Knowledge Environments, 386–394
Digital Medical Processing Environments, 394–401
Digital Object Processing Environments, 383–386
Digital Signal Processors (DSPs), 15, 172, 252, 355, 365, 382–383
digital subscriber loop (DSL), 298n10, 383
DIL&DIM, *See* Demand Intensities of Labor and Management
Dimensionality of the Nine variables A-I, 283–285
Direct memory access (DMA), 7, 16, 76, 94, 174, 370, 379, 372, 383, 385, 391, 411, 510
discrete multi tone (DMT), 510

Display/Graphics Processor Units (DPUs/GPUs), 380–381, 510
distance learning, 38–39, 219, 380, 429, 510
distillation of wisdom, 17, 22, 446
distributed queue dual bus (DQDB), 510
DM, *See* Data management/data manager
DMA, *See* Direct memory access
DMT, *See* discrete multi tone
DOD, *See* Department of Defense
Dollard, J, 222
domain of search, 35, 70, 154, 440, 448
Downward Migration, 471, 473
DPUs/GPUs, *See* Display/Graphics Processor Unit
DQDB, *See* distributed queue dual bus
DSL, *See* digital subscriber loop
DSLAM, 510
DSPs, *See* Digital Signal Processor
DSSs, *See* decision support systems
Dual KPU Human Interaction Model, 204–208, 388
duality of human nature, 112, 188, 193, 467, 469
duality of nature, 97, 102–108, 465
Dupont, 149n1
dynamic equilibrium, 124, 151, 165, 268, 286, 416, 456
Dynamics of Knowledge in Societies, 130–132

EC, *See* echo cancellation
echo cancellation (EC), 510
economic activity, 3n2, 41, 98, 233, 243, 496–497
economic motives, 14
Economics in Interactions, 420, 438, 447
economics of knowledge, 23–24
ecosystem, 197
Edison, 11
Education and Welfare, 40, 172, 430, 432, 441
education/distance learning, 38–39, 219, 380, 429, 510
education/library networks, 50–52
educational networks, 50–52
EDVAC, *See* Electronic Discrete Variable Automatic Computer
Effects of Intelligent Response, 197
EG, *See* electronic government
EG of a small nation, 41
EG platform, 39–40
EG-IT, *See* electronic government-IT
Einstein, 97, 106, 258, 276, 312, 314–315, 317–318, 479, 491, 496
EIS, *See* executive information system

elastic zero-sum game, 27
Electronic Discrete Variable Automatic Computer (EDVAC), 12
electronic government (EG), xxii, 36–37, 39–40, 57, 61, 367, 373, 429, 510
electronic government-IT (EG-IT), 37, 510
Electronic Numerical Integrator and Computer (ENIAC), 12
electronic switching systems (ESS), 13, 16, 373–379, 394, 456, 506–507, 510, 516–517
elegance, 27, 30–31, 33, 70, 88–89, 96, 98, 201, 259, 298, 341, 401, 411, 426, 462
Elemental Convolution and Knowledge Operations (kopcs), 108
elementary particles, 96, 126, 324
Elzas, M. S., 230
embedded intelligence, 39, 163, 423, 425,
Embedded NO's and VF's, 263
Employer resistance curves, 222, 227–228
Enhanced Knowledge Trail, 112–114
ENIAC, *See* Electronic Numerical Integrator and Computer
Enron, 73, 411, 445
Entropy, 20, 20n3, 101, 108, 141–145, 263n4, 320–353, 364
Entropy of knowledge, 141–145, 465
  change of entropy, 142
  duration for change of entropy, 142–143
  generalization of knowledge processes, 144–145
  no knowledge process, 143
  no prior knowledge, 143
epidemics, 98, 467
ESS, *See* electronic switching systems
Essence of Human Activity, 185–189
ethical dimensions, 359–360, 411
Ethical Issues and Role of Machines, 412–413
Ethics, 412, 417
Ethics and Values in High-Tech Society, 417–419
Evolution of Knowledge Networks, 58–59
Evolving Knowledge Environments, 19–22
EWSD, 378, 510
Examples of Prevalent Knowledge Processes, 143–144
executive information system (EIS), 40, 73, 78, 81, 516
expansion port network, 515
Expansionary Upward Movement (TVB), 470–471
Expert System Concepts, 231–233
Exponential Representation, 212

falsehood, distaste and corruption (FDC), 511
Faraday, 97
FCs, *See* functional components
FDC, *See* falsehood, distaste and corruption
fear and threat, 18, 195, 240
federal government, 14
feedback loop, 81, 222–223, 226, 231, 240–242, 452
FEPs, *See* front end processors
fiber optic, xxvii, 15, 18, 86, 107n4, 359, 434, 511, 516–517
fiber to the curb (FTTC), 511
fiber to the home (FTTH), 511
fifteen "B-Values", 508–509
Filter Template Based on TVB, 314–318
Filters, 246–318
financial networks, 441
FLAG, 511
flip-flop of attitude, 201, 511
Forces in Society and Operations on Objects, 150–152
Ford Motor Company, 498
Four Ports of Super Node-K, 476–477
Four Precursors of Modern Wisdom, 35
Fourier, 171, 249–250, 252, 258–259, 262, 278
Fragments of Overall Human Activity, 202–204
frame relay, xxvi, 16, 420, 436, 440–441
Framework of knowledge, 7, 103, 153–156, 466
  contours of knowledge, 155
  funicular polygon of concepts, 155–156
  hyper-dimensionality of knowledge, 153–154
  intertwined spaces of knowledge, 154–155
free market, 27
Freud, Sigmund, 70, 462
from CPU to KPU, 107–111, 197–198, 355–356
from node A_i to A_{i+}, 195–196
From Wisdom to Behavior, 95–97
Fromm, E., 218, 221, 424
front end processors (FEPs), 511
FTTC, *See* fiber to the curb
FTTH, *See* fiber to the home
Functional Approach, 103–104
functional components (FCs), 510
Functions, 22, 75–76, 101, 110, 148–179, 189n1, 254–256, 259n3, 265, 406, 421, 477
Funicular Polygon of Concepts, 155–156
Fuzziness, 239
Fuzzy Hamming Distance, 34, 511
fuzzy NPU, 511

G.DMT, 511
Galvani, 97
game theory, 23, 222, 230–231

INDEX **525**

Gauss, 29, 97, 106
general theory of knowledge, 101–145, 463, *See also* Entropy of knowledge; Origin and destination
　basis for the theory of knowledge, 102
　central processing and knowledge processing, 108–111
　comprehension, nature, and knowledge, 102–108
　　elemental convolution, 108
　　functional approach, 103–104
　　incremental changes, 104–108
　　knowledge operations, 108
　enhanced knowledge trail, 112–114
　information, knowledge, wisdom, accumulation of, 111–112
　inverse universe, 120–121
　sequencing of events at nodes, 114–116
　transition management at nodes, 117–120
　transitions at I, K, and C nodes, 116–117
Generality of Information Filtering, 310–311
Generalization of Knowledge Processes, 144–145
global concerns, human ethics, 430
Global Crossing, 73, 411, 445
Global packet radio services (GPRS), 511
Global Power of Knowledge, 6–9
global unrest and violence, 4, 98
GNP, *See* gross national product
GPRS, *See* Global packet radio services
GPU, *See* graphical processor unit
Gradation of KPU responses, 197–202
　artificial intelligence response, 199
　human response, 200–201
　humanistic response, 199–200
　life form response, 198
　other types of responses, 201–202
　rational response, 198–199
　traditional computer response, 198
graphical processor unit (GPU), 7, 322, 380–382, 384, 390, 393, 511
graphical user interface (GUI), 9, 43, 46, 162, 381, 421, 511
graphics processors, 380
greed, 3, 11, 417, 443, 449, 470
greed for personal gain, 443, 502
Green money, 442
gross national product (GNP), 149, 155, 511
GUI, *See* graphical user interface

Hamming, 29
Hamming distance, 34, 511
Hardware Configuration for the SCP of Intelligent Networks, 51, 59
hardware for KPS, 43
HDSL, *See* high-speed digital subscriber line
HDTV, *See* High Definition TV
Heisenberg's principle, 95, 156
Hicks, J. R., 219, 222, 226–227, 240–241
Hierarchy of Knowledge, 297, 451–453
Hierarchy of Motivation, 453–455
Hierarchy, 12, 37, 39, 150, 171–172, 223, 266n5, 352, 366, 451–455
High Definition TV (HDTV), 511
higher-level languages (HLLs), 9, 13, 58
higher-level needs, 6, 11, 71, 434, 436, 438
high-speed digital subscriber line (HDSL), 511
high-speed LANs, 371
high-speed lines, 51
Hindsight and Foresight, 489–493
HLLs, *See* higher-level languages
Hocquenghem, 29
hospital-based medical networks, 402
human activity, 185–189, 202–204
Human and Social Aspects, 423–424
human aspect, 249, 425n4
human behavior, 68, 121, 150–151, 189, 195, 197n7, 200, 206, 211, 267, 293, 451, 453
human beings, 4, 8, 32, 68, 72, 88, 104, 149, 186, 200, 266n5, 267, 411, 417, 425n4
Human bias, 18, 225, 257, 360, 446
Human-Concept Machines, 406
Human Element, 435–438
human errors, 130
human integrity, 462
human intellect, xxvi, 356, 364, 496
Human interactions, 204–208, 211–217
human needs, 6, 10, 12, 35, 70, 72, 307, 360, 424, 427, 433
Human Resources, 40, 57, 233, 478, 498
Human Response, 146, 198n8, 200–201, 206, 211
Human Search, 4, 6, 97–98, 256, 365
Humanistic and Semi-Human Systems, 182–244, *See also* Gradation of KPU responses; Human interactions; Knowledge machines for human interactions
　algorithmic representation, 193–197
　dual knowledge processor unit human interaction model, 204–208
　　bilateral human relations, 205–206
　　models of human interaction, 206–208
　effects of intelligent response, 197
　essence of human activity, 185–189
　　types of verb function commands, 188–189
　　varieties of noun objects, 186–188
　fragments of overall human activity, 202–204
　　major ordeals and complex programs, 203

Humanistic and Semi-Human Systems (*Continued*)
  microcode of the humanistic chip set, 203–204
  minor tasks and microcodes, 203
  humanistic chip sets, 183–185
  smart verbs and intelligent nouns, 189–192
    intelligent response, 192
    nonlinear response, 191
    qualitative synopsis of the response, 190–191
Humanistic Chip Set, 183–185, 203
Humanistic Response, 183, 199–200
Humanistic Systems that Refuse, 443
Humankind, 11, 18, 122, 124
human–machine interaction, 362
human-machine system, 34, 98, 256
Humans Machines and Networks, 416–458, *See also* Needs and networks, overlap of
  convergence of knowledge, 450
  ethics and values in high-tech society, 417–419
  knowledge machines on knowledge, 455–456
  knowledge rush, 440–443
  motivation hierarchies, 450–455
  needs pyramid of a society, 443–447
  needs that drive, 419–420
  networks that transport, 420–423
  rationality of humanistic machines, 431–440
    human element, 435–438
    open human inter-connect model, 438–440
    self-perpetuating power loops, 447–450
  wisdom domain, 440–443
Hype Knowledge, 247–249
hyperdimensional objects, 24, 387
Hyper-dimensionality of Knowledge, 153–154
Hypothesis, 76–77, 87, 103, 115–118, 168, 232, 405
Hysteresis effects, 213–215

I K C W and E Bases to Replace Nodes, 132–137
I K C W and E nodes, 112, 132, 137, 463
$I \lll K$, 253, 258, 266, 272, 275, 295
IAS, *See* Institute of Advanced Study
IBM, 82, 91, 379, 384, 386, 388, 397
IC, *See* integrated circuit
IKN, *See* intelligent knowledge network
I-KPS, 511
IKPS, *See* Intelligent Knowledge processing system
IKW, *See* intelligent-knowledgeware
IKWV chain, 511
Imaging, 2, 17, 114, 159, 266, 348, 390, 400, 402

IMF, *See* International Monetary Fund
Impasse in LM negotiations, 227–228
Implications of Knowledge Equations, 141
IN, *See* Intelligent network
IN/+, 512
in-band signaling, 59
INCAD environments, 512
INCM, *See* intelligent network conceptual model
Incremental Changes, 104–108
incremental gain, 32
incremental gradients, 131
Individual Needs, 424
Individuals Machines And Networks, 424–427
inertia/mass effect, 102
information age, 3, 64–65, 70, 182, 358–359, 413, 432–433, 461
Information and Knowledge Filters, 246–319, *See also* Application to three negative noun objects; Configuration of $(I \lll K)$ systems; practical information filter; Three-Stage 'SET' Filter (John Tyndall); two-port knowledge network; Two-Stage 'TR' Filter (Henry Higgins)
  $(I \lll K)$ filters, 253–258
    precise solutions and confident results, 257–258
    role of humans, 256–257
    symbols as objects and functions, 254–256
  contamination of $(I \lll K)$, 295–297
  decontamination of knowledge, 297–300
  design of $(I \lll K)$ filters, 258–266
    knowledge filter concepts, 259–265
    noun objects and verb functions, 265
    scanning for operative nouns and verbs, 266
  design of conventional signal filters, 249–252
    active filters, 251
    digital filters, 251–252
    types of traditional signal filters, 251
  junk information and hype knowledge, 247–249
  knowledge gates, 286–295
    knowledge-based AND gate, 289–290
    knowledge-based NOT gate, 291–292
    knowledge-based OR gate, 290–291
    use of knowledge gates, 292–295
  selection of noun objects, 273–274
  signal waves and knowledge flow, 252–253
  systems for $(I \lll K)$ filters, 275–277
  verb functions in samples, 273–274
information and knowledge, 23–24, 87, 161, 433
information filter, 8, 253, 259–260, 299, 304, 307, 312, 341
Information Knowledge and Wisdom, 111–112
information objects, 8, 24–25, 27–28, 30, 32

INDEX   527

information processing, 14, 25, 28, 122, 324, 343, 359, 503
Information processing systems, 179, 404
information processor unit (IPU), 116
Innovation, 286
Input/Output processors (IOPs), 379–380
Institute of Advanced Study (IAS), 12–13, 27
integrated circuit (IC), 36, 249–251, 260, 511, 518
integrated services digital network (ISDN), 4, 16, 376, 378, 512
intelligent agents, 71–72, 115, 163, 171, 205, 219, 281, 300, 490
intelligent communication systems, 285
intelligent decision support, 55, 61, 428
Intelligent hardware agents, 171–172
intelligent information-sourcing system, 111–112
Intelligent Internet and Knowledge Society, 34–58
  antiterrorism networks, 55–58
  educational networks, 50–52
  four precursors of modern wisdom, 35
  knowledge bases to derive wisdom, 36
  medical networks, 52–55
  role of national governments, 36–41
  universal knowledge-processing systems, 41–50
intelligent Internet, 34–58
intelligent-knowledgeware (IKW), 511
intelligent knowledge network (IKN), 511
Intelligent Knowledge processing system (IKPS), 79, 511
Intelligent Knowledge Two Port Networks, 277–286
intelligent medical Internet, 52–55
intelligent medical knowledge processing, 41–50
Intelligent network (IN), 58–59, 373, 386, 399
intelligent network conceptual model (INCM), 434, 512
intelligent networks, 16, 51–53, 58–60, 82, 373, 386, 427, 432, 434
Intelligent Nouns, 189–192
intelligent objects, 153, 186–195, 198, 201, 208, 214–218, 345
intelligent peripheral or Internet protocol (IP), 5, 35, 92, 186, 361
Intelligent Response, 192
intelligent robotics, 171, 184
intelligent systems, 150, 503
intensity of verb function, 20, 151, 159, 215–217
intermediate databases, 35, 39, 117, 140
Intern, 7, 111, 122, 451
International Monetary Fund (IMF), 172, 430, 432, 434

International Telecommunications Union (ITU), 59, 377, 434, 512
Internet environment, 58
Internet netware, 39
Internet platform, 14
Internet service provider (ISP), 54, 59–60, 88–89, 387, 393, 410–411, 512
Internet services, 14, 59–60, 88–89, 372, 411
Interoperability, 42
Interpersonal Relationships, 206
Intertwined Spaces of Knowledge, 154–155
Invention, 14–17, 23, 81, 113–114, 493
IOPs, *See* Input/Output processor
IP, *See* intelligent peripheral or Internet protocol
IPU, *See* information processor unit
ISDN, *See* integrated services digital network
ISP, *See* Internet service provider
IT aspects, 36–38, 412, 512
IT platform, 37–38, 81, 431
Iterative and Reflexive Processing, 347
Iterative Cyclic Social Processes, 359–364
ITU, *See* International Telecommunications Union

job control language, 512
Junk Information, 286, 497
Junk Information and Hype Knowledge, 247–249

KAND gate, 289–292, 512
Kavesh, R., 206n9
KBMS, *See* knowledge base management systems
KBs, *See* knowledge bases
KCO, *See* Knowledge Centric Object
KCP, *See* knowledge control point
KCPU, 512
KCS, *See* knowledge communication system
KD function, 486
KEE, *See* knowledge enhancement environments
Khayyám, Omar, 305–306
killer machine, 4, 71, 454
KIP, *See* knowledge intelligent peripheral
KIR, *See* knowledge instruction register
KM For Corporate Interactions, 229–244
KM for Human Interactive Processes, 219–221
KM for Labor Management (LM) Negotiation, 222–228
KM, *See* knowledge machine
KMM, *See* knowledge management module
KMs and WMs, *See* Knowledge machines and Wisdom machines
KMS, *See* Knowledge Maintenance System
KNOT, *See* Knowledge based NOT gate

Knowbula, 486–489
Knowbula Almost Fills the Cube of Knowledge, 493–495
Knowbula Becomes Insignificant in Cube of Knowledge, 496
Knowbula Loosely Fits the Cube of Knowledge, 495–496
Knowledge and Money, 496–498
Knowledge and Wealth, 17–19
knowledge and wisdom, 111–112
knowledge bank, 7, 26, 99, 184, 276–277, 419, 430, 503
knowledge base management systems (KBMS), 36, 512
Knowledge based NOT gate (KNOT), 291–292, 512
Knowledge based OR gate (KOR), 290–291
knowledge bases (KBs), 11, 36, 50–51, 54, 59, 92, 94, 264, 268, 358, 367, 479, 484, 512
Knowledge Bases in Primary K-Plane, 479
Knowledge Bases to Derive Wisdom, 36
knowledge cache, 281–283, 307
Knowledge Centric Objects (KCO), 321–326
knowledge communication system (KCS), 79, 512
knowledge control point (KCP), 50, 59–60, 393, 395, 512
knowledge dispensing, 33
knowledge domain, 2, 11, 20n3, 27, 29, 32–33, 155–156, 248n1, 263, 297, 321, 461n1, 494
knowledge enhancement environments (KEE), 512
knowledge environments, 1–61
knowledge equations, 141, 284
knowledge filter, 246–319
Knowledge Freedom and Creativity, 493–496
Knowledge gates, 286–295
Knowledge Hardware and Software Systems, 172–175
knowledge instruction register (KIR), 265, 333, 335, 512
knowledge instruction, 78, 153, 161, 265, 321, 335
knowledge intelligent peripheral (KIP), 512
Knowledge loop and its stability, 462–465
Knowledge Loops in X And Y Dimensions, 473–476
knowledge machine (KM), 78–80, 149, 160, 186, 207, 265, 320, 324, 358, 390, 440, 446, 501, 512
knowledge machine building blocks, 77–86
  bus configurations and switch locations, 84–86
  knowledge machines, 78–79

knowledge-machine-based wisdom machines, 79–80
sensor-scanner-based wisdom machines, 81–84
Knowledge machine code, 163
Knowledge Machine Software Hierarchy, 171–172
Knowledge machines and Wisdom machines (KMs and WMs), 9, 20–21, 64–98, 171–172, 219–221, 512
Knowledge Machines Building Blocks, 77–86
Knowledge Machines for Human Interactions, 218–221
Knowledge Machines on Knowledge, 455–456
Knowledge Maintenance System (KMS), 50–51, 60, 389, 395, 402
knowledge management module (KMM), 512
knowledge module, 367, 399–400, 404–406, 410
Knowledge Network Configuration, 59
knowledge network, 58–59
knowledge operation code (*kopc*), 19, 22, 108, 146, 164, 172, 180, 260, 321, 339, 386, 389, 392, 395, 488, 513
knowledge-oriented functions, 254, 373
knowledge-oriented problems, 22
knowledge processing environment (KPC), 513
knowledge-processing systems (KPSs), 41, 44, 60, 262–263, 308, 310
knowledge processing unit (KPU), 12, 78, 164, 166, 265, 320, 391–392, 395
knowledge processing, 41–50, 108–111
knowledge profile, 394
knowledge program, 78, 146, 148, 165, 168, 172, 180, 293, 323
knowledge ring, 394
knowledge services, 58–60
knowledge services control point (KSCP), 60, 513
knowledge service providers (KSP), 59–60
knowledge society, 34–58
knowledge software, 324
knowledge software or knowledge switch (KSW), 513
Knowledge system architectures, 355–414, *See also* Varieties of processors and machines
  from the CPU to KPU, 356–357
  iterative cyclic social processes, 359–364
    adjustment and relationships, 363–364
    awareness and knowledge, 359–361
    bargaining and negotiation, 363
    communications and interactions, 362
    compromise during transactions, 363
  philosophic dimension, 357–359
  recent trends toward wisdom, 406–412

artificial wisdom, 407–409
  instinct in humans, 411–412
  old and new wisdom, 409–411
  wisdom in machines, 411–412
scientific dimension, 364–365
  machine accuracy and human adaptability, 365
  search for precision and infallibility, 364–365
  trend of concepts, 401–406
    bridge between science and wisdom, 401–405
    concept processing, 406
    human concept machines, 406
  trend of values and ethics, 412–413
    conflictive roles of humans and machines, 413
    ethical issues and role of machines, 412–413
knowledge system management system (KSMS), 513
knowledge theory, 26–27, 33–34, 271, 285n8, 388
knowledge trail, 112–114, 122–126, 137
knowledge transfer point (KTP), 50–51, 513
Knowledge traps, 176
knowledge tree, 157, 275, 389
knowledge-ware, 173, 503
knowledge wisdom and culture, 124
Kompilation, 513
kopc, *See* knowledge operation code
kopcode, 336–340
koperand, 158, 189, 323–324
kopr, 19, 122, 145, 164, 166
KOR, *See* Knowledge based OR gate
KPE, *See* knowledge processing environment
KPSs, *See* knowledge-processing systems
KPU Architectures for *KOPC0*-Instructions, 348–353
KPU, *See* knowledge processing unit
Krichoff, 106
Krol, M., 52
KSCP, *See* knowledge services control point
KSMS, *See* knowledge system management system
KSP(i), 59
KSP, *See* knowledge service providers
KSW, *See* knowledge software or knowledge switch
KTP, *See* knowledge transfer point

labor-management, 222–228
Laplace, 249–250, 258
Laser, 249, 310, 310n14, 453

Laws of Nature, 97, 103–106, 108, 122, 146, 334, 484
leisure time, 224, 444
Lessons from Knowledge Lost, 483
Level I - Access and Administrative Functions, 76
Level II - Linkage Scientific and Statistical Functions, 76–77
Level III –Human Authentication, 77
Levels of information, 341
lexical analysis, 157, 418
Lexical Analysis for Knowledge, 157
Lexical analyzers, 513
Library of Congress, 36, 40, 50, 53, 59, 99, 159, 173, 175, 301, 323, 372, 461, 491–492
Life Form Response, 198
Life of Inventions, 7–8
Lightwave Systems, 310, 453
limited search, 99
LISP, 13, 386
LM Negotiations, 222–228
LM Relationships, 191, 222
LoC, *See* Library of Congress
longer-term benefits, 18, 442
Long-Term Movements within the K-Loop, 468–469
Low Pass and High Pass Information Filters, 271–272
lower level needs, 11, 71, 184, 434, 436
Luce, 222
Lueng, 52

Machine Accuracy and Human Adaptability, 365
Machine clusters, 86–95
  multiple-wisdom multiple-machine systems, 93–95
  multiple-wisdom single-machine systems, 91–92
  single-wisdom multiple-machine systems, 89–90
  single-wisdom single-machine systems, 86–89
machine-humanistic functions, 77, 203, 256, 412
Machines as Processors of Information, 445–447
Macro Instructions For KPU, 347–348
Mafia, 132, 134, 141, 450
magnetic resonance imaging (MRI), 2, 4, 7, 17, 54, 114, 266, 348, 380, 446
Major Ordeals and Complex Programs, 203
MAN, *See* Metropolitan Area Networks
Many Flavors of Wisdom, 65–67, 308
March, H. A., 70, 423
marginal utility, 18, 221, 230, 235–236, 268
Marshall, Alfred, 10, 221, 305–306
Marshall's marginal utility, 183, 479

530 INDEX

Martin, Norman, 206n9
Marx, Karl, 6, 11, 202, 357
Maslow's pyramid of needs, 443–447
materialistic wisdom, 68
Mathematical Implications of the Two Port Network, 283
Mathematical Representations, 235–237
Maxwell, 29, 97
Mead, G. H., 70, 424
medical computer system, 52, 397, 401
medical data bank, 52, 397, 399, 400
medical diagnostics, 451
medical knowledge networks (MKSs), 403, 514
medical knowledge transfer point (MKTP), 53, 514
Medical networks, 52–55
Medical processor, 365, 394–400, 514
medical services management system (MSMS), 514
Medical service providers, 52–55
Memory effects, 212
Metropolitan Area Networks (MAN), 51, 513
microcode, 17, 203–204, 388
microcodes, 203
Microcode of the Humanistic Chip Set, 203–204
microeconomics, 24, 202, 438
Microscopic or Basic Assembly Level Knowledge BALK Instructions, 164–165
microwave communications, 383, 393
Miller, N. E., 222
MIMD, *See* multiple instruction multiple data
ministries, 39–41, 57
Minor Tasks and Micro-codes, 203
MISD, *See* multiple instruction single data
MKNs, *See* medical knowledge networks
MKTP, *See* medical knowledge transfer point
Models Databases, 98, 206
Models of Human Interaction, 206–208
Modes of Wisdom, 64–98
Modular, 40, 42, 377–378, 389, 394
Mollah, N., 52
monetary theory, 98
moral content, 24, 92, 94, 471
MP-MCKO-MDO-MA, 340–341
MP-MCKO-MDO-NA, 340
MP-MCKO-MDO-SA, 340
MPMO, *See* multiple-process multiple-object
MPSO, *See* multiple process single object
MRI, *See* magnetic resonance imaging
MSMS, *See* medical services management system
MSP, *See* medical service providers
Multicast service, 41, 44, 47
multimedia traffic, 390, 420

Multiple Feedbacks, 126–130
multiple instruction multiple data (MIMD), 12, 15, 93, 164, 197, 350, 370–371, 388n3
multiple instruction single data (MISD), 12, 197, 350, 370–371, 388, 388n3
Multiple Process Instructions, 340–341
multiple-process multiple-object (MPMO), 173, 197, 348, 350
multiple process single object (MPSO), 173, 197, 348, 350
Multiple Wisdom Multiple Machine, 93–95
Multiple Wisdom Single Machine, 91–92
multiplier effects, 140
Multiplier Effects at Intermediate Nodes, 140–141
Mycin, 7, 111, 122, 162, 428, 451

national economy, 23, 58, 131, 144, 428
natural intelligence, 75–76, 92, 94, 98, 361, 404, 416, 493
Nature (N) Origin of Knowledge Trail, 122–124
NBES, *See* Network-based educational system
NCOM, *See* network communication unit
Nebula of Knowledge (Knowbula), 486–489
need extinction process, 269
Need Pyramid, 423–424
Need Pyramid of A Society, 443–447
Need to know, 2–11
  global power of knowledge, 6–9
  scientific aspect, 9–10
  wealth aspect, 10–11
    global power of nations, 11
    industrial development, 10–11
    materialistic achievement, 11
Needs and networks, overlap of, 423–431
  corporate needs, 427–428
  corporations, computers, and networks, 428–429
  human and social aspects, 423–424
  individual needs, 424
  individuals, machines, and networks, 424–427
  societal needs, 429–430
  societies, machines, and networks, 430–431
Needs and Wisdom, 70–71
Needs Knowledge and Innovations, 286
Needs That Drive, 419–420
negative use, 254
negative verbs, 194
negative wisdom, 67, 69, 73, 149, 356–357, 449
Negotiation Process and Validation, 225–227
negotiation, 222–228
NeoMycin, 7, 122, 162, 428, 451
netware programs, 39
Network-based educational system (NBSE), 514

network communication unit (NCOM), 514
network knowledge service providers (NKSP), 514
Network service, 378, 401
network terminations (NT), 41, 514
networks and machines, 413
Networks That Transport, 420–423
neural networks, 25
neutral verbs, 189
new breed of computer systems, 202
New Breed of Knowledge, 461–462
New Filter Template Based on TVB, 316–318
New knowledge environments, 1–61, See also Intelligent Internet and Knowledge Society; Need to know
   evolving knowledge environments, 19–22
      components of knowledge, 21–22
      processing of knowledge, 22
   knowledge and wealth, 17–19
   knowledge networks, 58–59
      evolution of knowledge networks, 58–59
      knowledge network configuration, 59
   role of technology, 12–17
      major contributions, 12–15
      peripheral contribution, 17
      string of secondary contribution, 15–17
   structure and communication of knowledge, 23–34, See also individual entry
Nirvana, 67n4, 74, 124–125, 258, 414
Nixon, 131n5, 411, 469, 472–473
NKSPs, See network knowledge service providers
No Knowledge Process –No Change in Entropy, 143
No Prior Knowledge –No Change in Entropy, 143
Nodal Forces in Societies, 137–139
node manager, 122
Noise, 247, 250, 297–298
noisy environments, 27
Non-linear Response, 384
Noun Objects and Verb Functions, 273–274
Novel, 1, 14, 25
NPU, See numeric processor units
NT, See network terminations
numeric processor units (NPU), 7, 144, 257–258, 349, 367, 371, 382, 388n3
Numeric Processor Units, 382
numerically controlled, 221

OA&M, See operations, administration and maintenance
object management, 226
object-oriented languages, 295
Object Oriented Languages (OOL), 7, 515
object-oriented programs, 174–175, 265, 295, 359, 383
object oriented programs (OOP), 113, 174–175, 265
object process, 197, 319, 338, 348, 350–351, 353, 383, 388, 394
objective function, 98, 199, 247, 263, 272, 275, 277, 304–305, 409
OC-, See Optical Carrier systems
OHI, See open human inter-connect model
Old and New Wisdom, 409–411
OOL, See Object Oriented Languages
OOP, See object oriented programs
open human inter-connect model (OHI), 440, 514
operand cache, 166
operand stacks, 166, 345
operations, administration and maintenance (OA&M), 514
OPNET, 515
opportunistic wisdom, 69–70
Optical Carrier systems (OC-), 378, 421
optical network terminations, 250, 357
optical switching, 5
Optimality, 20, 65, 73, 84, 173, 200, 249, 265, 383, 404, 456
Optimization of Wise Choices, 73–75
optimizing compliers, 344
Order Awareness and Search, 97–98
organizational behavior, 232, 301
Origin and destination, 122–141
   dynamics of knowledge in societies, 130–132
   I, K, C, W, and E bases to replace nodes, 132–137
   implications of knowledge equations, 141
   multiple feedbacks along the knowledge trail, 126–130
   multiplier effects at intermediate nodes, 140–141
   nature, origin of knowledge trail, 122–124
   nodal forces in societies, 137–139
   sociological forces on nodes, 139–140
   two destinations of knowledge trail, 124–126
oscillatory response, 215–217
OSI model, 35, 84, 361, 420–422, 440
Other Types of Responses, 201–202
Overlap of Needs And Networks, 423–431

packet-switched, 51–52
Parsing BOKS, 157
parsing knowledge, 157
Passive KOPCS, 341–343
Passive Objects (KCOs), 342–343
passive optical networks, 250, 357

patient databases, 52, 397, 400
patient information, 52
patient medical process cycle (PMPC), 515
pattern recognition, 8, 71, 148, 160, 168, 219, 248, 272–273, 281, 490
payload, 506
PBXs, *See* private branch exchanges
PCM, *See* pulse code modulation
PCP, *See* processor control point
PCU, *See* program control unit
Peaky response, 213
Pen, J., 206, 219, 222
PeopleSoft, 36, 40, 78, 270, 348, 428
Perfection, 35, 90, 124–125, 200, 318, 327, 426, 453
Peripheral Contributions, 17
personal intelligent network (PIN), 515
Philosopher, 11, 66, 72, 75, 91, 95, 97, 103, 199–200, 244, 357, 358n1, 359n2, 469, 484, 499
Philosophic, 18, 27–33, 64–65, 69, 88, 175–176, 357, 425n4
Philosophic Dimension, 357–359
philosophic validation, 29–32
photonic currents, 107
physical layer convergence protocol (PLCP), 515
Physicians, 54, 403
physician–patient, 54, 403
Physiological, 70–71, 171, 185, 223, 419
PIN, *See* personal intelligent network
Pinnacle of Mind, 498–500
Pipeline, 12, 172, 197, 348, 370, 383, 388
Pitfalls In Knowledge, 442–443
plain old telephone services (POTS/PSTN), 4, 421
Plato, 312, 314, 317–318
PLCP, *See* physical layer convergence protocol
PMPC, *See* patient medical process cycle
PMR, *See* private mobile radio
Political agenda, 500–502
Positive and Negative Social Forces, 149–153
positive side, 6, 491, 500
positive values, 121, 201, 417
positive verbs, 194
Positive Wisdom, 69, 149, 356–357, 447, 449, 468
POTS/ PSTN, *See* plain old telephone services
PPNs and EPNs, *See* processor port network and expansion port network
PR, *See* pattern recognition
Practical Information Filter (Plato And Einstein), 312–318
Practical Use of Knowledge Gates, 292–295

Precise Solutions And Confident Results, 257–258
predictive programming, 151, 347
primary driving force, 418
Primary K-Plane, 479
Primary K-Plane in XY Dimensions, 478–479
Priming of Wisdom Machines, 74–75
Primitive, 12, 44, 171, 257, 359, 417, 420, 458, 498
prior knowledge, 143
private branch exchanges (PBXs), 515
private mobile radio (PMR), 515
private virtual network, 390
Process and change of entropy, 320–353, *See also* Classification of knowledge-centric objects
  actions and entropy, 321
  active *kopcs*, 343–346
  clusters of complex knowledge-centric objects, 333–336
  execution of multiple-process instructions, 346–347
  iterative and reflexive processing, 347
  knowledge-centric objects, 321–326
    intelligent knowledge-centric objects, 322–326
    passive knowledge-centric objects, 322
  KPU architectures for *kopc01–12* instructions, 348–353
  macroinstructions for knowledge processor units, 347–348
  multiple-process instructions, 340–341
  passive *kopcs*, 341–343
  single-process *kopcodes* for generic knowledge-centric objects, 336–340
  single-process *kopc* for multiple complex objects, 339–340
  single-process *kopc* for single complex object, 337–339
  single-process *kopcodes* for simple objects, 336–337
Process hierarchy, 352
processing of knowledge, 22
processor control point (PCP), 395
processor port network and expansion port network (PPN and EPNs), 515
program control unit (PCU), 366, 368, 515
program databases, 395
program instructions, 347
programmed intelligence, 106, 271
PROLOG, 13, 162
Protocol, 5, 34, 41, 59, 186, 476
Psychological, 31, 185, 206, 257, 357, 424

INDEX **533**

psychotic fear, 208
pulse code modulation (PCM), 375, 515

QAM coding, 515
QoS, 52, 84, 515
Qualitative Synopsis of the Response, 190–191

RACE, 515
Raiffa, H, 222
Rational Response, 198–199
Rationality of Humanistic Machines, 431–440
Rationality, 26, 66, 68, 87, 96, 121, 186, 189, 198, 463n3, 473
raw data, 98, 301
RDTs, *See* remote data terminals
Reactive or Intelligent Objects (KCOs), 322–326
read only memory (ROM), 65, 515
Recent Trends towards Wisdom, 406–412
Recessive Downward Movement (DAH), 471–472
remote data terminals (RDTs), 515
Report Generation KOPCS on Dynamic KCOs, 343
Report Generation KOPCS on Passive KCOs, 342–343
Role of Humans, 256–257
Role of National Governments, 36–41
Role of OHI in National Settings, 440
Role of Technology, 12–17
ROM, *See* read only memory
Roman, E. G., 206, 219, 235, 240
Routers, 2, 51, 57, 84, 379
R-Zero Filtering and Output BOK-R, 304–305

SANs, *See* storage area networks
SAP, 36, 40, 78, 270, 348, 385, 428
SAT-/WASC/SAFE, 516
Satisficing, 232–233
Saturation, 212–213
Scanning For Operative Noun and Verbs, 266
scarcity theory, 98
SCE, *See* service creation environment
scenario identification, 25, 431, 435, 451, 502
Science and Wisdom, 401–405
Scientific Aspects, 9–10
Scientific Dimension, 364–365
Scientific Principles in Knowledge Domain, 32–33
scientific revolution, 356
SCP, *See* service control point
SDSL, *See* symmetric DSL
SEA-ME-WE, 516
Search for Precision and Infallibility, 364–365

search for wisdom, 90, 96–99
security and encryption, 403
Selection Of NOs And VFs, 273–274
self-adjusting, 454
self-destructive, 412
Self Perpetuating Power Loops, 447–450
selfishness, 3, 312, 418, 443–444, 501–502
self-learning, 99, 269
semantic analysis, 161–163
Semi-infinite Search Space, 462n2
sense and monitor, 55, 61, 357
sensing network, 81
Sensor-Scanner Based Wisdom Machines, 81–84
Sequencing of Events at Nodes, 114–116
service control point (SCP), 51, 59, 516
service creation environment (SCE), 516
service intelligent peripheral (SIP), 516
service management system (SMS), 51, 402, 516
Service network interface (SNI), 516
service provisioning, 53
service switching points (SSP), 51, 378
seven-level model, 54, 403
Shakespeare, 95, 305
Shannon, C, xvii, 29
short-term objective, 73
signal to noise ratio (SNR), 516
signal transfer points (STP), 51, 516
Signal Waves and Knowledge Flow, 252–253
SIMD, *See* single instruction multiple data
SIMO, *See* single-process multiple-object
Simon, H. A., 70
Simon, J. G., 423
simple network management protocol (SNMP), 516
Simplified Transition Diagram, 122
Single Complex Object Multiple Dependent Objects (ℓ) Multiple Attributes (n), 333
Single Complex Object Multiple Dependent Objects (ℓ) One Attribute, 332
Single Complex Object Multiple Dependent Objects (ℓ), 331
Single Complex Object Single Dependent Object Multiple Attributes, 330–331
Single Complex Object Single Dependent Object Single Attribute, 330
Single Complex Objects with Single Dependent Object, 329
single instruction multiple data (SIMD), 12, 15, 197, 370–371, 382, 388, 388n3
single object, 348–350
Single Process kopc for Multiple Complex Objects, 339–340
Single Process kopc for Single Complex Objects, 337–339

INDEX

Single Process Kopcodes for Generic KCOs, 336–340
Single Process Kopcodes for Simple Objects, 336–337
single-process multiple-object (SIMO), 197, 321, 387, 393
single process single object (SPSO), 173, 197, 348–350
Single Simple Knowledge Object (SSKO), 327–328
Single Simple Object with Multiple Attributes (-n), 328–329
Single Simple Objects with Single Attribute (SSKO-SA), 328
Single Wisdom Multiple Machine (SWMM) Systems, 89–90
Single Wisdom Single Machine (SWSM) Systems, 86–89
SIP, *See* service intelligent peripheral
SM, *See* switching module
small scale integration (SSI), 15
Smart Verbs, 189–192
SMDS, *See* switched multi-megabit digital service
Smith, Adam, 10–11, 23
SMS, *See* service management system
SNI, *See* Service network interface
SNMP, *See* simple network management protocol
SNR, *See* signal to noise ratio
social and cultural norms, 262, 408, 431
social benefit, 27, 301, 411, 470
Social Migration and Political Agenda, 500–502
social problems, 93, 102, 153–154, 165, 301, 307, 414, 431, 440
social values, 3n2, 88, 96, 98, 401, 407, 413, 419n2, 425
socially destructive human, 419n1
socially intelligent, 207
Societal Needs, 429–430
societal problems, 419
Societies Machines and Networks, 430–431
sociological forces on nodes, 139–140
software layers, 86, 174, 417, 456
SONET, *See* synchronous optical network
SONET/SDH, 516
SOPs, *See* standard operating procedures
SPC, *See* stored program control
special services logical interface (SSLI), 516
spread spectrum technique (SST), 516
SPSO, *See* single process single object
SSI, *See* small scale integration
SSLI, *See* special services logical interface
SSP, *See* service switching points

SST, *See* spread spectrum technique
Stacking of K-Loops, 479–482
standard operating procedures (SOPs), 146, 367, 411, 414, 418
status of noun object, 20n3, 163
STM-, 516
Stone, 74, 462
storage area networks (SANs), 516
stored program control (SPC), 12, 15, 346, 373, 375
STP, *See* signal transfer points
strategies, 232, 414, 423, 489, 497
String of Secondary Contributions, 15–17
Structure and Communication of Knowledge, 23–34
  aspects of knowledge, 33–34
  philosophic validation of knowledge, 29–32
  scientific principles in the knowledge domain, 32–33
  truisms in the knowledge domain, 29
  velocity of flow of knowledge, 23–29
structure of knowledge, 103–104, 106, 157, 161, 354, 473
Structure of the Model for LM Negotiations, 225
Structured and Predictive Knowledge, 491–493
STS, *See* synchronous transport signal
Successful Negotiation, 227
Successive K-Loops in T (Time) Dimension, 479–486
Supreme Court, 150, 172, 419, 432, 441, 447
switched multi-megabit digital service (SMDS), 516
switching centers, 16, 52
switching module (SM), 54, 60, 84, 309, 367, 371, 394–395, 400
switching modules, 83, 378, 399
switching system, 373–379
Symbols and Equations, 76–77, 92, 94
Symbols as Objects and Functions, 254–256
symmetric DSL (SDSL), 516
synchronous optical network (SONET), 16, 377–378, 421, 434
synchronous transport signal (STS), 516
Synergy, 2, 14, 64, 126, 168, 329, 462
syntactic analysis, 161
Syntactic Analysis of Knowledge, 161
System performance, 375
Systems for $(I \ll \gg K)$ Filters, 275–277
Systems Model of Interaction, 237–241
S-Zero Filtering and Output BOK-S, 310

T P and S, 517
T P S E, 517
T V and B, 298n10, 425, 517

INDEX   **535**

T/E lines, 517
TAT-, 517
TCP/IP, 5, 35, 41, 361, 372, 434
telecommunication networks, 4–5, 14, 50, 434
telecommunications industry, 59, 375
telemedicine, 37–39
telephone switching systems, 506
terrorists, 91, 98, 411
Tesla, 11
The *(I ≪≫ K)* Filters, 253–258
The Duration for Change of Entropy, 142–143
Theory of Knowledge, 101–145
Theory of Wages, 222
Thorstein Veblen, 202
Three Level Functions, 75–77
Three Major Contributions, 12–15
Three Negative Noun Objects, 312, 319
Three Orientations of Wisdom, 67–73
Three Positive Noun Objects, 312–313
Three-Stage 'SET' Filter (John Tyndall), 308–311
TINA, 517
TMM, *See* transaction management machine
torture training centers, xxvi
TPC, *See* Trans Pacific cables
Tracking of the K-loops, 469–472
   expansionary upward movement, 470–471
   recessive downward movement, 471–472
Traditional Computer Response, 198
Trans Pacific cables (TPC), 517
transaction management machine (TMM), 517
transfer function, 249, 253, 260, 262–264, 279, 285n8
Transition, 116–117, 176
Transitions at I K C Nodes, 116–117
Transmission, 298, 298n10
Trend of concepts, 401–406
Trend of values and ethics, 412–413
trigger condition, 59, 378
triple helix, 358, 479
Truisms in Knowledge Domain, 29
truth virtue and beauty (TVB), 18, 94, 125, 161, 200, 240, 298–299, 312, 366, 412, 425, 463n3
TR-Zero Filtering And Output, 305–307
Tsunami early warning, 46, 48–49
Tsunami warning system (TWS), 48, 517
TVB, *See* truth virtue and beauty
Two Destinations of Knowledge Trail, 124–126
Two Port Electrical Network, 278–279
Two-port knowledge network, 277–286
   change of operators (+, ×, /, and =), 283–285

configuration of a two-port knowledge network, 279–283
intelligent two-port knowledge networks, 286
mathematical implications of the two port network, 283
needs, knowledge, and innovations, 286
two-port electrical network theory, 278–279
Two-Stage 'TR' Filter (Henry Higgins), 303–307
TWS, *See* Tsunami warning system
Tyndall, John, 306, 306n12, 308–310
type A, 507
type B, 508
Types of Traditional Signal Filters, 251
Types of Verb Function Commands, 188–189
Tyrant kings the mafia and the thugs, 132
T-Zero filter, 304, 309
T-Zero Filtering and Output, 303–304

U T and S, 517
UKPS, *See* universal knowledge-processing system
UMTS, *See* universal mobile telephone system
UN, *See* United Nations
Uncertainty, 11, 48, 68, 106–107, 177, 203, 363, 435
unethical behavior, 254
unify, 97, 162, 288, 424, 428, 439
union concession curves, 222, 228
United Nations (UN), 172, 411, 430, 432–434
Universal Knowledge Processing Systems, 41–50
universal knowledge-processing system (UKPS), 41, 46–48
Universal Level (Knowledge) Programs ULKP, 168–171
universal mobile telephone system (UMTS), 517
universal services intelligent network (USIN), 517
Universality, 65, 173, 200, 298, 425
Unix, 13
unstable feedback, 145
Upward Migration, 472–473
USIN, *See* universal services intelligent network
utility array, 230, 234–235

valuation matrix, 234–235
values and ethics, 412–413
Varieties of Noun Objects (NOs), 186–188
Varieties of processors and machines, 365–401
   communications processor, 373–379
   CPU and generic computers, 366–373

Varieties of processors and machines (*Continued*)
  digital knowledge environments, 386–394
  digital medical-processing environments, 394–401
  digital object-processing environments, 383–386
  digital signal processors, 382–383
  display/graphics processor units, 380–381
  electronic switching systems, 373–379
  input/output processors, 379–380
  NPU and array processors, 382
VCs, *See* virtual circuits
VDSL, 518
velocity of flow of information, 23–24
Velocity of flow of knowledge, 23–29
velocity of money, 496
verb function, 188, 265, 273, 477
Verb Functions and Noun Objects, 148–179, *See also* Compilation of knowledge; Framework of knowledge
  classical migration path of knowledge, 175–179
    knowledge traps, 176
    transition at generic nodes, 176–179
  derivation of knowledge, 163–171
    application-level knowledge programs, 165–168
    basic assembly-level knowledge instructions, 164–165
    universal-level knowledge programs, 168–171
  knowledge hardware and software systems, 172–175
  knowledge machine software hierarchy, 171–172
  positive and negative social forces, 149–153
    assistive, neutral, and resistive objects, 152–153
    forces in society and operations on objects, 150–152
very high-speed DSL (VHDSL), 518
very large scale integration (VLSI), 2, 15, 355
VF&NO, 195
VHDSL, *See* very high-speed DSL
virtual circuits (VCs), 518

virtue/social benefit/benevolence, 173, 316, 426, 471
viscosity in the flow of information, 23
VLSI, *See* very large scale integration
voice over Internet deploying (VoIP), 518
VoIP, *See* voice over Internet deploying
von Neumann, 12, 74n5, 112, 159, 197n7, 200, 203, 349, 353, 356, 366, 447

Wafer/chip/configuration, 165, 367, 371, 396
war camps, xxvi
Waraporn, N., 52
Wealth Aspects, 10–11
White money, xxvi
WHO, *See* World Health Organization
Wisdom and ethics, 5, 132, 137, 266n5, 365, 463, 502
Wisdom Domain and Knowledge Rush, 440–443
Wisdom in Machines and Instinct in Humans, 411–412
Wisdom Machines, 64–98, *See also* Knowledge machine building blocks; Machine clusters
  flavors of wisdom, 65–67
  optimization of wise choices, 73–75
    derived axioms, 73–74
    priming of machine wisdom, 74–75
  order, awareness, and search, 97–98
  three orientations of wisdom, 67–73
    absolute wisdom, 68
    materialistic wisdom, 68
    needs and wisdom, 70–71
    opportunistic wisdom, 69–70
    wisdom machines, 71–73
  three-level functions, 75–77
    access and administrative functions, 76
    human authentication, 77
    linkage, scientific, and statistical functions, 76–77
wisdom to behavior, 95–97
WMs and KMs, 367
World Health Organization (WHO), 518

Zeigler, B. P., 219, 230
zones in knowledge domain, 112
Z-transform, 249–250, 253

AUTHOR BIOGRAPHY

Syed V. Ahamed received his Ph.D. and D. Sc. (E.E.) degrees from the University of Manchester and his MBA (Econ.) from the New York University. He taught at the University of Colorado for 2 years before joining AT&T Bell Laboratories in 1966. In 1982 he became a Professor of Computer Science at the City University of New York and a member of the Doctoral Faculty in 1985. He has taught at New York Polytechnic at Brooklyn as a visiting Professor of Computer Science from 1982 to 1986. Professor Ahamed has been a Telecommunications consultant to Bell Communications Research, AT&T Bell Laboratories, Lucent Technologies, Telecom Australia, ETH Zurich, and to Celcom, Malaysia. In Malaysia, he worked as a Management consultant to establish an IT platform and to design the high-speed backbone network. He is the co-recipient of the

1981 IEEE Communications Society, Leonard G. Abraham Prize Paper Award and the co-recipient of the 1964 Honorable Citation for the IEEE Aerospace Society paper. In 1992 he became a Fellow of the IEEE for his seminal contribution to the simulation and design studies of the High-speed Digital Subscriber Lines. The results published in the Bell System Technical Journal in 1982 form the basis for the later versions of the DSLs. He is listed in Marquis 1995, Who's Who in the World and holds many such honors. He has also coauthored two other books: "Intelligent Broadband Multimedia Networks," and "Design and Engineering of Intelligent Communications Systems," based on his investigations of subscriber networks around the world for high-speed data. His doctoral students continue to contribute to knowledge processing systems and wisdom machines proposed by him in 1999 and 2002. In 2004, he wrote the book "The Art of Scientific Innovation," for new doctoral students. In 2006, he wrote the precursor book, "Intelligent Internet Knowledge Networks, Processing of Concepts and Wisdom" dealing with the impact of the knowledge revolution on the human psyche and the subsequent enhancement of human needs. These two books are based on his teaching and mentoring scores of Ph.D. students at the Graduate Center of City University of New York. He holds over 20 American and European patents ranging from slip-meters for induction motors to medical networks for hospitals.